Communications
in Computer and Information Science　1182

Commenced Publication in 2007
Founding and Former Series Editors:
Phoebe Chen, Alfredo Cuzzocrea, Xiaoyong Du, Orhun Kara, Ting Liu,
Krishna M. Sivalingam, Dominik Ślęzak, Takashi Washio, Xiaokang Yang,
and Junsong Yuan

Editorial Board Members

Ana Paula Cláudio · Kadi Bouatouch ·
Manuela Chessa · Alexis Paljic ·
Andreas Kerren · Christophe Hurter ·
Alain Tremeau · Giovanni Maria Farinella (Eds.)

Computer Vision, Imaging and Computer Graphics Theory and Applications

14th International Joint Conference, VISIGRAPP 2019
Prague, Czech Republic, February 25–27, 2019
Revised Selected Papers

Editors
Ana Paula Cláudio
University of Lisbon
Lisbon, Portugal

Kadi Bouatouch
University of Rennes 1
Rennes, France

Manuela Chessa
University of Genoa
Genoa, Italy

Alexis Paljic
Mines ParisTech
Paris, France

Andreas Kerren
Linnaeus University
Växjö, Sweden

Christophe Hurter
French Civil Aviation University (ENAC)
Toulouse, France

Alain Tremeau
University Jean Monnet
Saint-Etienne, France

Giovanni Maria Farinella ⓘ
University of Catania
Catania, Italy

ISSN 1865-0929   ISSN 1865-0937   (electronic)
Communications in Computer and Information Science
ISBN 978-3-030-41589-1   ISBN 978-3-030-41590-7   (eBook)
https://doi.org/10.1007/978-3-030-41590-7

This Springer imprint is published by the registered company Springer Nature Switzerland AG
The registered company address is: Gewerbestrasse 11, 6330 Cham, Switzerland

Preface

The present book includes extended and revised versions of a set of selected papers from the 14th International Joint Conference on Computer Vision, Imaging and Computer Graphics Theory and Applications (VISIGRAPP 2019), held in Prague, Czech Republic, during February 25–27, 2019.

The purpose of VISIGRAPP is to bring together researchers and practitioners interested in both theoretical advances and applications of Computer Vision, Computer Graphics, Information Visualization, and Human-Computer Interaction. VISIGRAPP is composed of four co-located conferences, each specialized in at least one of the aforementioned main knowledge areas.

VISIGRAPP 2019 received 396 paper submissions from 49 countries, of which 6% were included in this book. The papers were selected by the event chairs and their selection was based on a number of criteria that included the classifications and comments provided by the Program Committee members, the session chairs' assessment, and also the program chairs' global view of all papers included in the technical program. The authors of selected papers were then invited to submit a revised and extended version of their papers, having at least 30% innovative material.

The papers selected to be included in this book contribute to the understanding of relevant trends of current research on Computer Vision, Imaging, and Computer Graphics Theory and Applications, including: network visualization, visual data analysis and knowledge discovery, volume rendering, visualization applications and technologies, 3D reconstruction and animation, virtual and augmented reality, image and video understanding, deep learning, first person (egocentric) vision, multi-modal, human-computer interaction, and user evaluation.

We would like to thank all the authors for their contributions and also the reviewers who helped ensure the quality of this publication.

February 2019

Ana Paula Cláudio
Kadi Bouatouch
Manuela Chessa
Alexis Paljic
Andreas Kerren
Christophe Hurter
Alain Tremeau
Giovanni Maria Farinella

Organization

Conference Chair

José Braz   Escola Superior de Tecnologia de Setúbal, Portugal

Program Co-chairs

GRAPP

Ana Paula Cláudio   BioISI, Universidade de Lisboa, Portugal
Kadi Bouatouch   IRISA, University of Rennes 1, France

HUCAPP

Manuela Chessa   University of Genoa, Italy
Alexis Paljic   Mines Paristech, France

IVAPP

Andreas Kerren   Linnaeus University, Sweden
Christophe Hurter   French Civil Aviation University (ENAC), France

VISAPP

Alain Tremeau   Université Jean Monnet, France
Giovanni Maria Farinella   Università di Catania, Italy

GRAPP Program Committee

Francisco Abad   Universidad Politécnica de Valencia, Spain
Marco Agus   King Abdullah University of Science and Technology,
                                     Saudi Arabia
Lilian Aveneau   University of Poitiers, France
Gérard Bailly   GIPSA-Lab, Université Grenoble Alpes, CNRS, France
Maria Beatriz Carmo   Universidade de Lisboa, Portugal
Gonzalo Besuievsky   Universitat de Girona, Spain
Carla Binucci   Università degli Studi di Perugia, Italy
Venceslas Biri   University of Paris-Est, France
Fernando Birra   UNL, Portugal
Kristopher Blom   Virtual Human Technologies, Czech Republic
Stephen Brooks   Dalhousie University, Canada
Dimitri Bulatov   Fraunhofer IOSB, Germany
Patrick Callet   Centre Français de la Couleur, France
L. G. Casado   University of Almeria, Spain

Eva Cerezo	University of Zaragoza, Spain
Teresa Chambel	Lasige, University of Lisbon, Portugal
Parag Chaudhuri	Indian Institute of Technology Bombay, India
Hwan-gue Cho	Pusan National University, South Korea
Teodor Cioaca	AID GmbH, Germany
António Coelho	Universidade do Porto, Portugal
Sabine Coquillart	Inria, France
Vasco Costa	INESC-ID, Portugal
Rémi Cozot	LISIC, University of Littoral, France
Luiz Henrique de Figueiredo	Impa, Brazil
Bailin Deng	Cardiff University, UK
Paulo Dias	Universidade de Aveiro, Portugal
John Dingliana	Trinity College Dublin, Ireland
Jean-Michel Dischler	Université de Strasbourg, France
Anastasios Drosou	Centre for Research and Technology, Hellas, Greece
Thierry Duval	IMT Atlantique, France
Elmar Eisemann	Delft University of Technology, The Netherlands
Marius Erdt	Fraunhofer IDM@NTU, Singapore
Petros Faloutsos	University of California, Los Angeles, USA
Jean-Philippe Farrugia	LIRIS Lab, France
Pierre-Alain Fayolle	University of Aizu, Japan
Francisco R. Feito	University of Jaén, Spain
Dirk Feldmann	University of Münster, Germany
Jie Feng	Peking University, China
Jie-Qing Feng	State Key Lab of CAD&CG, Zhejiang University, China
Leandro Fernandes	Universidade Federal Fluminense, Brazil
Carla Freitas	Universidade Federal do Rio Grande do Sul, Brazil
Ioannis Fudos	University of Ioannina, Greece
Alejandro García-Alonso	University of the Basque Country, Spain
Miguel Gea	University of Granada, Spain
Djamchid Ghazanfarpour	Xlim Laboratory (UMR CNRS 7252), University of Limoges, France
Enrico Gobbetti	CRS4, Italy
Stephane Gobron	HES-SO, Arc, Switzerland
Alexandrino Gonçalves	Polytechnic Institute of Leiria, Portugal
Marcelo Guimarães	Federal University of São Paulo, Brazil
James Hahn	George Washington University, USA
Vlastimil Havran	Czech Technical University in Prague, Czech Republic
Nancy Hitschfeld	University of Chile, Chile
Ludovic Hoyet	Inria Rennes, Centre Bretagne Atlantique, France
Andres Iglesias	University of Cantabria, Spain
Juan José Jiménez-Delgado	Universidad de Jaen, Spain
Xiaogang Jin	Zhejiang University, China
Robert Joan-Arinyo	Universitat Politecnica de Catalunya, Spain

Luís Romero	Instituto Politecnico de Viana do Castelo, Portugal
Isaac Rudomin	BSC, Spain
Wang Rui	Zhejiang University, China
Beatriz Santos	University of Aveiro, Portugal
Basile Sauvage	University of Strasbourg, France
Vladimir Savchenko	Hose University, Japan
Rafael J. Segura	Universidad de Jaen, Spain
Ari Shapiro	University of Southern California, USA
Frutuoso Silva	University of Beira Interior, Portugal
A. Augusto Sousa	FEUP/INESC TEC, Portugal
Jie Tang	Nanjing University, China
Gabriel Taubin	Brown University, USA
Matthias Teschner	University of Freiburg, Germany
Daniel Thalmann	École Polytechnique Fédérale de Lausanne, Switzerland
Juan Carlos Torres	Universidad de Granada, Spain
Alain Tremeau	Université Jean Monnet, France
Torsten Ullrich	Fraunhofer Austria Research GmbH, Austria
Kuwait University	Kuwait University, Kuwait
Anna Ursyn	University of Northern Colorado, USA
Cesare Valenti	Università degli Studi di Palermo, Italy
Thales Vieira	Universidade Federal de Alagoas, Brazil
Andreas Weber	University of Bonn, Germany
Burkhard Wuensche	University of Auckland, New Zealand
Ling Xu	University of Houston-Downtown, USA
Lihua You	Bournemouth University, UK
Jian Zhang	Bournemouth University, UK

GRAPP Additional Reviewers

Simone Balocco	University of Barcelona, Spain
Renan Bela	PUC-Rio, Brazil
Ehtzaz Chaudhry	Bournemouth University, UK
Joao Marcos da Costa	PUC-Rio, Brazil
Andrea Lins	PUC-Rio, Brazil
Franziska Lippoldt	Fraunhofer Singapore, Singapore
Francisco Daniel Pérez Cano	University of Jaén, Spain
Meili Wang	Northwest A&F University, China

HUCAPP Program Committee

Andrea Abate	University of Salerno, Italy
Leena Arhippainen	University of Oulu, Finland
Mirjam Augstein	University of Applied Sciences Upper Austria, Austria
Federica Bazzano	Politecnico di Torino, Italy

Yacine Bellik	LIMSI-CNRS Orsay, France
Cigdem Beyan	Istituto Italiano di Tecnologia, Italy
Leon Bodenhagen	University of Southern Denmark, Denmark
Rodrigo Bonacin	CTI Renato Archer, Brazil
Federico Botella	Miguel Hernandez University of Elche, Spain
Jessie Chen	U.S. Army Research Laboratory, USA
Manuela Chessa	University of Genoa, Italy
Yang-Wai Chow	University of Wollongong, Australia
José Coelho	University of Lisbon, Portugal
Cesar Collazos	Universidad del Cauca, Colombia
Alma Culén	University of Oslo, Norway
Damon Daylamani-Zad	University of Greenwich, UK
Lucia Filgueiras	Escola Politécnica da USP, Brazil
Juan Garrido Navarro	University of Lleida, Spain
Toni Granollers	University of Lleida, Spain
Barry Hughes	University of Auckland, New Zealand
Francisco Iniesto	The Open University and Centre for Research in Education and Educational Technology, UK
Jean-François Jégo	University of Paris 8, France
Chee Weng Khong	Multimedia University, Malaysia
Uttam Kokil	Kennesaw State University, USA
Heidi Krömker	Technische Universität Ilmenau, Germany
Fabrizio Lamberti	Politecnico di Torino, Italy
Chien-Sing Lee	Sunway University, Malaysia
Tsai-Yen Li	National Chengchi University, Taiwan, China
Régis Lobjois	IFSTTAR, France
Flamina Luccio	Università Ca' Foscari Venezia, Italy
Sergio Luján-Mora	University of Alicante, Spain
Malik Mallem	Université Paris Saclay, France
Santiago Martinez	University of Agder, Norway
Vincenzo Moscato	Università degli Studi di Napoli Federico II, Italy
Andreas Nearchou	University of Patras, Greece
Keith Nesbitt	University of Newcastle, Australia
Radoslaw Niewiadomski	University of Genoa, Italy
Nuno Otero	Linnaeus University, Sweden
Evangelos Papadopoulos	NTUA, Greece
James Phillips	Auckland University of Technology, New Zealand
Otniel Portillo-Rodriguez	Universidad Autonóma del Estado de México, Mexico
Nitendra Rajput	IBM Research New Delhi, India
Maria Lili Ramírez	Universidad del Quindío, Colombia
Juha Röning	University of Oulu, Finland
Paul Rosenthal	Germany
Andrea Sanna	Politecnico di Torino, Italy
Corina Sas	Lancaster University, UK
Berglind Smaradottir	University of Agder, Norway
Fabio Solari	University of Genoa, Italy

Daniel Thalmann   École Polytechnique Fédérale de Lausanne,
                                   Switzerland
Godfried Toussaint   New York University Abu Dhabi, UAE
Pauliina Tuomi   Tampere University of Technology, Finland
Kostas Vlachos   University of Ioannina, Greece
Gualtiero Volpe   Università degli Studi di Genova, Italy

IVAPP Program Committee

Vladan Babovic   National University of Singapore, Singapore
Juhee Bae   University of Skovde, Sweden
Maria Beatriz Carmo   Universidade de Lisboa, Portugal
Till Bergmann   Max Planck Society, Germany
David Borland   University of North Carolina at Chapel Hill, USA
Massimo Brescia   Istituto Nazionale di AstroFisica, Italy
Ross Brown   Queensland University of Technology, Australia
Guoning Chen   University of Houston, USA
Yongwan Chun   University of Texas at Dallas, USA
António Coelho   INESC-TEC, Universidade do Porto, Portugal
Danilo B. Coimbra   Federal University of Bahia, Brazil
Christoph Dalitz   Niederrhein University of Applied Sciences, Germany
Mihaela Dinsoreanu   Technical University of Cluj-Napoca, Romania
Georgios Dounias   University of the Aegean, Greece
Achim Ebert   University of Kaiserslautern, Germany
Danilo Eler   São Paulo State University, Brazil
Maria Ferreira de Oliveira   University of São Paulo, ICMC, Brazil
Chi-Wing Fu   The Chinese University of Hong Kong, Hong Kong
Mohammad Ghoniem   Luxembourg Institute of Science and Technology,
                                   Luxembourg
Randy Goebel   University of Alberta, Canada
Martin Graham   University of Edinburgh, UK
Naeemul Hassan   University of Mississippi, USA
Torsten Hopp   Karlsruhe Institute of Technology, Germany
Jie Hua   University of Technology Sydney, Australia
Jusufi Ilir   Linnaeus University, Sweden
Stefan Jänicke   University of Southern Denmark, Denmark
Mark Jones   Swansea University, UK
Bijaya Karki   Louisiana State University, USA
Jörn Kohlhammer   Fraunhofer Institute for Computer Graphics Research,
                                   Germany
Martin Kraus   Aalborg University, Denmark
Haim Levkowitz   University of Massachusetts Lowell, USA
Chun-Cheng Lin   National Chiao Tung University, Taiwan, China
Giuseppe Liotta   University of Perugia, Italy
Rafael Martins   Linnaeus University, Sweden
Krešimir Matkovic   VRVis Research Center, Austria

Kazuo Misue	University of Tsukuba, Japan
Steffen Oeltze-Jafra	Otto-von-Guericke-Universität Magdeburg, Germany
Benoît Otjacques	Luxembourg Institute of Science and Technology (LIST), Luxembourg
Jinah Park	KAIST, South Korea
Fernando Paulovich	Dalhousie University, Canada
Torsten Reiners	Curtin University, Australia
Philip Rhodes	University of Mississippi, USA
Patrick Riehmann	Bauhaus-Universität Weimar, Germany
Maria Riveiro	University of Skövde, Sweden
Adrian Rusu	Fairfield University, USA
Filip Sadlo	Heidelberg University, Germany
Beatriz Santos	University of Aveiro, Portugal
Giuseppe Santucci	University of Roma, Italy
Angel Sappa	ESPOL Polytechnic University, Ecuador, and Computer Vision Center, Spain
Falk Schreiber	University of Konstanz, Germany, and Monash University, Australia
Hans-Jörg Schulz	Aarhus University, Denmark
Celmar Silva	University of Campinas, Brazil
Juergen Symanzik	Utah State University, USA
Sidharth Thakur	Renaissance Computing Institute (RENCI), USA
Roberto Theron	Universidad de Salamanca, Spain
Günter Wallner	University of Applied Arts Vienna, Austria
Jinrong Xie	eBay Inc., USA
Anders Ynnerman	Linköping University, Sweden
Hongfeng Yu	University of Nebraska-Lincoln, USA
Lina Yu	Intel Corporation, USA
Jianping Zeng	Microsoft, USA
Yue Zhang	Oregon State University, USA
Jianmin Zheng	Nanyang Technological University, Singapore

IVAPP Additional Reviewers

Aris Alisandrakis	Linnaeus University, Sweden
Xiaopei Liu	ShanghaiTech University, China
Benjamin Mora	Swansea University, UK
Wei Zeng	Shenzhen Institutes of Advanced Technology, Chinese Academy of Sciences, China

VISAPP Program Committee

Amr Abdel-Dayem	Laurentian University, Canada
Ilya Afanasyev	Innopolis University, Russia
Palwasha Afsar	Algorithmi Research Center, Uminho, Portugal

Sotirios Diamantas	Tarleton State University, Texas A&M System, USA
Ernst Dickmanns	UniBw Munich, Germany
Yago Diez	Yamagata University, Japan
Mariella Dimiccoli	Insitut de Robòtica i Informàtica Industrial (CSIC-UPC), Spain
Jana Dittmann	Otto-von-Guericke-Universität Magdeburg, Germany
Aijuan Dong	Hood College, USA
Ulrich Engelke	CSIRO, Australia
Shu-Kai Fan	National Taipei University of Technology, Taiwan, China
Giovanni Maria Farinella	Università di Catania, Italy
Jean-Baptiste Fasquel	University of Angers, France
Jorge Fernández-Berni	CSIC, Universidad de Sevilla, Spain
Gernot Fink	TU Dortmund, Germany
David Fofi	ImViA, France
Gian Foresti	Unversity of Udine, Italy
Mohamed Fouad	Military Technical College, Egypt
Antonino Furnari	University of Catania, Italy
Claudio Gennaro	CNR, Italy
Przemyslaw Glomb	IITiS PAN, Poland
Seiichi Gohshi	Kogakuin University, Japan
Luiz Goncalves	Federal University of Rio Grande do Norte, Brazil
Manuel González-Hidalgo	Balearic Islands University, Spain
Levente Hajder	Eötvös Loránd University, Hungary
Xiyi Hang	California State University, USA
Daniel Harari	Weizmann Institute of Science, Israel
Walid Hariri	ETIS ENSEA, Université de Cergy-Pontoise, France
Aymeric Histace	ETIS UMR CNRS 8051, France
Wladyslaw Homenda	Warsaw University of Technology, Poland
Fay Huang	National Ilan University, Taiwan, China
Hui-Yu Huang	National Formosa University, Taiwan, China
Laura Igual	Universitat de Barcelona, Spain
Francisco Imai	Apple Inc., USA
Jiri Jan	University of Technology Brno, Czech Republic
Tatiana Jaworska	Polish Academy of Sciences, Poland
Xiaoyi Jiang	University of Münster, Germany
Luis Jiménez Linares	University of Castilla-La Mancha, Spain
Zhong Jin	Nanjing University of Science and Technology, China
Leo Joskowicz	The Hebrew University of Jerusalem, Israel
Paris Kaimakis	University of Central Lancashire, Cyprus
Martin Kampel	Vienna University of Technology, Austria
Etienne Kerre	Ghent University, Belgium
Anastasios Kesidis	National Center for Scientific Research, Greece
Nahum Kiryati	Tel Aviv University, Israel
Constantine Kotropoulos	Aristotle University of Thessaloniki, Greece

Arjan Kuijper	Fraunhofer Institute for Computer Graphics Research and TU Darmstadt, Germany
Mónica Larese	CIFASIS-CONICET, National University of Rosario, Argentina
Denis Laurendeau	Laval University, Canada
Sébastien Lefèvre	Université Bretagne Sud, France
Marco Leo	CNR, Italy
Daw-Tung Lin	National Taipei University, Taiwan, China
Huei-Yung Lin	National Chung Cheng University, Taiwan, China
Xiuwen Liu	Florida State University, USA
Giosue Lo Bosco	University of Palermo, Italy
Liliana Lo Presti	University of Palermo, Italy
Angeles López	Universitat Jaume I, Spain
Bruno Macchiavello	Universidade de Brasília, Brazil
Ilias Maglogiannis	University of Piraeus, Greece
Francesco Marcelloni	University of Pisa, Italy
Mauricio Marengoni	Universidade Presbiteriana Mackenzie, Brazil
Emmanuel Marilly	Nokia - Bell Labs France, France
Jean Martinet	University Cote d'Azur, CNRS, France
José Martínez Sotoca	Universitat Jaume I, Spain
Mitsuharu Matsumoto	The University of Electro-Communications, Japan
Radko Mesiar	Slovak University of Technology, Slovakia
Leonid Mestetskiy	Lomonosov Moscow State University, Russia
Cyrille Migniot	Université de Bourgogne, le2i, France
Dan Mikami	NTT, Japan
Steven Mills	University of Otago, New Zealand
Nabin Mishra	Stoecker & Associates, USA
Sanya Mitaim	Thammasat University, Thailand
Pradit Mittrapiyanuruk	Panasonic R&D Center Singapore, Singapore
Birgit Moeller	Martin Luther University Halle-Wittenberg, Germany
Thomas Moeslund	Aalborg University, Denmark
Ali Mohammad-Djafari	CNRS, France
Bartolomeo Montrucchio	Politecnico di Torino, Italy
Davide Moroni	ISTI-CNR, Italy
Kostantinos Moustakas	University of Patras, Greece
Dmitry Murashov	Federal Research Center "Computer Science and Control" of Russian Academy of Sciences, Russia
Feiping Nie	University of Texas at Arlington, USA
Mikael Nilsson	Lund University, Sweden
Nicoletta Noceti	Università di Genova, Italy
Yoshihiro Okada	Kyushu University, Japan
Gonzalo Pajares	Universidad Complutense de Madrid, Spain
Theodore Papadopoulo	Inria, France
Félix Paulano-Godino	University of Jaén, Spain
Felipe Pinage	Federal University of Parana, Brazil
Stephen Pollard	HP Labs, UK

Charalambos Poullis	Concordia University, Canada
William Puech	Université Montpellier, France
Giovanni Puglisi	University of Cagliari, Italy
Giuliana Ramella	CNR, Istituto per le Applicazioni del Calcolo "M. Picone", Italy
Huamin Ren	Inmeta Consulting AS, Norway
Phill Rhee	Inha University, South Korea
Joao Rodrigues	University of the Algarve, Portugal
Marcos Rodrigues	Sheffield Hallam University, UK
Bart Romeny	Eindhoven University of Technology (TU/e), The Netherlands
Ramón Ruiz	Universidad Politécnica de Cartagena, Spain
Silvio Sabatini	University of Genoa, Italy
Farhang Sahba	Sheridan Institute of Technology and Advanced Learning, Canada
Joaquin Salas	CICATA-IPN, Mexico
Ovidio Salvetti	CNR, Italy
Andreja Samcovic	University of Belgrade, Serbia
K. C. Santosh	University of South Dakota, USA
Jun Sato	Nagoya Institute of Technology, Japan
Ilhem Sboui	National School of Computer Sciences, Manouba University, Tunisia
Gerald Schaefer	Loughborough University, UK
Siniša Šegvic	University of Zagreb, Croatia
Kazim Sekeroglu	Southeastern Louisiana University, USA
Shishir Shah	University of Houston, USA
Caifeng Shan	Philips Research, The Netherlands
Lik-Kwan Shark	University of Central Lancashire, UK
Désiré Sidibé	Université Évry - Paris-Saclay, France
Bogdan Smolka	Silesian University of Technology, Poland
Ferdous Sohel	Murdoch University, Australia
Ömer Soysal	Southeastern Louisiana University, USA
Tania Stathaki	Imperial College London, UK
Mu-Chun Su	National Central University, Taiwan, China
Tamás Szirányi	MTA SZTAKI, Hungary
Ryszard Tadeusiewicz	AGH University of Science and Technology, Poland
Norio Tagawa	Tokyo Metropolitan University, Japan
Hiroki Takahashi	The University of Electro-Communications, Japan
Ricardo Torres	Norwegian University of Science and Technology (NTNU), Norway
Alain Tremeau	Université Jean Monnet, France
Vinh Truong Hoang	Ho Chi Minh City Open University, Vietnam
Yulia Trusova	Federal Research Center "Computer Science and Control" of the Russian Academy of Sciences, Russia
Du-Ming Tsai	Yuan Ze University, Taiwan, China

Aristeidis Tsitiridis	Universidad Rey Juan Carlos, Spain
Cesare Valenti	Università degli Studi di Palermo, Italy
Panayiotis Vlamos	Ionian University, Greece
Frank Wallhoff	Jade University of Applied Science, Germany
Tao Wang	BAE Systems, USA
Layne Watson	Virginia Polytechnic Institute and State University, USA
Quan Wen	University of Electronic Science and Technology of China, China
Laurent Wendling	LIPADE, France
Andrew Willis	University of North Carolina at Charlotte, USA
Christian Wöhler	TU Dortmund University, Germany
Stefan Wörz	Heidelberg University, Germany
Yan Wu	Georgia Southern University, USA
Pingkun Yan	Rensselaer Polytechnic Institute, USA
Guoan Yang	Xian Jiaotong University, China
Vera Yashina	Dorodnicyn Computing Center of the Russian Academy of Sciences, Russia
Hongfeng Yu	University of Nebraska-Lincoln, USA
Pietro Zanuttigh	University of Padova, Italy
Huiyu Zhou	Queen's University Belfast, UK
Yun Zhu	UCSD, USA
Zhigang Zhu	City College of New York, USA
Peter Zolliker	Empa-Swiss Federal Laboratories for Materials Science and Technology, Switzerland
Ju Zou	University of Western Sydney, Australia

VISAPP Additional Reviewers

Eman Ahmed	University of Luxembourg, Luxembourg
Gerasimos Arvanitis	ECE, National Technical University of Athens, Greece
Renato Baptista	University of Luxembourg, Luxembourg
Júlio Batista	Brazil
Romain Dambreville	IRISA, France
Rig Das	University of Luxemburg, Luxembourg
Konstantinos Delibasis	University of Central Greece, Greece
Enjie Ghorbel	University of Luxembourg, Luxembourg
Andrei Kopylov	Tula State University, Russia
Arnau Mir	University of Balearic Islands, Spain
Deependra Mishra	IBM, USA
Mohammad Abdul Mukit	University of Oklahoma, USA
Konstantinos Papadopoulos	University of Luxemburg, Luxembourg
Zoltan Rozsa	Hungarian Academy of Sciences Institute for Computer Science and Control (MTA SZTAKI), Hungary
Oleg Seredin	Tula State University, Russia

Ismael Serrano	Vicomtech, Spain
Nikos Stagakis	University of Patras, Greece
Zhihao Zhao	USA

Invited Speakers

Daniel McDuff	Microsoft, USA
Diego Gutierrez	Universidad de Zaragoza, Spain
Jiri Matas	Czech Technical University in Prague, Czech Republic
Dima Damen	University of Bristol, UK

Contents

Computer Vision Theory and Applications

Computer Graphics Theory and Applications

Synthesis and Validation of Virtual Woodcuts Generated with Reaction-Diffusion

Davi Padilha Mesquita$^{(\boxtimes)}$ (iD) and Marcelo Walter (iD)

Universidade Federal do Rio Grande do Sul, Porto Alegre, Brazil
davipadilhamesquita@gmail.com
http://www.inf.ufrgs.br/

Abstract. Although woodcuts are a traditional artistic technique, in which a woodblock is carved and then printed into paper, few works have attempted to synthesize woodcuts in the context of Non-Photorealistic Rendering (NPR). We previously presented a woodcut synthesis mechanism based on Turing's reaction-diffusion. In that work, an input image is preprocessed in order to gather information which will be used to control the reaction-diffusion processing. In this article, we expanded our previous work by the addition of noise to improve the appearance of wood and a higher degree of user control toward the final result. We also validated our results by comparison with actual woodcuts and performed a qualitative evaluation with users. This work expands the range of artistic styles which can be generated by NPR tools.

Keywords: Reaction-Diffusion (RD) · Woodcuts · Non-Photorealistic Rendering (NPR) · Evaluation of NPR

1 Introduction

Traditional computer graphics (CG) intends to simulate the higher possible degree of similarity with the real world, or, in other words, to be photorealistic. On the other hand, since the 90s, many published works followed a different path, deviating from photorealism to instead simulate many artistic styles, such as oil painting [12,46], black-and-white drawings [21,24] and stained glass [43]. These works compose a research area appropriately known as *Non-photorealistic rendering* (NPR) [20].

One of these artistic styles is the woodcut (Fig. 1), a technique which prints an image based in an engraving in a woodblock. Woodcuts are generated through the carving in the surface of the woodblock, where the carvings correspond the lines in the final image. Then paint is rolled on the block, and a paper is pressed over it, so only the non-carved areas will be painted in the paper. This results in a two color image, where the carvings maintain the paper color (usually white), while the non-carved areas have the same tone as the paint

© Springer Nature Switzerland AG 2020
A. P. Cláudio et al. (Eds.): VISIGRAPP 2019, CCIS 1182, pp. 3–29, 2020.
https://doi.org/10.1007/978-3-030-41590-7_1

(often black), although some woodcut styles repeat the process with more colors of paint, resulting in multi-color prints [23].

Despite the multitude of NPR works in the literature, few works have tried to synthesize woodcuts. In a previous work [29], we presented a new method to generate woodcuts through reaction-diffusion (RD), a mechanism to generate textures originally proposed by Alan Turing [48]. In this paper, we expand our previous work by the addition of mechanisms to simulate the aspect of wood, new options to allow a finer user-defined control over the final result, new experiments, and a validation through a qualitative evaluation study and comparison with an actual woodcut derived from the same input image.

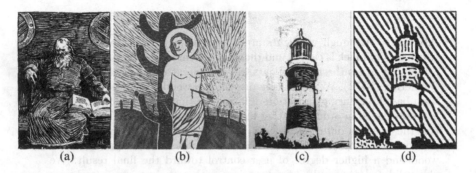

    (a)   (b)   (c)   (d)

Fig. 1. Example of woodcuts. (a) *Snorre Sturluson*, by Krogh [19]. (b) *São Sebastião*, by Francorli [10]. (c) Real lighthouse woodcut by de Jesus [16]. (d) A result of our method (for details regarding the parameters used, see Fig. 5d).

2   Literature Review

In this section we will discuss the previous works regarding the generation of woodcuts, reaction-diffusion, and evaluation of NPR works.

2.1   Synthesis of Woodcuts

The first work which tried to generate woodcut-like images in the context of CG applications was Mizuno et al. [32] (Fig. 2a). In his work, the real process to generate a woodcut is emulated in a virtual environment, where the user carve a 3D model of a woodblock in a manner analogous to carving an actual woodblock, by the addition or removal of shapes from the model surface. This block will receive paint over its surface and finally it will be pressed onto a virtual sheet of paper, yielding the final woodcut image. However, it is not a fully automated system, since a direct user input is demanded to produce the result. Another issue of Mizuno's system is the need of some artistic skills to properly carve the woodblock.

Their system was improved in further works, with the automation of parts of the carving process [30]. There was also a focus on *Ukiyo-e*, a typical woodcut style from Japan [33, 34]. Finally, the carving system was implemented using a pressure sensitive pen over a tablet, easing how the user controls the virtual carving tool, since a pen is a more direct analog of the chisels used to carve actual woodblocks [31].

The first work to generate woodcuts from a base image, bringing this artistic style into the context of image-based artistic rendering [20], was Mello, Jung, and Walter's [28] (Fig. 2c). Their work used a series of image processing operations over an input image, producing an image resembling an actual woodcut. Basically, these operations were done in four steps: image segmentation, calculation of carving directions, distribution of carvings according to this calculation, and post-processing operations to render the image. Although their results have a certain degree of similarity with some styles of actual woodcuts, the simulated carvings did not show a good variation on their appearance, with the system only able to produce strokes resembling short lines.

The next attempt to produce a result resembling woodcuts from a base image was Winnemoller's [51]. They proposed the extension of the Difference-of-Gaussians operator, traditionally used to detect edges (Fig. 2b). It should be emphasized that this work did not intend to generate only woodcuts, but it is focused on general black-and-white drawing styles, such as hatching and pencil drawing.

One of the most recent attempts, showing a considerable improvement in the quality of the final results, was Li et al. approach to simulate the woodcuts typical from the Chinese region of Yunnan [22, 23] (Fig. 2d). Their work is able to produce multi-color woodcuts, a feature from this style of woodcuts. Although in this work the direction of the carvings is calculated using a similar algorithm also used by Mello et al. [28], they innovate by adding them in the image through different textures collected from real woodcuts, simulating the aspect of actual carvings.

It should be highlighted that in all of these works the only types of carvings present are in the form of short lines or variations (such as broken stripes), unlike real woodcuts, where other structures such as dots or continuous stripes can be found. They also lack a mechanism to set different properties for each part of the image, except for absence or presence in lines (in opposition to pure black and white areas) and varying the direction of carvings. For example, there is no way to set different sizes or styles for the carvings in each region.

To address these issues, in [29] we presented a new method to generate woodcuts using Reaction-Diffusion, a widely used pattern generation mechanism. To present our methodology, first we need to briefly introduce RD and its usage on CG applications.

2.2 A Brief Review of Reaction-Diffusion

In 1952, Alan Turing [48] proposed a new mechanism to explain how biological patterns are formed, from the zebra stripes to the development of fingers during

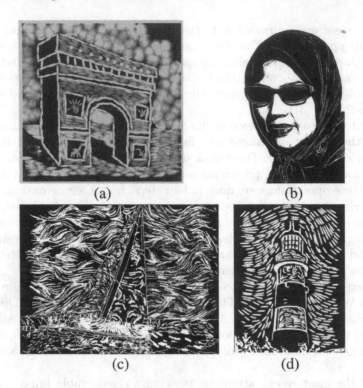

Fig. 2. Woodcut synthesis methods. (a) Mizuno et al. [32]. (b) Mello et al. [28]. (c) Winnemoller [51]. (d) Li et al. [23].

embryonic development. This mechanism, denominated reaction-diffusion (RD), works through a set of chemical substances, named morphogens, whose concentrations vary in space and time due to reactions among themselves (so the substances can be synthesized or decomposed) and diffusion. The morphogens react and diffuse until their concentrations achieve a stationary state. The local variations of the morphogen concentration form a pattern, with some regions with higher and other with lower concentrations. The morphogen concentrations can be translated into a color value, so RD can be used to generate textures.

Two works published in 1991 introduced RD into the world of CG. Turk [49] used RD to generate patterns resembling the different types of spots and stripes found in mammalian coat patterns, such as zebra's stripes and leopard's spots, simulating the already proposed application of RD as the mechanism behind these biological patterns by Bard [2] and Murray [37]. Witkin and Kass [52] applied RD to generate different kinds of textures, being able to control the direction of stripes and stretched spots by using a rotation matrix. This approach was expanded by Sanderson et al. [41]. More recent works show RD being used for distinct tasks, such as image restoration [53], simulation of rotting in fruits and vegetables [8,18] and simulation of human skin pigmentation disorders [4].

Recent works introduced RD approaches to NPR goals. In 2016, Chi et al. [5] published a work which uses RD for image stylization. They used a modified anisotropic RD system as part of a workflow which processes an input image to yield a final stylized result. This anisotropic RD system is guided by a flow field calculated from the input image. It is also able to turn the circular spots into triangles, squares or drop-like figures, allowing the final result to simulate different artistic styles, as paper-cut and halftone.

Jho and Lee [17] used RD as a halftoning mechanism. They use a linear RD system, where a parameter is correlated with the general color intensity of the generated patterns. Thus, by changing the value of this parameter, it is possible to produce a darker or lighter pattern. To improve the speed of their method, allowing real-time processing, a RD system is previously generated and then used as a mask. However, neither Chi et al. nor Jho and Lee works synthesize woodcuts.

2.3 Our Previous Work: Reaction-Diffusion Woodcuts

In our previous work [29] we developed a mechanism to synthesize woodcuts with RD from an input image, being able to generate woodcuts with a distinct aspect from previous methods. We will briefly describe our algorithm here (for a more detailed explanation, consult [29]).

Our method was subdivided into three steps (Fig. 3): preprocessing, processing and post-processing. In the preprocessing step, we gather information from the original image to control the RD system, which runs in the processing step. Post-processing applies filters to enhance the appearance of the RD result, to

Table 1. Parameters used in our system and their default values. Adapted from [29].

Parameter	Default value
T_{size}	500 pixels
k	{1, 5, 9, 13, 17}
T_{diff}	500 pixels
S_{min}	0.005
S_{max}	0.015
D_{ah}	0.125
D_{av}	0.125
D_{bh}	0.030
D_{bv}	0.025
δ_{min}	0.5
δ_{max}	1.5
M_{white}	8.0
M_{black}	0.0

generate the final woodcut-like image. Table 1 lists the parameters which are relevant for the simulations in this article. These parameters will be explained in this section. For the complete list of parameters, see [29].

Preprocessing. The preprocessing step receives as input, besides the image to be used as basis for the woodcut, a segmentation of it, where each region is marked as a different color. First, we expand the borders of the segmented regions and merge the smallest regions (with less than T_{size} pixels) into the borders, which will be marked as black in the final image. Since in real woodcuts there are regions with plain black and white color and no cravings, our algorithm marks the brighter and darker regions (determined by the average gray-scale color of the region) as respectively pure white and black.

We wanted to allow different properties to each part of image, starting from the size of carvings. We observed that smaller structures in the RD system allow more details from the original image to be preserved in the final woodcut, so we developed a mechanism to compute the detail level of each region through comparisons between the original image and blurred versions of it, in a similar way as Hertzmann [13]. The detail level was mapped to a parameter of our RD system, S, related to the size of spots and stripes - the higher the value of S, the smaller the structures will be, so regions with more details will receive a larger value of S and vice-versa. Basically, for each integer k_i in k, we do a Gaussian blur over the original image with a $k_i \times k_i$ kernel. For each pixel we compare the average difference between the original and blurred version for its neighborhood, and if this difference is smaller than T_{diff} we consider that this pixel is in a less detailed region (since the blurred and the original versions are similar for it), and consequently it will have a lower S. More blurred images make the difference between pixels higher, so the higher the blur the less detailed a region would need to be to have an average difference below T_{diff} and then it will receive a lower value of S. The parameters S_{max} and S_{min} set the interval of accepted values for S.

Next, we calculated the orientation of the carvings. Our system allows three possibilities for that: (a) *orientation by region*, (b) *orientation by pixel*, and (c) *adaptive orientation*. The orientation will be saved as an angle, θ, for each pixel, which will direct the stripes and spots generated by the RD system.

The orientation by region option sets for each region a single direction for the cuts, generating parallel stripes inside the region. This direction is the one from the brightest pixel in the region border to its center, since in most woodcuts the carvings tend to start in the lighter area of a region [28]. This option allows a high level of coherence of the cuts in a single region, easing the identification of each image part. On the other hand, there is no way to preserve details internal to a region by this kind of orientation.

The second option, orientation by pixel, calculates for each pixel a specific orientation angle, through the gradient of variations in luminance in the neighbourhood of that pixel. This option preserves in the final woodcut the contour details present in the original image, allowing a higher degree of fidelity in relation to it.

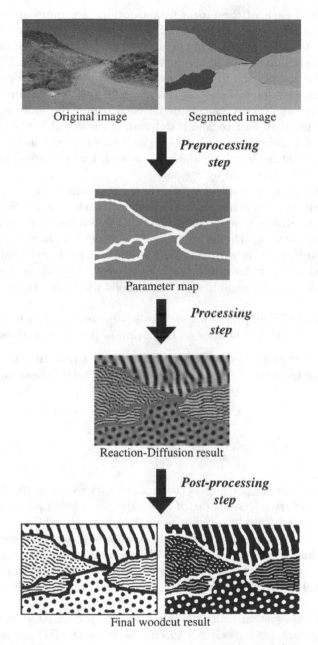

Original image   Segmented image

Preprocessing
step

Parameter map

Processing
step

Reaction-Diffusion result

Post-processing
step

Final woodcut result

Fig. 3. Summary of the steps in our woodcut synthesis method (image adapted from [29]). Our system receives the input image together with its segmentation. These two are used during preprocessing to extract information about the base image, and this information is encoded into a parameter map. Processing through RD uses this parameter map to generate the raw pattern of stripes and spots. This raw result is filtered in the post-processing step to generate the final woodcut-like image.

The results of each of these orientation options are more appropriated to different parts of the image: in foreground areas, specially in regions representing more complex objects such as an human face, orientation by pixel performs better by showing the original details in our final result. On the other hand, background areas where few details are present, are better represented by orientation by region, allowing viewers to focus on the more detailed foreground. The third option, adaptive orientation, calculates for each region the more appropriated method and apply it to that region. It determines the detail level of each region by its value of S, previously calculated - regions with higher S are more detailed and so uses orientation by pixel, and regions with lower S are considered less detailed, receiving orientation by region.

Previous methods of woodcut synthesis only produced stripe-like carvings, but many woodcuts show other kinds of structures, such as spots. So we added a parameter to our RD system, δ, which controls the tendency of a given pixel to be part of stripes or spots. Higher values of δ yield spots or short, broken stripes, while lower values allow the stripes to be continuous, crossing the entirety of the regions in which they are localized. In our work we calculated this parameter by the average color of each region, so darker regions will show stripes and lighter regions spots.

The output of the preprocessing step is a parameter map, where each pixel has a value for S, θ and δ, besides an indication of presence or absence of RD.

Processing. After preprocessing, our system is ready to execute the RD system. In our work, we used the Turing non-linear system [3,48], whose equations are shown below:

$$\frac{\partial a}{\partial t} = S(16.0 - ab) + D_a \nabla^2 a$$

$$\frac{\partial b}{\partial t} = S(ab - b - \beta) + D_b \nabla^2 b$$

This system has two morphogens, whose concentrations are a and b, with temporal variations of, respectively, $\partial a/\partial t$ and $\partial b/\partial t$ for a given point in the space. These morphogens diffuse in a bidimensional surface with diffusion coefficients D_a and D_b. The reaction parameter S represents how fast the reactions occur and β the randomness of the system. To allow computational simulation of this RD process, we divided the space in a grid of pixels, and the time in discrete steps. Each pixel has a different value for β, generated randomly during program initialization.

We control this system using the parameter map calculated in the preprocessing step, so each pixel receives a specific value for the RD parameter S and a value for the parameters θ and δ. Also, we mark which pixels will not participate in the RD processing (areas with plain black and white colors in the final woodcut). By setting a pixel-specific value for S, we can change the size of stripes and spots generated by the RD system in the proximity of that pixel. By setting a single S value for a segmented region, that region will show thinner stripes and smaller spots (high S) or thicker stripes and larger spots (low S).

With the parameter θ, it is possible to control the direction of stripes. This parameter works through the introduction of anisotropic diffusion in a way similar to Witkin and Kass work [52]. First, we set a different value for the diffusion coefficient in the horizontal and vertical directions for the general RD system (D_{ah} and D_{bh} for the former and D_{av} and D_{bv} for the latter). This makes the system to form stripes in the direction of the bigger diffusion coefficient. Then, we particularize the direction of these stripes for a given pixel by rotating the system by an angle equal to the θ value for that pixel, so the stripes will have the same direction as this angle.

Finally, to allow stripes, spots and intermediate forms (as broken stripes in the form of slight elongated spots) in the same image, we use the parameter δ to change the difference between the horizontal and vertical diffusion coefficients for each pixel. Larger values for δ reduce the difference between these two diffusion coefficients, forming spots, while smaller values increase this difference, producing stripes. δ_{min} and δ_{max} sets the maximum and minimum values for δ.

After processing, we need to translate the concentrations into a grayscale image. We use the value of b for this, translating its concentrations into the interval between 0 (black) and 255 (white). To do this, we have two options: dynamic visualization and static visualization. Dynamic visualization consists in considering the maximum value of b as white and the minimum as black, with linear interpolation for intermediate values. For most cases this option works, but for a few ones an outlier (a point whose value for b is too higher than other pixels) makes the image too dark, preventing the post-processing binarization to form a good black and white woodcut image. The static option deals with this situation by using two thresholds, M_{white} and M_{black}, which are the b concentration values corresponding to white and black, respectively. Values beyond this interval are considered the corresponding threshold for the purposes of conversion of concentration to color.

Now we have a processed raw image with the pattern of wood carvings produced by the RD system. This raw image will undergo post-processing operations in order to yield the desired woodcut image.

Post-processing. During RD processing, the color of regions without carving patterns was not considered, so we start the post-processing by applying pure black or pure white colors for the regions without RD, as defined in the preprocessing operations.

A median filter with a 3×3 kernel is executed over the image to remove unwanted noise from it. Finally, an Otsu filter is applied, resulting in a black and white binarized image whose aspect resembles an actual woodcut with black lines and a white background. We also generate the inversion of this image (white lines with a black background) to reproduce white-line woodcuts, a style in which the carvings are white over a dark background.

2.4   Evaluation in NPR

One of the goals of NPR is to generate images in several artistic styles. Consequently, the quality of the results is somewhat hard to judge, since the goal is not some defined, objective property, but the aesthetic appeal instead, which is subject to the personal preferences of viewers. This makes validation of NPR works a challenging task.

Isenberg [14] mentioned several questions to be answered by NPR validation procedures, such as how people answer to NPR images, if a given technique is accepted by the final users, and how NPR rendered images compare with hand-made artworks. The last question can be formulated as the so-called *NPR Turing test*, i.e., if it is possible to create a non-photorealistic image which is impossible for a regular person to distinguish from an actual similar artwork, one of the seven grand challenges for NPR according to Salesin [40].

The first to propose an evaluation of NPR methods was Schumann et al. [42] work, in which the authors compared NPR sketches with CAD plot images and shaded images regarding communication of information in architectural design projects. This comparison was done by showing the images to architects and architectural students and asking them to rate the images. They found that the NPR sketch was preferable for presenting early drafts, while CAD plots and shaded images were considered more appropriated to final presentations.

An example of quantitative evaluation was shown by Maciejewski et al. [25, 26] works, which evaluated the differences regarding area and distribution of stipple dots between hand-made stippling drawings and NPR using a statistical analysis, concluding that even adding randomness NPR results can be tell apart each other because of other regularities. However, Martín et al. [27] applied the same statistical analysis to study a halftoning-based stippling method containing both randomness regarding location of stippling and example-based stipple dots, and they found that, using this example-based approach, they were able to generate stipple drawings with the same characteristics as actual hand-made drawings.

Isenberg et al. [15] provided an interesting qualitative study regarding how people compare hand-made artworks with NPR-generated ones. In their study, a set of images, containing both hand-drawn and computer generated in the pen-and-ink style, were shown to the participants, which should group these images in piles. The quantity, size and nature of these piles was not determined previously, so participants were unconstrained in how to group the images. Interestingly, no participant separated the images into computer-generated and hand-made images, preferring to form groups by the drawing style (such as the usage of lines and dots or tone) or the level of realism and/or detail present in the illustration. Also, they usually were able to correctly guess if an image was hand-made or generated through computational means.

Mould [35] argued for subjective evaluations as a better way to validate NPR works. According to him, quantitative measurements are harder to design and slower to conduct. Subjective evaluation is specially suitable when the goal of a NPR method is aesthetic, such as when trying to emulate a specific artistic style or simply creating beautiful images, instead of serving for a specific purpose. More recently, Mould and Rosin [36] proposed a benchmark image set to evaluate NPR results.

Spicker et al. [45] compared the number of primitives (such as spots in stippling or small lines) with the perceived abstraction quality of computer-generated images. They found a logarithmic relationship between the amount of primitives in an image and its perceived quality, finding related to the Weber-Fechner law proposed to quantify the human reaction to sensory stimulus, which states a logarithmic relationship between a stimulus and how the human mind perceives it (in other words, the more the stimulus grow, the lesser is the effect of each differential increase in the observer).

In this article, we combined some of the above ideas on validation to design our own evaluation method, as we will explain in the Sect. 5.

3   Methodology

For this work, we expanded our woodcut synthesis method, explained in the Subsect. 2.3. A more detailed description is available in [29]. In this section we will describe the additions over our previous work.

3.1   Input Images and Manual Control

To enable a higher level of user control over the result, we added two optional image inputs, which allows the manual setting of which regions will have plain black and white color and the orientation of wood carvings (i.e., the direction of RD stripes and stretched spots).

The black and white regions image (Fig. 4c) consists in a gray-scale image where black and white pixels denote the pixels which will have a plain color in the final result, while any shade of gray will show the craving pattern as determined by RD. If this image is being used, then we do not calculate black and white regions by its average gray-scale color, neither we merge the smaller regions into the border. We also added a parameter, *Border Color*, to manually set the color of region borders, instead of the default black.

The orientation image (Fig. 4d) is a full-color image map, in which the color of each pixel is related to the orientation angle θ. This image map is read with the HSV color mode, and, for each pixel with saturation equals to 1.0 (i.e., maximum saturation), we calculate its orientation angle θ by the following linear equation:

$$\theta_{i,j} = \pi h_{i,j}.$$

where $\theta_{i,j}$ is the orientation angle for the pixel (i, j) and $h_{i,j}$ is its hue value (in HSV) from the orientation image. This yields angles between 0 and π. In our RD system, two opposites directions (i.e., θ and $\theta + \pi$) produces the same orientation for the stripes, so we do not need to consider angles larger than π. For pixels with saturation lower than the maximum, the orientation angle is calculated by any of the three automatic methods, so this option still allows automatic calculation of θ for any area in which this is desired.

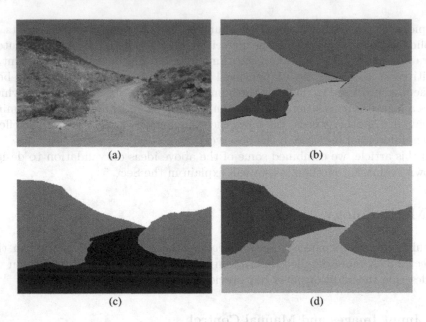

Fig. 4. The image inputs used in our system. (a) The original base image. (b) The segmentation image, where each color represents a segmented region. (c) The black and white regions image. Black and white regions will have the respective pure color in the final image, with only gray areas having carving patterns due to RD. (d) The orientation image, in which each color corresponds to an orientation angle for the carvings.

3.2 Noise Addition

In order to simulate the aspect of woodgrain, and improve the final rendering, we added a new substep during post-processing to add noise to the result after Otsu binarization. It should be emphasized that this noise has a distinct nature than the one removed by the median filter, since it intends to reproduce the effect of wood irregularities on the printing process, which prevents ink transmission to the paper for some points in the painted area. We apply this noise only on the black areas of the resulting image, since we suppose for our work a white paper and black ink (both for the standard woodcut and the white-line one). There are three noise types available in our system: Gaussian, Poisson and Contoli.

The noise types based on statistical distributions (Gaussian and Poisson) generate a point-like noise. In these methods, we alter the color value of each black pixel (color equals to 0) to a sample coming from the distribution in question (for the Gaussian noise, which works with real numbers instead as integers as Poisson, the value is converted to an integer to allow representation in the color space). Any values below 0 and 255 are considered as, respectively, 0 and 255 to avoid overflow issues. Since in both methods the sample results are mostly nearer to 0 than 255, the overall aspect is still dark, with points of lighter colors meaning areas where paint failed to pass to the paper. Table 2 summarize the parameters for these two noise generation methods.

Table 2. Parameters used in Gaussian and Poisson noise.

Parameter	Meaning	Default value
μ_{Gauss}	Mean value	0
σ_{Gauss}	Standard deviation	50
$\sigma_{Poisson}$	Standard deviation	6
$\lambda_{Poisson}$	Expected number of occurrences	5

The Contoli noise uses a different mechanism, adapted from [6], which produces a line-like noise. In this method, semi-transparent objects are iteratively added to random positions in the image, until a noisy texture is generated. In our case, our object is an horizontal line with 1 pixel of width and a randomly determined length $2l + 1$, where l is a random number between $l_{Contoli_{min}}$ and $l_{Contoli_{max}}$ and corresponds to each "arm" of the line, starting from its central point. We add $N_{Contoli}$ lines to the image, randomly choosing a central point for each line and "growing" it l pixels to each horizontal side, increasing the color intensity of these pixels by $Color_{Contoli}$ (this has the same effect of a transparent white line placed over the image). Any value above 255 is capped to 255 (so already white pixels will not change). Table 3 summarizes the parameters for the Contoli noise.

Table 3. Parameters used in Contoli noise.

Parameter	Meaning	Default value
$N_{Contoli}$	Quantity of lines added to the image	3000
$Color_{Contoli}$	Color added to each pixel in the line	40
$l_{Contoli_{min}}$	Minimum length of the line "arm"	2
$l_{Contoli_{max}}$	Maximum length of the line "arm"	8

4 Results

For our experiments, we used the image dataset ADE20k [54], which contains several different kinds of natural images, from landscapes to indoor scenes, and each image is semantically segmented (i.e., the many elements are segmented in a meaningful way for human viewers instead of by some visual property). We also used other images, such as the lighthouse image used in previous woodcut works (Fig. 5a). In these cases, we performed a manual segmentation over it (Fig. 5b).

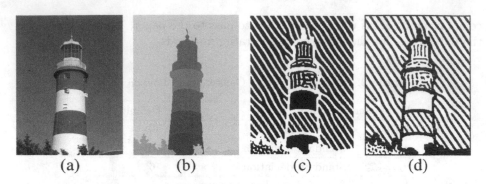

(a)   (b)   (c)   (d)

Fig. 5. Lighthouse image used in our tests (adapted from [29]). (a) Original image from [28]. (b) Manual segmentation of this image. (c) Result for adaptive orientation (with $k = \{1, 11, 21\}$, $T_{diff} = 10.0$, $D_{bh} = 0.040$, $D_{bv} = 0.020$, and inverted colors). (d) Result for adaptive orientation (with $k = \{1, 11, 21\}$, $T_{diff} = 10.0$, $D_{bh} = 0.040$, $D_{bv} = 0.020$ and Contoli noise with default values).

By default our results used orientation by region, no noise, and the default value for the parameters, unless stated otherwise. To translate the processing result into image, we use the static visualization option, since its results are almost equal to the dynamic option in most cases and it prevents the few cases where no pattern can be formed due to an outlier concentration of b in a pixel.

Figure 6 shows how for the same base image it is possible to obtain different patterns in the final woodcut, as combinations of areas with stripes and spots in different sizes. When using orientation by pixel or adaptive (Fig. 6c, d, j, k, l) some details internal to the regions are shown in the engraving patterns of the final image. Figure 7 shows the effect of different sizes for the stripes and spots in the figure by enlarging the difference between S_{min} and S_{max}, allowing more details of the hill contours to be preserved into the final image.

Variations of the diffusion coefficient D_b was used as the mechanism to differentiate the pattern into stripes and spots by the usage of the parameter δ [29] (Fig. 6e and i). So we investigated what happens when we change the value of the other diffusion coefficient, D_a. The first row of Fig. 8 shows the results for the same parameters used in Fig. 5c, with long, straight stripes. As we increased the value of D_a, lines seems a bit more thicker and more bifurcations appear, but overall there is not a big difference between these images. In the second row, we reduced the difference between D_{bh} and D_{bv}, reducing the tendency to form straight, continuous stripes. Now the effect of D_a variation is more noticeable: lower values of this coefficient generates thin, broken stripes mixed with some stretched spots (specially inside the lighthouse building), while higher values yield thicker, sinuous stripes with more bifurcations. For all of these images, $D_{ah} = D_{av}$. Simulations with $D_{ah} \neq D_{av}$ did not produced interestingly different patterns than the ones in Fig. 8, and they altered the orientation angle of the stripes due to changing how diffusion works on each direction, and so reduced the ability to control our system using the pre-processing procedures.

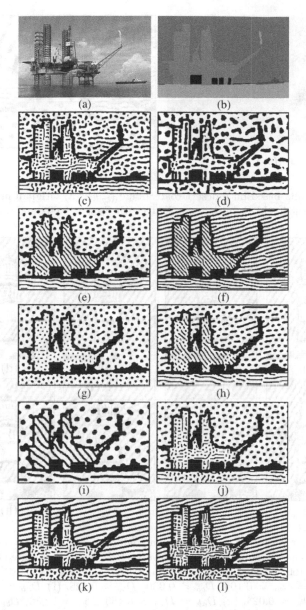

Fig. 6. Our results with variation of the parameters for the same image of an oil platform (adapted from [29]). (a) Original image from the ADE20k dataset [54]. (b) Segmentation of the image from the ADE20k dataset. (c) Result for θ calculated by pixel ($D_{bh} = 0.040$, $D_{bv} = 0.020$). (d) Result for θ calculated by pixel ($D_{bh} = 0.040$, $D_{bv} = 0.020$, $S_{min} = 0.002$, $S_{max} = 0.005$). (e) Result using the default value for parameters. (f) Result for $\delta_{max} = 0.5$. (g) Result for $\delta_{min} = 1.2$ and $\delta_{max} = 1.2$. (h) Result for $\delta_{min} = 0.8$ and $\delta_{max} = 1.2$. (i) Result for $S_{min} = 0.002$ and $S_{max} = 0.005$. (j) Result for adaptive orientation. (k) Result for adaptive orientation ($D_{bh} = 0.040$ and $D_{bv} = 0.020$). (l) Result for adaptive orientation ($D_{bh} = 0.040$, $D_{bv} = 0.020$ and $\delta_{max} = 0.5$).

Fig. 7. Our results based on a desert road image ([29]). (a) Original image from the ADE20k dataset [54]. (b) Resulting woodcut (with $S_{min} = 0.002$ and $S_{max} = 0.018$).

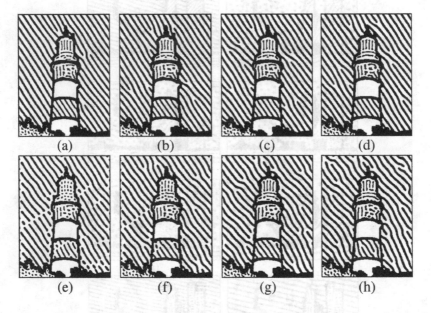

Fig. 8. Results for variations in the diffusion coefficient D_a. All these results used the same parameters as Fig. 5c with normal, not inverted color, unless stated otherwise. (a) $D_{ah} = D_{av} = 0.100$. (b) $D_{ah} = D_{av} = 0.125$. (c) $D_{ah} = D_{av} = 0.150$. (d) $D_{ah} = D_{av} = 0.175$. (e) $D_{ah} = D_{av} = 0.100$, $D_{bh} = 0.030$, $D_{bv} = 0.025$. (f) $D_{ah} = D_{av} = 0.125$, $D_{bh} = 0.030$, $D_{bv} = 0.025$. (g) $D_{ah} = D_{av} = 0.150$, $D_{bh} = 0.030$, $D_{bv} = 0.025$. (h) $D_{ah} = D_{av} = 0.175$, $D_{bh} = 0.030$, $D_{bv} = 0.025$.

The results after noise addition are shown in Fig. 9. Figures 9a to b show the line-like noise generated by Contoli noise, while the point-like noise produced by Gaussian and Poisson noise are shown in Figs. 9c and d, respectively. In Contoli noise, a larger value for $Color_{Contoli}$ together with a smaller amount of noise elements added ($N_{Contoli}$) (Fig. 9a) resulted in clearly visible line-like irregularities in the wood, while decreasing $Color_{Contoli}$ while increasing $N_{Contoli}$

(Fig. 9b) turns the noise into a subtle variation on the black background color of the painted part. The main difference which can be seen between Gaussian and Poisson noise is that the former produced a higher contrast between the noise elements and the black background, being more prone to yield white or light gray points (Fig. 9c). In real woodcuts, both dots and short stripes can be found as irregularities in the wood, holding a similar aspect to our noise methods (Fig. 10).

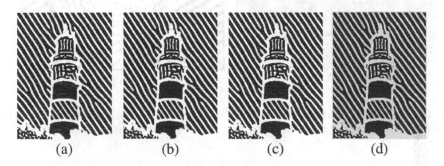

(a)   (b)   (c)   (d)

Fig. 9. Results with noise added in the post-processing step. These simulations have the same parameters as Fig. 5c. (a) Contoli Noise with $N_{Contoli} = 3000$, $l_{Contoli_{min}} = 2$, $l_{Contoli_{max}} = 8$ and $Color_{Contoli} = 40$. (b) Contoli Noise with $N_{Contoli} = 30000$, $l_{Contoli_{min}} = 2$, $l_{Contoli_{max}} = 8$ and $Color_{Contoli} = 4$. (c) Gaussian Noise with $\sigma_{Gauss} = 50$. (d) Poisson Noise with $\sigma_{Poisson} = 12$, $\lambda_{Poisson} = 2$.

(a)   (b)

Fig. 10. Real Brazilian Northeast woodcuts with irregularities in the wood. The images were digitally treated to black and white, with the original woodcut (in a green-stained paper color) in the upper left corner of each image. (a) *A Rendeira*, by Ciro Fernandes [9] showing line-like irregularities similar to our Contoli noise, (b) detail of *Os bóias frias*, by da Silva [44] showing point-like irregularities like our Gaussian or Poisson noise.

Our previous tests were all based on natural images as input (i.e., photographs). To make our mechanism more versatile, it should be able to produce woodcuts from other types of images, such as 3D rendering scenes. We used two

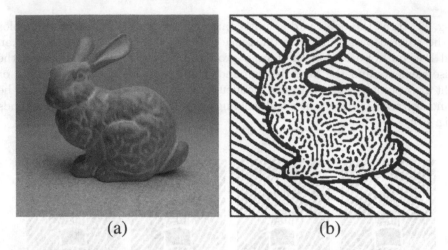

(a)   (b)

Fig. 11. Our results based on a rendering of the Stanford Bunny. (a) Original rendering from [55]. (b) Resulting woodcut (with $D_{bh} = 0.040$, $D_{bv} = 0.020$, $S_{min} = 0.002$ and $S_{max} = 0.018$).

(a)   (b)

Fig. 12. Our results based on a 3D model of the battle cruiser Dunkerque. (a) Original rendering (available at [7]). (b) Resulting woodcut (with orientation set by image for the ocean region, black and white region image to set the sky as black, $D_{bh} = 0.040$, $D_{bv} = 0.020$, $S_{min} = 0.002$, $S_{max} = 0.018$, inverted colors and Poisson noise with $\sigma_{Poisson} = 5$).

images, one from the Stanford Bunny (Fig. 11) and the other from a 3D model of the French battle cruiser Dunkerque (Fig. 12). These tests show that both for natural images and 3D rendering scenes our system is able to generate woodcuts consistent with the input image. We point out that the many contour details of the inner region of the bunny body are present in the final woodcut, as the eyes and the separation between head and main body.

We observed a certain degree of similarity between our results and the wood-cuts typical from the Brazilian Northeast region (Fig. 13). Both our results and these woodcuts commonly have areas with pure black and white color, the usage of different geometric figures to express distinct textures (such as dots, continuous lines and broken lines), an abstraction of the form of figures and the already mentioned irregularities in the woodgrain (Fig. 10).

5   Evaluation

As stated before, evaluation is a challenging task in the context of NPR or other mechanisms to generate artwork. To validate our results, we first compared them with methods already known in the literature and with an actual woodcut of one of the input images we used in our simulations. Then we proceeded to a qualitative evaluation through a questionnaire where the participants qualified both our results and a set of similar images.

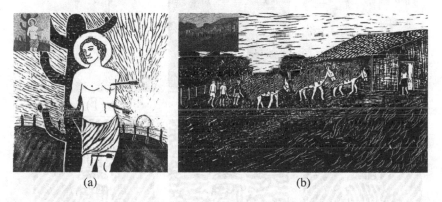

(a)   (b)

Fig. 13. Real Brazilian Northeast woodcuts (adapted from [29]). The images were digitally treated to black and white, with the original woodcut (in a green-stained paper color) in the upper left corner of each image.(a) *São Sebastião*, by Francorli [10]. (b) *Meninos Cambiteiros*, by Gonzaga [11].

5.1   Visual Comparison

To properly compare our results with previous works from the literature we used the lighthouse image (Fig. 5), since it was used on Mello et al. [28] (Fig. 14a) and Li's et al. [23] (Fig. 14b). Our results are shown on the last row (Fig. 14g to i). The three mechanisms produced distinct styles of woodcuts, with ours producing continuous stripes (Figs. 14g and i), which are not present in the other works, where the wood carvings are more similar to small cuts. This shows that our method is able to expand the set of artistic styles that can synthesize by computational means, since it presents different features than the previous works, such as spots and stripes in different regions of the same image with the

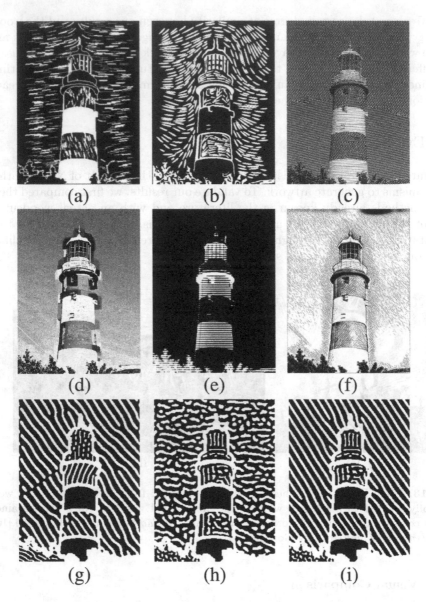

(a)   (b)   (c)

(d)   (e)   (f)

(g)   (h)   (i)

Fig. 14. Comparison between different woodcut synthesis methods (expanded from [29]). (a) Mello's result [28]. (b) Li's result [23]. (c) Xylograph plugin [1], from [28]. (d) Simplify 3 Preset List (Adobe Photoshop plugin) for the Wood Carving option [47]. (e) A GIMP pipeline for woodcut-like images [50]. (f) Prisma (smartphone app) with the Light Summer Reading filter with sharpen +100, contrast +100 and gamma −100 [38]. (g) Our result (with $k = \{1, 11, 21\}$, $T_{diff} = 10.0$ and inverted colors). (h) Our result for θ calculated by pixel (with $k = \{1, 11, 21\}$, $T_{diff} = 10.0$, $D_{bh} = 0.040$, $D_{bv} = 0.020$, and inverted colors). (i) Our result for adaptive orientation (for the parameters used see Fig. 5c).

possibility of different sizes for them and continuous stripes that travel from one side to another of a given region.

Besides academic works, we also tested some graphic design applications, such as the Adobe Photoshop plugin Simply 3 Preset List (Fig. 14d), a GIMP pipeline which intends to produce woodcts (Fig. 14e) and the smartphone app Prisma (Fig. 14f), besides the Adobe Photoshop plugin Xylograph (Fig. 14c) already tested by Mello et al. [28]. None of these were able to produce an image resembling actual woodcuts.

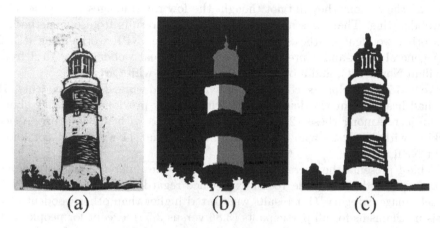

Fig. 15. Comparison of our result with an actual woodcut. (a) Real woodcut by Guilherme León Berno de Jesus [16]. (b) Black and white regions image. (c) Our results using setting of black and white parts and orientation by an image loading, Border Color = 255 (i.e., white border color), $T_{size} = 600$, $k = \{1, 11, 21\}$, $T_{diff} = 10.0$, $D_{bh} = 0.040$, $D_{bv} = 0.020$ and inverted colors.

Another comparison we did was with an actual woodcut produced to resemble one of our input images, specifically the lighthouse image. This actual woodcut was done by the artist Guilherme León Berno de Jesus [16] without any specifications about how our system operates, except for the presence of spots and stripes, so the woodcut would be in a style which our system can replicate (Fig. 15a). To reproduce this image using our system, we used the two optional manual inputs (black and white regions and orientation angles). The black and white regions image is shown in (Fig. 15b), while the final result is in Fig. 15c. With the manual inputs, we were able to replicate the overall appearance of the manual woodcut, but not specific details such as the windows in the lighthouse top.

5.2 Qualitative Evaluation

The qualitative evaluation was performed using a questionnaire, present to several participants. After collecting etnographic data (such as previous experience with woodcuts and NPR), we explained the process to create actual woodcuts. Then we showed a set of images in a random order, containing both our results, results of other techniques, actual woodcuts and other manual and computer-generated black and white images. For each image the questionnaire asked the participants to give it a rating regarding how plausible they consider that the image could be produced by the manual woodcut technique, from "1" (very implausible) to "5" (very plausible). That way we could assess the ability of our method to replicate the woodcut style, instead of measuring subjective aesthetic preference as simply asking if they like the images. Finally, we asked the participants the reasons they did not thought the low rated images were generated by woodcutting. There was a total of 23 images: 7 results from our method, 3 from other woodcut synthesis methods, 2 from other NPR works using RD, 2 from general black-and-white NPR procedures, 6 real woodcuts (with 3 from Brazilian Northeast), and 3 other manual black-and-white artworks.

We had 72 participants, containing only 2 who had worked with woodcuts, 51 who had heard about woodcuts, and 19 without any previous knowledge about the subject. Among these 72 participants, we had 3 with previous experience working with NPR, 25 who had heard about it, and 44 which did not know about NPR.

Table 4 presents the average ratings given by the different types of participants (separated by their degree of experience regarding woodcuts and NPR), to each image category. Our results were rated higher than other woodcut synthesis mechanisms for all participants (3.86 versus 3.55), except for people with previous experience working with NPR (2.90 versus 3.78), although our rating was slightly below average (2.93 versus 2.17) for people with experience in woodcutting. Interestingly, actual woodcuts received a lower rating than the average given to all images and our results for both people who have worked with woodcuts (3.00) and the average of all participants (3.64), but the ones who had worked with NPR gave a higher rating not only for the real woodcuts but for other woodcut synthesis methods, being more critical about our results. Ratings given by people without experience working with either of the techniques mentioned, but had heard about them (the last two columns) are similar to overall ratings (the second column).

About the reasons stated to give a lower rating, participants who had worked with NPR mentioned the excess of details which would be hard to manually produce in a woodcut, aliasing, and photorealistic appearance for some images, which denotes their digital origin. The ones with experience in woodcuts did not gave specific reasons, but stated that perceived these images as different from actual woodcuts.

While most participants considered relatively plausible that our images could be produced by the woodcut process, more than other woodcut synthesis mechanisms, the lower rating given by people with experience in NPR suggests that we need to improve our method to make more realistic woodcuts. However, the small sample size of experienced people indicates that we should read the validation results with caution.

Table 4. Qualitative evaluation results.

Image category	All participants	Worked w/woodcuts	Worked w/NPR	Know woodcuts	Know NPR
All images	3.67	3.04	3.65	3.63	3.61
Our results	3.86	2.93	2.90	3.82	3.65
Other woodcut synthesis	3.55	2.17	3.78	3.53	3.51
Other RD	3.19	3.75	3.67	3.13	3.26
Other NPR	3.64	3.50	3.67	3.51	3.52
Actual woodcuts	3.64	3.00	4.17	3.67	3.83
Other manual artworks	3.74	3.50	4.22	3.65	3.44

6 Conclusion and Future Work

This article expanded on our previous woodcut synthesis by RD with the addition of noise and new input images to better control our system. We also did new experiments, both using the new functionalities and the previous ones. To validate our results, we performed a comparison with an actual woodcut generated from the same input image we used in our tests and a qualitative evaluation through a questionnaire to assess how participants consider plausible that our results could be produced by the woodcutting technique, i.e., if our system is able to replicate the style of actual woodcuts.

Our results hold a different aspect than the ones from previous methods, being similar to certain styles of woodcuts, particularly from the Brazilian Northeast region. We have therefore expanded the set of artwork styles that can be produced through computer-based tools. This work also confirms the potential of RD for NPR tasks. The questionnaire responses show that, on one hand most participants considered plausible that our results could be generated by the same process as actual woodcuts, but on the other hand the few participants who were more experienced with the technique gave a lower rating. Given that only 2 respondents had worked with woodcuts, validation should be taken with caution.

Among the future work which can be done to improve our results, the possibility to use more than one type of noise in the same image could improve the replication of the woodgrain aspect. Manual segmentation is demanding, and most segmentation techniques, which relies in color variations, are limited in their ability to separate the objects inside an image in a meaningful way. Using trainable segmentation methods could generate a more semantically significant segmentation of image regions [39]. Finally, improvements on the usability and system interface could ease the usage for non-programmers, allowing artists without advanced computational skills to produce their own artworks.

References

1. AMPHISOFT: Xylograph plugin for photoshop (2007). http://photoshop.msk.ru/as/ngrave.html. Accessed 7 Feb 2019
2. Bard, J.: A model for generating aspects of zebra and other mammalian coat patterns. J. Theor. Biol. **93**(2), 363–385 (1981). https://doi.org/10.1016/0022-5193(81)90109-0
3. Bard, J., Lauder, I.: How well does turing's theory of morphogenesis work? J. Theor. Biol. **45**(2), 501–531 (1974). https://doi.org/10.1016/0022-5193(74)90128-3
4. Barros, R.S., Walter, M.: Synthesis of human skin pigmentation disorders. In: Computer Graphics Forum, vol. 36, pp. 330–344. Wiley Online Library (2017). https://doi.org/10.1111/cgf.12943
5. Chi, M.T., Liu, W.C., Hsu, S.H.: Image stylization using anisotropic reaction diffusion. Vis. Comput. **32**(12), 1549–1561 (2016). https://doi.org/10.1007/s00371-015-1139-2
6. Contoli, A.: Procedural seamless noise texture generator (2015). https://www.codeproject.com/Articles/838511/Procedural-seamless-noise-texture-generator. Accessed 31 Jan 2018
7. Cordier, C.: Croiseur de bataille Dunkerque. Rendering available at Wikimedia Commons (2006). https://commons.wikimedia.org/wiki/File:Dunkerque_3d.jpeg; model available in http://le.fantasque.free.fr/. Accessed 2 July 2019
8. Fan, D., Liu, S., Wei, Y.: Fruit ring rot simulation based on reaction-diffusion model, pp. 199–205, September 2013. https://doi.org/10.1109/ICVRV.2013.38
9. Fernandes, C.: A Rendeira. Available at the Centro Nacional de Folclore e Cultura Popular (CNFCP) site (2018). http://www.cnfcp.gov.br/interna.php?ID_Secao=64. Accessed 14 Feb 2019
10. Francorli: São Sebastião. Available at the Centro Nacional de Folclore e Cultura Popular (CNFCP) site (2018). http://www.cnfcp.gov.br/interna.php?ID_Secao=64. Accessed 10 Dec 2018
11. Gonzaga, J.L.: Meninos Cambiteiros. Available at the Centro Nacional de Folclore e Cultura Popular (CNFCP) site (2018). http://www.cnfcp.gov.br/interna.php?ID_Secao=64. Accessed 10 Dec 2018
12. Haeberli, P.: Paint by numbers: abstract image representations. In: ACM SIGGRAPH Computer Graphics, vol. 24, pp. 207–214. ACM (1990). https://doi.org/10.1145/97880.97902
13. Hertzmann, A.: Painterly rendering with curved brush strokes of multiple sizes. In: Proceedings of the 25th Annual Conference on Computer Graphics and Interactive Techniques, pp. 453–460. ACM (1998). https://doi.org/10.1145/280814.280951
14. Isenberg, T.: Evaluating and validating non-photorealistic and illustrative rendering. In: Rosin, P., Collomosse, J. (eds.) Image and Video-Based Artistic Stylisation. CIVI, vol. 42, pp. 311–331. Springer, London (2013). https://doi.org/10.1007/978-1-4471-4519-6_15
15. Isenberg, T., Neumann, P., Carpendale, S., Sousa, M.C., Jorge, J.A.: Non-photorealistic rendering in context: an observational study. In: Proceedings of the 4th International Symposium on Non-photorealistic Animation and Rendering, NPAR 2006, pp. 115–126. ACM, New York (2006). https://doi.org/10.1145/1124728.1124747
16. de Jesus, G.L.B.: Lighthouse. Artwork generated to this project (2018)
17. Jho, C.W., Lee, W.H.: Real-time tonal depiction method by reaction-diffusion mask. J. R. Time Image Process. **13**(3), 591–598 (2017). https://doi.org/10.1007/s11554-016-0643-6

18. Kider Jr., J.T., Raja, S., Badler, N.I.: Fruit senescence and decay simulation. Comput. Graph. Forum **30**(2), 257–266 (2011). https://doi.org/10.1111/j.1467-8659. 2011.01857.x
19. Krogh, C.: Snorre Sturluson - Illustration for Heimskringla 1899-edition (1890). Available at Wikimedia Commons. https://en.wikipedia.org/wiki/File:Snorre_ Sturluson-Christian_Krohg.jpg. Accessed 17 Feb 2019
20. Kyprianidis, J.E., Collomosse, J., Wang, T., Isenberg, T.: State of the "art": a taxonomy of artistic stylization techniques for images and video. IEEE Trans. Vis. Comput. Graph. **19**(5), 866–885 (2013). https://doi.org/10.1109/TVCG.2012.160
21. Lai, Y.K., Rosin, P.L.: Non-photorealistic rendering with reduced colour palettes. In: Rosin, P., Collomosse, J. (eds.) Image and Video-Based Artistic Stylisation. CIVI, vol. 42, pp. 211–236. Springer, London (2013). https://doi.org/10.1007/978-1-4471-4519-6_11
22. Li, J., Xu, D.: A scores based rendering for Yunnan out-of-print woodcut. In: 2015 14th International Conference on Computer-Aided Design and Computer Graphics (CAD/Graphics), pp. 214–215. IEEE (2015). https://doi.org/10.1109/ CADGRAPHICS.2015.42
23. Li, J., Xu, D.: Image stylization for Yunnan out-of-print woodcut through virtual carving and printing. In: El Rhalibi, A., Tian, F., Pan, Z., Liu, B. (eds.) Edutainment 2016. LNCS, vol. 9654, pp. 212–223. Springer, Cham (2016). https://doi.org/ 10.1007/978-3-319-40259-8_19
24. Lu, C., Xu, L., Jia, J.: Combining sketch and tone for pencil drawing production. In: Proceedings of the Symposium on Non-Photorealistic Animation and Rendering, NPAR 2012, Goslar Germany, Germany, pp. 65–73. Eurographics Association (2012). http://dl.acm.org/citation.cfm?id=2330147.2330161
25. Maciejewski, R., Isenberg, T., Andrews, W.M., Ebert, D., Costa Sousa, M.: Aesthetics of hand-drawn vs. computer-generated stippling, pp. 53–56, January 2007. https://doi.org/10.2312/COMPAESTH/COMPAESTH07/053-056
26. Maciejewski, R., Isenberg, T., Andrews, W.M., Ebert, D., Costa Sousa, M., Chen, W.: Measuring stipple aesthetics in hand-drawn and computer-generated images. IEEE Comput. Graph. Appl. **28**, 62–74 (2008). https://doi.org/10.1109/MCG. 2008.35
27. Martín, D., Arroyo, G., Luzon, M., Isenberg, T.: Scale-dependent and example-based stippling. Comput. Graph. **35**, 160–174 (2011). https://doi.org/10.1016/j. cag.2010.11.006
28. Mello, V., Jung, C.R., Walter, M.: Virtual woodcuts from images. In: Proceedings of the 5th International Conference on Computer Graphics and Interactive Techniques in Australia and Southeast Asia, pp. 103–109. ACM (2007). https://doi. org/10.1145/1321261.1321280
29. Mesquita., D.P., Walter., M.: Reaction-diffusion woodcuts. In: Proceedings of the 14th International Joint Conference on Computer Vision, Imaging and Computer Graphics Theory and Applications - Volume 1: GRAPP, pp. 89–99. INSTICC, SciTePress (2019). https://doi.org/10.5220/0007385900890099
30. Mizuno, S., Kasaura, T., Okouchi, T., Yamamoto, S., Okada, M., Toriwaki, J.: Automatic Generation of Virtual Woodblocks and Multicolor Woodblock Printing. Comput. Graph. Forum (2000). https://doi.org/10.1111/1467-8659.00397
31. Mizuno, S., Kobayashi, D., Okada, M., Toriwaki, J., Yamamoto, S.: Creating a virtual wooden sculpture and a woodblock print with a pressure sensitive pen and a tablet. Forma **21**(1), 49–65 (2006)
32. Mizuno, S., Okada, M., Toriwaki, J.i.: Virtual sculpting and virtual woodcut printing. J. Vis. Comput. **14**, 39–51 (1998). https://doi.org/10.1007/s003710050122

33. Mizuno, S., Okada, M., Toriwaki, J.i., Yamamoto, S.: Improvement of the virtual printing scheme for synthesizing Ukiyo-e, vol. 3, pp. 1043–1046, January 2002. https://doi.org/10.1109/ICPR.2002.1048217

34. Mizuno, S., Okada, M., Toriwaki, J.i., Yamamoto, S.: Japanese traditional printing "Ukiyo-e" in a virtual space (2002)

35. Mould, D.: Authorial subjective evaluation of non-photorealistic images. In: Proceedings of the Workshop on Non-Photorealistic Animation and Rendering, NPAR 2014, pp. 49–56. ACM, New York (2014). https://doi.org/10.1145/2630397.2630400

36. Mould, D., Rosin, P.L.: A benchmark image set for evaluating stylization. In: Proceedings of the Joint Symposium on Computational Aesthetics and Sketch Based Interfaces and Modeling and Non-Photorealistic Animation and Rendering, Expresive 2016, Aire-la-Ville, Switzerland, Switzerland, pp. 11–20. Eurographics Association (2016). http://dl.acm.org/citation.cfm?id=2981324.2981327. https://doi.org/10.2312/exp.20161059

37. Murray, J.: A pre-pattern formation mechanism for animal coat markings. J. Theor. Biol. 88(1), 161–199 (1981). https://doi.org/10.1016/0022-5193(81)90334-9

38. PRISMA LABS: Prisma photo editor (2016). https://prisma-ai.com/. Accessed 7 Feb 2019

39. Ronneberger, O., Fischer, P., Brox, T.: U-Net: convolutional networks for biomedical image segmentation. In: Navab, N., Hornegger, J., Wells, W.M., Frangi, A.F. (eds.) MICCAI 2015. LNCS, vol. 9351, pp. 234–241. Springer, Cham (2015). https://doi.org/10.1007/978-3-319-24574-4_28

40. Salesin, D.H.: Non-photorealistic animation & rendering: 7 grand challenges. In: Keynote Talk at Second International Symposium on Non-Photorealistic Animation and Rendering, NPAR 2002, Annecy, France, 3–5 June 2002. http://www.research.microsoft.com/~salesin/NPAR.ppt

41. Sanderson, A.R., Kirby, R.M., Johnson, C.R., Yang, L.: Advanced reaction-diffusion models for texture synthesis. J. Graph. GPU Game Tools 11(3), 47–71 (2006). https://doi.org/10.1080/2151237X.2006.10129222

42. Schumann, J., Strothotte, T., Laser, S., Raab, A.: Assessing the effect of non-photorealistic rendered images in cad. In: Proceedings of the SIGCHI Conference on Human Factors in Computing Systems, CHI 1996, pp. 35–41. ACM, New York (1996). https://doi.org/10.1145/238386.238398

43. Setlur, V., Wilkinson, S.: Automatic stained glass rendering. In: Nishita, T., Peng, Q., Seidel, H.-P. (eds.) CGI 2006. LNCS, vol. 4035, pp. 682–691. Springer, Heidelberg (2006). https://doi.org/10.1007/11784203_66

44. da Silva, E.F.: Os bóias frias. Available at the Centro Nacional de Folclore e Cultura Popular (CNFCP) site (2018). http://www.cnfcp.gov.br/interna.php?ID_Secao=64. Accessed 14 Feb 2019

45. Spicker, M., Götz-Hahn, F., Lindemeier, T., Saupe, D., Deussen, O.: Quantifying visual abstraction quality for computer-generated illustrations. ACM Trans. Appl. Percept. 16(1), 51–520 (2019). https://doi.org/10.1145/3301414

46. Stuyck, T., Da, F., Hadap, S., Dutré, P.: Real-time oil painting on mobile hardware. In: Computer Graphics Forum, vol. 36, pp. 69–79. Wiley Online Library (2017)

47. TOPAZ LABS: Simplify plugin for photoshop (2018). https://topazlabs.com/simplify/. Accessed 23 Sept 2018

48. Turing, A.M.: The chemical basis of morphogenesis. Philos. Trans. R. Soc. Lond. B 237(641), 37–72 (1952). https://doi.org/10.1098/rstb.1952.0012

49. Turk, G.: Generating textures for arbitrary surfaces using reaction-diffusion. In: Computer Graphics (Proceedings of SIGGRAPH 1991), pp. 289–298, July 1991. https://doi.org/10.1145/127719.122749
50. Welch, M.: Gimp tips - woodcut (2018). http://www.squaregear.net/gimptips/wood.shtml. Accessed 23 Sept 2018
51. Winnemöller, H.: XDoG: advanced image stylization with extended difference-of-Gaussians. In: Proceedings of the ACM SIGGRAPH/Eurographics Symposium on Non-Photorealistic Animation and Rendering, pp. 147–156. ACM (2011). https://doi.org/10.1016/j.cag.2012.03.004
52. Witkin, A., Kass, M.: Reaction-diffusion textures. In: Computer Graphics (Proceedings of SIGGRAPH 1991), pp. 299–308, July 1991. https://doi.org/10.1145/127719.122750
53. Yu, J., Ye, J., Zhou, S.: Reaction-diffusion system with additional source term applied to image restoration. Int. J. Comput. Appl. **147**(2), 18–23 (2016). https://doi.org/10.5120/ijca2016910998. http://www.ijcaonline.org/archives/volume147/number2/25625-2016910998
54. Zhou, B., Zhao, H., Puig, X., Fidler, S., Barriuso, A., Torralba, A.: Scene parsing through ade20k dataset. In: Proceedings of the IEEE Conference on Computer Vision and Pattern Recognition, vol. 1, p. 4. IEEE (2017). https://doi.org/10.1109/CVPR.2017.544
55. Zzubnik: Computed generated render of the "Stanford Bunny" (2019). Available at Wikimedia Commons: https://commons.wikimedia.org/wiki/File:Computer_generated_render_of_the_%22Stanford_Bunny%22.jpg. Accessed 2 July 2019

Synthesising Light Field Volume Visualisations Using Image Warping in Real-Time

Seán K. Martin[✉], Seán Bruton, David Ganter, and Michael Manzke

School of Computer Science and Statistics, Trinity College Dublin, Dublin 2, Ireland
{martins7,sbruton,ganterd,manzkem}@tcd.ie

Abstract. We extend our prior research on light field view synthesis for volume data presented in the conference proceedings of VISIGRAPP 2019 [13]. In that prior research, we identified the best Convolutional Neural Network, depth heuristic, and image warping technique to employ in our light field synthesis method. Our research demonstrated that applying backward image warping using a depth map estimated during volume rendering followed by a Convolutional Neural Network produced high quality results. In this body of work, we further address the generalisation of Convolutional Neural Network applied to different volumes and transfer functions from those trained upon. We show that the Convolutional Neural Network (CNN) fails to generalise on a large dataset of head magnetic resonance images. Additionally, we speed up our implementation to enable better timing comparisons while remaining functionally equivalent to our previous method. This produces a real-time application of light field synthesis for volume data and the results are of high quality for low-baseline light fields.

Keywords: Light fields · View synthesis · Volume rendering · Depth estimation image warping · Angular resolution enhancement

1 Introduction

In our previous work we demonstrated a method for high quality light field synthesis on a specific volume and transfer function [13]. Lacking a baseline for our particular problem, we evaluated multiple different strategies for each step of our method. To be specific, multiple CNNs, depth heuristics and warping methods were experimented with. Although this provided detailed results for our specific dataset, we had minimal tests for generalisation of the method to other volumes and transfer functions. We demonstrated that training the CNN on a heart Magnetic Resonance Imaging (MRI) would not be sufficient to then use the CNN on a head MRI. But, it is possible that the CNN could be trained on head MRIs and generalise to work on other unseen MRIs, or be trained on a set of transfer functions and generalise to work on similar transfer functions. To investigate this idea of generalisability, we performed additional experiments on the best

© Springer Nature Switzerland AG 2020
A. P. Cláudio et al. (Eds.): VISIGRAPP 2019, CCIS 1182, pp. 30–47, 2020.
https://doi.org/10.1007/978-3-030-41590-7_2

performing strategy from our previous work. We present results for a dataset of multiple head MRIs with different transfer functions, a far more variable case than our original dataset of a heart MRI. Furthermore, the implementation has been rewritten to take better advantage of shader functions, leading to far more efficient approach and more accurate timings than the previous work.

As before, we aim towards low baseline light field displays with a fast light field synthesis approach for volumetric data. Although light field display technology is still in its infancy, devices such as The Looking Glass [2] and a near-eye light field display [6] show promising results. Generating a light field allows for intuitive exploration of volumetric data. In gross anatomy, studies have shown that students respond better to three dimensional (3D) anatomical atlases for rapid identification of anatomical structures [16], but is not necessary for deep anatomical knowledge. In this sense, a light field synthesis approaches fits very well for rapid exploration of anatomical structure. For instance, on the microscopic level, histology volume reconstruction [24] could be readily visualised in 3D without the need for glasses. However, it is impossible to render a light field from volumetric data using ray tracing techniques at interactive rates without an extensive and expensive hardware setup. To move towards light field volume visualisation at interactive rates, we use a view synthesis method from a single sample view to avoid rendering every view in the light field.

Unlike conventional images, which record a two dimensional (2D) projection of light rays, a light field describes the distribution of light rays in free space [33] described by a four dimensional (4D) plenoptic function. Capturing an entire light field is infeasible, so a light field is generally sampled by capturing a large uniform grid of images of a scene [7]. However, capturing this 4D data leads to an inherent resolution trade off between dimensions. As such, the angular resolution, the number of images in the grid, is often low compared to the spatial resolution, the size of each image in the grid. Because of this, angular resolution enhancement, or view synthesis, from a limited number of reference views is of great interest and benefit to a light field application.

Angular resolution enhancement for light fields of natural images (images taken by a real camera) has many strong solutions [4,14,34]. However, for light fields from volumetric data, we face some unique challenges. Natural image approaches generally assume a unique depth in the scene, but volume renderings frequently feature semi-transparent objects, resulting in an ill-defined depth. Additionally, as opposed to natural light field imaging where the limitation is the resolution of the sensor, for volumetric data the limitation is time. Speed of the approach is vital to us, as an exact light field rendering can be produced with enough time. As such, many existing approaches are not relevant without heavy modification. For instance, the recent state of the art approach by [32] takes 2.57 s to produce a single 541 × 376 view, while an entire light field from a volume could be produced in a tenth of this time with a powerful Graphics Processing Unit (GPU) for a small volume.

Our proposed method synthesises a light field from a single volume rendered sample image, represented as a 2D grid of images captured by a camera moving along a plane. As the first step in this process, a depth heuristic is used to estimate a depth map during volume ray casting [36]. This depth map is converted to a disparity map using the known virtual camera parameters. Backward image warping is performed using the disparity map to shift information from the single sample image to all novel view locations. Finally, the warped images are passed into a CNN to improve the visual consistency of the synthesised views. This method is demonstrated for an 8×8 angular resolution light field (a grid of 64 images) but there is no inherent limitation on this other than GPU memory.

We previously demonstrated that a CNN increases the visual consistency of synthesised views, especially for those views at a large distance from the sample reference view [13]. However, by testing on a large dataset of head MRIs, we further back up our previous results on the lack of generalisation of the CNN. The CNN must be retrained to be used on a specific volume and transfer function, its performance does not transfer between sets. This is a significant limitation, but the CNN could be ommitted for particular use cases if a lower quality estimate is acceptable. Alternatively, for use cases such as education, a library of specifically trained models could be produced.

The re-implemented method is very fast, an order of magnitude faster than rendering each view in the light by traditional volume ray casting. For low baseline light fields, the results are of high quality, as the single sample view provides sufficient information to estimate the light field. The method is particularly beneficial for large volumes because the time to synthesise a light field is independent of the size and complexity of the volume and rendering techniques once a sample view has been produced. In high precision fields where exact results are required, the synthesis could be used for exploratory purposes and a final render performed when the camera is held fixed. In this manner, inevitable errors in the visualisation technique will not lead to medical misdiagnosis or other serious problems.

2   Background

2.1   Light Field Representation

The first description of the light field was the 7D plenoptic function recording the intensity of light rays travelling in every direction (θ, ϕ) through every location in space (x, y, z) for every wavelength λ, at every time t [1]. As a simplification, the plenoptic function is assumed to be mono-chromatic and time invariant, reducing it to a 5D function $L(x, y, z, \theta, \phi)$. In a region free of any occluders, the radiance along a ray remains constant along a straight line, and the 5D plenoptic function reduces to the 4D function $L(x, y, \theta, \phi)$ [3,7]. This 4D plenoptic function is termed the 4D light field, or simply the light field, the radiance along rays in empty space.

The most common light field representation is to parameterise a ray by its intersection with two planes, a uv plane and an st plane [7]. As such the 4D light field maps rays passing through a point (u, v) on one plane and (s, t) on another plane to a radiance value:

$$L : \mathbb{R}^4 \to \mathbb{R}, \quad (u, v, s, t) \mapsto L(u, v, s, t)$$

With this parameterisation, a light field can be effectively sampled using arrays of rendered or digital images by considering the camera to sit on the uv plane, and capture images on the st plane. It was later shown that if the disparity between neighbouring views in a light field sampling is less than one pixel, novel views can be generated without ghosting effects by the use of linear interpolation [9]. Since sampling rates are rarely this high, to achieve such a densely sampled light field, view synthesis is a necessity.

2.2 Light Field View Synthesis

Light field view synthesis is the process of producing unseen views in the light field from a set of sample views. This is a well studied problem [18,30,35], see [33] for a review of light field view synthesis, as well as multiple other light field image processing problems. However, because speed is essential to our problem, much of the existing literature is unusable without significant modifications. Nonetheless, we recount prominent literature here and elucidate why the majority of it can not be used directly.

A wide range of view synthesis approaches take advantage of the inherent structure of light fields. A set of state of the art structure based approaches revolve around using the properties of Epipolar-Plane Images (EPIs) [32,34]. In this context an EPI is a 2D slice of the 4D light field in which one image plane co-ordinate is fixed and one camera plane co-ordinate is fixed. For instance, fixing a horizontal co-ordinate t^* on the image plane, and v^* on the camera plane produces an EPI:

$$E : \mathbb{R}^2 \to \mathbb{R}, \quad (u, s) \mapsto L(u, v^*, s, t^*)$$

A line of constant slope in an EPI corresponds to a point in 3D space where the slope of the line is related to the depth of the point. For Lambertian objects, this means that the instensity of the light field should not change along a line of constant slope [30]. This strong sense of structure is the basis of accurate view synthesis using EPIs. Other structure based approaches transform the light field to other domains with similar strong contraints [23,28]. Despite the good results from these approaches, the necessary domain transfer or slicing transforms are too slow for our purpose. As an example, [34] takes roughly 16 min to synthesise a $64 \times 512 \times 512$ light field which could be rendered in under a second.

For large-baseline light fields, such as that required for the Looking Glass [2], methods based on multiplane images [22] represent the state of the art [14,38]. These revolve around the principle of splitting reference views into multiple images at different depth planes to gain more information and simplify the aggregation process with great success. In volume rendering, a similar idea has been applied to view synthesis [10], by subdividing the original ray-casting into a layered representation. The layered information can be effectively transformed to a novel view, and combined via compositing an efficient manner. Although this produces fast single view synthesis for a low number of layers, and would be very effective for stereo magnification, the number of views required for a light field adds significant difficulty to the fast application of this paradigm.

As such, fast depth-based warping approaches are most relevant to our problem of synthesising low-baseline light fields, as we can cheaply estimate some form of depth during a volume rendering. With this in mind, the most relevant body of work to the problem at hand is by Srinivasan et al. [26]. This is fast, synthesising a $187 \times 270 \times 8 \times 8$ light field in under one second on a NVIDIA Titan X GPU. For comparison, a state of the art depth based approach [4] takes roughly 12.3 s to generate a single novel view from four input images of 541×376 resolution. Srinivasan et al.'s [26] method is fast, uses deep learning to account for specular highlights, and only requires a single input view to synthesise a full light field. This is very relevant for volume rendering, as speed is essential, surfaces are often anisotropically shaded, and rendering sample views is especially slow.

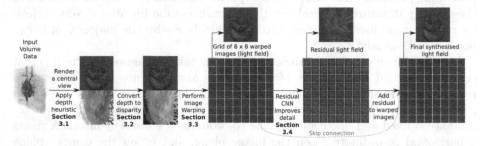

Fig. 1. Our proposed light field synthesis method can be broken down into distinct stages, including an initial depth heuristic calculation stage and a final CNN stage acting as a residual function to improve fine-grained detail. Extracted from our previous work [13].

A single image does not carry enough information to truthfully reconstruct the light field, so at best a good estimate will result from this process. By taking advantage of redundant information in the light field representation, the Srinivasan et al. [26] achieve a high quality estimate, and a method summary follows here. First, a 3D CNN is applied to a single sample view to estimate the depth in the scene for a full light field. Backward warping is then applied to the sample view using the depth maps to estimate a Lambertian light field. Finally, the Lambertian light field is passed through a 3D CNN to help account for specular highlights. This produces high quality results for objects from specific categories, such as flowers. We base our approach on the method of [26], but provide modifications to increase suitability for volume rendering.

Firstly, we avoid the expensive CNN based depth estimation step by estimating depth during volume rendering for a single view. Warping information from sample volume rendered views to synthesise new views [12, 15] is feasible when rendered images do not change dramatically between viewpoints. Zellmann et al. [36] proposed to warp images based on depth heuristics. Due to alpha compositing resulting in transparent surfaces without single depth values, the authors present multiple depth heuristics for image warping. Returning depth value at the voxel where the accumulated opacity along the ray reaches 80% during ray tracing achieved the best balance between speed and quality.

Secondly, we propose to apply a 2D CNN to improve the quality of the novel views as the 3D CNN from [26] is too slow for this problem. Although we have volumetric information in a light field, remapping the 3D volume to a 2D structure is faster and current deep learning architectures are often unable to fully exploit the power of 3D representations [20]. Due to limitations of 3D CNNs, Wang et al. [29] demonstrate how to map a 4D light field into a 2D VGG network [25] instead of using a 3D CNN. This is beneficial as the weights of a pre-trained 2D model can be updated. Additionally, although the 4D filters in 3D CNNs are intuitive to use on a 4D light field, the number of parameters quickly explode.

3   Light Field Synthesis

Our goal is to quickly synthesise a low-baseline light field for visualisation purposes from volumetric data. We demonstrate this for an 8×8 angular resolution light field, but there is no inherent limitation on this size other than GPU memory. The pipeline of the method has not changed from our original description in [13]. The following steps are involved in our light field synthesis (Fig. 1).

1. Render a reference view by direct volume rendering and use a depth heuristic to estimate a depth map during ray casting.
2. Convert the depth map to a disparity map using the intrinsic camera parameters.
3. Apply backward image warping to the reference view using the disparity map to approximate a light field.
4. Apply a CNN to the warped images to improve visual consistency. This is modelled as a residual function which is added to the approximate light field from the previous step.

3.1   Volume Depth Heuristics

To apply image warping based on a depth map, we quickly estimate depth values during the ray casting process. Due to the semi-transparent structures that frequently occur in volume rendering, there is no well defined depth and a heuristic is a necessity. An obvious approach is to use the depth of the first non-transparent voxel along the ray, but this is often ineffective due to near transparent volume information close to the camera. Using isosurfaces gives a good view of depth, but these must be recalculated during runtime if the volume changes.

In our previous work [13], we proposed to modify the best performing single pass depth heuristic from [36]. The heuristic from [36] is to save the depth of the voxel at which the opacity accumulated along a ray exceeds a pre-defined opacity threshold. A depth map formed in this way frequently lacks information at highly transparent regions as the opacity threshold needs to be set to a large value. To counteract this limitation, we applied a two-layered approach. A depth value is saved when a ray accumulates a low threshold opacity and overwritten if the ray accumulates the high threshold opacity. The layered approach can be repeated multiple times, but two layers was enough to capture the volume in most cases. This improved the quality of the depth map over isosurfaces and the best single pass method from Zelmann et al. [36]. A more detailed comparison of different depth heuristics is provided in our previous work [13].

3.2   Converting Depth to Disparity

This mathematical conversion is the same process as our previous paper [13]. During rendering, a depth value from the Z-buffer $Z_b \in [0, 1]$ is converted to a pixel disparity value using the intrinsic camera parameters as follows. The depth buffer value Z_b is converted into normalised device co-ordinates, in the range $[-1, 1]$, as $Z_c = 2 \cdot Z_b - 1$. Then, perspective projection is inverted to give depth in eye space as

$$Z_e = \frac{2 \cdot Z_n \cdot Z_f}{Z_n + Z_f - Z_c \cdot (Z_f - Z_n)} \tag{1}$$

Where Z_n and Z_f are the depths of the camera's near and far clipping planes in eye space, respectively. Note that Z_n should be set as close to the visualised object as possible to improve depth buffer accuracy, while Z_f has negligible effect on the accuracy. Given eye depth Z_e, it is converted to a disparity value d_r in real units using similar triangles [31] as

$$d_r = \frac{B \cdot f}{Z_e} - \Delta x \tag{2}$$

Where B is the camera baseline, or distance between two neighbouring cameras in the grid, f is the focal length of the camera, and Δx is the distance between two neighbouring cameras' principle points. Again, using similar triangles, the disparity in real units is converted to a disparity in pixels as

$$d_p = \frac{d_r W_p}{W_r} \tag{3}$$

Where d_p and d_r denote the disparity in pixels and real world units respectively, W_p is the image width in pixels, and W_r is the image sensor width in real units. If the image sensor width in real units is unknown, W_r can be computed from the camera field of view θ and focal length f as $W_r = 2 \cdot f \cdot \tan(\frac{\theta}{2})$.

3.3 Disparity Based Image Warping

The mathematical description of this operation is identical to our previous work [13]. To synthesise a novel view, a disparity map $D : \mathbb{R}^2 \mapsto \mathbb{R}$ is used to relate pixel locations in a novel view to those in the reference view. Let $I : \mathbb{R}^2 \mapsto \mathbb{R}^3$ denote a reference Red Green Blue (RGB) colour image at grid position (u_r, v_r) with an associated pixel valued disparity map D. Then a synthesised novel view I' at grid position (u_n, v_n) can be formulated as:

$$I'(x + d \cdot (u_r - u_n), \ y + d \cdot (v_r - v_n)) = I(x, \ y)$$
$$\text{where} \quad d = D(x, \ y) \tag{4}$$

As opposed to the non-surjective hole producing forward warping operation, we apply a surjective backward warping. For each pixel in the novel view, information is inferred from the reference view. This results a hole free novel view but it is generally an oversampled from the reference view. Pixels in the novel view that require information from outside the border of the reference view instead used the closest border pixel. This strategy effectively stretches the border of the reference view in the absence of information. Since are dealing with low baseline light fields, this strategy is not overly restrictive.

3.4 Convolutional Neural Network

To improve the fine-grained details of the synthesised light field, we apply a CNN to the resulting images from image warping. The network is framed as a residual function that predicts corrections to be made to the warped images to reduce the synthesis loss in terms of mean squared error. The residual light field has full range over the colour information to allow for removal of predicted erroneous information and addition of predicted improvements. Srinivasan et al. [26] applied 3D convolutions to achieve a similar goal. Because the light field is 4D, 3D CNNs which use 4D filters are intuitive to apply to this problem. However, using 2D convolutions is advantageous as less parameters are required for a 2D CNN. Additionally, more pretrained models and better optimisation tools exist for 2D CNNs than 3D CNNs. Wang et al. [29] previously demonstrated strong evidence that 3D CNNs can be effectively mapped into 2D architectures.

To test remapping the 3D network from [26] into a 2D network, we compared four CNNs architectures to improve the synthesised light field [13]. The CNNs were tested on a single heart MRI with a fixed transfer function from different camera positions. Each network was evaluated based on the difference to the baseline result of pure geometrical image warping in terms of Peak Signal to Noise Ratio (PSNR) and Structural Similarity (SSIM). In the experiments, all networks improved the resulting light field quality measured by SSIM. As such, the 3D CNN Srinivasan et al. [26] could be effectively remapped into a 2D architecture. In our tests, the best performing network in terms of quality was a slightly modified Enhanced Deep Super-Resolution (EDSR) network [8] taking an angular remapped input from [29], achieving an average of 0.923 SSIM on the validation set [13].

We briefly describe the 3D occlusion prediction network from Srinivasan et al. [26], and the modified EDSR network [8] we use in its place. The 3D network is structured as a residual network with $3 \times 3 \times 3$ filters that have access to every view. The input to the 3D network is all warped images and colour mapped disparity maps. Our EDSR network is the same as the original EDSR network from [8], bar removal of spatial upscaling at the last layer and application of tanh activation at the final layer. Spatial upscaling is removed as we only require angular resolution enhancement, while tanh is applied to allow residual output to add and remove information. To map the light field input into the three colour channel RGB input required for EDSR, angular remapping from [29] is applied. Angular remapping transforms an $n \times m$ angular resolution light field with $x \times y$ spatial resolution into an $(n \cdot x) \times (m \cdot y)$ image. In this remapped image, the uppermost $n \times m$ pixels contain the upper-left pixel from each of the original $n \times m$ light field views. There are two significant differences in this EDSR network from the network of Srinivasan et al. [26], besides being a 2D network. Firstly, the disparity map is not input to the network, reducing the number of input channels. Secondly, the 3×3 filters used in this network only consider the nearest neighbours to a view and this local connectivity is very efficient.

4   Implementation

Note that these test system specifications differ from our previous work [13]. All new experiments were performed on a computer with 8 GB memory, an Intel(R) Xeon(R) CPU E5-1620 v3 @ 3.50 GHz Central Processing Unit (CPU), and a NVIDIA GeForce RTX 2080 GPU running on Ubuntu 16.04. For deep learning, the PyTorch library [17] was used with Cuda 10.0, cuDNN 7.1.2, and NVIDIA driver version 410.104.

4.1   Speed Improvements

Compared to our previous work [13], we have improved the implementation and speed of multiple aspects of this research, but the result is functionally identical. Firstly, the light field capturing approach was improved over rendering each view in the light field separately. This enables faster data capture times, and more importantly allows for a fairer baseline comparison of our method's speed. This was achieved by creating two large textures containing the entry and exit positions into the volumetric data for each ray in the light field, and casting all of these rays in one step. Traversing the rays in a single pass made better advantage of the highly parallel nature of GPUs. The resulting method is roughly twice as fast as rendering views individually, leading to a fairer speed comparison.

Additionally, the image warping procedure was improved considerably. Instead of performing the image warping on the CPU using NumPy and PyTorch, fixed function GPU shader code was implemented. First, the depth map from the reference view is converted to a disparity in a fragment shader, and saved to a depth texture. Then, the reference view colour texture and depth texture are passed to another fragment shader which performs backward warping for each novel view, saving the result into multiple viewports of a large colour texture. Unsurprisingly, this is far faster than the CPU warping, close to two orders of magnitude faster.

4.2 Datasets

Using a 2D array of outward looking (non-sheared) perspective views with fixed field of view [7], synthetic light field datasets were captured in the volume visualisation framework Inviwo [27]. The cameras were shifted along an equidistant grid to keep their optical axes parallel, removing the need to rectify the images to a common plane. Sampling was performed uniformly, with the camera fixed to lie within a certain distance of the central object. Additionally, a plane with its normal vector aligned with the camera view direction is used to clip the volume, revealing detailed structures inside the volume and demonstrating the accuracy of the depth heuristic. See Fig. 2 for the central sub-aperture image of five captured light fields for each dataset.

In our prior research [13], the validity of our method was demonstrated for one specific dataset, an MRI of a heart with visible aorta and arteries with a resolution of $512 \times 512 \times 96$ [21]. Each of the 2000 sample training light fields and 100 validation light fields were captured at 512×512 spatial resolution. This was a difficult dataset because the heart has a rough surface, and the aorta and arteries create intricate structures which are difficult to reconstruct. However, the colours in the transfer function were quite dark with low contrast. Additionally, the CNN could learn a specific single case and generalisation was not fully tested. To address the limitations in these experiments, we tested a new dataset with far more variety.

Using multiple head MRIs from a large scale neurological study [19] we captured a highly varying dataset. Many hand designed functions are applied to these images, as well as randomly generated transfer functions. The density values in each MRI volume were normalised to a common data range, so the transfer functions acted similarly across the volumes. The generalisation of the deep learning could be evaluated by running the network against unseen MRIs volumes and unseen transfer functions during training. In this manner, we could test if the volume or transfer function had more of an effect on the generalisation of learning. The head MRI volumes are smaller than the heart dataset, so these were captured at 256×256 spatial resolution. 2000 training light fields are captured, with 400 validation light fields for non-training volumes and 400 light fields for non-training transfer functions.

4.3 Training Procedure

The training procedure for the convolutional network is identical to that described in [13]. The CNNs are trained by minimising the per-pixel mean squared error between the ground truth views and the synthesised views. For the heart dataset, to increase training speeds and the amount of available data,

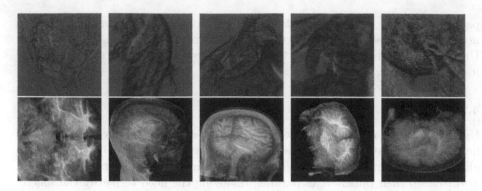

Fig. 2. Sample training light field central sub-aperture views. Heart MRI images are from our previous work [13]. It should be immediately apparent that the dataset of head MRIs has far more variation.

four random spatial patches of size 128×128 were extracted from each light field at every training epoch. Training colour images for both datasets have a random gamma applied as data augmentation. Network optimisation was performed with Stochastic gradient descent and Nesterov momentum. An initial learning rate of 0.1 was updated during learning by cosine annealing the learning rate with warm restarts [11]. Gradients were clipped based on the norm at a value of 0.4 and an L2 regularisation factor of 0.0001 was applied. Training takes about 14 h using the 2D CNN architecture with eight CPU cores used for data loading and image warping.

5   Experiments

5.1   Generalisation of Deep Learning

We previously tested multiple CNNs on a single volume and transfer function, with different camera positions [13]. As discussed in Sect. 3, the best performing network in terms of quality was the EDSR network [8] taking an angular remapped input. This resulted in a residual function which improved the visual quality of novel views, especially for those far from the reference view, see Fig. 3. Additional evaluation performed with PSNR and the Learned Perceptual Image Patch Similarity (LPIPS) metric [37] using the deep features of AlexNet [5] to form a perceptual loss function agrees with the per image values for SSIM. See Fig. 4 for the bottom right sub-aperture view of this light field from a validation set along with difference images to visualise the effect of the CNN.

Although the learnt residual function worked well for a single volume, our initial tests indicated that this learnt function would not generalise for multiple transfer functions and volumes [13]. In contrast, the image warping procedure and depth heuristic did generalise well across volumes and transfer functions. Using our dataset consisting of multiple different head MRIs and transfer functions from [19], we found that the learning certainly did not generalise. Due to

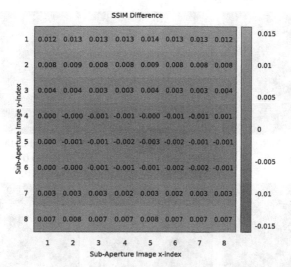

Fig. 3. The difference in SSIM per image location after applying the EDSR network to the warped images. Images far away from the reference view exhibited lower loss, but the CNN caused a degradation in quality of the reference image. Position (5, 5) is the location of the reference view. Extracted from our previous work [13].

the lack of a clear pattern in this widely varying data, the CNN quickly learnt to regress the residual function to a blank output. This demonstrates that the CNN could not predict effective changes for multiple volumes and transfer functions. Although this a significant limitation of this method, it is an extremely important result. In particular, it is necessary to train the CNN to handle specific volumes with fixed transfer functions. For some cases, such as an educational demonstration, this would be feasible as a specific set of volumes could be selected that elucidate a topic.

5.2   Example Synthesised Light Fields

To investigate the method performance, examples of a low, middling, and high quality synthesised light fields from the validation sets are presented in Figs. 5 and 6. For the head MRI dataset, Fig. 5(b) is a poor reconstruction due to a large number of cracks appearing in the wavy semi-transparent structure of the cerebral cortex. Figure 5(e) is a reasonably well synthesised view, but skull and brain's borders are inaccurately estimated. Figure 5(h) is accurate synthesis as the information is moved very well to the novel view, though the image is blurry. For the heart MRI dataset, Fig. 6(b) is a poor reconstruction due to the opaque structure that should be present in the centre of the view causing a large crack in the image. Figure 6(e) is a reasonably well synthesised view, though some arteries lose their desired thickness and the image is not very sharp. Figure 6(h) is an accurate synthesis, although some errors are seen around object borders, such as on the arch of the aorta.

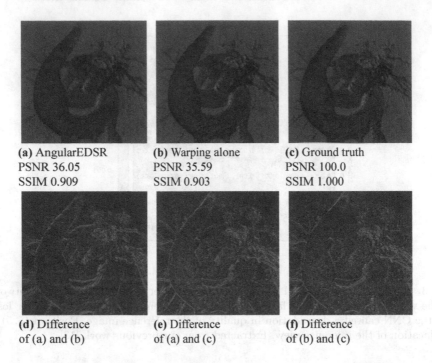

(a) AngularEDSR
PSNR 36.05
SSIM 0.909

(b) Warping alone
PSNR 35.59
SSIM 0.903

(c) Ground truth
PSNR 100.0
SSIM 1.000

(d) Difference
of (a) and (b)

(e) Difference
of (a) and (c)

(f) Difference
of (b) and (c)

Fig. 4. The bottom right view in the light field which Fig. 3 presents results for. Figure (d) visualises the residual applied by the CNN to the warped images to improve visual quality. The CNN detects broad edges to improve, such as the central arch of the aorta, but fails to improve finer details such as the arteries in the top right of the image. Extracted from our previous work [13].

5.3   Time Performance

All timing performances are reported on 300 $512 \times 512 \times 8 \times 8$ light fields captured in Inviwo of the heart MRI discussed in Sect. 4.2. In our previous work [13], we assumed that the light field was rendered view by view. With this assumption, rendering such a light field takes 1.25 s on average with 93 ms std dev. With the improved method taking advantage of large textures, rendering a light field takes 612 ms on average, with standard deviation of 20 ms. Using our method is an nearly an order of magnitude faster than this, taking 72 ms on average, with negligible standard deviation of less than 0.1 ms. This is 72 ms is broken down in Table 1, with average values given over all 300 light fields. The timing for the CNN includes the time to transform the values in the GPU texture to a PyTorch CUDA tensor, which could potentially be alleviated by reusing the GPU memory. The GLSL shader code implementation of our image warping operation is two orders of magnitude faster than our previous pytorch implementation [13] with a time reduction from 2.77 s to 10 ms. Note that the whole pipeline takes only 32 ms to synthesise a light field without applying a CNN. As such, a 512×512

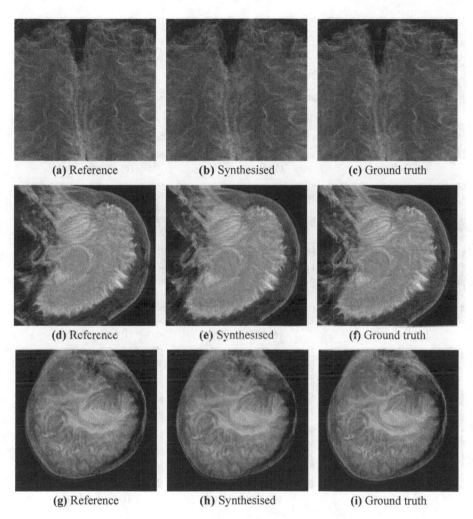

(a) Reference   (b) Synthesised   (c) Ground truth

(d) Reference   (e) Synthesised   (f) Ground truth

(g) Reference   (h) Synthesised   (i) Ground truth

Fig. 5. Example synthesised upper-left images for the head MRI. The first row has low performance (25.8 PSNR, 0.60 SSIM) due to the abundant translucent structures. The second row has middling performance (25.3 PSNR, 0.75 SSIM) since the background has erroneously been picked up in the novel view information. The third row has high performance (28.0 PSNR, 0.86 SSIM) with small inaccuracies, including a general blurriness and splitting on the lower edge of the head.

volume could be visualised in a $512 \times 512 \times 8 \times 8$ light field at interactive rates of 30 frames per second.

As aforementioned, the time for light field synthesis has far less deviation than directly volume rendering a light field, because the latter depends heavily on the complexity of the scene. A CNN performs the same operations regardless of input volume size and complexity, which results in steady performance. The fixed function backward warping and disparity conversion shaders also perform

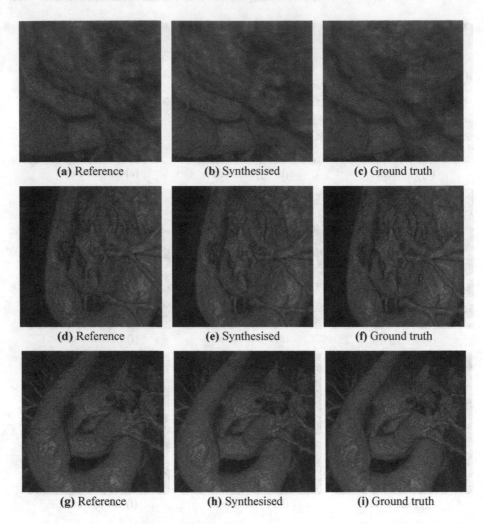

(a) Reference   (b) Synthesised   (c) Ground truth

(d) Reference   (e) Synthesised   (f) Ground truth

(g) Reference   (h) Synthesised   (i) Ground truth

Fig. 6. Example synthesised upper-left images for the heart MRI. The first row has low performance (29.0 PSNR, 0.90 SSIM) due to the translucent structures at the front of the view. The second row has middling performance (31.7 PSNR, 0.86 SSIM) since the arteries are not perfectly distinguished from the aorta. The third row has high performance (35.3 PSNR, 0.91 SSIM) with small inaccuracies, such as on the lower right edge of the aorta.

the same operation regardless of the volume size and complexity. For a fixed spatial resolution, the only time variable is rendering the single sample view. Accordingly, this method could be applied to very large complex volumes with expensive rendering techniques.

Table 1. Average timing values in ms.

Operation	Average time (ms)
Rendering the reference view	19 ms
Computing the depth heuristic	1 ms
Converting depth to disparity	2 ms
Image warping	10 ms
CNN	40 ms

6 Conclusion

Synthesising light fields by image warping produces a useful visualisation method that runs in real-time producing a $512 \times 512 \times 8 \times 8$ light field in an eight of the time to render the light field by ray casting. Additionally, the rendering time for our method is fixed after rendering the central view regardless of the size of the volume or the transfer function applied. A signification limitation of this approach is that the CNN must be retrained for every combination of volume and transfer function. However, a purely image warping based visualisation would still be useful for exploratory purposes as the exact result can be rendered when the camera is held fixed. There is still significant limitations in the quality of the result beyond low-baseline light fields, as a single sample view is not sufficient to extrapolate for a wide-baseline. For wide-baselines, we believe that a fruitful avenue could be to combine multiplane image approaches [10,14,38] for a new volume targeted approach.

Acknowledgements. This research has been conducted with the financial support of Science Foundation Ireland (SFI) under Grant Number 13/IA/1895.

References

1. Adelson, E.H., et al.: The plenoptic function and the elements of early vision. In: Computational Models of Visual Processing, pp. 3–20. MIT (1991)
2. Frayne, S.: The looking glass (2018). https://lookingglassfactory.com/. Accessed 22 Nov 2018
3. Gortler, S.J., Grzeszczuk, R., Szeliski, R., Cohen, M.F.: The lumigraph. In: Proceedings of the 23rd Annual Conference on Computer Graphics and Interactive Techniques, pp. 43–54. SIGGRAPH 1996. ACM (1996). https://doi.org/10.1145/237170.237200, http://doi.acm.org/10.1145/237170.237200
4. Kalantari, N.K., Wang, T.C., Ramamoorthi, R.: Learning-based view synthesis for light field cameras. ACM Trans. Graph. **35**(6), 193:1–193:10 (2016). https://doi.org/10.1145/2980179.2980251, http://doi.acm.org/10.1145/2980179.2980251
5. Krizhevsky, A., Sutskever, I., Hinton, G.E.: Imagenet classification with deep convolutional neural networks. In: Advances in Neural Information Processing Systems, pp. 1097–1105 (2012)

6. Lanman, D., Luebke, D.: Near-eye light field displays. ACM Trans. Graph. (TOG) **32**(6), 220 (2013)
7. Levoy, M., Hanrahan, P.: Light field rendering. In: Proceedings of the 23rd Annual Conference on Computer Graphics and Interactive Techniques, pp. 31–42. SIGGRAPH 1996. ACM (1996)
8. Lim, B., Son, S., Kim, H., Nah, S., Lee, K.M.: Enhanced deep residual networks for single image super-resolution. In: The IEEE Conference on Computer Vision and Pattern Recognition (CVPR) Workshops, vol. 1, p. 4 (2017)
9. Lin, Z., Shum, H.Y.: A geometric analysis of light field rendering. Int. J. Comput. Vis. **58**(2), 121–138 (2004). https://doi.org/10.1023/B:VISI.0000015916.91741.27
10. Lochmann, G., Reinert, B., Buchacher, A., Ritschel, T.: Real-time novel-view synthesis for volume rendering using a piecewise-analytic representation. In: Vision, Modeling and Visualization. The Eurographics Association (2016)
11. Loshchilov, I., Hutter, F.: SGDR: stochastic gradient descent with warm restarts. In: International Conference on Learning Representations (2017)
12. Mark, W.R., McMillan, L., Bishop, G.: Post-rendering 3D warping. In: Proceedings of the 1997 Symposium on Interactive 3D Graphics, pp. 7–16. ACM (1997)
13. Martin, S., Bruton, S., Ganter, D., Manzke, M.: Using a Depth Heuristic for Light Field Volume Rendering, pp. 134–144, May 2019. https://www.scitepress.org/PublicationsDetail.aspx?ID=ZRRCGeI7xV8=&t=1
14. Mildenhall, B., et al.: Local light field fusion: Practical view synthesis with prescriptive sampling guidelines. arXiv preprint arXiv:1905.00889 (2019)
15. Mueller, K., Shareef, N., Huang, J., Crawfis, R.: Ibr-assisted volume rendering. In: Proceedings of IEEE Visualization, vol. 99, pp. 5–8. Citeseer (1999)
16. Park, S., Kim, Y., Park, S., Shin, J.A.: The impacts of three-dimensional anatomical atlas on learning anatomy. Anat. Cell Biol. **52**(1), 76–81 (2019). https://doi.org/10.5115/acb.2019.52.1.76, https://www.ncbi.nlm.nih.gov/pmc/articles/PMC6449593/
17. Paszke, A., et al.: Automatic differentiation in pytorch (2017)
18. Penner, E., Zhang, L.: Soft 3D reconstruction for view synthesis. ACM Trans. Graph. **36**(6), 235:1–235:11 (2017). https://doi.org/10.1145/3130800.3130855
19. Poldrack, R.A., et al.: A phenome-wide examination of neural and cognitive function. Sci. Data **3**, 160110 (2016)
20. Qi, C.R., Su, H., Nießner, M., Dai, A., Yan, M., Guibas, L.J.: Volumetric and multi-view CNNs for object classification on 3D data. In: Proceedings of the IEEE Conference on Computer Vision and Pattern Recognition, pp. 5648–5656 (2016)
21. Roettger, S.: Heart volume dataset (2018). http://schorsch.efi.fh-nuernberg.de/data/volume/Subclavia.pvm.sav. Accessed 15 Aug 2018
22. Shade, J., Gortler, S., He, L.W., Szeliski, R.: Layered depth images. In: Proceedings of the 25th Annual Conference on Computer Graphics and Interactive Techniques, pp. 231–242. SIGGRAPH 1998. ACM, New York (1998). https://doi.org/10.1145/280814.280882, http://doi.acm.org/10.1145/280814.280882
23. Shi, L., Hassanieh, H., Davis, A., Katabi, D., Durand, F.: Light field reconstruction using sparsity in the continuous fourier domain. ACM Trans. Graph. **34**(1), 1–13 (2014). https://doi.org/10.1145/2682631
24. Shojaii, R., et al.: Reconstruction of 3-dimensional histology volume and its application to study mouse mammary glands. J. Vis. Exp.: JoVE **89**, e51325 (2014). https://doi.org/10.3791/51325
25. Simonyan, K., Zisserman, A.: Very deep convolutional networks for large-scale image recognition. arXiv preprint arXiv:1409.1556 (2014)

26. Srinivasan, P.P., Wang, T., Sreelal, A., Ramamoorthi, R., Ng, R.: Learning to synthesize a 4D RGBD light field from a single image. In: IEEE International Conference on Computer Vision (ICCV), pp. 2262–2270, October 2017. https://doi.org/10.1109/ICCV.2017.246
27. Sundén, E., et al.: Inviwo - an extensible, multi-purpose visualization framework. In: IEEE Scientific Visualization Conference (SciVis), pp. 163–164, October 2015. https://doi.org/10.1109/SciVis.2015.7429514
28. Vagharshakyan, S., Bregovic, R., Gotchev, A.: Light field reconstruction using shearlet transform. IEEE Trans. Pattern Anal. Mach. Intell. **40**(1), 133–147 (2018). https://doi.org/10.1109/tpami.2017.2653101
29. Wang, T.-C., Zhu, J.-Y., Hiroaki, E., Chandraker, M., Efros, A.A., Ramamoorthi, R.: A 4D light-field dataset and CNN architectures for material recognition. In: Leibe, B., Matas, J., Sebe, N., Welling, M. (eds.) ECCV 2016. LNCS, vol. 9907, pp. 121–138. Springer, Cham (2016). https://doi.org/10.1007/978-3-319-46487-9_8
30. Wanner, S., Goldluecke, B.: Variational light field analysis for disparity estimation and super-resolution. IEEE Trans. Pattern Anal. Mach. Intell. **36**(3), 606–619 (2014). https://doi.org/10.1109/TPAMI.2013.147
31. Wanner, S., Meister, S., Goldluecke, B.: Datasets and benchmarks for densely sampled 4D light fields. In: Vision, Modeling, and Visualization (2013)
32. Wu, G., Liu, Y., Dai, Q., Chai, T.: Learning sheared EPI structure for light field reconstruction. IEEE Trans. Image Process. **28**(7), 3261–3273 (2019). https://doi.org/10.1109/TIP.2019.2895463
33. Wu, G., et al.: Light field image processing: an overview. IEEE J. Sel. Top. Sig. Process. **11**(7), 926–954 (2017). https://doi.org/10.1109/jstsp.2017.2747126
34. Wu, G., Zhao, M., Wang, L., Dai, Q., Chai, T., Liu, Y.: Light field reconstruction using deep convolutional network on EPI. In: IEEE Conference on Computer Vision and Pattern Recognition (CVPR), pp. 1638–1646, July 2017. https://doi.org/10.1109/CVPR.2017.178
35. Yoon, Y., Jeon, H.G., Yoo, D., Lee, J.Y., So Kweon, I.: Learning a deep convolutional network for light-field image super-resolution. In: Proceedings of the IEEE International Conference on Computer Vision Workshops, pp. 24–32, December 2015. https://doi.org/10.1109/ICCVW.2015.17
36. Zellmann, S., Aumüller, M., Lang, U.: Image-based remote real-time volume rendering: decoupling rendering from view point updates. In: ASME 2012 International Design Engineering Technical Conferences and Computers and Information in Engineering Conference, pp. 1385–1394. ASME (2012)
37. Zhang, R., Isola, P., Efros, A.A., Shechtman, E., Wang, O.: The unreasonable effectiveness of deep features as a perceptual metric. In: Proceedings of the IEEE Conference on Computer Vision and Pattern Recognition (CVPR) (2018)
38. Zhou, T., Tucker, R., Flynn, J., Fyffe, G., Snavely, N.: Stereo magnification: Learning view synthesis using multiplane images. arXiv preprint arXiv:1805.09817 (2018)

Motion Capture Analysis and Reconstruction Using Spatial Keyframes

Bernardo F. Costa$^{(\boxtimes)}$ and Claudio Esperança

PESC/COPPE, UFRJ, Av. Horácio Macedo 2030, CT, H-319 21941-590,
Rio de Janeiro/RJ, Brazil
{bfcosta,esperanc}@cos.ufrj.br

Abstract. Motion capturing is the preferred technique to create real-istic animations for skeleton-based models. Capture sessions, however, are costly and the resulting motions are hard to analyze for posterior modification and reuse. In this paper we propose several tools to analyze and reconstruct motions using the concept of *spatial keyframes*, ideal-ized by Igarashi et al. [19]. Captured motions are represented by curves on the plane obtained by multidimensional projection, allowing the ani-mator to associate regions on that plane with regions in pose space so that representative poses can be located and harvested. The problem of reconstruction from representative poses is also investigated by conduct-ing experiments that measure the error behavior with respect to different multidimensional projection and interpolation algorithms. In particular, we introduce a novel multidimensional projection optimization that min-imizes reconstruction errors. These ideas are showcased in an interactive application that can be publicly accessed online.

Keywords: Motion capture · Keyframing · Spatial keyframe ·
Dimension reduction · Multidimensional projection · Visualization

1   Introduction

The synthesis of motion is an intensively studied area of computer graphics. The most common way of creating animated sequences for characters is through *keyframing*, a technique widely used in traditional animation. The idea is to manually configure a model skeleton to a number of different poses that are assigned to given moments in time. Interpolation functions can then be used to generate poses for any moment in time.

A related idea is the concept of *spatial keyframing* (SK), proposed by Igarashi et al. [19]. The main idea is to associate keyframe poses to carefully placed points on a plane, rather than to points in time as is done in standard keyframing. Using an ingenious space interpolation algorithm based on radial basis functions, each point on the plane can be associated with a different pose in a continuous fashion. Thus, a trajectory traced on this bidimensional pose space can be translated into animated motion. SK aims at producing casual animations with relatively few

© Springer Nature Switzerland AG 2020
A. P. Cláudio et al. (Eds.): VISIGRAPP 2019, CCIS 1182, pp. 48–70, 2020.
https://doi.org/10.1007/978-3-030-41590-7_3

keyframes and skeletons containing few degrees of freedom, since each key pose has to be manually crafted and its representative on the plane must also be positioned by hand.

Obtaining a realistic animation, however, is still a challenge, given the fact that the human eye is specially good at spotting artificially engineered animations that try to mimic the movement of humans. This has led to the development of what is known as *motion capture* (mocap, for short), a process whereby the movement of actors is registered digitally using cameras, sensors and other equipment. The result of a mocap session is a file containing a complete description of the actor's movement in the form of a set of skeleton poses and positions sampled with high precision, both in time and space. More specifically, it contains a description of the skeleton's bones and joints, together with a set of poses, each containing a timestamp, the rotation of each joint and a translation vector for the root joint with respect to a standard rest pose.

Since motion capture is a particularly expensive endeavor, much effort has been spent in the development of techniques to repurpose available mocap files. In the present work, we propose adapting the SK framework to the task of reusing mocap data. It extends the work of Costa and Esperança [12] by proposing new tools for visual analysis and motion reconstruction. The idea is to employ algorithms that automate much of the manual animation workflow of SK so that, rather than authoring individual poses, a subset of meaningful poses can be harvested from mocap files and placed on a plane using multidimensional projection and keyframe selection techniques. As a side-effect, the trajectory defined by projecting all poses on that plane can be used for motion analysis. For instance, a walking cycle animation is usually projected onto a closed curve. This research is supported by a publicly accessible online tool called the *Mocap Projection Browser*, that can be used to view and analyze original and reconstructed motions.

We also investigate whether motion can be realistically reconstructed from a small set of poses using SK interpolation rather than the traditional time-based interpolation scheme. As a byproduct, we developed a projection optimization algorithm that substantially reduces reconstruction error for a class of two-step projectors used for interactive visualization. Empirical support for this investigation has been gathered by conducting experiments where files from a mocap database are subjected to several types of projection and keyframe selection algorithms and evaluated with respect to reconstruction error using variants of temporal and spatial interpolation.

The remainder of this paper is organized as follows. Section 2 presents a short review of techniques for pose sampling and reconstruction. Section 3 discusses some of the most important work on multidimensional projection. Section 4 describes the parameters of our study and the organization of the software used in the experiments. Section 5 reports on the uses of our proposal for mocap visualization and describes how the mocap projection browser can be used for these tasks. Section 6 presents the projection optimization algorithm. Section 7 reports on several reconstruction error experiments used to assess how the

SK-based schemes fare with respect to the traditional temporal interpolation of mocap data. Finally, Sect. 8 contains some concluding remarks and suggestions for the continuation of this research.

2   Pose Sampling and Reconstruction

A mocap file consists of an uniform sampling of the poses assumed by a skeleton during the capture session. The shape of the skeleton, i.e., the length of each bone and the way bones are joined with each other, is usually fixed throughout the session. Thus, each time sample (or frame) is a pose expressed by the translation of a central bone and a list of joint rotations. For instance, all mocap files used in our experiments have 38 bones and poses are sampled at a rate of 120 frames per second.

The high frame rates used in mocap files mean that poses vary only slightly from one frame to the next. This suggests that a coarser sampling might still yield natural motions by interpolating between sufficiently dissimilar poses. The poses selected from the file are called *keyframes*, whereas the interpolated poses are also known as *in-between frames* or simply *tweens*. Two related problems then arise: (1) how to select important poses (keyframes), and (2) how to interpolate them.

2.1   Keyframe Selection

The most naive approach for selecting keyframes is to use uniform time sampling, but at a lower rate. This, however, may include many similar poses from when the movement is slow, and fail to include poses from rapidly varying sections of the animation.

Since motion can be regarded as a parametric curve on the time domain, a common idea is to use curve simplification techniques. For instance, Lim and Thalmann [25] propose adapting the Ramer-Douglas-Peucker curve simplification algorithm [15,27] for mocap data, yielding a scheme known as *simple curve simplification* (SCS). This idea was later enhanced by Xiao et al. [37] by employing an optimized selection strategy, which they named *layered curve simplification* (LCS). On the other hand, Togawa and Okuda [35] propose starting with the full set, repeatedly discarding the frame which least contributes to the interpolation. This algorithm is called *position-based* (PB) selection.

Other possible approaches include clustering methods and matrix factorization. Clustering methods divide frames into clusters and search for a representative frame in each group. The works of Bulut and Capin [9] and Halit and Capin [16] fall into this category. Both also consider dynamic information in the clustering metrics. Matrix factorization uses linear algebra methods to reconstruct mocap data represented in matrix format. Examples of such algorithms are the work of Huang et al. [18], called *key probe*, and that of Jin et al. [21].

In this paper, we propose adapting an algorithm called regularized orthogonal least squares (ROLS) [11], which is tailored for use with reconstructions performed with radial basis functions (RBFs) [14]. The idea is to first compute a dense RBF interpolation matrix and rank the poses with respect to a reconstruction error measure, so that the first k best ranked poses are chosen.

For completeness, it is also worth mentioning the work of Sakamoto et al. [30] that uses of self-organizing maps [23] to select representative poses in the context of a mocap analysis application.

2.2 Interpolation

Reconstructing motion from a few selected poses requires the use of interpolating functions for the rotations associated with each joint of the skeleton. The position of the central joint must also be interpolated, but this problem is less challenging, since it consists of interpolating points in 3-space.

Animation software usually expresses poses as functions of time. If rotations are expressed by Euler angles, these can be interpolated by linear or higher order polynomials. Euler angles, however, are known to be subject to a problem known as "gimbal-lock", where rotations can lose a degree of liberty. A better alternative is to use spherical linear interpolation (*slerp*) [32]. We refer to schemes that use time as the interpolation parameter as *temporal interpolation*.

Rather than using time as the interpolating parameter, Igarashi et al. [19] propose associating poses with points in 2-dimensional space. This scheme, known as *spatial keyframes* (SK), was used to obtain simple animations from a few key poses modeled by hand. By associating each key pose with a point on a plane (also chosen manually), reconstruction becomes a scattered data interpolation problem, i.e., each point on the plane can generate a different pose using the distances to the key poses as interpolation parameters. The authors solve this problem using radial basis functions (RBFs) to interpolate the rotation matrices of each joint. The interpolated matrices are then orthonormalized using a relaxation algorithm, which the authors claim produces a more natural result than the well-known Gram-Schmidt method. Alternatively, RBFs can also be used to interpolate Euler angles or quaternion coefficients directly. We use the term *spatial interpolation* to refer to schemes which use points in two or more dimensions as the interpolation parameter.

3 Multidimensional Projection and Unprojection

The poses sampled in a motion capture session can be seen as points in multidimensional space, which makes them amenable to analysis by multidimensional projection, also known as *dimension reduction*. The idea is to associate each pose with a point on the plane so that similar poses fall close to each other. Referring to the work of Igarashi et al. [19], we notice that one could harvest poses from mocap files rather than creating keyframe poses by hand, making spatial keyframe selection and placement fully automated.

3.1 Projection Techniques

Multidimensional projection techniques have been intensely researched in the last years for use in several areas of application. Although all of these can be used for visualizing multidimensional data in general, many works employ them in mocap compaction and analysis.

Principal Component Analysis (PCA) is a standard technique used in several works. Its advantages include low sensitivity to noise, easy implementation and good performance. Arikan [6], Halit and Capin [16], Safonova et al. [29] and Jin et al. [21] all employ PCA for mocap compression, motion analysis and synthesis.

Locally Linear Embedding (LLE) is an "unsupervised learning algorithm that computes low dimensional, neighborhood preserving embeddings of high dimensional data" [28]. Zhang and Cao [38] and Jin et al. [21] use LLE to ease their search for keyframes in the frame set. LLE tries to find a projection where relative distances between each point and their nearest neighbors in lower dimension space is preserved in the least squares sense. The number of nearest neighbors is a parameter of the algorithm.

Multidimensional Scaling (MDS) [13] produces projections where the relative distances in lower dimension space are as close as possible to the corresponding distances in the original space in the least squares sense. The result depends on the distance metric, being equivalent to PCA if the euclidean norm is used. For instance, Assa et al. [7] use MDS to project the motion curve onto a 2D space to find keyframe candidates.

Isomap [34] is related to MDS in the sense that it uses the same least-squares formulation, but distances are defined by geodesic paths imposed by a weighted graph. For instance, Jenkins and Matarić [20] use isomap to reduce human motion to a smaller dimension for clustering purposes.

Force-based approaches used in graph drawing applications can also be used for dimension reduction. For instance, Tejada el al. [33] developed a fast force-based iterative approach on graphs defined by distance relationships, producing interesting results for data visualization applications.

t-distributed Stochastic Neighborhood Embedding (t-SNE) [26] is an iterative optimization technique for dimensionality reduction that is particularly well suited for the visualization of high-dimensional datasets. It has become increasingly popular in the last few years for its almost magical ability to create compelling two-dimensonal "maps" from data with hundreds or even thousands of dimensions [36]. A key parameter in this approach is called *perplexity*, and it controls the number of neighbors analyzed for each point for each iteration. In our experiments on mocap compression (see Sect. 7) it has yielded the lowest overall error rates for the tested datasets.

3.2 Two-Step Projection Schemes

Data visualization often resorts to interaction to enhance the exploration experience. In the context of multidimensional projection this means that the projection of the dataset should be subject to interactive fine-tuning. The idea is to first project a small representative subset (control points). The remaining points then find their place on the projective plane using some interpolation method. By moving control points around, a different projection of the whole set is produced.

Local Affine multidimensional Projection (LAMP) [22] is a two-step scheme where control points are randomly picked from the original set and projected using some projection algorithm, whereas the remaining points are projected by means of an affine map expressed in terms of their distance to the control points in the original space.

The *Radial Basis Function Projection* (RBFP) scheme of Amorim et al. [5] uses ROLS to pick control points and replaces the affine map of LAMP by RBFs, which provide a smoother interpolation.

3.3 Unprojection

Unprojection or backward projection is a less developed field compared to dimension reduction. It aims at producing points in the original multidimensional space which do not belong to the input data set.

Inverse LAMP (iLAMP) [31] uses the affine maps of LAMP [22] to get a backward projection, swapping high dimension with lower dimension.

RBF unprojection is proposed in [4] where radial basis functions are used to transport information from the reduced dimension to the original space in an application to explore facial expressions. We note that this is completely analogous to the spatial keyframe interpolation scheme [19] used for skeleton animation.

4 Proposal and Implementation

In this work we investigate the use of multidimensional projection and spatial keyframing techniques for analyzing and reconstructing motion capture data. Two main aspects are of special interest. The first concerns how well multidimensional projection behaves in mapping pose space into a bidimensional domain, i.e., its usefulness as a visual analytics tool for mocap data. The second aspect concerns the question of whether spatial keyframing is a proper tool for reconstruction of motion from a small set of selected poses.

To help address these questions, we have implemented a software that processes mocap files and rebuilds them in several ways as depicted in Fig. 1. Here are the tested algorithms and their parameters as used in the experiments:

(a) The keyframe selection strategies (box labeled "Selection" in Fig. 1) are: PB, SCS, ROLS, and Uniform Sampling (US), which selects keyframes at regular time intervals (see Sect. 3.1).
(b) For experiments using spatial keyframes we tested 6 main projection algorithms, namely: Force, MDS, LLE, PCA, Isomap and t-SNE (boxes labeled "Projection" and "1st step projection" in Fig. 1). Our LLE implementation

uses the 15 nearest neighbors to reconstruct its surrounding areas. Force uses 50 iterations to reach its final projection with a step size of 0.125. Experiments with t-SNE use a perplexity value of 50 and a low learning rate[1]. The projection algorithms correspond to multidimensional projection of frames using the strategies discussed earlier. In addition to these single-step schemes, we also tested the two-step RBFP-based projection scheme where the first step (keyframe projection) is performed with one of the former projection algorithms and the remaining frames are projected with RBFs (see Sect. 3.1).

(c) The interpolation algorithm. For temporal schemes we use Spherical linear (Slerp) interpolation of rotations expressed as quaternions. All spatial schemes use the spatial keyframes back-projection algorithm (see Sect. 2.2). In Sect. 5, we also describe a novel hybrid temporal/spatial interpolation and explain its uses and limitations in visualization. In Sect. 7, we also show tests with Linear and Hermitian interpolation of rotations expressed with Euler angles.

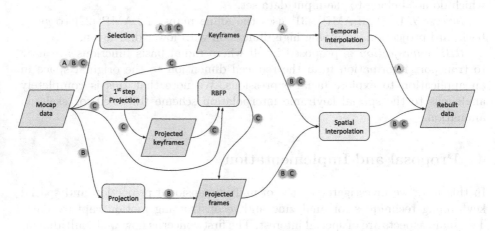

Fig. 1. Mocap analysis and reconstruction pipelines. Arrows labeled (A) mark the standard temporal reconstruction workflow used in commercial animation software. Arrows labeled (B) show the pipeline required to reconstruct motion from frames associated with points on the plane obtained by single-step projection approaches. In pipeline shown with arrows labeled (C), projection is achieved with RBFP, a two-step projection approach.

We used as input data a subset of the CMU Mocap database [1] with 70 files containing various types of motions (walking, running, ballet sequences, etc.) and ranging in size from 129 to 1146 frames. Among the many versions and formats publicly avalailable that database, we chose those in BVH format converted for

---

[1] The learning rate appears in [26] as η in its simple version of the t-SNE algorithm. We used the value of 10 which produces less projection artifacts.

the 3DMax animation software [2]. These files, along with projections created with our software are available in https://github.com/bfcosta/mocapskf. Projections are stored as plain text files, where each line corresponds to a frame and contains three numbers: the x and y coordinates of the projection, and a boolean indicator k, which is 1 if this frame was selected as a keyframe or 0 otherwise.

4.1   Pose Distance and Error Estimation

Projection algorithms and reconstruction fidelity estimation require some way of measuring the similarity between any two given poses. This is a distance measure, i.e., two poses have distance zero if, and only if, they are identical poses. There are several ways of defining such a measure, such as the sum of distances between corresponding bone extremity positions, or the sum of rotation differences for all joints. In our work, we use a weighted sum of rotation differences as defined in Eq. (1). A and B are skeleton poses, r is a joint taken from the set of all joints R, A_r and B_r are the local rotation matrices corresponding to joint r in A and B, respectively, $\text{Tr}[\cdot]$ is the trace operator, and w_r is a weight assigned to joint r.

$$D_q(A, B) = \sum_{r \in R} w_r \cos^{-1}(\frac{\text{Tr}[A_r B_r^\top] - 1}{2}).\tag{1}$$

The rationale for weights w_r is that they should lend more importance to rotation dissimilarities with more visual impact. Clearly, a rotation closer to the root of the skeleton hierarchy should have more weight than a rotation closer to one of the leaves. Thus, we set w_r to the sum of the lengths all branches stemming from r. The length of a branch stemming from r is the length of the path between joint r and the most distant leaf in the branch.

5   Visualization

By its very nature, recorded motion must be played in order to be understood. Thus, animation software must provide means for playing the motion from several angles, or examining time intervals of interest. For this reason, we have built the *Mocap Projection Browser*, an interactive online application which can load and exhibit motions from our selected database, as well as projection files. This software can be used directly from any modern web browser by accessing the address https://observablehq.com/@esperanc/mocap-projection-browser. It was written in the form of an interactive notebook, so that the interested reader can not only play with the interface, but also inspect the code directly. The interface allows the user to select one of the 70 database mocap files and any of the projection variants computed for it by selecting keyframe ratio (3, 5, 7 or 9%), projection algorithm, keyframe selection algorithm, and whether the projection optimization algorithm (discussed in Sect. 6) was applied or not. If desired, the user can also upload any mocap or projection file for inspection using the browser.

The main visualization panel of the browser (see Fig. 2) shows a perspective rendering of a pose from the original file (red skeleton) together with reconstructions created with spatial keyframes and with temporal interpolation (blue and green skeletons, respectively). The inset panel on the lower left shows the selected projection where red circles mark keyframes and the small outlined square marks the current frame. Using the interface, it is possible to select other frames or display a synchronous animation of all skeletons. Optionally, the panel can also render the trajectory of a particular set of bones throughout the animation. In Fig. 2, the three colored surfaces show the trajectory of the bone connecting the left knee to the left ankle.

5.1 Pose Inspection

Mere observation may not be adequate when one must compare details of poses separated in time or when comparing several capture sessions of the same motion. This has led to several attempts to enhance the analysis of mocap data by using additional visualization tools. One idea is to represent poses as points on a plane where similar poses are clustered together. A curve then can be defined connecting poses for successive frames of the animation. Sakamoto et al. [30] propose an efficient motion data retrieving scheme based on image-based keys for semantic-based retrieval. A self-organizing map [23] is built for several mocap files in order

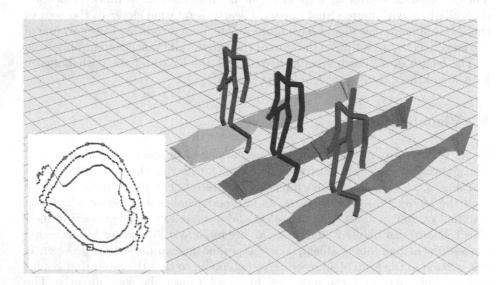

Fig. 2. Screen capture of the interactive mocap projection and reconstruction browser for mocap file 02_01.bvh with 3% keyframe ratio. The red skeleton shows a pose from the original mocap file, while the blue and green skeletons show reconstructions created with spatial and temporal interpolation, respectively. The coloured surfaces are renderings of the trajectory of the left tibia/fibula for each skeleton. Red dots on the projection plane represent the keyframes. (Color figure online)

to offer recognizable icons of poses to the user, who then able to pick a desirable motion segment by selecting two points – begin and end – on the image. Hu et al. [17] propose a novel visualization approach to depict variations of different human motion data. Their idea is to represent the time dimension as a curve in two dimensions called the *motion track*. They argue that such visualization framework is able to highlight latent semantic differences between motion variations. Other forms of motion analysis use interactive visualization tools. Alemi et al. [3] and Bernard et al. [8] are examples of interactive visualizations focused in time comparison. We refer the reader to the work of Li et al. [24] for a broader discussion of mocap static visualizations.

Visual inspection functionality is also supported in our prototype. In addition to inspecting poses already present in the dataset, spatial keyframe interpolation provides the user with means to interactively examine reconstructed poses for any point in the projection plane. This is illustrated in Fig. 3, which shows 4 different poses taken from different points in the projection plane.

Another pose inspection feature supported by the browser is the visualization of bone trajectories. One can select any subset of bones of the skeleton and have the browser render a surface connecting successive positions of these bones for the complete duration of the animation. Figure 2 shows an example of this for the bone connecting the left knee and the left ankle. This helps spot unwanted irregularities in the movement of the original data as well as reconstruction problems.

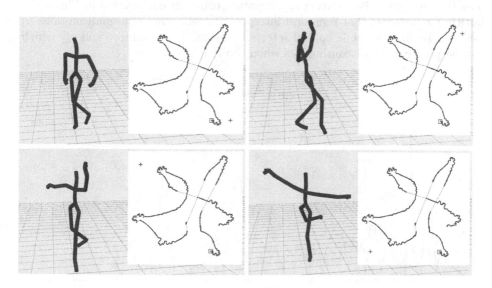

Fig. 3. Inspecting the pose space by reconstruction. Four poses created by SKF-reconstruction based on file `05_10.bvh`, where 3% of the frames were selected as keyframes using ROLS and projected using t-SNE. The blue crosses on the projection plane mark the corresponding points in pose space. (Color figure online)

5.2   Reconstruction Error Visualization

Reconstruction error is one of the key aspects that need to be assessed when considering a given pose interpolation strategy. The Browser automatically computes and displays the reconstruction error averaged over all frames for temporal and SKF-based interpolations. Clearly, both depend on the keyframe to frame ratio, yielding smaller errors for higher ratios. For temporal interpolation, keyframe selection is restricted to uniform sampling and rotations are always interpolated with *slerp*. Spatial interpolation depends on a wider range of factors, namely the projection, keyframe selection and rotation interpolation algorithm.

It is also useful to understand how reconstruction error behaves during the whole animation. The Browser automatically computes and displays a line chart showing the reconstruction error for each frame in the animation for both temporal and spatial interpolations. Figure 4 shows an example of this chart for mocap file 05_10.bvh. The average errors for spatial and temporal reconstructions are 0.05566 and 0.05729, respectively.

Low reconstruction errors, however, may mask problems that become apparent when the whole animation is played. This is due to the fact that a good projection algorithm might yield a very convoluted curve which is not smooth in time, i.e., the distance between two consecutive frames in the projection plane can be very irregular, as observed in Fig. 3. Another way to detect discontinuities in the reconstruction is to observe the distance between consecutive frames in the original data and compare these to the reconstructed skeleton behaviors. The Browser plots these curves in a separate chart, as exemplified in Fig. 5. We notice, for instance, that temporal interpolation smoothes out small motions in between keyframes, while spatial interpolation might introduce "spikes" which can be perceived as discontinuities when the animation is played.

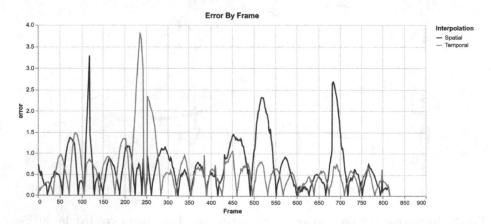

Fig. 4. Plot of error by frame for file 05_10.bvh using t-SNE projection with 3% keyframe ratio, ROLS selection and SKF rotation interpolation.

This suggests that by mixing spatial and temporal interpolation we may obtain smaller errors while increasing the movement smoothness. A simple way of achieving this hybridization is to add time as a third dimension in the domain of the RBFs used for interpolating rotations, i.e., the domain consists of points of the form $(x, y, \alpha \frac{frame}{120})$, since CMU mocap files are sampled with 120 frames per second, where α is a weighting factor for the temporal component. This idea is implemented in the Browser and corresponds to selecting the *Hybrid SKF+time* radio button and adjusting the weight box in the interface. Figure 5 (bottom) shows the effect of using hybrid interpolation with weight 5. This setting reduces the average error of the spatial interpolation from 0.05566 to 0.04940. We note that hybrid interpolation does not necessarily reduces the error but almost always produces a smoother movement.

6   Projection Optimization

Since radial basis functions interpolate the control points, SKF reconstructs the 2D projection of a keyframe pose exactly, but non-key poses are reconstructed only approximately. Surely, for a given keyframe selection, and once the projection of the keyframe poses has been established, there must be some *optimal* projection for each non-frame pose, i.e., one whose reconstruction by the back-projection algorithm will yield the smallest error.

This observation led us to develop an optimization algorithm that will help us to find a better projection of non-key poses starting from some initial projection. A pseudo-code for it is shown Algorithm 1, which takes as input the mocap data, the selected keyframes and the initial projected frames (refer to Fig. 1). The algorithm produces an optimized projection by comparing the back-projected poses with the original poses, moving each projected point whenever it finds a neighboring location that yields a back-projected pose with a smaller error. The algorithm is an application of gradient descent [10] optimization where, at each iteration, the gradient of a given function is estimated and steps in the gradient direction are taken to reach a local minimum.

An important component of the process is the estimation of the gradient at a given point. The numerical process used in our algorithm estimates the gradient around a particular 2D point by back-projecting four neighbor points within a disk of small radius and computing the error with respect to the original pose. If all points yield a bigger error than the original point, then that point is a local minimum. Otherwise, a gradient direction is computed based on the error variation. Parameters α and δ in the algorithm are typically chosen in the interval $[0, 1]$ and were set to 0.5 in our experiments.

Figure 6 shows the effect of applying our proposed optimization algorithm on projections of mocap file 05_10.bvh obtained with RBFP, where the first step was done with each of the other implemented single-step projectors, namely: Force, Isomap, LLE, MDS, PCA and t-SNE. All use the same set of keyframes selected with ROLS (3% keyframe ratio). We note that the optimization tends to scatter non-key frames with respect to the non-optimized projections. Clearly,

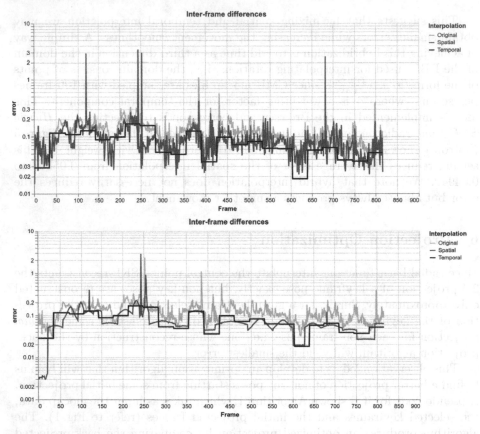

Fig. 5. Errors between consecutive frames exhibited by the original data and temporal and spatial reconstructions. Data and parameters are the same used for Figs. 3 and 4. The top chart shows the standard SK reconstruction, but the chart on the bottom uses a hybrid spatial-temporal rotation interpolation with weight 5 for the temporal component.

cases where greater differences exist between the optimized and original projections can be attributed to the original algorithm choosing poorer positions in the first place.

Table 1 summarizes the reconstruction error improvement obtained by optimization for the projections shown in Fig. 6. In all cases, optimization significantly improved the original error. In a sense, this is expected since the RBFP approach, although the only one that supports interactive placement of control points, is not a particularly good projector for mocap data (see Sect. 7). For reference, the temporal reconstruction error for this file, using uniform sampling and 3% keyframe ratio was 0.60901. Thus, Isomap and t-SNE obtained better results, while LLE and MDS yielded slightly worse results.

Algorithm 1. Gradient descent optimization for pose projection.

1: **function** GETOPTIMIZEDPROJECTION($\mathbb{F}$, $\mathbb{Q}$, $\mathbb{G}$) ▷ Optimizes projection $\mathbb{Q}$. Where $\mathbb{F}$ - original mocap data, $\mathbb{Q}$ - projected frames, $\mathbb{G}$ - projeced keyframes.

2:      $\mathbb{O} \leftarrow \emptyset$

3:      **for** $i \leftarrow 1$, *maxtries* **do**

4:          $done \leftarrow$ **true**

5:          **for all** $f_i \in \mathbb{F} \setminus \{\mathbb{G} \cup \mathbb{O}\}$ **do**

6:              **if** $gradientDescent(f_i, \mathbb{F}, \mathbb{Q}, \mathbb{G})$ **then**

7:                  $\mathbb{O} \leftarrow \mathbb{O} \cup \{f_i\}$

8:                  $done \leftarrow$ **false**

9:              **end if**

10:         **end for**

11:         **if** $done$ **then**

12:             **break**

13:         **end if**

14:     **end for**

15: **end function**

16: **function** GRADIENTDESCENT(f_i, $\mathbb{F}$, $\mathbb{Q}$, $\mathbb{G}$) ▷ Updates $q_i \in \mathbb{Q}$ and returns if q_i has changed. Where f_i - frame, $\mathbb{F}$ - frame set, $\mathbb{Q}$ - projected pose set, $\mathbb{G}$ - keyframe set.

17:     $h \leftarrow (0, 0)$

18:     $v_x \leftarrow (1, 0)$

19:     $v_y \leftarrow (0, 1)$

20:     $q_i \leftarrow$ projection of f_i stored in $\mathbb{Q}$

21:     $q_{i+1} \leftarrow$ projection of f_{i+1} stored in $\mathbb{Q}$

22:     $q_{i-1} \leftarrow$ projection of f_{i-1} stored in $\mathbb{Q}$

23:     $r \leftarrow (\|q_i - q_{i-1}\| + \|q_i - q_{i+1}\|)/2$

24:     $e_o \leftarrow Err(f_i, BackProj(q_i, \mathbb{F}, \mathbb{Q}, \mathbb{G}))$

25:     $e_{xp} \leftarrow Err(f_i, BackProj(q_i + \delta r v_x, \mathbb{F}, \mathbb{Q}, \mathbb{G}))$

26:     $e_{xn} \leftarrow Err(f_i, BackProj(q_i - \delta r v_x, \mathbb{F}, \mathbb{Q}, \mathbb{G}))$

27:     $e_{yp} \leftarrow Err(f_i, BackProj(q_i + \delta r v_y, \mathbb{F}, \mathbb{Q}, \mathbb{G}))$

28:     $e_{yn} \leftarrow Err(f_i, BackProj(q_i - \delta r v_y, \mathbb{F}, \mathbb{Q}, \mathbb{G}))$

29:     **if** $(e_o > e_{xp}) \vee (e_o > e_{xn})$ **then**

30:         $h \leftarrow h + v_x(e_{xn} - e_{xp})$

31:     **end if**

32:     **if** $(e_o > e_{yp}) \vee (e_o > e_{yn})$ **then**

33:         $h \leftarrow h + v_y(e_{yn} - e_{yp})$

34:     **end if**

35:     **if** $\|h\| > 0$ **then**

36:         $h \leftarrow rh/\|h\|$

37:         $q_i' \leftarrow q_i + \alpha h$

38:         replace q_i by q_i' in $\mathbb{Q}$

39:         **return true**

40:     **else**

41:         **return false**

42:     **end if**

43: **end function**

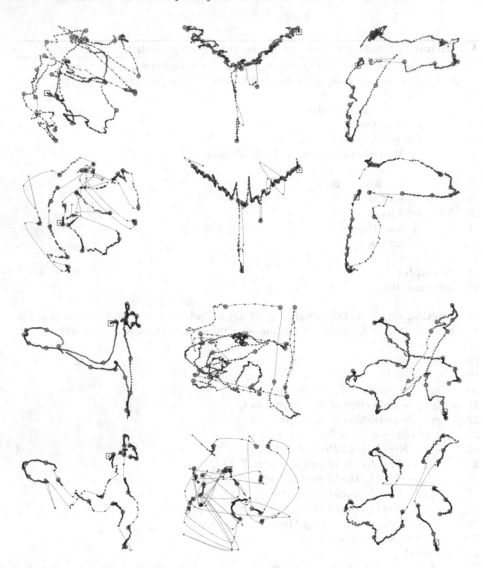

Fig. 6. Projections of file `05_10.bvh` using RBFP with several first step projectors. First/second rows: non-optimized/optimized Force, Isomap and LLE. Third/fourth row: non-optimized/optimized MDS, PCA and t-SNE.

Table 1. SKF reconstruction errors for file `05_10.bvh` with 2-step RBF projection, 3% keyframe ratio, ROLS selection, SKF rotation interpolation. (Projection names are for the first step).

	Force	Isomap	LLE	MDS	PCA	t-SNE
Non-optimized	1.81834	1.52071	1.43434	1.08269	2.72115	0.91568
Optimized	0.81384	0.47108	0.65679	0.63223	1.18766	0.57173

6.1   Optimization for Other Projection Applications

The optimization algorithm presented above can also be adapted to other uses of multidimensional projection for which a back-projection algorithm is available and an error metric is defined. In this paper, we are mostly interested in mocap reconstruction, for which the optimization algorithm has shown to be largely beneficial (see Sect. 7). For other uses of multidimensional projection, however, these gains might not be verified.

When multidimensional projection is used, for instance, in data and information visualization, it usually is evaluated with respect to its capacity to preserve distance and neighborhood relations in the low dimension space. The evaluation of the projection quality can be conducted subjectively by visual inspection, but an objective assessment requires the use of some metric such as the *stress*. The stress function (see Eq. 2) tries to measure how distance relations in high dimension space given by $d_{ij} = \|X_i - X_j\|$ are similar to the corresponding distances in low dimension space $\delta_{ij} = \|Y_i - Y_j\|$. In general, as the projection gets better, the stress becomes lower.

$$s(X,Y) = \frac{\sum_{i=1}^{n} \sum_{j=i+1}^{n} (\delta_{ij} - d_{ij})^2}{\sum_{i=1}^{n} \sum_{j=i+1}^{n} \delta_{ij}^2} \tag{2}$$

We conducted several experiments in order to establish how stress is affected by applying the optimization algorithm to the mocap data in our database. We tested all 6 implemented single-step projection algorithms and 4 keyframe selection strategies, as well as the RBFP 2-step projection algorithm, with the first step performed with each of the same 6 single-step projection algorithms and keyframes selected with ROLS. Back-projected poses were reconstructed with spatial keyframes. The average results for all 70 data sets used in our compression experiments are shown in Table 2, which plots average stress ratio, i.e., the between the stress of the optimized projection and the stress of the original projection. Gains for the optimization algorithm appear when the ratio is less than one. High-dimension distance is computed using Eq. 1, while low-dimension distance is calculated using Euclidean distance. It is possible to observe consistent stress gain for RBFP, except when Isomap is the seed projection heuristic. For all others projectors, there is no real gain in terms of stress. This suggests that the projection optimization algorithm 1 does not, in general, influence the stress measure for these data sets at least when only 3% of the input is used as keyframes. The exception is the RBFP algorithm, with consistent gains for all first-step projectors, except Isomap. This suggests that using optimization in interactive projections with RBFP is highly beneficial.

Table 2. Average stress ratio for 3% keyframe ratio. Each cell shows the ratio between the calculated stress of the optimized projection with respect to that of the original projection. Ratio values below one indicate stress gains only for the RBFP algorithm, except when Isomap is used to project the seeds.

Selection Algorithms	1-step projection algorithms					
	Force	Isomap	LLE	MDS	PCA	t-SNE
PB	1.145	1.07	1.065	1.173	1.245	1.008
ROLS	1.154	1.057	1.033	1.072	0.926	0.969
SCS	1.043	1.085	1.004	1.121	1.046	0.966
US	1.11	1.098	1.049	1.098	1.247	1.215
Selection Algorithms	2-step RBFP - seed projection algorithms					
	Force	Isomap	LLE	MDS	PCA	t-SNE
ROLS	0.396	1.013	0.441	0.585	0.518	0.58

7   Motion Capture Compression Evaluation

In this section we report the results of experiments conducted in order to obtain a quantitative evaluation of the factors that influence SKF reconstruction, trying to contrast this method with the more traditional time-based approach.

All experiments consist of selecting keyframes from mocap data, employing some reconstruction method and measuring the error of the reconstructed mocap with respect to the original. Although subsampling and reconstructing can be viewed as a scheme for mocap compression, we are mostly interested in evaluating which combination of pose selection and multidimensional projection algorithm produces the least amount of error. These issues help assessing the value of the technique for both motion analysis and synthesis. As mentioned in Sect. 5, mocap and projection files for these experiments can be downloaded from https://github.com/bfcosta/mocapskf or visualized with the Mocap Projection Browser.

A preliminary batch of tests was conducted in order to investigate how the compaction schemes fare with respect to the desired compaction ratio. Since this ultimately depends on the ratio between keyframes and total frames (KF ratio), we selected 9 representative animations and four compaction schemes, two temporal (Linear and Slerp) and two spatial (MDS and t-SNE), all run with SCS keyframe selection strategy, and measured the obtained error for ratios between 1 and 10%. This experiment reveals that error decreases sharply until reaching a KF ratio of about 4%, after which the error decreases at a slower rate. We used this observation to restrict further comparison tests to a KF ratio of 3%, since a smaller ratio would probably lead to bad reconstructions and a larger ratio would not reveal much about advantages of one scheme over another.

Having set the keyframe ratio baseline to 3%, Fig. 7 summarizes the behaviour of the average error with respect to the six single-step projectors, and four keyframe selection algorithms, where reconstructions are obtained with SK interpolation. That figure shows average reconstruction error per frame per joint over all 70 mocap files of our database. For reference, the lowest average error obtained with temporal interpolation was 0.8262, using slerp interpolation and SCS keyframe selection, whereas the highest error was 0.8759 using linear interpolation of Euler angles with keyframes selected by uniform sampling. The top chart, which refers to the original projector output, indicates that SK reconstruction does not fare better than temporal approaches, except for Isomap and t-SNE which yield error values below 0.8. However, the bottom chart in Fig. 7 bottom graph, which shows projection error rates after using our optimization algorithm, indicates that SK beats temporal reconstruction for all projectors when keyframe selection is done with US or ROLS. Some projectors however, notably Force and PCA, might show no error reconstruction gain if PB or SCS are used as the keyframe selector. This is probably related to their poor performance when compared to the other projectors.

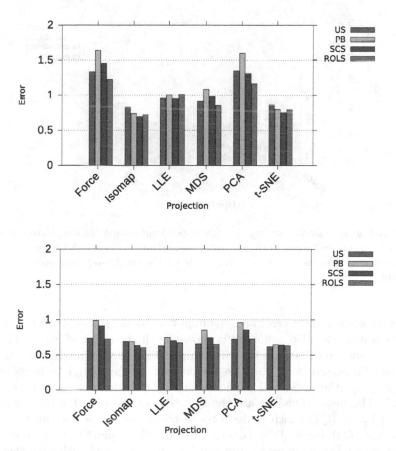

Fig. 7. Average reconstruction error per frame per joint for single-step projections. Abscissa groups correspond to projection algorithms. Second graph shows the same measure after running our proposed optimization algorithm. Bars group values by keyframe selectors.

Figure 8 shows a chart summarizing the average reconstruction errors obtained with variants of the RBFP two-step projection scheme, relating them to the analogous results obtained with single-step projections. Each bar group corresponds to a different projector used either in the single step variant (bars labeled '1s') or in the first step of the RBFP scheme (bars labeled '2s'). Bars labeled 'ori' correspond to the original non-optimized projections, whereas those labeled 'opt' corresponds to errors obtained after optimization. Since the seminal paper on RBF projection [5] proposed ROLS selection, this was used for the experiments. We can see from Fig. 8 that the first two bars are always taller than the last two, indicating a consistent gain in the reconstruction error through the use of optimization. Note also that the difference between single-step and two-step bars is smaller when the optimization algorithm is used, indicating that optimization tends to fix poor projections more than good ones.

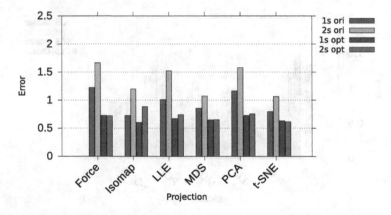

Fig. 8. Average reconstruction error per frame per joint comparing single-step/two-step ('1s'/'2s') projections before ('ori') and after optimization ('opt'). All tests used ROLS as the keyframe selector. Abcissa corresponds to the single-step/first-step projection algorithm.

Finally, since average error analysis might be biased towards special kinds of motions, a more detailed summary is shown in the box plot of Fig. 9. The first three boxes show results obtained with temporal interpolation schemes – namely hermitian, linear and slerp – and, in the last six, to the spatial keyframe reconstruction using the optimized version of the six single step projectors shown in Fig. 7. The boxes divisions are the quartiles of the distribution, except for the top whisker ($Q4$) which is the top error instance below the limit given by $Q3+1.5(Q3-Q1)$, being $Q1$ and $Q3$ the first and third quartile, respectively, and $Q2$ the median. Points above the top whisker are considered outliers in these distributions. In Fig. 9, we can see that not only the medians of reconstructions done with spatial keyframes are consistently lower than those of temporal schemes, but they also exhibit a lower incidence of outliers.

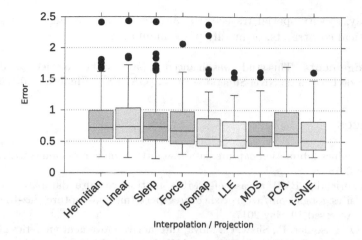

Fig. 9. Boxplot of Error distribution per interpolation scheme. This graph shows the error distribution divided into quartiles and grouped by interpolation or interpolation and projection schemes. Last six groups use the same interpolation scheme – Spatial Keyframe – with projections that used our optimization algorithm for mocap reconstruction. Points over the top whisker are considered outliers.

8 Final Remarks

In this work we have examined several aspects related to the idea of applying spatial keyframing to both analysis and reconstruction of motion captured animations. Although we tried to cover several reasonable combinations of projection and keyframe selection algorithms as well as reconstruction variations, there are still many other variations which could conceivably yield better results. We have confirmed that t-SNE is probably the best single-step projection algorithm for mocap data and that the alliance of RBF projection with our optimization algorithm yields competitive results. Of course, since RBFP is intended for use in interactive visualizations, error rates might change widely if the user moves keyframe projections away from the initial positions prescribed by the first-step projector.

A more consistent evaluation of our work for mocap visualization requires a more thorough investigation. For instance, the comparison of two takes of the same motion would involve the capability of projecting multiple files, something our Mocap Browser is not equipped to do at the moment. Also, the utility of such a tool can only be judged by professional animators who would know what to look for.

Lastly, it should be stressed that one of the initial goals of this study has not yet been achieved, i.e., the combined use of mocap data and keyframing in a real motion synthesis context. Since our browser tool already permits inspecting the continuous pose space associated to the SK plane, adding the functionality of tracing a trajectory on that plane would be a simple extension. Nonetheless, implementing a real motion authoring platform would require many additional tools to, say, specify spatial trajectories, author motion timings, care for real-world physical constraints, or modify poses manually.

Acknowledgements. This study was financed in part by the Coordenação de Aperfeiçoamento de Pessoal de Nível Superior - Brasil (CAPES) - Finance Code 001.

References

1. CMU graphics lab motion capture database. http://mocap.cs.cmu.edu/. Accessed 30 June 2015
2. The daz-friendly BVH release of CMU's motion capture database - cgspeed. https://sites.google.com/a/cgspeed.com/cgspeed/motion-capture/daz-friendly-release. Accessed 19 May 2017
3. Alemi, O., Pasquier, P., Shaw, C.: Mova: Interactive movement analytics platform. In: Proceedings of the 2014 International Workshop on Movement and Computing, MOCO 2014, pp. 37:37–37:42. ACM, New York (2014). https://doi.org/10.1145/2617995.2618002. http://doi.acm.org/10.1145/2617995.2618002
4. Amorim, E., et al.: Facing the high-dimensions: Inverse projection with radial basis functions. Comput. Graph. **48**, 35–47 (2015). https://doi.org/10.1016/j.cag.2015.02.009. http://www.sciencedirect.com/science/article/pii/S0097849315000230
5. Amorim, E., Vital Brazil, E., Nonato, L.G., Samavati, F., Sousa, M.C.: Multidimensional projection with radial basis function and control points selection. In: Pacific Visualization Symposium (PacificVis), 2014 IEEE, Pacific Vis 2014, pp. 209–216 (2014). https://doi.org/10.1109/PacificVis.2014.59
6. Arikan, O.: Compression of motion capture databases. ACM Trans. Graph. **25**(3), 890–897 (2006). https://doi.org/10.1145/1141911.1141971. http://doi.acm.org/10.1145/1141911
7. Assa, J., Caspi, Y., Cohen-Or, D.: Action synopsis: pose selection and illustration. ACM Trans. Graph. **24**(3), 667–676 (2005). https://doi.org/10.1145/1073204.1073246. http://doi.acm.org/10.1145/1073204.1073246
8. Bernard, J., Wilhelm, N., Krüger, B., May, T., Schreck, T., Kohlhammer, J.: MotionExplorer: Exploratory search in human motion capture data based on hierarchical aggregation. IEEE Trans. Vis. Comput. Graph. **19**(12), 2257–2266 (2013). https://doi.org/10.1109/TVCG.2013.178
9. Bulut, E., Capin, T.: Key frame extraction from motion capture data by curve saliency. In: Computer Animation and Social Agents, p. 119 (2007)
10. Burges, C., et al.: Learning to rank using gradient descent. In: Proceedings of the 22Nd International Conference on Machine Learning, ICML 2005, pp. 89–96. ACM, New York (2005). https://doi.org/10.1145/1102351.1102363. http://doi.acm.org/10.1145/1102351.1102363
11. Chen, S., Chng, E.S., Alkadhimi, K.: Regularized orthogonal least squares algorithm for constructing radial basis function networks. Int. J. Control **64**(5), 829–837 (1996). https://doi.org/10.1080/00207179608921659
12. Costa., B.F., Esperança., C.: Enhancing spatial keyframe animations with motion capture. In: Proceedings of the 14th International Joint Conference on Computer Vision, Imaging and Computer Graphics Theory and Applications - Volume 1: GRAPP, pp. 31–40. INSTICC, SciTePress (2019). https://doi.org/10.5220/0007296800310040

13. Cox, M.A.A., Cox, T.F.: Multidimensional Scaling, pp. 315–347. Springer, Heidelberg (2008). https://doi.org/10.1007/978-3-540-33037-0_14
14. Dinh, H.Q., Turk, G., Slabaugh, G.: Reconstructing surfaces by volumetric regularization using radial basis functions. IEEE Trans. Pattern Anal. Mach. Intell. **24**(10), 1358–1371 (2002). https://doi.org/10.1109/TPAMI.2002.1039207
15. Douglas, D.H., Peucker, T.K.: Algorithms for the reduction of the number of points required to represent a digitized line or its caricature. Cartogr. Int. J. Geogr. Inf. Geovis. **10**(2), 112–122 (1973). https://doi.org/10.3138/FM57-6770-U75U-7727
16. Halit, C., Capin, T.: Multiscale motion saliency for keyframe extraction from motion capture sequences. Comput. Animat. Virtual Worlds **22**(1), 3–14 (2011). https://doi.org/10.1002/cav.380
17. Hu, Y., Wu, S., Xia, S., Fu, J., Chen, W.: Motion track: Visualizing variations of human motion data. In: 2010 IEEE Pacific Visualization Symposium (PacificVis), pp. 153–160, March 2010. https://doi.org/10.1109/PACIFICVIS.2010.5429596
18. Huang, K.S., Chang, C.F., Hsu, Y.Y., Yang, S.N.: Key probe: a technique for animation keyframe extraction. Vis. Comput. **21**(8), 532–541 (2005). https://doi.org/10.1007/s00371-005-0316-0
19. Igarashi, T., Moscovich, T., Hughes, J.F.: Spatial keyframing for performance-driven animation. In: Proceedings of the 2005 ACM SIGGRAPH/Eurographics Symposium on Computer Animation, SCA 2005, pp. 107–115. ACM, New York (2005). https://doi.org/10.1145/1073368.1073383
20. Jenkins, O.C., Matarić, M.J.: A spatio-temporal extension to isomap nonlinear dimension reduction. In: Proceedings of the Twenty-First International Conference on Machine Learning, ICML 2004, p. 56. ACM, New York (2004). https://doi.org/10.1145/1015330.1015357. http://doi.acm.org/10.1145/1015330.1015357
21. Jin, C., Fevens, T., Mudur, S.: Optimized keyframe extraction for 3D character animations. Comput. Anim. Virtual Worlds **23**(6), 559–568 (2012). https://doi.org/10.1002/cav.1471
22. Joia, P., Paulovich, F., Coimbra, D., Cuminato, J., Nonato, L.: Local affine multidimensional projection. IEEE Trans. Vis. Comput. Graph. **17**(12), 2563–2571 (2011). https://doi.org/10.1109/TVCG.2011.220
23. Kohonen, T.: The self-organizing map. Proc. IEEE **78**(9), 1464–1480 (1990)
24. Li, W., Bartram, L., Pasquier, P.: Techniques and approaches in static visualization of motion capture data. In: Proceedings of the 3rd International Symposium on Movement and Computing, MOCO 2016, pp. 14:1–14:8. ACM, New York (2016). https://doi.org/10.1145/2948910.2948935. http://doi.acm.org/10.1145/2948910.2948935
25. Lim, I.S., Thalmann, D.: Key-posture extraction out of human motion data by curve simplification (2001). swiss Federal Inst. Technol. (EPFL), CH-1015 Laussane, Switzerland, http://infoscience.epfl.ch/record/98933
26. van der Maaten, L., Hinton, G.: Visualizing data using t-SNE. J. Mach. Learn. Res. **9**, 2579–2605 (2008). http://www.jmlr.org/papers/v9/vandermaaten08a.html
27. Ramer, U.: An iterative procedure for the polygonal approximation of plane curves. Comput. Graph. Image Process. **1**(3), 244–256 (1972). https://doi.org/10.1016/S0146-664X(72)80017-0. http://www.sciencedirect.com/science/article/pii/S0146664X72800170
28. Roweis, S.T., Saul, L.K.: Nonlinear dimensionality reduction by locally linear embedding. Science **290**(5500), 2323–2326 (2000). https://doi.org/10.1126/science.290.5500.2323. http://science.sciencemag.org/content/290/5500/2323

29. Safonova, A., Hodgins, J.K., Pollard, N.S.: Synthesizing physically realistic human motion in low-dimensional, behavior-specific spaces. ACM Trans. Graph. **23**(3), 514–521 (2004). https://doi.org/10.1145/1015706.1015754. http://doi.acm.org/10.1145/1015706.1015754

30. Sakamoto, Y., Kuriyama, S., Kaneko, T.: Motion map: image-based retrieval and segmentation of motion data. In: Proceedings of the 2004 ACM SIGGRAPH/Eurographics Symposium on Computer Animation, SCA 2004, pp. 259–266. Eurographics Association, Goslar Germany, Germany (2004). https://doi.org/10.1145/1028523.1028557

31. dos Santos Amorim, E.P., Brazil, E.V., II, Daniels, J., Joia, P., Nonato, L.G., Sousa, M.C.: iLamp: Exploring high-dimensional spacing through backward multidimensional projection. In: IEEE VAST, pp. 53–62. IEEE Computer Society (2012). http://dblp.uni-trier.de/db/conf/ieeevast/ieeevast2012.html#AmorimBDJNS12

32. Shoemake, K.: Animating rotation with quaternion curves. In: ACM SIGGRAPH Computer Graphics, vol. 19, pp. 245–254. ACM (1985)

33. Tejada, E., Minghim, R., Nonato, L.G.: On improved projection techniques to support visual exploration of multidimensional data sets. Inf. Vis. **2**(4), 218–231 (2003). https://doi.org/10.1057/palgrave.ivs.9500054. https://doi.org/10.1057/palgrave.ivs.9500054

34. Tenenbaum, J.B., Silva, V.D., Langford, J.C.: A global geometric framework for nonlinear dimensionality reduction. Science **290**(5500), 2319–2323 (2000). https://doi.org/10.1126/science.290.5500.2319. http://science.sciencemag.org/content/290/5500/2319

35. Togawa, H., Okuda, M.: Position-based keyframe selection for human motion animation. In: 11th International Conference on Parallel and Distributed Systems, 2005, Proceedings, vol. 2, pp. 182–185, July 2005. https://doi.org/10.1109/ICPADS.2005.239

36. Wattenberg, M., Viégas, F., Johnson, I.: How to use t-sne effectively. Distill (2016). https://doi.org/10.23915/distill.00002. http://distill.pub/2016/misread-tsne

37. Xiao, J., Zhuang, Y., Yang, T., Wu, F.: An efficient keyframe extraction from motion capture data. In: Nishita, T., Peng, Q., Seidel, H.-P. (eds.) CGI 2006. LNCS, vol. 4035, pp. 494–501. Springer, Heidelberg (2006). https://doi.org/10.1007/11784203_44

38. Zhang, Y., Cao, J.: A novel dimension reduction method for motion capture data and application in motion segmentation. In: 11th International Conference on Parallel and Distributed Systems, 2005, Proceedings, vol. 12, pp. 6751–6760 (2015). https://doi.org/10.12733/jics20150024

Human Computer Interaction Theory and Applications

Involving Hearing, Haptics and Kinesthetics into Non-visual Interaction Concepts for an Augmented Remote Tower Environment

Maxime Reynal[1]([✉]) [iD], Pietro Aricò[2] [iD], Jean-Paul Imbert[1] [iD],
Christophe Hurter[1], Gianluca Borghini[3], Gianluca Di Flumeri[2],
Nicolina Sciaraffa[2], Antonio Di Florio[4], Michela Terenzi[3], Ana Ferreira[3],
Simone Pozzi[3] [iD], Viviana Betti[2] [iD], Matteo Marucci[2], and Fabio Babiloni[2]

[1] University of Toulouse (ENAC), Toulouse, France
maxime.reynal@enac.com
[2] Sapienza University, Roma, Italy
[3] Deep Blue, Roma, Italy
[4] Fondazione Santa Lucia, Roma, Italy

Abstract. We investigated the contribution of specific HCI concepts to provide multimodal information to Air Traffic Controlers in the context of Remote Control Towers (i.e. when an airport is controlled from a distant location). We considered interactive spatial sound, tactile stimulation and body movements to design four different interaction and feedback modalities. Each of these modalities have been designed to provide specific solutions to typical Air Traffic Control identified use cases. Sixteen professional Air Traffic Controllers (ATCos) participated in the experiment, which was structured in four distinct scenarios. ATCos were immersed in an ecological setup, in which they were asked to control (i) one airport without augmentations modalities, (ii) two airports without augmentations, (iii) one airport with augmentations and (iv) two airports with augmentations. These experimental conditions constituted the four distinct experimental scenarios. Behavioral results shown a significant increase in overall participants' performance when augmentation modalities were activated in remote control tower operations for one airport.

Keywords: Human-Computer Interaction · Kinesthetic interaction · Vibrotactile feedback · Interactive spatial sound · Air Traffic Control · Remote control tower

1 Introduction

Airport and aerodrome's Air Traffic Control (ATC) is evolving towards a concept that is both in line with the digital age and innovative for the discipline. This concept is called *Remote and Virtual Tower* (RVT), and consists in controlling an airport from a distant location. The present study is inscribed in this context,

© Springer Nature Switzerland AG 2020
A. P. Cláudio et al. (Eds.): VISIGRAPP 2019, CCIS 1182, pp. 73–100, 2020.
https://doi.org/10.1007/978-3-030-41590-7_4

and aims at designing and evaluating new Human-Computer Interaction (HCI) techniques for RVT environments. Several midsize to small airports in Europe and worldwide are enduring high operating costs while still having low traffic density. This kind of problematic situation generates difficulties to provide Air Traffic Services (ATS). They are often located in sparsely populated areas, poorly accessible or with harsh meteorological conditions. Some solutions have already been proposed to solve this issue [44]. One would be to move their ATS to places where human resources would be grouped together and thus better managed.[1] RTCs would be located in areas which are already equipped with ATS facilities such as international airports. In this regard, the Air Navigation Authorities and some laboratories have made some specific recommendations in order to frame future developments [4, 10, 23]. The study reported in this paper is part of this approach, and aims at developping and testing new HCI concepts for RVT environments, having still in mind the big image goal of improving safety.

Remote Tower Service (RTS) is basically providing ATS remotely, and allows Air Traffic Controllers (ATCos) to work from a different location than the airport which they are actually controlling. In such a case, ATCos' working task would be operated from a RTC that can be everywhere. They would work in a RVT room [22], with a safety level that should be equivalent to the related physical control tower. RTS would be executed by streaming a real-time view from high definition cameras located in the distant controlled area. Airport vicinity would be replicated onto LCD screens in the RVT room to make an out-the-window image. Since the airport environment is digitally reproduced through screens, the way is open for developers to integrate visual augmentations. Night vision, for example, zooming in on a particular area, or holographic projections [52] are some of the forms of visual interaction that can already be found in existing RVT solutions.

Therefore, since many concepts based on vision have already been developed, this study focuses on non-visual augmentations. In order to propose new methods to provide useful information to the user and thus relieve ATCos' visual channel, we have focused our work on the development of HCI concepts based on the senses of touch and hearing. Hence, four interaction and feedback modalities were designed and tested, making it possible to address some problematic situations that had been isolated through brainstorming with ATC professionals. Unlike Single RVT concept, which is the fact of controlling one airport at a time, Multiple RVT concept allows an ATCo to be responsible for the operations at several airports simultaneously. It requires multiple ratings for each ATCo and careful staffing schedules. This concept is relatively new compared to Single RVT and still require to be studied because of its potential huge impact on ATCos' working task. However, it allows to reduce drastically operating costs by dividing human ressources. If such a system is not well designed, the ATCos' workload may increase, which could lead to an increase in the number of errors and therefore to a decrease in safety level. In this context, the work reported in

---

[1] In such a case we talk about Remote Tower Centers (RTC), which should emerge at the operational level in the relatively near future.

this study is also to consider certain aspects of Multiple RVT concept through behavioural and subjective measures.

This paper is an addition to a study published at HUCAPP conference in 2019 [49]. We enriched the section related to the state-of-the-art and the theoritical framework for the design of HCI techniques is presented in detail. Furthermore, we enhanced behavioral diagrams. Finally, we give more details concerning the use of specific design spaces and guidelines (i.e. for kinesthetic interaction [21], earcons [6,7] and tactons [5]) into our design process.

2   Theoritical Backgrounds

2.1   HCI for Remote Control Towers

It appeared that for Single RVT environment, sight is the most studied of the sensory channels [14,33,58]. Furthermore, ATCos' working audio environment appears to be quite poor but sufficient for them to have a mental auditory representation of the controlled airspace. In addition to radio communications based on hearing and speech, the auditory senses can be stimulated in a physical control tower by relatively discrete events. For example, the wind may be felt on the walls, or some engines may be heard if they are located relatively close to the tower. These auditory cues are involved in ATCos' immersion and should consequently be considered for the design of ecological RVT solutions.

Our problematic is about the interaction with sound, which is therefore an augmentation regarding the common control tower auditory environment. There are few examples of sound-based interactive solutions for ATC. An innovative method of sound spatialization using binaural stereo in order to discriminate the communications of enroute ATCos[2] is reported in [25]. Enroute ATCos work in pairs, usually sitting on separate chairs in front of the same desk. They both have separate tasks, which complement each other for the smooth running of the enroute ATC task. This system allows different audio channels to be broadcast simultaneously and without affecting the audio environment of the other ATCo.

However, and to the best of our knowledges, few works exists that consider the sense of touch in control towers. A concrete example of research works considering tactile sensations in ATC are DigiStrips [42] and Strip'TIC [33] initiatives. In this latter, the authors introduced a system that aims at replacing actual tangible paper strips (and therefore there associated tactile aspects) used by ATCos to track the state of aircraft. The research is based on the observation that using this kind of paper tool the system cannot be easily updated with real-time aircraft-related information. Therefore, the authors proposed a system combining vision-based tracking, augmented rear and front projection, radar visualisation, and electronical paper strips wich is made usable, graspable with a digital pen. Finnally, as this is a relatively recent field, multiple RVT

---

[2] *Enroute* ATC must be distinguished from *approach* ATC, which is considered for this study. It concerns aircraft in the cruise phase, while approach ATC concerns aircraft in the approach, descent, landing or take-off phases.

contexts are increasingly studied in the literature, but this is still an area to be explored. However, we can report here a study published in 2010 demonstrating the feasibility of the control of two airports simultaneously [43, 45].

2.2 Hearing as a Way to Interact

The process of weakly modifying (e.g. simply amplifying) a sound played by the system is called *audification*. Audification is a particular type of sonification— actually the simplest type [31]. Gregory Kramer gave a definition in [37]: *"the direct translation of a data waveform into sound"*. Audification can be useful in particular context, for example when raw data can already be sonified without modification. The technique used in the design of Audio Focus interaction modality can be qualified as *interactive spatial audification*, since engine sounds are played while user's head movement modulate their loudness to have a clue on related aircraft position (see next section).

There are relatively few concepts of sound interaction similar to the Audio Focus paradigm (see next sections). The oldest that can be mentioned could be the one published in 1981 [3]. Bolt, inspired by the work of Keen and Scott Morton [35], proposed his contribution to HCI with an anticipation of what he expected for the everyday future interactions. He imagined that we will be quickly exposed to different simultaneous videos and sound sources, and gave a name to this particular environment ("world-of-windows"). The goal was to focus user's attention on sounds while facing a wall of screens operating simultaneously and broadcasting different images (e.g. approximately 20 simultaneous images and sounds). The sounds emanating from the televisions were amplified according to the users' gaze ("eyes-as-output"). More recently, we can also mention the OverHear system [55], which provides a method for remote sound sources amplification based on the user gaze direction using directional microphones. Applied to the field of video conferencing, this tool allows the users' to be remotely focused on a particular interlocutor speech, in an environment that can be noisy. The system used an integrated eye-tracker and a webcam. Finally, a concept of 3D sound interaction presented in [16,17,51] resembles in some aspects to the present research. In this study, the user is surrounded by interactive sound sources organized in a "ring" topology. They can select specific sound sources with 3D-pointing, gestures and speech recognition inputs. The goal is to provide a way to explore the auditory scene, which is provided using head-related transfer functions [8,13,60], with the use of direct manipulation [34] via this ring metaphor mapping of a structured environment.

The third concepts which was considered for this study (i.e. for Spatial Sound Alert) was the earcons. They are synthetic auditory icons but not ecological nor environmental sounds. The first definition was given by Blattner et al. in 1989: *"[earcons are] non-verbal audio messages used in the user-computer interface to provide information to the user about some computer object, operation, or interaction"* [2]. Later on in 1994, Stephen Brewster refined this definition: *"[...] abstract, synthetic tones that can be used in structured combinations to create auditory messages"* [7]. Unlike auditory icons [24], earcons do not require the user to be previously accustomed to the sounds: the relationship between earcons and the event that they are linked with need to be learned by the user. A design space with a set of parameters to consider is given by Brewster et al. in [2, 6, 7]:

- *Timber*: the timber should be made of multiple harmonics; however, the resulted sound should not be too complex to be easily understandable;
- *Register*: if users are required to make judgments in their interaction with the system, the semantic register of the sounds used should not be to much meaningfull;
- *Pitch*: the frequencies used for earcons must be fairly simple and few in number; when several earcons are designed, their differentiation by users is efficient if the frequency differences between them are aligned with their rhythmic differences;
- *Rythm, duration and tempo*: in order for the earcons to be as differentiable as possible from each other, while remaining relatively short, the temporal dimensions considered must be different;
- *Intensity*: this parameter is the most common source of discomfort, and must remain audible enough so that the earcons are sufficiently perceptible and understandable, while keeping a relatively low sound volume to fully integrate into the HCI's sound environment;
  Spatial location: the term *spatial* refers to both a sound played in stereo and a sound having a predefined position in space; this parameter should be taken in consideration when several earcons are intended to be played simultaneously.

Several types of earcons have been formalized (*one-element, compound, transformational* and *hierarchical* earcons). In the present study, we considered the abovementioned set of parameters for the design of spatial sound alerts as compound earcons. This type of earcons are simply made of a collection of one-element earcons, which are basically synthetic sounds (e.g. one musical note) with rythmic properties.

2.3   Providing Information Using Touch

Devices providing tactile feedback to users use one of the following three perceptions coming from our somatosensory system: *cutaneous* or *passive* perception, *tactile-kinesthetic* or *active* perception, and *haptic* perception. The one which is considered in this study is *passive perception*, which occurs when a tactile stimulation is applied to the surface of the skin while the concerned part of the body is immobile. This form of perception is involved in the design space considered in this study, namely the formal framework of tactons, which was considered for the vibrotactile patterns provided to the users.

Haptic technologies are numerous and contribute to many areas of HCI, and tactile stimulation is a broad topic within the field. The most part of techniques for creating stimulation under the skin rely on mechanical devices. However, tactile perception can also be artificially obtained using mid-air ultrasound technology [11, 32], or even using other forces like sustion effect [28]. Tactile transducers devices can also be considered, which are in some ways almost infrabass speakers. They should be fixed on the support we want to make vibrate (preferably made of wood to provide a better medium to carry the signal). Numerous studies and systems have been presented in recent years to communicate spatial information to users using vibrotactile patterns. For example in the context of driving, several studies use vibrotactile cues to indicate directions or obstacles to be avoided to the driver. In their study, Schwalk et al. [53] worked subjective and behavioral measurements regarding recognition, adequacy and workload of tactile patterns presented to participants through a driving seat. Likewise, Petermeijer et al. shown that vibrotactile feedback provided in the driver seat within highly automated driving task can convey a take-over request [47]. They also shown that static vibrotactile stimuli bring to faster reaction times than dynamic ones. This study and their results are usefull for our purpose because the authors postulate that the drivers are not focused on their visual sense since in the framework of the autonomous driving, studies have shown that they would be acceded to other tasks than the driving or the monitoring of the driving (such as eating, talking, listening to music, etc). Hence, they considerate the case in which drivers are deprived of their visual sense (and also hearing), and therefore, better able to "listen" to their other sensations (such as touch). Their are also many studies in the aeronautical field. Van Erp et al. designed a system providing vibrotactile feedback to the pilots through a belt to give waypoint directions, and thus orientation cues [20,57]. Another study investigated helicopter pilots to help them performing hover maneuvers [48].

Different HCI concepts built on tactle sensations to provide structured information to users have been designed. *Haptic* (or *tactile*) icons and the family of underlying concepts are precisely designed for that purpose. MacLean defined *haptic icons* in 2003 as *"brief computer-generated signals, displayed to a user through force or tactile feedback to convey information such as event notification, identity, content or state [...] Synthetic icons may be experienced passively or actively"* [41]. They was refined a year later with the more formalized concept of *tactons*, which have been introduced in 2004 by Brewster and Brown. They are defined as *"structured, abstract messages that can be used to communicate complex concepts to users non-visually"*. The research shown that when visual icons are good to provide spatial information to the users, their tactile equivalent, namely tactons, are good to convey temporal information. Their design is quite similar to the one used for earcons because they are both made with sound signals. They are designed through the parameters of cutaneous (passive) perception, i.e. [5]:

- *Frequency*: the perceivable frequencies on the skin are located in the 20–1000 Hz bandwidth [26], and the maximum perception potential is approximately around 250 Hz. In order to further optimise cutaneous perception, no more than nine frequencies should be haptified simultaneously;
- *Amplitude*: only one amplitude-related parameter should be considered because a large number of interferences with the frequency parameter have

been observed; the maximum amplitude to be considered for a skin stimulus is 28 dB [15];

- *Waveform*: this parameter is equivalent to that of the *timber* for earcons; due to our limited skin perception, we cannot perceive finer differences than those between a sinusoidal, triangular or square wave;
- *Duration*: it is recommended to not use stimuli of less than 100 ms because they are perceived as bites on the skin; in order to not make the interaction unpleasant it is therefore recommended to consider longer skin stimuli [27];
- *Rythm*: this parameter refers to a combinations of stimuli with different *duration*;
- *Body Location*: different locations should be considered on the body when possible. The fingers are the most sensitive part of the body in terms of cutaneous perception. However, hands are often used for other interactions. The back, tigh or abdomen can then be considered with good results [15];
- *Spatiotemporal Patterns*: this parameter refers to a combination of *rythm* and *body location* parameters. Transducers arrays, for example, can be used to display spatially and temporally coordinated haptic patterns.

As for earcons, tactons can be of *one-element*, *compound*, *hierarchical*, or *transformational* type. However, their learning time is often quite short. Using the abovementionned design space, we considered the concept of one-element tactons for our study.

2.4 Considering the Body to Communicate with the Systems

Kinesthetic interaction (KI) refers to engaging our body, or a part of, into an interaction with a system. This term was defined for the first time in 2008 by Fogtmann et al. [21]. It is based on previous work, including the Gesture-based Interaction introduced in 1999 by Long et al. [40] that consists in considering the body as an input device for interaction, and the Full-Body Interaction [46]. Fogtmann also draws on the work of Dourish [19] and Klemmer [36] on *embodied cognition*, which is a model in which the representation of one's own space depends on our perceptual and motor sensations.[3] More precisely, there is a recent consensus in the cognitive science community that the perception of one's own body in space critically depends on multisensory integration [38]. Embodiment can be significantly increased in an immersive environment by considering sensations from the somatosensory system. KI could also enhance this feeling by permitting the users to act as natural as possible, and therefore, in an ecological perspective, enhance the feeling of immersion. It is based on our kinesthetic sensations to create interactive frameworks improving user's implication and, therefore, serenpidity. In their study [21], the authors identified the following seven parameters to define a framework to consider for the design of kinesthetic interaction:

- *Engagement*: this parameter describes a KI concept that can be easily memorized by users and that facilitates serenpidity through their body movements;

---

[3] The terms *embodiment* [18] and *sense of presence* can also be used.

- *Sociality*: describes the fact of considering one body among others in the case of a multi-user system;
- *Movability*: this parameter must be considered to be aware, within the design phase, of the fact that the body, during a KI with a system, can be free or spatially constrained in its capacity of movements;
- *Explicit motivation*: this parameter means that the system must explicitly describe to users how to interact;
- *Implicit Motivation*: this parameter must be taken into account when the KI is well opened, without restriction on the movements themselves (not only spatially);
- *Expressive Meaning*: the expressivity parameter is well taken into account when the KI is perceptibly related to the expected result of the interaction;
- *Kinesthetic Empathy*: is where specific and controlled movement patterns are affected by the relation to other people and stimuli from the surrounding environment.

3   Identification of Relevant ATC Situations

3.1   Aircraft Location in Poor Visibility Conditions

Situations of poor or with no visibility are likely to be common in approach ATC. It also appeared that this kind of situation seem to be problematic for ATCos. Indeed, within approach ATC, ATCos often use their vision to physically search for aircraft with which they are in communication. This allows them to have a better idea of their positions within the controlled area, but also to anticipate their passage times and, thus, to improve their mental representation of the current control configuration. In the case of heavy fog, which prevents such visual information on aircraft from being available, or simply at night when the concerned aerodrome is minimally equipped, the ATCos with whom we worked found it useful to have a tool that could allow them to act as (or approximately as) if they were in good visibility conditions. Hence, a use case we considered was the lack of visibility on the controlled area. This has been the subject of a separate study; the results are presented in [1] and [50].

3.2   Abnormal Events Location

One situation to which ATCos referred several times in our interviews was that of abnormal events. This term refers to events that are abnormal from the ATCo point of view, typically unforeseen events. Most of the time, these are actions performed by pilots without prior authorization. In general, they may start their engines, perform engine tests or even move on the ground without previously having ask for permission. Under normal circumstances, ATCos respond to pilot requests and authorize or deny them the clearance to perform this type of action. This situation is the first use case considered for our study.

3.3   Runway Incursion Awareness

One of the most dangerous situations that an ATCo may have to deal with is the runway incursion. This situation occurs when an aircraft enters the runway while a second one is on short final and is about to land (or conversely, when an aircraft enters the runway while another is taking off, see Fig. 1). The runway incursion was the cause of several air incidents and crashes in the past. In particular, it is partly involved in the most significant crash in the history of aviation, that occurred in Tenerife in 1977 [59]. Many airports are already equipped with systems to prevent such situations. However, this is very often not the case at small airports and aerodromes, to which our study is related. That is why we have chosen the runway incursion as the second use case to consider for the present study.

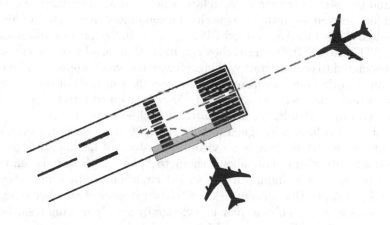

Fig. 1. A schematic view on a runway incursion situation. The aircraft on the right is about to land in the next seconds while another one, at the bottom, is about to cross the line between taxi routes and runway (holding point, rectangle in transparent red). (Color figure online)

3.4   Multiple Radio Calls in Case of Multiple RVT

The concept of multiple RVT generates a large number of situations that do not yet have solutions. One of the most common is the communication with several pilots simultaneously. Each airport uses its own radio frequency, and pilots use it for communicating with related ATC. However, in the case of multiple RVT, ATCos must manage as many radio-communications as they are controlling airports (i.e. radio frequencies). Therefore, they may receive a radio message from an airport A while they are already in communication with a pilot from an airport B. As radio frequency management is an important topic in ATC, we have chosen this situation, which is specific to multiple RVT, for a third and final use case.

4   HCI Techniques for Specific ATC Use Cases

4.1   Using Kinesthetic Interaction and Audification to Provide Interactive Spatial Sound

We called the interaction modality we designed to overcome the visibility issue *Audio Focus (AF)*. This interaction modality allows ATCos to have a way to get a mental representation of the airport vicinity. More precisely, it makes it possible to assess the spatial position of aircraft by making head movements to search not visually for aircraft, but audibly. A sound source playing a generic aircraft engine sound is assigned to each aircraft in the airport vicinity[4], and spatially positioned according to the actual position of the related aircraft. Since vision is no longer stimulated, hearing takes over and becomes the sensory channel on which ATCos rely on. AF principle is similar to the one proposed by Savidis et al. [51], and is based on the strong correlation between head position and visual attention [56], as well as on the increase in the sound level associated to aircraft that are aligned with the sagittal axis. When the visibility is poor and does not allow the ATCos to see the aircraft, they can move their head to make the sound volume associated to each aircraft varying. Hence, if a sound appears to be much louder than the other ones, it means that an aircraft is in front of them (Fig. 2). Also, the distance between aircraft and the user is mapped into the gain of the related sound source (louder when the aircraft is close to the tower).

This interaction form is designed to help the ATCos to locate the aircraft in the airport vicinity when there is no visibility, instead of having an impossible visual contact with them. This allows them to "play" with sounds, and thus to have a better understanding of the sound environment in which they are immersed. In addition, this concept belongs to the category of *enactive interfaces* because it allows users to obtain new knowledge by simply moving their bodies ("act of doing") [39]. AF was studied in more detail in a previous experiment and the results were discussed in [1] from a neurophysiological point of view, and in [50] from a behavioral and HCI point of view. Anyway, KI guidelines was considered for its design:

- **Engagement:** the body language is natural and therefore the interaction is easily memorable, since the principle of the AF interaction only asks the users to move their head to search for object;
- **Sociality:** RVT (single and multiple) are managed by a single operator, so this parameter is not considered for the AF interaction;
- **Movability:** the head movements required for the AF interaction are completely free;
- **Explicit Motivation:** we will see later in this document that users had a training phase before starting the experiment to learn the different interaction modalities. However, AF interaction requires only natural movements: from the moment the participants understand the link between the increase in sound levels and their head movements, the interaction is learned;

---

[4] It should be noted that this type of sound cannot normally be heard in a physical control tower; it is therefore an augmentation.

- **Implicit Motivation:** users can do very small movements to accurately appreciate the location of aircraft. They can also move their head forward or backward to perceive relative distances more precisely;
- **Expressive Meaning:** AF interaction have a good reactivity. Actually, head movements and related sound level increases are well synchronized (this synchronization seems to be the millisecond range and no participants have alerted us to a potential desynchronization in the sound increase);
- **Kinesthetic Empathy:** for the same reason as for parameter related to the *sociability* aspect, this parameter have not been taken into account for AF interaction.

4.2 Spatial Alert as Earcon to Provide Location of Abnormal Events

To warn ATCos from unauthorized events that could be dangerous for the smooth running of operations, we have designed the *Spatial Sound Alert* modality (*SSA*). Its principle is based on an audible warning that is spatially triggered towards the azimuth to which the abnormal event occured. In this way, SSA attracts ATCos' attention to a specific area. When their head is aligned with the alert azimuth, the alert stops (Fig. 2). This principle has already been studied in the past and generally allows a finer and faster reaction, especially when the visual sense is unavailable [54]. SSA modality was made of 3 A musical tones (880 Hz) with a duration of 50 ms, in the form of sine waves separated with 50 ms silences (Fig. 3). This sequence was repeated two times with 100 ms of silence to make a compound earcon. This earcon was looped until the ATCos' head was aligned with the associated event azimuth. During the experimental phase, this modality was used in two specific cases: when an aircraft was taxiing without prior authorization, and in the case of runway incursions. We applied the earcon design guidelines [6]:

- **Timber:** the timber is made of sine wave at 880 Hz, which is not too complex, easily to understand and distinguish in the RVT auditory environement;
- **Register:** this parameter is important here since the concern earcon is an auditory alert. Therefore, its semantic implies urgency to tell the ATCos' that a potentially dangerous situation is occuring at a speific azimuth;
- **Pitch:** we used an A musical tone because this is a frequency that users are accustomed to hear;
- **Rythm, Duration and Tempo:** we used one earcon with a rythmic based on 50 ms blocks of A musical tone and silences: A—silence—A—silence—A—silence—silence, repeated twice. No other earcon was used so we had no need to use this parameter as a mean of differentiation;
- **Intensity:** SSA was audible enough to catch ATCos' attention, while being not too much disruptive sa as not to be too uncomfortable;
- **Spatial location:** by nature, SSA was spatializded; however, the envisaged scenarios did not trigger two alerts simultaneously.

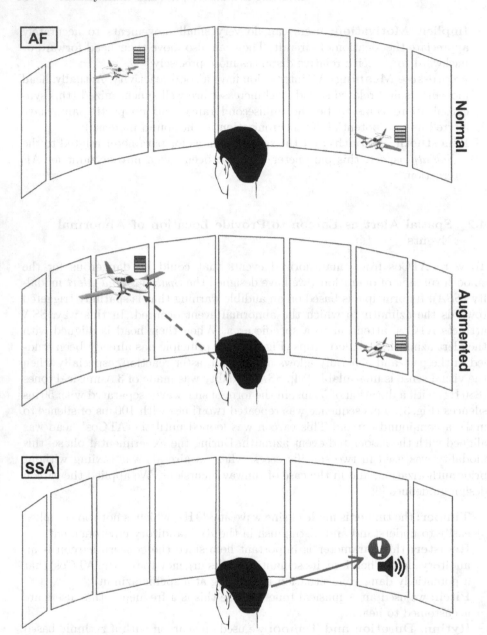

Fig. 2. *Top*: Audio Focus (AF) modality principle. When not activated (top), aircraft are associated to engine sounds but does not interact with user's movements. When activated (bottom), the interaction modality increases the loudness of sound sources (i.e. aircraft) that are aligned with the head of the user. *Bottom*: Spatial Sound Alert (SSA) modality principle. An event that requires ATCo attention occurs on an azimuth that is not currently monitored. A spatial sound alert is raised on this azimuth to attract the ATCo's attention.

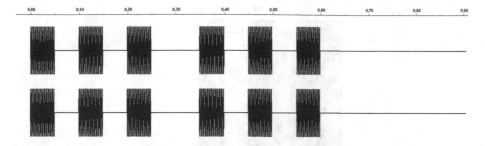

Fig. 3. The wave used for the earcon related to SSA modality. The sound signal is made of two sets of three 800 Hz sinusoid waves of 10 ms separated from each other with a 10 ms silence block. The two blocks are separated with a 100 ms silence block. The scale on top is in milliseconds.

4.3   Vibrotactive Feedback to Alert from Critical Events

Because the runway incursion is associated with the highest degree of danger, the related feedback must be highly disruptive. When anticipating this kind of situation, ATCos shut down all the tasks they are currently performing in order to manage quickly and efficiently the runway incursion situation and avoid an accident at all costs. Runway incursion is a priority situation for most control tasks. We call the related HCI modality *RWY*.

The tactile feedback involved in the RWY modality have been presented to ATCos through a so-called "haptic" chair (as for the next modality, see Fig. 4). It is basically a wooden chair on which two tactile transducers have been attached, one behind the back and the other under the seat. Hence, when a runway incursion situation is detected by the system, the seat of the chair began to vibrate continuously through a sinusoidal signal at 50 Hz. We chose the seat of the chair for this modality with the idea to associate it with the ground, which is the location of the aicraft that crosses the holding point. From the moment the chair starts to vibrate, the ATCos must manage the situation. To do so, they order the aircraft entering the runway to stop suddenly and without delay, give a go-around order to the aircraft ready to land, while managing the current in-flight situation (i.e. other aircraft that may be in the airport environment). This modality also asks the ATCos to inform the system that they are effectively taking the situation into account by validating a button located in front of them on the radar HMI. This button appears when the runway incursion is detected. At the same time, an SSA is also triggered when the runway incursion is detected.

4.4   Other Vibrotactile Feedback to Distinguish Between Radio Calls

In order to avoid the confusing situation that could arise when multiple and simultaneous conversations occur from two separate airports in the case of multiple RVT, a particular type of vibrotactile feedback was used and evaluated. We call

Fig. 4. The wooden chair equipped with two transducers used for Runway Incursion and Call from Secondary Airport alerts.

this modality *Call from Second Airport*, or *CSA*. Its principle consists in spreading a specific and recognizable vibratory pattern through the haptic chair backrest when a message is received by the ATCo from the second controlled airport. We differentiated this vibratory pattern from the one used for RWY modality: it is located in the back of the chair, while for RWY modality the seat vibrates. In addition, it is not continuous (Fig. 5) as for the runway incursion, and consists of intermittent vibrations that are looped as long as the remote pilot is speaking. More precisely, it is made of 55 Hz sinusoid signals of 10 ms followed by 10 ms of silence. In this way, the second airport is associated to this vibratory pattern. We used the guidelines for the design of tactons [5] to formalize it:

- **Frequency:** we used only one signal with a frequency of 55 Hz because of the pseudo infra-bass nature of the tactile transducers;
- **Amplitude:** the amplitude of the signal was configured to be as perceivable as possible, and it was also adjusted from a participant to another by directly varying the intesnity level of the transducers;
- **Waveform:** considering the low number of signal types which are perceivable by the skin, and whishing to propose a smooth signal to the participants, we chose sinusoid waveform for the tacton;
- **Duration:** one element of the tacton only last 10 ms, and they are looped until the pilot from the second airport stop to talk;
- **Rythm:** the alternance of 10 ms of sinusoid signal and 10 ms of silence are creating the rythmic dimention of the tacton;
- **Body Location:** since the ATCos use their fingers for other tasks, the abdomen was not possible with the equipment we had, and the thighs were already considered for RWY modality, we retained the back to play the vibrations (which is one of the most receptive part of the body for cutaneous stimulation);

– **Spatiotemporal Patterns:** the tacton is played only on a single location on the body (between the two shoulder blades) and does not have more than one spatial dimension, so this parameter is not considered in its design.

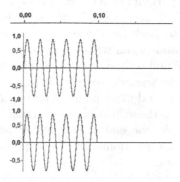

Fig. 5. The wave used for the designed of the tacton involved in the CSA modality. The sound signal is made of a 55 Hz sinusoid wave of 10 ms and 10 ms of silence. The scale on top is in milliseconds.

5   Method

5.1   Hypothesis

As a result of our workshops with expert ATCos during the design process to well design scenarios and solutions, we have isolated some situations that seemed to require new tools in order for them to provide ATS more comfortably. In this way, we have designed specific interaction and feedback modalities, which we have therefore tested using the experiment reported in this paper. More precisely, the previous terms "more comfortably" are used here to describe a more flexible working task, giving better results in terms of cognitive workload, response times, accuracy or, more generally, trying to give ATCos a general impression of ease compared to their current working conditions. Therefore, our work aims to improve immersion and sense of presence through interactions based on under-exploited sensory channels (e.g. hearing, touch), in order to improve safety in RVT environment. More formally, our hypotheses can be summarized as follows: *User performances, in terms of reaction times and perceived workload, are improved when all the augmentation modalities (i.e. AF, SSA, RWY, and CSA) are activated.".*

5.2   Experimental Protocol

The protocol presented was designed to record behavioural and subjective values in relation to the tested modalities, namely SSA, RWY and CSA, in well-defined situations embodied in the scenarios mentioned above. The experiment was divided into two parts. The first was the training phase, in which a general presentation of the experiment was made to the participants, as well as a 30 min specific training scenario to make them familiar with the experimental platform, employed technologies and interaction modalities. Once this training phase was completed, the participant has the opportunity to continue to familiarize with the test bench if they wished to. Otherwise, they was considered ready and would be invited back for the second part of the experiment. This latter lasted about 1h30, and began with a welcome of the participants and a reminder of the experience and technologies used. The experimental phase itself begins with a briefing of the scenario to the participant. The air configuration (number, position, direction, and flight phase of current and future aircraft) was thus accurately described to the ATCo before starting each of the 4 scenarios. Then the scenario began. At the end of the four scenarios, participants were asked to complete a post-run questionnaire. The experiment ended once all the scenarios had been presented.

5.3 Participants

Sixteen professional ATCos participated in the experiment. Genres were equally distributed: 8 women and 8 men, all of French nationality. The average age was 39.4 years ($SD = 7$). All of them reported no hearing problems, which is consistent with their profession that requires them to be tested regularly. After providing them with a detailed explanation of the study and experimentation, all of them gave their consent to participate in accordance with the revised Declaration of Helsinki.

5.4 Scenarios and Experimental Conditions

Four different configurations were used to assess the different remote ATC conditions: Single RVT context without any augmentation (SRT), Single RVT context with augmentations ($SART$), Multiple RVT context without augmentation (MRT), and Multiple RVT context with augmentations ($MART$). Each of these configurations had to be presented once to each participant, and correspond to the four scenarios considered. In order to decrease the learning effect, 4 different scripts were designed for a total of 8 scenarios. In this way, 4 different scenarios using different scripts could be used for each participant. The scripts were designated by *Single 1*, *Single 2*, *Multiple 1*, and *Multiple 2*. They were all different scripts while including equivalent operational events, with the aim to decrease potential learning effects. The events, at minimum, were raised for each scenario. The augmentation modalities (i.e. AF, SSA, RWY and CSA) were activated only during augmented scenarios (i.e. SARTs and MARTs). Single 1 was used for SRT1 and SART1, Single 2 for SRT2 and SART2, Multiple 1 for MRT1 and MART1, and logically Multiple 2 for MRT2 and MART2. Three types of events were used: *Spatial*, *Runway Incursion* and *Call from the Secondary Airport*. AF modality was always activated during augmented scenarios. During the passes, 4 different scenarios of each configuration were randomly presented to each participant, whose were divided into two groups: G_1 was composed of SRT1, SART2, MRT1 and MART2 scenarios, and G_2 was composed of SRT2, SART1, MRT2 and MART1 scenarios. With this experimental design, each participant had to pass through 4 distinct scripts (see Fig. 6 for a direct understanding).

These scripts had to be designed to be both plausible from the volunteers point of view, and comparable between them to allow us to analyze the different modalities once the experiment was over. These various considerations taken upstream mean that the experimental environment offered to participants could be described as ecological and comparable to the real environments that few have now encountered in operational RVT around the world. In general, ATCos have had to deal with common situations to which they are accustomed. All these scenarios were written in collaboration with an ATC expert, former ATCo, in low visibility conditions in which no aircraft were visible. Within this design, 2 exogenous parameters have been considered as independent variables: *Context* that evolves in the set of values [Single, Multiple], and *Augmented* which could be [No, Yes]. Single and Multiple scripts were made similar using comparable traffic complexity and events to raise. Fog was added to the visuals to make poor visibility conditions all along the experiment; no aircraft were visibile throughout the entire experiment, expect aicraft which were on the parking area until the holding point, juste before entering the runway.

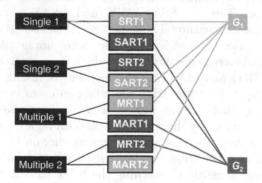

Fig. 6. The repartition of scripts and scenarios between the two experimental groups.

5.5 Apparatus

The experimental apparatus was quite substantial with the goal to provide realistic working conditions to participants (see Fig. 7). Two locations were needed: the ATCo position and two pseudo-pilots positions located in another room. The apparatus was composed of 8 UHD 40 in. Iiyama Prolite X4071 screens, arranged vertically and semi-circularly to provide panoramic view on the controlled airport[5]. Main airport visuals were made using RealTower ad-hoc solution developped under Unity IDE, which provide a photorealistic view on the controlled

---

[5] Muret aerodrome, near Toulouse, South of France.

airport using photo mapping coupled with 3D animated airplane models. The secondaty airport[6] (only for MRTs and MARTs scenarios) was displayed using a 40 in. Eizo Flexscan screen. Secondary airport visuals were made using Flight Gear open flight simulator software. A Wacom 27 in. tablet was used for ground radar and a 19 in. Iiyama screens was used for air radar display. Radio communication with pseudo-pilots were carried out using Griffin PowerMate buttons used as push-to-talk actuators coupled with two microphones (one for each controlled airport). Participants were ask to seat on the haptic chair, located in front of the table on witch the two radars were placed (see Fig. 4). The two tactile transducers (back of the haptic and under its seat), were manufactured by Clark Synthesis. Spatial sound sources associated with aircraft were put in place using the Iyama Prolite screens' embedded audio (i.e. 8 pairs of stereo speakers) to provide the participants with a physical sound spatialization instead of using binaural headphone.[7] ATCos' head orientation was retrieved using a Microsoft HoloLens AR headset.[8] Finally, the different software modules were written in C# language under Microsoft .Net 4.6 framework. Network communications were developped using ad-hoc Ivy bus software, which provides a simple communication mean using string messages and a regular expression binding mechanism [9,12].

5.6   Measurements

Behavioral Variables. Response times were measured for each of the interaction modalities studied (i.e. SSA, RWY and CSA) in order to have a behavioural performance measure. These data could then be compared between non-augmented scenarios, in which modalities were not available (SRTs and MRTs scenarios), and increased scenarios, in which these modalities were usable by participants (SARTs and MARTs scenarios). For both augmented and non-augmented scenarios, a timer was triggered when an event occurred (unauthorized movement on ground, runway incursion or call from the second airport), until the event was managed by the participants. With regard to the SSA modality, the timer was started when an aircraft was moving on the ground without any prior clearance, and then stopped when the participants' heads were aligned with the related event azimuth. Concerning the RWY modality, the trigger was started when an aircraft crossed the holding point while another one was about to land, and stopped when the participants pressed the "Runway Incursion" button located on the ground radar HMI in front of them. Finally, the trigger associated with CSA modality was triggered when a radio communication was

---

[6] Lasbordes aerodrome, South of France.

[7] This solution was chosen because we had demonstration constraints; this way, the platform could be seen and heard by several people at the same time when it was demonstrated.

[8] This can be seen as unsuitable and expensive, however we could not use an inertial unit because of the magnetic field induced by the two transducers. The HoloLens was only used to measure the participants' head orientation, and its augmented reality features were not considered for this experiment.

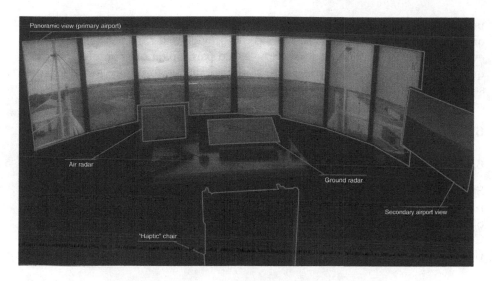

Fig. 7. The appartus used for the participants' position during the experiment.

initiated by a pilot located at the second airport, and then stopped when participants answered by clicking on the associated radio communication button to initiate their response.

All scripts are written to trigger at least one event of each type, hence average values have been computed for all events, both for Single and Multiple contexts, as well as when augmentations were activated or not. In total, 4 average values have been obtained per participant for the SSA and RWY modalities, i.e. for all the scenarios. The CSA modality could only be tested for Multiple RVT scenarios, therefore 2 mean values could be measured for each participant.

Subjectives Variables: Workload Measurements and Questionnaires. Participants had to complete a questionnaire after each scenario so that we had subjective data about the interaction modalities tested. Perceived mental workload (i.e. mental, physical and temporal demands, performance, effort and frustration) was first measured using a NASA Task Load Index (NASA-TLX) questionnaire [29,30]. At the very end of the experiment, we invited each participant to a guided interview using another questionnaire. We were able to ask them to evaluate the perceived contribution in terms of utility, accuracy, situational awareness, sense of presence and cognitive workload for each of the three modalities. In a second time, we asked the participants to assess the suitability of these modalities for particular operational contexts.

Ratings from Subject Matter Expert. During the experiment, and for each of the four scenarios presented to each participant, another direct and non intrusive evaluation was carried out by a Subject Matter Expert (SME, a former professional approach control ATCo) to obtain an additional measure of performance. This measurement was carried out using a dedicated tool launched on a tactile tablet, on which the SME could provide real-time ratings about the evaluated performance level. Finally, he asked participants to complete another questionnaire to subjectively measure workload and performance felt after each scenario treated on a scale from 0 (very low) to 10 (very high).

6   Experimental Results

6.1   Performance Analysis

Statistical analysis is reported in Table 1 below. A two-way ANOVA with repeated measures was conducted on Response times measurements. Concerning SSA modality, the analysis revealed a main effect of Augmented factor and a Context × Augmented interaction. However, no results were found concerning the two other modalities (see Fig. 8).

Table 1. Results from 2×2 ANOVA (CI = .95) with repeated measures (Context [Single, Multiple] × Augmented [No, Yes]) conducted on Response times measurements for SSA modality.

	ddl	F	p	η_p^2
Context	1,14	.85	.37	.06
Augmented	1,14	7.27	<.05	.34
Context × Augmented	1,14	6.92	<.05	.33

Post-hoc analysis was conducted using Tukey's HSD tests when a main effect or an interaction between two factors was revealed. At first sight, it seems that SSA modality allowed ATCos to resolve unauthorized movements on ground more quickly: under augmented scenarios, they were faster to manage these specific situations ($M = 18455.54$; $SD = 4217.22$) than when SSA modality was not activated ($M = 41438.31$; $SD = 7733.1$). More precisely, post-hoc analysis conducted after Context × Augmented interaction revaled that when they were under Single RVT environment, ATCos were significantly faster to resolve these kind of events using SSA modality ($M = 9820.64$; $SD = 1276.44$) than without ($M = 58103.2$; $SD = 14813.18$, see Fig. 8).

This behavioral measurements analysis of performance was then supported by a subjective measurements analysis of performance, which was conducted to confront i) the perceived performance P_{TLX} by each participant using NASA-TLX factors, and ii) the SME performance ratings both during each scenario P_{SME}, and after each scenario $P_{SME_{post}}$. Two-way 2×2 ANOVA (CI = .95) with repeated measures (Context [Single, Multiple] × Augmented [No, Yes]) on the normalized performance values has been conducted using Arcsin transform [62] were conducted for each of these variables.[9] P_{TlX} (resp. $P_{SME_{post}}$) did not show any significant difference $[F = .00034; p = .98]$ (resp. $[F = 1.53; p = .02]$). However, a significant trend on P_{SME} highlighted a decrease of performances when no modality was activated $[F = 5.98; p < .05]$. Duncan post-hoc analysis showed that augmentation modalities induced a significant decrement in overall performane $(p < .05)$ under Multiple airport contexts (i.e. MARTs scenarios). Besides, no result was highlight under Single airport contexts (i.e. SARTs scenarios).

To conclude on performance results, despite finding no result concerning RWY and CSA modality, SSA modality seems to enhance the operators' performance by reducing the response times when ATCos had only one airport to manage. In addition, no significant result were found concerning subjective measurements. In particular, the only observed result is that SME highlighted a significant decrease in performance under multiple airport contexts when augmentation modalities were activated. No results were found regarding the perceived performance (P_{TLX}) and SME real-time performance ratings (P_{SME}).

6.2 Perceived Workload Analysis

NASA-TLX subjective workload scores W_{TLX}) and post-run SME ratings W_{SME}) were considered for the evaluation of perceived workload. As before, two-way ANOVAs 2 × 2 (CI = .95) with repeated measures (Context [Single, Multiple] × Augmented [No, Yes]) on the normalized workload values with the same Arcsin transformation [62] were performed. A significant main effect was highlighted for each factor. It means that for each factor, the same results were found both for W_{TLX}) and W_{SME}: the perceived workload was lower when the interaction modalities were disabled $(p < .05)$. Moreover, in multiple airport context (i.e. MRTs and MARTs scenarios), the perceived workload was higher with respect to single airport contexts (i.e. SRTs and SARTs scenarios) $(p < .05)$. However this second result was not replicated with NASA-TLX values in which no significant difference were found regarding multiple and single airport contexts. No other significant main effect was found.

To sum up, cognitive workload measurements showed that i) multiple airport context (i.e. MRTs and MARTs scenarios) induced higher workload with respect to the single airport context (i.e. SRTs and SARTs scenarios), and ii) the augmentation modalities generate a higher perceived workload. Finally, no significant difference have been highlighted elsewhere.

---

[9] These transformation consists in computing the Arcsin of the square root of the value considered between 0 and 1.

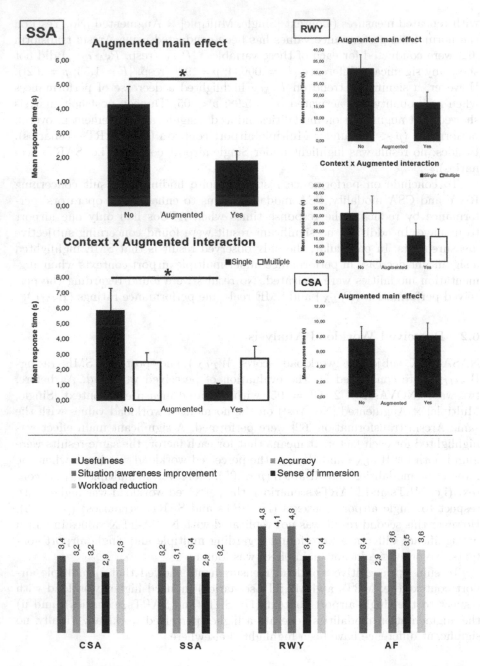

Fig. 8. Top: Results from inferential analysis conducted on Response times data for SSA modality (left diagrams). Right top diagram is for RWY modality (no significant result found). Right bottom diagram is for CSA modality (no significant result found). Error bars are standard errors. Bottom: Descriptive results from post-experiment interview. The scale goes from 1 (*Strongly Disagree*) to 5 (*Strongly Agree*).

6.3   Post-experiment Interview Analysis

In the post-experiment questionnaire, participant ATCos were asked to rate the four augmentation modalities (including AF) along a scale going from 1 (stronly disagree) to 5 (strongly agree), regarding the 5 following assertions: *Perceived contribution to usefulness* ("The current augmentation modality is a useful aid for RT operations"), *Perceived contribution to accuracy* ("The current augmentation modality is accurate enough to support you during the RT operations"), *Perceived contribution to situation awareness improvement* ("The current augmentation modality improves your situation awareness in RT operations"), *Perceived contribution to the sense of immersion* ("The current augmentation modality improves your sense of immersion in RT operations"), and *Perceived contribution to workload reduction* ("The current augmentation modality does not have a negative impact on your workload in RT operations").

By reading these post-experiment interview results, we can tell that from a general point of view the different augmentation modalities have been well received by the participant ATCos. In particular, the one that was felt the more usefull, accurate and providing the better support to SA improvement was the RWY modality (see Fig 8). However, from a decriptive point of view, on average the different interaction modalities received a score higher than 2.5 over 5 but we can not strictly tell that they were perceived to greatly contribute to the sense of immersion and to workload reduction. Nevertheless, AF modality received the highest scores concerning these two points, with respectively 3.5 and 3.6 on average.

7   Discussion

This paper was intended to enrich a study that was presented at the VIS-AGRAPP conference in 2019 [49]. The clearest contribution compared to the previous version is essentially theoretical. Actually, we focused on the state-of-the-art, which is now more deeply rooted in the literature, the use of specific guidelines and design spaces related to the concepts of earcons, tactons and kinesthetic interaction, as well as their application in the design process of the interaction and feedback modalities experimentally studied. Besides, the work reported in these two articles aimed at studying the impact of specific interaction modalities and feedback on a RVT environment. In addition, one of the aims of this work was also to design novel forms of interaction for the context of Single RVT, but also for the one of Multiple RVT, which is still poorly studied in HCI literature. Moreover, knowing that the most commonly used sense among ATCos is vision [61], we have focused our work on interaction and feedback modalities that used the senses of touch, hearing, and kinesthetics, in order to prevent potential cognitive overload related to the visual channel. These choices were made following previously published studies [1,50].

The results obtained show that the interaction and feedback modalities are well received by the end users. Designed to respond to specific problematic or even dangerous situations (use cases), we have experimentally shown that some of these interaction modalities are well adapted, in the sense that they allow better performance in terms of response time. This is indeed the case for the SSA modality, which allows ATCos to be more responsive when an unauthorized event occurs on ground (e.g. when pilots begin to move their aircraft to taxi without prior authorization). In this particular case, participants were 3 times faster to react when the SSA modality was audible. However, we note that the scores assigned by the SME as well as the participants' perception of their own overall performance were clearly reduced when the augmentation modalities were activated. Similarly, a trend could be observed regarding a potential increase in cognitive workload when the interaction and feedback modalities were activated. These results, which were anticipated, can be explained by a lack of experience from the ATCos. They are not accustomed to work in an environment in which their senses of touch and hearing are almost as stressed as their sense of vision. In addition, studies show that, taken separately, this type of interaction concept (at least based on sound) could significantly improve performance [50]. This may therefore be a familiarization issue. Anyway, this remains hypothetic at this stage of the study.

In fact, our exchanges with ATCos during the design process allowed us to observe that they do not consciously integrate the information that can reach them through sensory channels other than sight. Control towers are generally sound-proofed, making them almost airtight on the outside. However, some elements can still be perceived from a control room (e.g. aircraft engines, wind, rain). The state-of-the-art of HCI in RVT environment shows that hearing remains the most considered sensory channel for designing new interaction modalities. The addition of new interaction modalities based on sensory channels other than vision would give RVT environments more immersion and sense of presence, making communication with the system more natural and embodied. However, these new forms of interaction require time to adapt, and their learning needs to be further explored. Even if participants have been thoroughly trained before participating in the experiment, this type of interaction modality is still too far from their normal working environment, which can be disturbing. Incidentally, once the experience was over, some participants expressly told us that the vibrotactile feedback felt through the haptic chair could become embarrassing at times. In any case, regarding the non-significant results, we can assume that the interaction modalities and feedback tested did not increase the cognitive workload of the participants. Moreover, performance has been improved in the case of the SSA modality in Single RVT environment.

8   Conclusion

This experiment allowed us to partially validate our hypotheses. In particular, one of the main result is that some modalities can improve performance, at least in terms of user responsiveness. This was clearly the case for SSA modality. Besides, the underlying message is that the introduction of new interaction modalities, based on the senses of hearing and touch, does not necessarily improve ATCos' workload. It is not significantly increased in Single RVT environment, however we have not been able to demonstrate an hypothetic decrease. Anyway, it seems that these types of interaction modalities do not affect performance. To conclude, further studies are needed to investigate more precisely the positive and negative impacts of the introduction of novel interaction concepts in Single and Multiple RVT context.

Fundings and Acknowledgment. This work was co-financed by the European Commission as part of the Horizon 2020 project Sesar-06-2015 "the embodied reMOte TOwer", namely MOTO, GA n. 699379. We would like to thank all the ATCos who participated in this experiment during their working time, as well as all the people involved in the MOTO project, in France, Italy and in the Netherlands.

References

1. Aricó, P., et al.: Human-machine interaction assessment by neurophysiological measures: a study on professional air traffic controllers. In: 2018 40th Annual International Conference of the IEEE Engineering in Medicine and Biology Society (EMBC), pp. 4619–4622, July 2018. https://doi.org/10.1109/EMBC.2018.8513212
2. Blattner, M.M., Sumikawa, D.A., Greenberg, R.M.: Earcons and icons: their structure and common design principles (abstract only). SIGCHI Bull. **21**(1), 123–124 (1989). https://doi.org/10.1145/67880.1046599
3. Bolt, R.A.: Gaze-orchestrated dynamic windows. SIGGRAPH Comput. Graph. **15**(3), 109–119 (1981). https://doi.org/10.1145/965161.806796
4. Braathen, S.: Air transport services in remote regions. International Transport Forum Discussion Paper (2011)
5. Brewster, S., Brown, L.M.: Tactons: Structured tactile messages for non-visual information display. In: Proceedings of the Fifth Conference on Australasian User Interface, AUIC 2004, vol. 28, pp. 15–23. Australian Computer Society Inc., Darlinghurst (2004)
6. Brewster, S.A., Wright, P.C., Edwards, A.D.N.: Experimentally derived guidelines for the creation of earcons (1995)
7. Brewster, S.A.: Providing a structured method for integrating non-speech audio into human-computer interfaces. Technical report (1994)
8. Brungart, D.S., Simpson, B.D.: Auditory localization of nearby sources in a virtual audio display. In: Proceedings of the 2001 IEEE Workshop on the Applications of Signal Processing to Audio and Acoustics (Cat. No. 01TH8575), pp. 107–110, October 2001. https://doi.org/10.1109/ASPAA.2001.969554
9. Buisson, M., et al.: Ivy: un bus logiciel au service du développement de prototypes de systèmes interactifs. In: IHM 2002, 14ème Conférence Francophone sur l'Interaction Homme-Machine, p. 223 (2002)
10. Calvo, J.: SESAR solution regulatory overview - single airport remote tower. Technical report (2009)
11. Carter, T., Seah, S.A., Long, B., Drinkwater, B., Subramanian, S.: Ultrahaptics: multi-point mid-air haptic feedback for touch surfaces. In: Proceedings of the 26th Annual ACM Symposium on User Interface Software and Technology, UIST 2013, pp. 505–514. ACM, New York (2013). https://doi.org/10.1145/2501988.2502018
12. Chatty, S.: The ivy software bus. White paper. www.tls.cena.fr/products/ivy/documentation/ivy.pdf (2003)
13. Cheng, C.I., Wakefield, G.H.: Introduction to head-related transfer functions (HRTFs): representations of hrtfs in time, frequency, and space. J. Audio Eng. Soc. **49**(4), 231–249 (2001)

14. Cordeil, M., Dwyer, T., Hurter, C.: Immersive solutions for future air traffic control and management. In: Proceedings of the 2016 ACM Companion on Interactive Surfaces and Spaces, ISS Companion 2016, pp. 25–31. ACM, New York (2016). https://doi.org/10.1145/3009939.3009944

15. Craig, J.C., Sherrick, C.E.: Dynamic tactile displays. In: Tactual Perception: A Sourcebook, pp. 209–233 (1982)

16. Crispien, K., Fellbaum, K., Savidis, A., Stephanidis, C.: A 3D-auditory environment for hierarchical navigation in non-visual interaction. Georgia Institute of Technology (1996)

17. Crispien, K., Würz, W., Weber, G.: Using spatial audio for the enhanced presentation of synthesised speech within screen-readers for blind computer users. In: Zagler, W.L., Busby, G., Wagner, R.R. (eds.) ICCHP 1994. LNCS, vol. 860, pp. 144–153. Springer, Heidelberg (1994). https://doi.org/10.1007/3-540-58476-5_117

18. Csordas, T.J.: Embodiment as a paradigm for anthropology. Ethos 18(1), 5–47 (1990)

19. Dourish, P.: Where the Action Is. MIT Press, Cambridge (2001)

20. Erp, J.B.F.V., Veen, H.A.H.C.V., Jansen, C., Dobbins, T.: Waypoint navigation with a vibrotactile waist belt. ACM Trans. Appl. Percept. 2(2), 106–117 (2005). https://doi.org/10.1145/1060581.1060585

21. Fogtmann, M.H., Fritsch, J., Kortbek, K.J.: Kinesthetic interaction: revealing the bodily potential in interaction design. In: Proceedings of the 20th Australasian Conference on Computer-Human Interaction: Designing for Habitus and Habitat, OZCHI 2008, pp. 89–96. ACM, New York (2008). https://doi.org/10.1145/1517744.1517770

22. Fürstenau, N. (ed.): Virtual and Remote Control Tower: Research, Design, Development and Validation. RTA. Springer, Cham (2016). https://doi.org/10.1007/978-3-319-28719-5

23. Fürstenau, N., Schmidt, M., Rudolph, M., Möhlenbrink, C., Papenfuß, A., Kaltenhäuser, S.: Steps towards the virtual tower: remote airport traffic control center (raice). Reconstruction 1(2), 14 (2009)

24. Gaver, W.W.: The sonicfinder an interface that uses auditory icons (abstract only). SIGCHI Bull. 21(1), 124 (1989). https://doi.org/10.1145/67880.1046601

25. Guldenschuh, M., Sontacchi, A.: Application of transaural focused sound reproduction. In: 6th Eurocontrol INO-Workshop 2009 (2009)

26. Gunther, E., O'Modhrain, S.: Cutaneous grooves: composing for the sense of touch. J. New Music Res. 32(4), 369–381 (2003). https://doi.org/10.1076/jnmr.32.4.369.18856

27. Gunther, E.E.L.: Skinscape: a tool for composition in the tactile modality. Ph.D. thesis, Massachusetts Institute of Technology (2001)

28. Hamdan, N.A.h., Wagner, A., Voelker, S., Steimle, J., Borchers, J.: Springlets: expressive, flexible and silent on-skin tactile interfaces. In: Proceedings of the 2019 CHI Conference on Human Factors in Computing Systems, CHI 2019, pp. 488:1–488:14. ACM, New York (2019). https://doi.org/10.1145/3290605.3300718

29. Hart, S.G.: Nasa-task load index (nasa-tlx); 20 years later. Proc. Hum. Factors Ergon. Soc. Annu. Meet. 50(9), 904–908 (2006). https://doi.org/10.1177/154193120605000909

30. Hart, S.G., Staveland, L.E.: Development of NASA-TLX (task load index): results of empirical and theoretical research. In: Hancock, P.A., Meshkati, N. (eds.) Human Mental Workload, Advances in Psychology, vol. 52, pp. 139–183. North-Holland (1988). https://doi.org/10.1016/S0166-4115(08)62386-9

31. Hermann, T., Hunt, A., Neuhoff, J.G.: The Sonification Handbook. Logos Verlag Berlin, Germany (2011)
32. Hoshi, T., Takahashi, M., Iwamoto, T., Shinoda, H.: Noncontact tactile display based on radiation pressure of airborne ultrasound. IEEE Trans. Haptics 3(3), 155–165 (2010). https://doi.org/10.1109/TOH.2010.4
33. Hurter, C., Lesbordes, R., Letondal, C., Vinot, J.L., Conversy, S.: Strip-TIC: exploring augmented paper strips for air traffic controllers. In: Proceedings of the International Working Conference on Advanced Visual Interfaces, AVI 2012, pp. 225–232. ACM, New York (2012). https://doi.org/10.1145/2254556.2254598
34. Hutchins, E.L., Hollan, J.D., Norman, D.A.: Direct manipulation interfaces. Hum. Comput. Inter. 1(4), 311–338 (1985). https://doi.org/10.1207/s15327051hci0104_2
35. Keen, P.G., Scott Morton, M.S.: Decision support systems; an organizational perspective. Technical report (1978)
36. Klemmer, S.R., Hartmann, B., Takayama, L.: How bodies matter: five themes for interaction design. In: Proceedings of the 6th Conference on Designing Interactive Systems, DIS 2006, pp. 140–149. ACM, New York (2006). https://doi.org/10.1145/1142405.1142429
37. Kramer, G.: Auditory Display: Sonification, Audification, and Auditory Interfaces. Perseus Publishing, New York (1993)
38. Lackner, J.R., DiZio, P.: Vestibular, proprioceptive, and haptic contributions to spatial orientation. Annu. Rev. Psychol. 56, 115–147 (2005)
39. Loftin, R.B.: Multisensory perception: beyond the visual in visualization. Comput. Sci. Eng. 5(4), 56–58 (2003). https://doi.org/10.1109/MCISE.2003.1208644
40. Long Jr., A.C., Landay, J.A., Rowe, L.A.: Implications for a gesture design tool. In: Proceedings of the SIGCHI Conference on Human Factors in Computing Systems, CHI 1999, pp. 40–47. ACM, New York (1999). https://doi.org/10.1145/302979.302985
41. Maclean, K., Enriquez, M.: Perceptual design of haptic icons. In: In Proceedings of Eurohaptics, pp. 351–363 (2003)
42. Mertz, C., Chatty, S., Vinot, J.L.: The influence of design techniques on user interfaces: the DigiStrips experiment for air traffic control. In: HCI-Aero (2000)
43. Moehlenbrink, C., Papenfuss, A.: ATC-monitoring when one controller operates two airports: research for remote tower centres. In: Proceedings of the Human Factors and Ergonomics Society Annual Meeting, vol. 55, pp. 76–80. Sage Publications, Los Angeles (2011)
44. Nene, V.: Remote tower research in the United States. In: Fürstenau, N. (ed.) Virtual and Remote Control Tower. RTA, pp. 279–312. Springer, Cham (2016). https://doi.org/10.1007/978-3-319-28719-5_13
45. Papenfuss, A., Friedrich, M.: Head up only - a design concept to enable multiple remote tower operations. In: 2016 IEEE/AIAA 35th Digital Avionics Systems Conference (DASC), pp. 1–10, September 2016. https://doi.org/10.1109/DASC.2016.7777948
46. Parés, N., Carreras, A., Soler, M.: Non-invasive attitude detection for full-body interaction in mediate, a multisensory interactive environment for children with autism. In: VMV, pp. 37–45. Citeseer (2004)
47. Petermeijer, S., Cieler, S., de Winter, J.: Comparing spatially static and dynamic vibrotactile take-over requests in the driver seat. Accid. Anal. Prev. 99, 218–227 (2017). https://doi.org/10.1016/j.aap.2016.12.001
48. Raj, A.K., Kass, S.J., Perry, J.F.: Vibrotactile displays for improving spatial awareness. Proc. Hum. Factors Ergon. Soc. Ann. Meet. 44(1), 181–184 (2000). https://doi.org/10.1177/154193120004400148

49. Reynal, M., et al.: Investigating multimodal augmentations contribution to remote control tower contexts for air traffic management. In: Proceedings of the 14th International Joint Conference on Computer Vision, Imaging and Computer Graphics Theory and Applications, HUCAPP, vol. 2, pp. 50–61. INSTICC, SciTePress (2019). https://doi.org/10.5220/0007400300500061

50. Reynal, M., Imbert, J.P., Aricó, P., Toupillier, J., Borghini, G., Hurter, C.: Audio focus: interactive spatial sound coupled with haptics to improve sound source location in poor visibility. Int. J. Hum. Comput. Stud. **129**, 116–128 (2019). https://doi.org/10.1016/j.ijhcs.2019.04.001

51. Savidis, A., Stephanidis, C., Korte, A., Crispien, K., Fellbaum, K.: A generic direct-manipulation 3D-auditory environment for hierarchical navigation in non-visual interaction. In: Proceedings of the Second Annual ACM Conference on Assistive Technologies, Assets 1996, pp. 117–123. ACM, New York (1996). https://doi.org/10.1145/228347.228366

52. Schmidt, M., Rudolph, M., Werther, B., Fürstenau, N.: Remote airport tower operation with augmented vision video panorama HMI. In: 2nd International Conference Research in Air Transportation, pp. 221–230. Citeseer (2006)

53. Schwalk, M., Kalogerakis, N., Maier, T.: Driver support by a vibrotactile seat matrix - recognition, adequacy and workload of tactile patterns in take-over scenarios during automated driving. Procedia Manuf. **3**, 2466–2473 (2015). https://doi.org/10.1016/j.promfg.2015.07.507. 6th International Conference on Applied Human Factors and Ergonomics (AHFE 2015) and the Affiliated Conferences, AHFE 2015

54. Simpson, B.D., Brungart, D.S., Gilkey, R.H., McKinley, R.L.: Spatial audio displays for improving safety and enhancing situation awareness in general aviation environments. Technical report, Wright State University, Department of Psychology, Dayton, OH (2005)

55. Smith, D., et al.: Overhear: augmenting attention in remote social gatherings through computer-mediated hearing. In: CHI 2005 Extended Abstracts on Human Factors in Computing Systems, CHI EA 2005, pp. 1801–1804. ACM, New York (2005). https://doi.org/10.1145/1056808.1057026

56. Stiefelhagen, R., Yang, J., Waibel, A.: Estimating focus of attention based on gaze and sound. In: Proceedings of the 2001 Workshop on Perceptive User Interfaces, pp. 1–9. ACM (2001)

57. Van Erp, J., Jansen, C., Dobbins, T., Van Veen, H.: Vibrotactile waypoint navigation at sea and in the air: two case studies. In: Proceedings of EuroHaptics, pp. 166–173 (2004)

58. Van Schaik, F., Roessingh, J., Lindqvist, G., Fält, K.: Assessment of visual cues by tower controllers, with implications for a remote tower control centre (2010)

59. Weick, K.E.: The vulnerable system: an analysis of the tenerife air disaster. J. Manag. **16**(3), 571–593 (1990). https://doi.org/10.1177/014920639001600304

60. Wenzel, E.M., Arruda, M., Kistler, D.J., Wightman, F.L.: Localization using non-individualized head-related transfer functions. Acoust. Soc. Am. J. **94**, 111–123 (1993). https://doi.org/10.1121/1.407089

61. Wickens, C.D., Mavor, A.S., McGee, J., Council, N.R., et al.: Panel on human factors in air traffic control automation. In: Flight to the Future: Human Factors in Air Traffic Control (1997)

62. Wilson, E., et al.: The arcsine transformation: has the time come for retirement (2013)

Virtual Reality System for Ship Handling Simulations: A Case Study on Nautical Personnel Performance, Observed Behaviour, Sense of Presence and Sickness

Chiara Bassano[1], Manuela Chessa[1(✉)] ⓘ, Luca Fengone[2], Luca Isgrò[2], Fabio Solari[1] ⓘ, Giovanni Spallarossa[2], Davide Tozzi[2], and Aldo Zini[2]

[1] Department of Informatics, Bioengineering, Robotics, and Systems Engineering, University of Genoa, Genoa, Italy
manuela.chessa@unige.it
[2] CETENA S.p.A., Genoa, Italy
aldo.zini@cetena.it

Abstract. In this paper we introduce the virtual reality ship simulator we designed and the results of an experimental session, in which a manoeuvring task was proposed. In particular, we considered three conditions: (i) the kind of visualisation setup, i.e. a non-immersive system based on standard monitors and an immersive system using a head mounted display; (ii) the path users were instructed to follow, Elliptic or Eight shape path; (iii) the boat type, Slow and Fast. We analyzed three different aspects: performances, defined as the correctness of the followed path; cybersickness, assessed by the Simulator Sickness Questionnaire and physiological measurements (heart rate and skin conductance); sense of presence, determined through the Igroup Presence Questionnaire and the participants' head rotation. In order to evaluate the proposed system from the point of view of experts, tests were conducted on 20 volunteer skilled users, specifically students of a naval academy. Results show that: (i) expert users are able to follow the predefined path in a quite accurate manner; (ii) both visualization systems do not introduce serious undesired effects or stress and the use of immersive virtual reality itself does not explain the increase of user malaise state; (iii) immersive virtual reality systems allow users to feel more involved and present in the simulation scenario; (iv) there are no appreciable differences with respect to the degree of knowledge of virtual reality systems, thus indicating that such simulator can be also used for training users without specific technological skills.

This work was supported by the Innovation Project "LEADERSHIP TECNOLOGICA - Acquisizione di nuove conoscenze propedeutiche a future applicazioni navali e sviluppo delle tecnologie abilitanti per la leadership tecnologica di Fincantieri", co-funded by the Italian Ministry of Transport MIT - Ministero delle Infrastrutture e dei Trasporti.

A. P. Cláudio et al. (Eds.): VISIGRAPP 2019, CCIS 1182, pp. 101–127, 2020.
https://doi.org/10.1007/978-3-030-41590-7_5

Keywords: Presence · Immersivity · Cybersickness · Task
performance · User experience · Gamification · Navigation
performance · Ship simulator

1   Introduction

Simulation is an important field of research and application in all the domains, where complex and demanding tasks are required. Ship handling simulators have always taken advantages from computer-based environments representing a replica of the real world, in which the ship is operating. Such a kind of systems can be used for both design assessment and for training purposes [2,27,28]. Simulation is one of the key aspect of the concepts behind Industry 4.0 [30]. The diffusion of immersive virtual reality (VR) head mounted displays (HMDs) gives the users the possibility of training in synthetic environments for more realistic experiences.

The goal of the current study is the assessment of a VR technological system for ship handling simulation, which can be used both with an immersive HMD and with standard displays, developed in the context of the project MIT - Leadership Tecnologica[1]. In particular, we aim to analyze the possible undesired effect (i.e. simulator sickness), the involvement and the sense of presence experienced by professional users. Indeed, the prolonged use of this type of systems might produce on users different negative effects related both to a decrease of performances and an increase of sickness. Moreover, we are interested into understanding how expert users behave in such an environment. The long term goal of the study is the assessment of the use of such systems in professional applications.

The considered ship handling simulator is a simulation framework designed with different targets on mind: training, virtual prototyping and virtual test bed. Major strengths of the framework are the high detailed real-time physical behavior reproduction of any type of ships (from small boat to big ships) and a powerful visualization system using up to date gaming technologies for the best cost effective virtual reality environment available nowadays.

In this paper, we present the results of an experimental session lead on students of the Genoa naval academy, by testing how the operator can feel using different types of immersive experience during navigation activities. We specifically addressed expert naval users as a target of our study, in order to analyze the different visualization modalities, in a non-trivial situation, nevertheless without taking care of the complexity of the task. In particular, we evaluate and compare different technological solutions and techniques for the implementation of an interactive VR system: on the one hand, a simulation system, composed of a monitor for visualization; on the other hand, an immersive system composed of a HMD for VR (the Oculus Rift). Interaction is done through the physical

---

[1] The involved partners are the Company Cetena S.p.A. and the Department DIBRIS of the University of Genoa, Italy.

reproduction of a ship command panel, which, in the first case, is completely visible to the user, while, in the second case, has to be substituted by a schematic virtual representation in the virtual environment and synchronized with it.

A preliminary version of the results here analyzed have been presented in [1]. Here, we extend the analysis, by including the differences among different population of volunteers, i.e. different expertise in ship handling in the real world, and different attitudes to the use of VR systems.

The paper is organized as follows: in Sect. 2 we briefly discuss the state of the art of the analysis of cybersickness and presence in VR simulators; in Sect. 3 we describe the materials and the methods of the experiment. In Sect. 4 we present and discuss the obtained results, and in Sect. 5 we conclude and discuss the further developments and implication of our research.

2   State of the Art

Cybersickness, defined as a state of malaise and unpleasant side effects associated to use of immersive simulations, is still a common problem, affecting 60–80 % of the users, and it is a potential issue for the broader adoption of these technologies, although its minor, short term health risk [16]. Typically, cybersickness varies between individuals but common symptoms are nausea, eyestrain, dizziness, apathy, sleepiness, disorientation, fatigue and general discomfort. It can occur immediately after training or even up to 5–12 h later [10,13]. It is also worth noting that this state of malaise can also cause cognitive impairment and negatively affect user's performance while accomplishing a task [16].

Causes of cybersickness are still over debate, but three prominent theories are: poison theory [3], postural instability theory [20] and sensory conflict theory [18]). In particular, the latter thesis suggests that the mismatch of vestibular and visual sensory systems could cause sickness, due to the absence of inertial displacement and could explain why higher Visually Induced Motion Sickness (VIMS) levels were reported in passive exploration compared to active exploration of virtual environment [21].

Factors influencing cybersickness include individual, device and task differences [5,6,12,16]. It can be quantify and assessed using both subjective and objective measures, e.g. see [9,16].

Subjective scales of evaluation include questionnaires which are usually filled in before and after the trial, and where the user is asked to rate the severity level of different symptoms. Some example of symptoms questionnaires are the Simulator Sickness Questionnaire [8], the Motion Sickness Assessment Questionnaire [7], the Pensacola Motion Sickness Questionnaire [11], and the Nausea Profile [14].

Objective measures, instead, include physiological measurements, some of which have been proven to be correlated with VIMS [6]. [15] studied the effect of motion sickness on thermoregulation, using provocative visual stimuli (immersion into the virtual reality simulating rides on a rollercoaster). They registered in participants an increase of vasodilation, sweating, skin conductance, skin warming and tachycardia, which are related to the activity of the sympathetic nervous system, as a defensive reaction against the sensation of nausea [17] related to cybersickness.

Another important aspect to be considered is the sense of presence, or spatial presence, which is defined as the psychological state where virtual experiences and computer-generated environments feel authentic, similar to the corresponding physical experience. In other words, presence is the sense of "being physically there" [22]. A related term, often confused with presence, is immersion, which depends essentially on the type of technology used (immersive versus non-immersive) and can be objectively described. On the contrary, the sense of presence is primarily subjective and it is linked to the user experience. Although its subjective nature, however, several factors influence spatial presence [26]: the degree of interaction user has with the virtual environment, as the presence of the player in the virtual world implicitly implies his ability to act in it and interact with it; the proper implementation of an action-effect loop; the high resolution of information displayed, in a manner that it does not indicate the existence of the display; the consistency of the displayed information across different sensory modalities; the presence of a first person avatar, as a self-representation of the user in the virtual world, which should be similar in appearance or functionality to the individual's body.

Again, it is difficult to measure the sense of presence. In the literature, there are some questionnaire commonly used to measure the sense of presence are the IGroup Presence Questionnaire [19], the Presence Questionnaire [29] and the Slater, Usoh and Steed (SUS) Questionnaire [23, 25]. Moreover, we can rely on some indirect measurements, e.g. postural stability, users' movements and heart rate [4].

3 Materials and Methods

Here, we describe how the software application has been designed, the two different hardware solutions taken into account and the parameters measured to evaluate the experience of expert users in the two different setups.

3.1 Software

The simulation system is conceived as a gamification of a ship handling experience. Gamification is a process consisting in the introduction of techniques, methods and strategies typical of entertainment world in educational contexts, which, otherwise, would be deficient in induced interactivity. From the user point of view, he is encouraged to familiarize with a new experience bypassing the specific physical interface and constraints, due to a lack of knowledge or experience. From the experimenter point of view, instead, it is possible to set two different parameters, the boat dimension and the path the user is asked to follow. This way, the trainer is able to control the difficulty of the task according to the user's ability and pre-defined goals of the simulation activity.

When the application starts, both with the monitor and the VR setup, participant finds himself on a boat offshore, some kilometers far from the coast. The scenario is inspired to actually existing coastal areas, a surface of about 50 km × 50 km around Genoa harbour, taken from a 3D graphics database. Rendering is not detailed in order to avoid to distract the user with an excess of visual information, while maintaining a good level of involvement, determined by the recognition of familiar scenes.

User has to steer the ship following one of the predefined paths: the Ellipse, the simplest one, and the Eight path, more complex because of the frequent changes in direction, (see Fig. 1). The Eight path implies a greater freedom of movement and a higher probability of losing the original route. Moreover, participants are required to have a better spatial mapping and control over the ship, being aware of its turning rate, i.e. time and space required to veer.

The path is fixed and visible to the user as a route just above the sea (visible in Fig. 3).

Weather conditions are an additional feature of the system: in the Eight path simulation, sea is calm and plain, while in the elliptic path simulation, sea is more rough, in order to counterbalance the simplicity of the route. The system can be further personalized with different weather conditions.

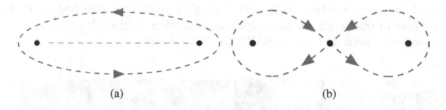

(a)   (b)

Fig. 1. The two paths: (a) ellipse and (b) eight.

Finally, participants have to accomplish the task with two different ships: a patrol boat and a coastguard, afterwards referred to as fast and slow boat, respectively (see Fig. 2). The patrol boat is a small light vessel (15 m long and 20 t of weight), that can reach 50 n of speed. The coastguard, instead, is a larger ship (20 m long and 50 t of weight) that can reach a maximum speed of 20 n. The aesthetics and the operating mechanisms of these two ships have been modeled and implemented as faithfully as possible.

3.2 Hardware

Two different setups have been designed and implemented: a traditional simulation non-immersive virtual reality system, based on 3 standard monitors, and an immersive VR system, based on a HMD (see Fig. 3).

In both cases, interaction (i.e. steering the vessel) is done through the physical reproduction of a ship command panel, so people can drive ships rolling a real

Fig. 2. (a) Slow ship: a coastguard (b) Fast ship: a patrol boat.

rudder and moving the accelerator knobs. In the first case, the command panel is completely visible to the user, while, in the second case, the user has the haptic and force feedback of the panel, but he cannot see it. For this reason, it has to be substituted by a schematic representation in the virtual environment, a slider showing rudder rotation, and synchronized with it.

In the monitor setup, three 27 in. monitors are disposed vertically side by side, in order to mimic the view from the ship command bridge. The user is required to wear a safety helmet, with a HTC Vive tracker attached on it, for the purpose of tracking his head position and rotation. In the VR setup, instead, monitors are substituted with a HMD, the Oculus Rift.

Fig. 3. The two visualization modalities: (a) visualization with monitors and (b) visualization in immersive VR (the user only sees what is shown inside the VR headset).

3.3 Parameters

The goal of our work is evaluating participant performance; cybersickness and comfort in general; sense of presence, in the two different hardware solutions.

Performance. Performance, in this specific scenario, is considered as the ability of the user to follow the proposed paths, so boat latitude and longitude are recorded during the experimental session.

Cybersickness and Comfort. We quantify cybersickness by using both subjective valuations, i.e. the Simulator Sickness Questionnaire (SSQ) [8], and objective parameters, i.e. physiological measurements. The SSQ is an instrument commonly used to quantify this state of malaise. It is composed by 16 questions in a 4-points Likert scale evaluating three main aspects: Nausea, Oculomotor disorders and Disorientation. Skin conductance, i.e. the continuous variations of the electrical characteristics of the skin caused by variations of the sweating, and heart rate are, instead, physiological parameters strictly linked to the emotional and mental state of the user: variations of these two parameters from the baseline could be a consequence of stress, fatigue, excitement or cybersickness.

We used the Mindfield eSense Skin Response sensors, to measure skin conductance, and the Scosche Rhythm armband, for the heart rate. The first sensor is connected to a smartphone Galaxy S4 by wire and uses a proprietary software to record and send 5 samples/second, while the second sensor is connected to the same phone via Bluetooth and exploits the BLE Heart Rate Monitor software in order to memorize and send 1 sample/second.

Sense of Presence. In order to quantify sense of presence we decided to use the Igroup Presence Questionnaire (IPQ), one of the standard questionnaire for measuring presence [19]. It is composed of 13 7-points Likert scale questions evaluating three different aspects: the Spatial Presence; the Involvement, intended both as attention during the interaction with the virtual world and as perceived involvement; the Experienced Realism, which measures the perceived realism of the VR experience. An additional question rates the sense of presence from the original definition on [24].

User rotation and 3D position are recorded during the test, and used as a measure of the tendency of people to explore the surrounding environment in a natural way [4].

3.4   Procedure

We consider three different independent variables: boat type, visualisation modality and path shape. Considering the two first variables we use a repeated measure experimental design, as all participants accomplish the task both with the slow and fast vessel with either the monitor and the HMD. While considering the latter parameter, we adopt a between group experimental design, half of the participant has to follow the Ellipse route and half the Eight one.

The order of execution was fixed: participants start with the monitor simulation and the slow boat (Monitor Slow), then the fast boat (Monitor Fast); after this, they wear the Oculus Rift and accomplish the task in the immersive virtual environment with the slow (HMD Slow) and fast (HMD Fast) vessel. Each trial lasts 5 min. A further development of this work will take into account learning effects and so the execution order will be shuffled in the experiment. Prior to the experiment execution, the experimenter explains participant the modality and the purpose of the test, the different tasks, the setup and the instrumentation used. Then subjects have to sign a written consent and the privacy policy. Participants are told that they could interrupt the experiment whenever they wanted.

After this, they have to fill in an anonymous module giving personal information (age, genre, previous experience in simulation, VR environments, and boat driving).

Afterward, the experimenter attaches the different sensors to the participant (Scosche Rhythm armband and Mindfield eSense Skin Response sensors) and give him an armband for the smartphone. Sensor choice and position have been thought in order not to interfere with participants movements or cause discomfort, reducing sense of presence.

Next, participant is asked to fill in the first SSQ (SSQ Pre) and, once finished, he accomplishes the four tasks. After each trial, he has to complete a separate SSQ, in order to monitor his state of malaise. A brief introductory tutorial phase precedes the use of each new hardware setup. Guided by the experimenter, users start familiarizing with the interface and visualization system and try driving the ship for 1 min.

Finally, volunteers are asked to fill in an IPQ for each trial accomplished.

The experiment has a total duration of 45 min, 20 min of which for simulation.

3.5 Participants

20 volunteer healthy male subjects aged between 20 and 24 years (21.8 ± 1.1 years) participated to the experiment. They had normal or corrected to normal vision. The majority of them were naive towards Virtual Reality (74 %), while 5 % had already took part to experiments involving VR systems. All of them were expert boat drivers (see Table 1), in fact they were students from the Genoa naval academy. On one side, we wanted to evaluate the reaction of experts to specific stressors; on the other hand, we wanted to understand their propensity to use VR technologies, which could be perceived more as a game than a serious tool for learning.

Table 1. Boats usually driven by participants, from [1].

Kind of boat	Number of people
Fast smaller than 15 m	8
Fast longer than 15 m	2
Slow smaller than 40 m	3
Slow longer than 40 m	2
None	4

4   Results

In this section, we present results obtained from the analysis of the recorded data: trajectories for performances; SSQ, skin conductance and heart rate for cybersickness; IPQ and head rotation for the sense of presence, comfort and naturalness of the experience.

4.1   Analysis of Performances

The latitude and longitude of the virtual boat during task execution have been recorded. Figure 4 shows that participants trajectories seems are accurate and the original path shape is easily identifiable, especially concerning the Eight path data.

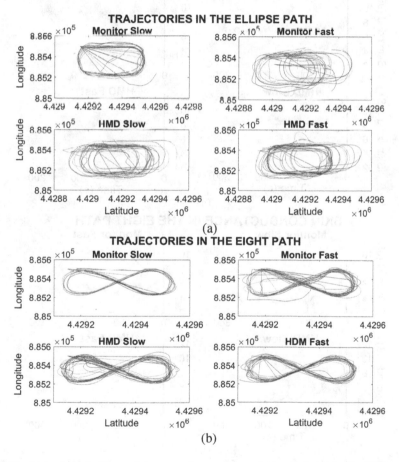

Fig. 4. Trajectories of all participants who performed the task with the Ellipse path (a) and Eight path (b). Figure from [1].

4.2 Analysis of Cybersickness

Skin Conductance. Skin conductance was recorded as a measure of change in participant emotional state and well-being. Measurements were analyzed considering the kind of trial (Monitor Slow, Monitor Fast, HMD Slow, HMD Fast) and then taking into account both the kind of trial and the path (Ellipse or Eight). Additionally, we decided to consider participants previous experience with VR setups and handling ships, in order to investigate their influence. In particular, we expect that people who have already tried Virtual Reality will be less excited or anxious, at least in the beginning of the trial, as they already know

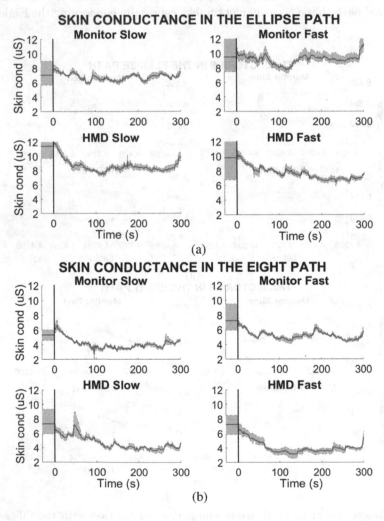

Fig. 5. Average skin conductance considering the trial and the path: Ellipse (a) or Eight (b). Figure from [1].

how an immersive VR environment should look like. Nonetheless, students with practical experience in handling ships should be more confident about the task.

Results obtained from the first analysis are shown in Fig. 5. In the monitor case, in general, skin conductance is stable and constant, even if slightly higher in the trial with the faster ship, probably because of the greater difficulty of the task. In the HMD case, it initially fast decreases and then settles around a stable value, similar to the one recorded during the simulation with the monitor. This descending trend could be associated to people expectation over the trial difficulty, or the excitement of using a new hardware system. However, when they realize that the task consists in an activity they are used to, skin conductance decreases.

Moreover, skin conductance in the Monitor Fast trial is in average the highest one in both cases. This can be due to the fact that participants face for the first time the fast task and still have little confidence with the setup, hardware and software. This effect, however, is attenuated in the following trials. Future analysis with a counterbalanced presentation of the setups could better explain this trend.

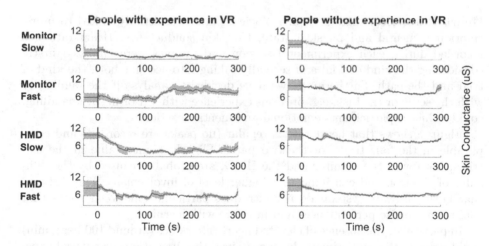

Fig. 6. Average skin conductance of participants with (left column) or without (right column) previous experience in VR based on the trial.

The fact of having already tried VR has a mild influence on participants' skin conductance levels, both in the baseline and general trend. As shown in Fig. 6, subjects new to Virtual Reality show higher value of skin conductance, probably due to their state of excitement.

Finally, results obtained grouping participants based on their previous experience on ship handling (Fig. 7) suggest an influence of the task confidence on skin conductance values.

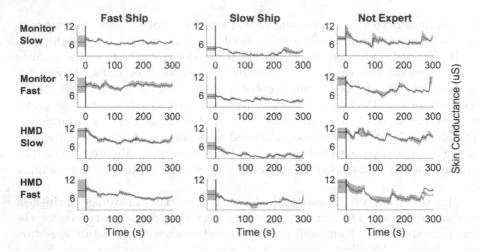

Fig. 7. Average skin conductance of participants having experience with different kind of ships based on the trial.

Heart Rate. Heart rate is a physiological parameter strictly linked to users' emotional, mental and physical state, like skin conductance. Heart rate was recorded during each experimental session and samples were analyzed firstly considering the kind of trial and secondly taking into account both the kind of trial and the path. Furthermore, we have decided to analyse if the familiarity with the setup or the task, e.g. previous experience with VR or in ship handling, could influence participants emotional and mental reaction.

Figure 8 shows that heart rate is regular (no peaks are recorded) and comparable in the four trials for the two paths. Slightly higher value in the fast trials, both with the monitor and the HMD, are probably caused by the difficulty of the task and can indicate a higher level of involvement. This indicate that both hardware systems do not introduce particular emotional states that could compromise performances and interfere with learning.

In particular, the absence of elevated heart rate values (around 100 beat/min) could indicate that participants have perceived the simulations as natural experiences, comparable to real life driving experiences.

Similarly to what described for the analysis of skin conductance, we further analysed data considering participants previous experience with VR setups. As shown in Fig. 9, people who have never experienced Virtual Reality systems, tend to have slightly higher heart beat rate, especially in the initial phase of the simulations, probably because of the excitement/anxiety due to the fact of using a new device. However, both VR naive and non-naive subjects do not show elevated heart rate values (around 100 beat/min), excluding the presence of cybersickness. This confirms the trend of skin conductance.

Finally, we organised data, considering participants previous experience on handling ships of different velocity. While expert students maintain standard heart rate values, with the only exception of the first trial with the fast boat,

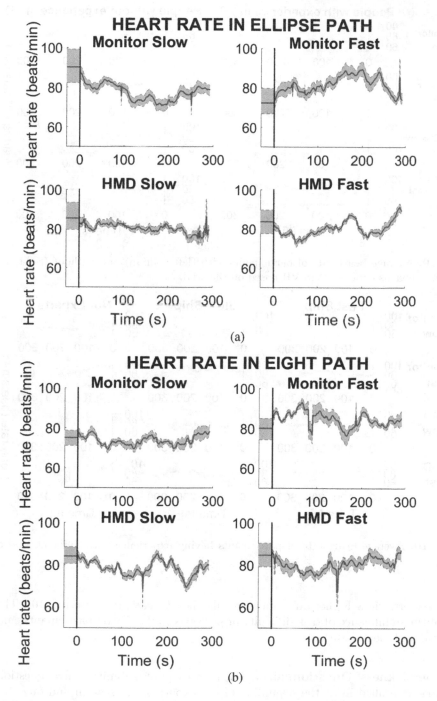

Fig. 8. Average heart rate considering the trial and the path: Ellipse (a) or Eight (b). Figure from [1].

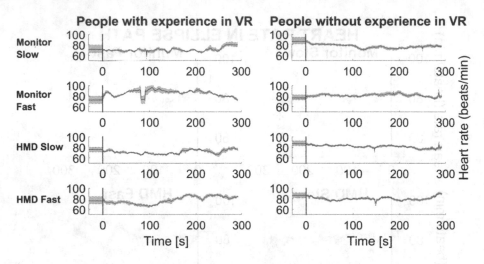

Fig. 9. Average heart rate of participants with (left column) and without (right column) previous experience in VR based on the trial.

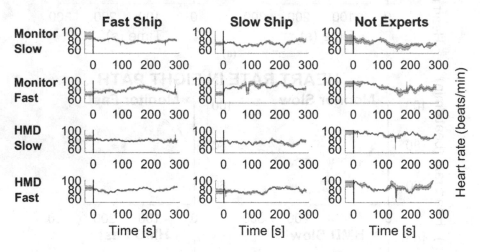

Fig. 10. Average heart rate of participants having experience with different kind of ship.

not-experts show higher and more irregular heart beat rates (see Fig. 10). This confirm an influence of task difficulty or perceived difficulty on participants emotional and mental state.

Cybersickness Questionnaire. Each participant submitted five questionnaires, one filled in at the beginning of the experimental session and one after each trial. We analyzed the answers given by all the volunteer subjects, firstly, considering the kind of trial and, secondly, taking into account also the path.

We performed a between group Wilcox test, in order to evaluate if the increment of cybersickness is statistically significant. Figure 11a highlights that differences between results obtained in the Pre questionnaire and in the two questionnaires referred to the monitor and those referred to the Monitor Slow trial and all the following experiments are statistically significant. This suggests that cybersickness can be caused either by the hardware system and by the boat velocity. Wilcox test performed between group, points out that results obtained with the total grade can be extended to the three subcategories (Table 2, [1]).

Table 2. P-value obtained making a between group Wilcox test and comparing Nausea (N), Oculomotor (O) and Disorientation (D) grades in the five questionnaires. Table from [1].

	N	O	D
Pre-Post HMD S			0.0030
Pre-Post HMD F	0.0029	0.0119	0.0059
Post L S-Post M F	0.0111	0.0438	
Post M S-Post HMD S	0.0123	0.0032	0.0004
Post M S-Post HMD F	0.0006	0.0009	0.0026
Post M F-Post HMD S			0.0147
Post M F-Post HMD F		0.0475	0.0124

The analysis on SSQ questionnaire grades referred to the Ellipse and Eight paths (see Fig. 12) shows an increase in the cybersickness final values between consecutive trials. It is worth noting, however, that trial had fix order of execution (monitor first and HMD second), so the worst grade in HMD trials could be due also to fatigue. Cybersickness is higher in the trial with elliptic path, probably because the sea was more rough than in the simulation with the Eight path. The worst sickness value, so, could depend on a combined influence of sea conditions, kind of hardware used (immersive or non-immersive) and ship velocity, and the use of the HMD alone does not explain the increase of malaise.

A Wilcox test was performed in order to determine the statistical significance of these results. In particular, in the between groups analysis the null hypothesis was never rejected, while the within group analysis revealed interesting correlations shown in Fig. 12. In the elliptic path case, the differences between the total values of sickness in the Monitor Slow trial and in the HMD Fast trial are statistically significant ($p < 0.02$), as the differences between the initial and final total grades ($p < 0.05$). While in the Eight path case, only the results obtained in the first and final questionnaire ($p < 0.05$) and in the trial with the monitor and the slow ship and the following tests are statistically significant ($p < 0.02$). Therefore, in the first case, the factors majorly influencing sickness seem to be the boat velocity and the hardware used for simulation, with the HMD negatively affecting participants well-being. Whereas, in the second case, the velocity

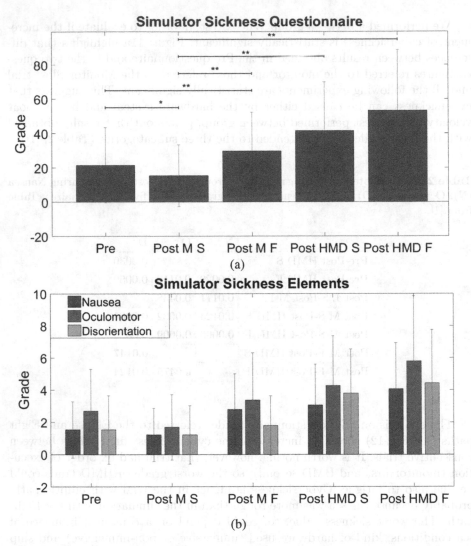

Fig. 11. SSQ results of the five questionnaires submitted. (a) Total grade. (b) Results divided in the three subcategories of the SSQ (Nausea, Oculomotor and Disorientation). M = monitor, HDM = head mounted display, S = slow, F = fast. * p-value < 0.05 e ** p-value < 0.02. Figure from [1].

of the boat plays a fundamental role: curved and irregular trajectories and sudden direction changes, notwithstanding, cause malaise more easily than regular linear path.

These consideration are confirmed by the evaluation of the three major symptoms of cybersickness. If we consider separately the three major symptoms of cybersickness (Nausea, Oculomotor e Disorientation), in general, they tend to increase during the experimental session, in particular Oculomotor grades, which is consistent with results found in the literature.

4.3   Analysis of the Sense of Presence

IGroup Presence Questionnaire. At the end of the experimental session, participants were asked to fill in four IPQ questionnaires, one for each trial they had accomplished. Again data collected have been analyzed, firstly, considering the kind of trial, secondly, taking into account also the path, thirdly considering people's previous experience with VR and ship handling.

The rates given to the three subscales that compose the questionnaire (Fig. 13a) are better for the HMD, indicating a higher sense of presence. Moreover, the trials with the fast boat have less Spatial Presence but greater Experienced Realism if compared to the trials with the slow ship, maybe because of the realism and response speed of the vessel to user's commands.

A between group Wilcox test was performed and only the difference of Involvement and Spatial Presence parameters in the monitor and HMD trials has been found to be statistically significant. This means that the use of the Oculus Rift allows the user to feel more involved and present in the virtual simulated environment.

If we consider the Presence Factor (Fig. 13b), trials with the HMD obtained better results and this difference is statistically significant: in particular, the average grade in the Monitor Slow and Fast trials with the HMD Fast.

Figure 14a and Table 3 show results organized based on the path shape. Differences between monitor and HMD are more evident in the Eight case than in the Ellipse case, where there is a clear distinction of grades only for Involvements. In fact, in the elliptic path only the difference between Monitor Fast and

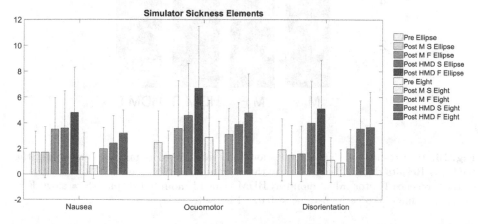

Fig. 12. SSQ results of the five questionnaires submitted organized considering the path. Results are divided in the three subcategory of the SSQ (Nausea, Oculomotor and Disorientation). M = monitor, HDM = head mounted display, S = slow, F = fast.

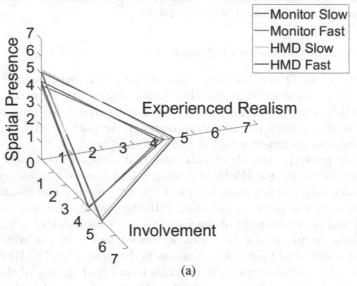

(a)

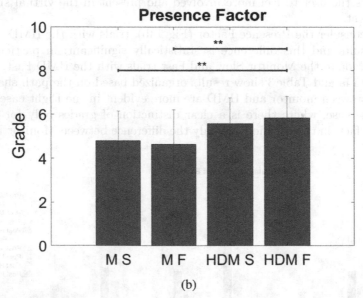

(b)

Fig. 13. IPQ results of the four questionnaires submitted organized considering the trial. (a) Results divided based on the three evaluation subscales. (b) Mean of grades of the Presence Factor. M = monitor, HDM = head mounted display, S = slow, F = fast. * p-value < 0.05 e ** p-value < 0.02. Figure from [1].

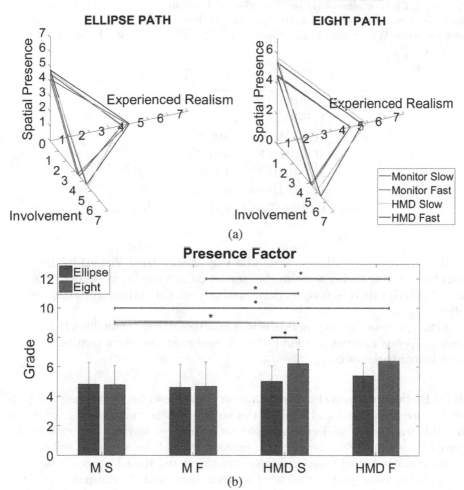

Fig. 14. IPQ results of the four questionnaires submitted organized considering the trial and the path. (a) Results divided based on the three evaluation subscales. (b) Mean of grades of the Presence Factor. M = monitor, HDM = head mounted display, S = slow, F = fast. * p-value < 0.05 e ** p-value < 0.02. Figure from [1].

HMD Slow and Fast for Involvement is statistically significant (p < 0.05). In the Eight path, instead, the differences in Spatial Presence between monitor and headset are statistically significant.

The Presence Factor shown in Fig. 14b is better with the HMD in both paths. This trend is more evident in the Eight path case, were results are also statistically significant. All these data confirm an actual increasing of sense of presence in VR.

Table 3. P-value obtained making a within group Wilcox test and comparing Spatial Presence (SP), Involvement (I) and Experienced Realism (ER) grades in the four questionnaires considering the two path separately (Ellipse and Eight). Cross refers to the between group Wilcox test, made comparing Ellipse path trials and Eight path trials. Table from [1].

		SP	I	ER
Ellipse	M F-HMD S		0.0254	
	M F-HMD F		0.0409	
Eight	M S-HMD S	0.0030		
	M S-HMD F	0.0427		
	M F-HMD S	0.0110		
	M F-HMD F	0.0472		
Cross	HMD S-HMD S	0.0178		0.0261

We further noticed that, people who had already tried Virtual Reality systems tends giving higher score to the simulation systems under analysis. Differences between expert and not expert, however, are not statistically significantly different.

Finally, we considered people previous experience in ship handling. In general, grades depends more on the kind of trial, confirming previous results, than on participants background.

Head Rotation Analysis. Head rotation angles describe the tendency of people to turn their head and explore the scenario. It is worth noting, however, that this tendency is subjective: people can be more or less prone to explore the virtual environment. So the following considerations have a relative value.

We extracted head rotation angles and calculated their histogram, in order to highlights users' preferential head rotation angle and distribution across the trial. In our analysis, yaw is the rotation around the vertical axis (Y), pitch is the rotation around lateral axis (X) and roll is the rotation around the sagittal axis (Z), referring to Unity coordinate system.

Figure 17 shows results referred to two participants who accomplished the task with Ellipse (a) and Eight (b) paths both using the monitor and the Oculus Rift. Considering each subject, in the monitor case, we can notice that rotations are scattered around a central value, which corresponds to the initial head rotation when the application starts and the user looks at the horizon in front of him; in the HMD case, instead, especially in the Eight path, participants tended to turn their head more, in order to receive more information from the surrounding environment and better follow the path. This is due to the higher difficulty of the task, in fact the Eight path has more changes in direction while the elliptic one can be considered quite linear. This trend is even more evident in the trial with the fast boat (Figs. 15 and 16).

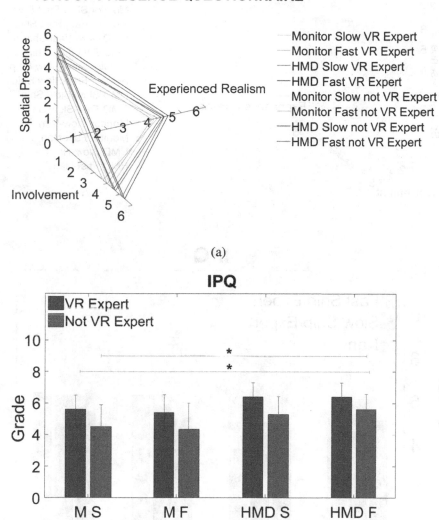

(a)

(b)

Fig. 15. IPQ results of the four questionnaires submitted organized considering people experience with VR setups. (a) Results divided based on the three evaluation subscales. (b) Mean of grades of the Presence Factor. M = monitor, HDM = head mounted display, S = slow, F = fast. * p-value < 0.05.

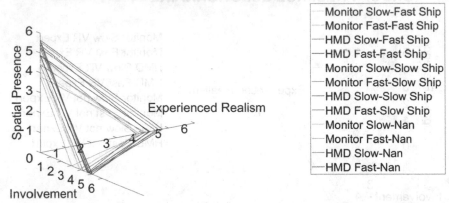

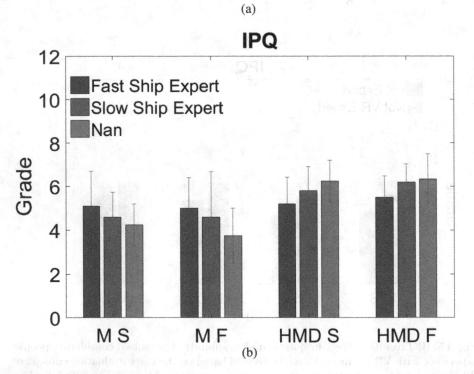

Fig. 16. IPQ results of the four questionnaires submitted organized considering people experience with ship handling. (a) Results divided based on the three evaluation subscales. (b) Mean of grades of the Presence Factor. M = monitor, HDM = head mounted display, S = slow, F = fast.

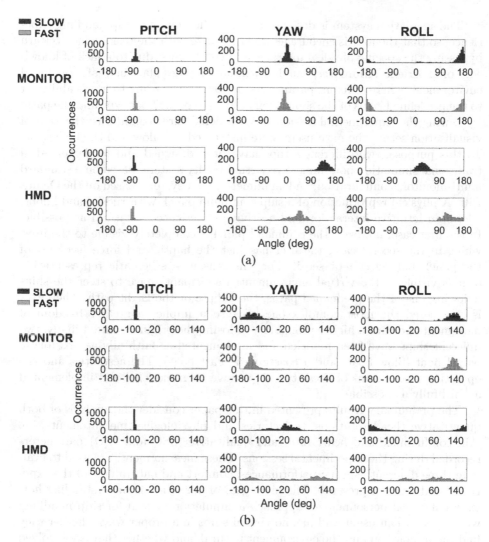

Fig. 17. Non normalized histograms of Pitch, Yaw and Roll head rotation angles of participant 1 (a), who did the trial with the Ellipse path, and participant 2, who did the trial with the Eight path (b). Results are divided based on the system used (monitor or HMD) and the boat velocity (slow or fast). Figure from [1].

5   Conclusions

The aim of this paper is assessing the actual usability of our virtual reality ship simulator, conceived for training nautical personnel, asking expert users to perform a manoeuvring task. Actually, only 4 of the students had no previous experience in driving boats.

The simulation system is designed as a gamification of a ship handling experience, so that the user is encouraged to familiarize with the task he has to learn bypassing the specific physical interface and constraints, due to a lack of knowledge or experience. Moreover, handling simulation takes place in a safe controlled environment, where different parameters can be set according to user's ability or to the predefined goals of the simulation activity, e.g. the boat type and the path.

During the experimental session we considered three conditions: the kind of visualisation setup, the path users' were instructed to follow and the boat type. To this purpose, two different setups have been designed and implemented: a traditional simulation non-immersive virtual reality system, based on a standard multi-monitors configuration, and an immersive VR system, based on the Oculus Rift. A physical reproduction of a ship command panel, with rudder and knobs, allows an intuitive interaction and a driving experience as natural as possible. In the first case, however, the command panel is completely visible to the user, while, in the second case, the user has just the haptic and force feedback of the panel, but he cannot see it. For this reason, a schematic representation of it is visible in the virtual environment. Participants have to steer the ships following one of the predefined paths, the Ellipse or the Eight path. The latter is considered the most complex one because it implies a greater freedom of movement and thus a higher probability of losing the original route. Finally, two different ships have been modeled and implemented as faithfully as possible: a patrol boat (Slow boat) and a coastguard (Fast boat). The aesthetics and the operating mechanisms of these two ships have been modeled and implemented as faithfully as possible.

The preliminary results presented in this paper consider the analysis of both quantitative (head rotations, trajectories and physiological measurements, i.e. skin conductance and heart rate) and qualitative (SSQ and IPQ) parameters recorded during the experimental sessions. These measurements are used to evaluate three different aspects: performances, comfort and naturalness of the experience and sense of presence. The main goal of our study is understanding how expert nautical personnel can approach a simulation system for ship handling: whether they can use it and handle virtual ships in a proper way, whether they find the interaction and the environment natural, and whether they are involved in task execution rather than considering it like a game.

- **Analysis of Performances.** Even if the analysis on the raw trajectories presented in this paper may be qualitative and coarse, it clearly shows that expert drivers follow the predefined path in a quite accurate manner. In future, we intend to calculate the actual accuracy between participants and real path.
- **Comfort and Naturalness of the Experience.** Skin conductance and heart rate remain constant and stable in the four trials, even if slightly higher in the fast vessel tests, which is the most complex one. Furthermore, it is worth noting, that people naive to VR or who have never handled a ship, at least at the beginning of the test, have a higher level of skin conductance and heart rate, probably because they are more anxious about the task and the interaction system.

In general, therefore, both setups do not introduce anxiety, stress or particular emotional or malaise states that could compromise performances and, eventually, learning of new skills. These results on physiological responses could also indicate that participants have perceived simulations as natural experiences, comparable to the real one.

From the analysis of answers given to the SSQ, it has been observed an increase of cybersickness, especially in the virtual reality setup, with a general increment of all the parameters (Nausea, Disorientation and Oculomotor). As the order of the trials was fixed, however, it can not be stated if this increase was due to the setup used, immersive or not-immersive, or to a physiological fatigue as the experimental session proceeds. In the Eight path, an influence of boat velocity over sickness has been noticed; whereas in the Ellipse case, the rise of sickness could depend on a combine action of game settings (calm or rough sea), hardware setup (immersive or non-immersive) and speed of the ship.

– **Sense of Presence.** Sense of presence is higher for the immersive setup and this is more evident in the Eight path. Moreover, the trials with the slow ship have a higher Spatial Presence but a lower Experienced Realism with respect to trials with the fast vessel, probably because of the better realism and response speed of the fast ship. Furthermore, there is a statistically significant difference between the Presence Factor in immersive and non-immersive simulations. So we can suppose that VR simulations are better for the user in terms of involvement and presence.

Also the propensity of people to explore the virtual space by turning their head could be considered as a manner to evaluate the degree of immersion in the simulated environment. While head rotations in the monitor case are scattered around a central value, in the HMD case, in general, we have observed that people have a greater propensity to turn their head, especially in the Eight path with the fast boat, probably because the task is more difficult and they need to collect more information from the environment. Anyway, the analysis of head rotation angles is subjective, in fact people can be more or less prone to explore the virtual world surrounding them.

In conclusion, the two setups have been proven to be equivalent in terms of performances. Considering cybersickness, the non-immersive system seems better, while considering the sense of presence the immersive setup has higher rates. Further acquisitions, on non expert drivers or senior drivers, are required in order to better understand the usability of the system on a large scale and its actual usefulness in providing long term learning of new skills. In this work, non expert drivers who took part to this experimental session were not completely naive to boat driving as they were students at Genoa Naval Academy. This could explain why the comparison of physiological measurements (skin conductance and heart rate), taken from expert and non expert drivers, showed an influence of the task on the emotional state, compatible with the self-confidence level of participants, while no appreciable differences on performances were found.

In future, however, we plan to include in the experimental sessions totally naive users and cadets without driving experience, in order to distinguish between them, and to define a proper baseline for the evaluation of the usability of the simulation system and its actual usefulness in the process of learning how to handle a ship. In the naive users case, in fact, we expect both higher level of skin conductance and heart rate, worse performances and a slow learning rate, i.e. their ability to follow the path as precisely as possible.

References

1. Bassano, C., et al.: Evaluation of a virtual reality system for ship handling simulations. In: Proceedings of the 14th International Joint Conference on Computer Vision, Imaging and Computer Graphics Theory and Applications - Volume 2: HUCAPP, pp. 62–73. INSTICC, SciTePress (2019). https://doi.org/10.5220/0007578900620073
2. Benedict, K., et al.: Simulation augmented manoeuvring design and monitoring-a new method for advanced ship handling. TransNav, Int. J. Mar. Navig. Saf. od Sea Transp. **8**(1), 131–141 (2014)
3. Bouchard, S., Robillard, G., Renaud, P., Bernier, F.: Exploring new dimensions in the assessment of virtual reality induced side effects. J. Comput. Inf. Technol. **1**(3), 20–32 (2011)
4. Chessa, M., Maiello, G., Borsari, A., Bex, P.J.: The perceptual quality of the Oculus Rift for immersive Virtual Reality. Hum. Comput. Interact. **34**(1), 51–82 (2019)
5. Davis, S., Nesbitt, K., Nalivaiko, E.: A systematic review of cybersickness. In: Proceedings of the 2014 Conference on Interactive Entertainment, pp. 1–9. ACM (2014)
6. Davis, S., Nesbitt, K., Nalivaiko, E.: Comparing the onset of cybersickness using the Oculus Rift and two virtual roller coasters. In: Proceedings of the 11th Australasian Conference on Interactive Entertainment (IE 2015), vol. 27, p. 30 (2015)
7. Gianaros, P.J., Muth, E.R., Mordkoff, J.T., Levine, M.E., Stern, R.M.: A questionnaire for the assessment of the multiple dimensions of motion sickness. Aviat. Space Environ. Med. **72**(2), 115 (2001)
8. Kennedy, R.S., Lane, N.E., Berbaum, K.S., Lilienthal, M.G.: Simulator sickness questionnaire: an enhanced method for quantifying simulator sickness. Int. J. Aviat. Psychol. **3**(3), 203–220 (1993)
9. Keshavarz, B., Hecht, H.: Validating an efficient method to quantify motion sickness. Hum. Factors **53**(4), 415–426 (2011)
10. Kim, Y.Y., Kim, H.J., Kim, E.N., Ko, H.D., Kim, H.T.: Characteristic changes in the physiological components of cybersickness. Psychophysiology **42**(5), 616–625 (2005)
11. Lawson, B., Mead, A.: The sopite syndrome revisited: drowsiness and mood changes during real or apparent motion. Acta Astronaut. **43**(3–6), 181–192 (1998)
12. McGill, M., Ng, A., Brewster, S.: I am the passenger: how visual motion cues can influence sickness for in-car VR. In: Proceedings of the 2017 Chi Conference on Human Factors in Computing Systems, pp. 5655–5668. ACM (2017)
13. Munafo, J., Diedrick, M., Stoffregen, T.A.: The virtual reality head-mounted display Oculus Rift induces motion sickness and is sexist in its effects. Exp. Brain Res. **235**(3), 889–901 (2017)

14. Muth, E.R., Stern, R.M., Thayer, J.F., Koch, K.L.: Assessment of the multiple dimensions of nausea: the nausea profile (NP). J. Psychosom. Res. **40**(5), 511–520 (1996)
15. Nalivaiko, E., Davis, S.L., Blackmore, K.L., Vakulin, A., Nesbitt, K.V.: Cybersickness provoked by head-mounted display affects cutaneous vascular tone, heart rate and reaction time. Physiol. Behav. **151**, 583–590 (2015)
16. Nesbitt, K., Davis, S., Blackmore, K., Nalivaiko, E.: Correlating reaction time and nausea measures with traditional measures of cybersickness. Displays **48**, 1–8 (2017)
17. Ohyama, S., et al.: Autonomic responses during motion sickness induced by virtual reality. Auris Nasus Larynx **34**(3), 303–306 (2007)
18. Reason, J.T., Brand, J.J.: Motion Sickness. Academic Press, New York (1975)
19. Regenbrecht, H., Schubert, T.: Real and illusory interactions enhance presence in virtual environments. Presence Teleop. Virt. Environ. **11**(4), 425–434 (2002)
20. Riccio, G.E., Stoffregen, T.A.: An ecological theory of motion sickness and postural instability. Ecol. Psychol. **3**(3), 195–240 (1991)
21. Sharples, S., Cobb, S., Moody, A., Wilson, J.R.: Virtual reality induced symptoms and effects (VRISE): comparison of head mounted display (HMD), desktop and projection display systems. Displays **29**(2), 58–69 (2008)
22. Sheridan, T.B.: Musings on telepresence and virtual presence. Presence Teleop. Virt. Environ. **1**(1), 120–126 (1992)
23. Slater, M., McCarthy, J., Maringelli, F.: The influence of body movement on subjective presence in virtual environments. Hum. Factors **40**(3), 469–477 (1998)
24. Slater, M., Usoh, M., Steed, A.: Depth of presence in virtual environments. Presence Teleop. Virt. Environ. **3**(2), 130–144 (1994)
25. Usoh, M., et al.: Walking > walking-in-place > flying, in virtual environments. In: Proceedings of the 26th annual conference on Computer graphics and interactive techniques, pp. 359–364. ACM Press/Addison-Wesley Publishing Co. (1999)
26. Usoh, M., Catena, E., Arman, S., Slater, M.: Using presence questionnaires in reality. Presence Teleop. Virt. Environ. **9**(5), 497–503 (2000)
27. Varela, J., Soares, C.G.: Interactive 3d desktop ship simulator for testing and training offloading manoeuvres. Appl. Ocean Res. **51**, 367–380 (2015)
28. Varela, J.M., Rodrigues, J., Soares, C.G.: 3D simulation of ship motions to support the planning of rescue operations on damaged ships. Procedia Comput. Sci. **51**, 2397–2405 (2015)
29. Witmer, B.G., Singer, M.J.: Measuring presence in virtual environments: a presence questionnaire. Presence **7**(3), 225–240 (1998)
30. Xu, J., Huang, E., Hsieh, L., Lee, L.H., Jia, Q.S., Chen, C.H.: Simulation optimization in the era of industrial 4.0 and the industrial internet. J. Simul. **10**(4), 310–320 (2016)

A Process Reference Model for UX

Suzanne Kieffer[1]([envelope])[iD], Luka Rukonić[1][iD], Vincent Kervyn de Meerendré[1][iD], and Jean Vanderdonckt[2][iD]

[1] Institute for Language and Communication, Université Catholique de Louvain, 1348 Louvain-la-Neuve, Belgium
{suzanne.kieffer,luka.rukonic,vincent.kervyn}@uclouvain.be
[2] Louvain Research Institute in Management and Organizations, Université Catholique de Louvain, 1348 Louvain-la-Neuve, Belgium
jean.vanderdonckt@uclouvain.be

Abstract. We propose a process reference model for UX (UXPRM), which includes a description of the primary UX lifecycle processes within a UX lifecycle and a set of supporting UX methods. The primary UX lifecycle processes are refined into objectives, outcomes and base practices. The supporting UX methods are refined into related techniques, specific objectives and references to the related documentation available in the literature. The contribution of the proposed UXPRM is three-fold: conceptual, as it draws an accurate picture of the UX base practices; practical, as it is intended for both researchers and practitioners and customizable for different organizational settings; methodological, as it supports researchers and practitioners to make informed decisions while selecting UX methods and techniques. This is a first step towards the strategic planning of UX activities.

Keywords: User experience · UX base practice · UX methods · Process reference model · Capability/maturity model · Agile methods

1 Introduction

This paper elaborates on our previous work entitled *Specification of a UX process reference model towards the strategic planning of UX activities* [40] and consolidates it by updating the set of considered references. This work aims at supporting and improving the integration and establishment of UX practice in software organizations. The notable changes from the previous version of this work include:

- The addition of a background section focused on capability/maturity models and their application in the human-computer interaction (HCI) field;
- The adoption of ISO24744 [28] to specify the UXPRM according to software modeling practices;
- The increased level of detail with which the UXPRM is specified (base practices and related references);
- The improvement of the classification of supporting UX methods;
- The perspective of the contribution in relation to the related work.

© Springer Nature Switzerland AG 2020
A. P. Cláudio et al. (Eds.): VISIGRAPP 2019, CCIS 1182, pp. 128–152, 2020.
https://doi.org/10.1007/978-3-030-41590-7_6

2   Background

2.1   Process Reference Model

A process reference model (PRM) describes a set of primary processes and their interrelations within a process lifecycle in terms of process objectives and process outcomes [31]. A primary process is a group of processes that belong to the same category and that are associated with the same objectives. The term process outcomes refers to work products, which are the artifacts associated with the execution of a process. Process reference models are refined into base practices that contribute to the production of work products [31].

Usually, a process reference model is associated with a process assessment model (PAM), which is a measurement structure for the assessment of the capability of organizations to achieve process objectives and to deliver process outcomes consistently [31]. Together, a process reference model and a process assessment model constitute a capability/maturity model.

2.2   Capability/Maturity Model

A capability/maturity model (CMM) formalizes the software development processes and recommends base practices to enhance software development and maintenance capability [67]. Especially, a capability/maturity model serves as a tool to help organizations select process-improvement strategies by determining their current process maturity and identifying the issues most critical to improving their software quality and process [67]. The purpose of such models is to support organizations moving from lower to higher capability/maturity level.

Typically, CMMs include five or six levels that describe the level of capability/maturity of a process (Table 1). Process attributes, which include methods, techniques, tools, base practices, data collection and analysis, goal setting and work products, serve as indicators of the capability/maturity level of processes.

2.3   Capability/Maturity Models in HCI

In HCI, capability/maturity models, which are referred to as usability capability/maturity models (UCMMs) or user experience capability/maturity models (UXCMMs), aim to help software organizations establish user-centered processes in order to improve software quality and increase organizational efficiencies [72,75]. Among the most often cited in the literature, there are Jonathan Earthy's 1998 Usability Maturity Model [18], Eric Schaffer' 2004 model for Institutionalization of Usability [76], and Jakob Nielsen's 2006 model for Corporate Usability Maturity [63]. The two latter have been renamed Institutionalization of UX and Corporate UX Maturity respectively, as a result of the paradigm shift within the HCI community from usability to user experience [40].

Since the late 90s, the International Organization for Standardization (ISO) actively contributes the documentation of UCMMs/UXCMMs.

Table 1. Capability/maturity levels in CMMs.

ISO/IEC 15504 [31]	Software CMM [67]
0. Incomplete: There is general failure to attain the purpose of the process. There are few or no easily identifiable work products or outputs of the process	
1. Achieved: The achievement may not be rigorously planned and tracked	1. Initial: The software process is characterized as ad hoc, and occasionally even chaotic
2. Managed: The process delivers work products according to specified procedures and is planned and tracked	2. Repeatable: Basic project management processes are established to track cost, schedule, and functionality
3. Established: The process is performed and managed using a defined process based upon good software engineering principles	3. Defined: The software process for both management and engineering activities is documented, standardized, and integrated into a standard software process for the organization
4. Predictable: The defined process is performed consistently in practice within defined control limits, to achieve its defined process goals	4. Managed: Detailed measures of the software process and product quality are collected. Both the software process and products are quantitatively understood and controlled
5. Optimizing: Performance of the process is optimized to meet current and future business needs, and the process achieves repeatability in meeting its defined business goals	5. Optimizing: Continuous process improvement is enabled by quantitative feedback from the process and from piloting innovative ideas and technologies

- ISO/IEC 15504 [31], which has been revised in 2015 by ISO/IEC 33004 [32], sets out the requirements for process reference models, process assessment models, and maturity models for software engineering;
- ISO 13407 [26], which has been revised in 2019 by ISO 9241-210 [29], provides requirements and recommendations for human-centred design principles and activities throughout the lifecycle of computer-based interactive systems;
- ISO/TR 18529 [33], which has been revised in 2019 by ISO 9241-220 [30], describes the processes and specifies the outcomes by which human-centred design is carried out within organizations;

– ISO18152 [34] presents a human-systems (HS) model for use in ISO/IEC 15504-conformant assessment of the maturity of an organization in performing the processes that make a system usable, healthy and safe.

Table 2. Summary findings of systematic literature review of UCMMs/UXCMMs [42] (AUCDI stands for agile user-centered design integration, HF for human factors, HS for human-systems, U for usability, UX for user experience).

Year	Reference	Process assessment model	Process reference model	Domain
1998	[18]	6 tailored levels		U
2000	[19]	ISO 15504 [31]	ISO 13407 [26]	HF
2000	[33]	ISO 15504 [31]	ISO 13407 [26]	U
2001	[86]	CMM [67]		U
2003	[36]	ISO 15504 [31]	ISO 13407 [26]	U
2007	[78]	CMM [67]		UX
2009	[55]	CMM [67]		U/UX
2009	[84]	5 tailored levels		UX
2010	[34]	ISO 15504 [31]	ISO 13407 [26]	HS
2011	[24]	5 tailored levels		U
2012	[70]	5 tailored levels		U
2013	[60]	6 tailored levels		AUCDI
2014	[12]	5 tailored levels		UX
2014	[68]	CMM [67]		agile
2016	[41]	3 tailored levels		U

Table 2 summarizes the findings of a 2018 systematic literature review (SLR) of UCMMs/UXCMMs [42] that identified and analyzed 15 UCMMs/UXCMMs in total. The observed trends can be synthesized as follows:

– Regarding measurement scales, half of the process assessment models (8 out of 15) conform either with ISO 15504 [31] or with the software CMM [67], while the other half informs a tailored capacity/maturity scale (column 3);
– Only a quarter of the UCMMs/UXCMMs (4 out of 15) explicitly defines ISO 13407 [26] as process reference model, the remaining 75% not providing any such information (column 4);
– A large majority of UCMMs/UXCMMs (13 out of 15) focuses on usability, user experience and related disciplines, while only 15% are tailored for agile software development or agile user-centered design integration (column 5);
– There is a clear split between the 2000s when most UCMMs/UXCMMs were compliant with or based on existing models and the 2010s when user-centered design is tailored, in all likelihood, to fit more specific domains and contexts.

The 2018 SLR [42] also indicates the following:

- There is a lack of information on how most UCMMs/UXCMMs (9 out of 15) were evaluated, which leaves their validity questionable;
- Two-thirds of the UCMMs/UXCMMs (10 out of 15) provide no or limited guidance for the use of their process assessment model, which hinders their application or adoption in practice; the five remaining UCMMs/UXCMMs [18,33,34,36,60] provide a glossary of key terms and detailed information on how to carry out the process assessment;
- Two-thirds of the process assessment models (10 out of 15) focus managing the UX process (defining, scoping, controlling); one focuses only on executing the UX process (performing UX activities); the remaining four PRMs [34,36, 68,86] focus on both process management and process execution.

3   Contribution

We argue that planning activities (e.g. scoping, defining, benchmarking) and execution activities (e.g. monitoring, performing, controlling) take place within two distinct processes, namely planning and execution, as they do not have the same purpose and do not involve the same stakeholders [27]. While the planning process intends to specify the activities to be performed to achieve the project goals, the execution process consists in actually performing these activities. While the planning process involves a wide range of stakeholders (steering committee, sponsor, special interest groups, customers, project team), the execution process involves the project team and the users.

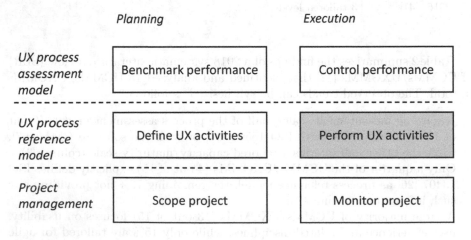

Fig. 1. Contribution of the paper (grey rectangle).

We propose a process reference model for UX (UXPRM), which focuses on the execution of UX activities (Fig. 1). The UXPRM includes a description of the primary UX lifecycle processes within a UX lifecycle and a set of supporting UX methods. The primary UX lifecycle processes are refined into objectives, outcomes and base practices. The supporting UX methods are refined into related techniques, specific objectives and references to the related documentation available in the literature.

The contribution of the proposed UXPRM is conceptual, practical and methodological. The conceptual contribution lies in the choice to leave the planning of UX processes aside, and to focus on the execution of UX processes. This conceptual choice is original compared to existing UCMMs/UXCMMs which either define ISO 13407 [26] as PRM and therefore include planning, or do not explicit define any such PRM. This conceptual choice also better matches what happens in practice, for example when the planning activities occur months ahead of the project launch, as it is the case during seeking funding. This conceptual choice has practical implications, as it makes the PRM customizable for different settings (e.g. agile vs. waterfall, in-house team vs. group of organizations). The methodological contribution of the proposed UXPRM lies in the detailed description of the supporting UX methods. This detailed description draws an accurate picture of the UX methods currently available and allows, we believe, the reader to make more informed decision while selecting UX methods and techniques.

4 Methodology

4.1 Comparison of the Related Work

To identify the primary UX lifecycle processes and organize them within a UX lifecycle, we compared the lifecycles in user-centered methodologies. The user-centered methodologies are usability engineering (UE), human-centered design (HCD), user-centered design (UCD) and agile user-centered design integration (AUCDI). We have selected these four methodologies as they involve user-centered methods articulated across a lifecycle.

UE is a set of activities that take place throughout a lifecycle and focus on evaluating and improving the usability of interactive systems [57,62]. UE base practices include conducting analysis activities early in the product development lifecycle, before design activities in order to specify the needs and the requirements of users. In line with this recommendation, additional references demonstrate the significance of such early stages activities [6,24].

HCD [29] and UCD [36] both aim to develop systems with high usability by incorporating the user's perspective into the software development process. In this regard, the user requirements specification (URS) is critical to the success of interactive systems and user requirements (UR) are refined iteratively throughout the lifecycle [51,52].

AUCDI is concerned with the integration of UCD into agile software development methods. The scientific consensus on AUCDI is that agile and UCD should be iterative and incremental, organized in parallel tracks, and continuously involve users [8,74]. The barriers to AUCDI are the lack of time for upfront analysis and the difficulty optimizing the work dynamics between developers and designers [15,22]. Several models have been proposed for supporting the management of the AUCDI process [20,49].

Figure 2 compares the user-centered lifecycles found in the related work. We conformed to ISO 24744 [31] in order to ensure a consistent, rigorous and up-to-date approach while expressing the processes involved in the user-centered lifecyles. ISO 24744 concerns method engineering and is aimed at expressing software processes involved in a workflow for any particular goal in software engineering. Figure 3 depicts the legend extracted from ISO 24744.

As can be seen from Fig. 2, user-centered lifecycles are very similar: they are iterative; they use a similar terminology (i.e., requirements, analysis, design, testing or evaluation); except for the AUCDI lifecycle, they follow a similar sequence of processes: analysis, design and evaluation. The main difference between these lifecycles lies in the perspective on the requirements. Requirements correspond to a primary process in UE (see requirement analysis), HCD (see requirements specification) and in UCD (see user requirements). By contrast, in AUCDI, requirements correspond to a process outcome fed throughout the development lifecycle by the primary processes.

4.2 Targeted Literature Review

To identify the supporting UX methods, we ran a targeted literature review (TLR) which is a non-systematic, in-depth and informative literature review. A TLR is expected to guarantee keeping only the references maximizing rigorousness while minimizing selection bias. Contrastingly, systematic literature reviews usually aim at addressing a predefined research question by extensively and completely collecting all the references related to this question and by considering absolute inclusion and exclusion criteria. Inclusion criteria retain references that fall in scope of the research question, while exclusion criteria reject irrelevant or non-rigorous references.

We chose to run a TLR for the following four reasons:

1. Translating our research question into a representative syntactical query to be applied on digital libraries is not straightforward and may lead to many irrelevant references [54];
2. If applied, such a query may result into a very large set of references that actually use a UX method, but which do not define any UX method or contribution to such a UX method;
3. The set of relevant references obtained through TLR is quite limited and stems for a knowledgeable selection of high-quality, easy-to-identify references on UX methods, as opposed to an all-encompassing list of irrelevant references;
4. A TLR is better suited at describing and understanding UX methods one by one, at comparing them, and at understanding the trends of the state of the art.

The TLR allowed us to identified 59 relevant references documenting UX methods, UX techniques and UX artifacts (Table 3).

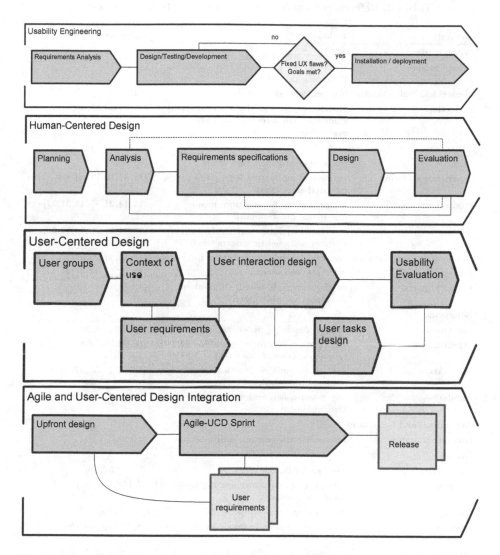

Fig. 2. Comparison of the lifecycles in user-centered methodologies specified according to ISO 24744 [28,77].

Fig. 3. Legend extracted from ISO 24744 [31,77].

Table 3. References identified by the targeted literature review.

Method	Techniques	References
Without users		
GOMS models	CMN-GOMS, CPM-GOMS, NGOMSL, Keystroke-Level Model	[10,35]
hierarchical task analysis	hierarchical task analysis	[14,59]
inspection	abstract tasks; cognitive walkthrough; guideline/expert review; heuristic evaluation	[45,46,53,57,64]
Attitudinal		
cards	cards; emocards; emotion cards	[22,88]
experience sampling	automated interaction logs; daily diary; repeated-entry diary	[11,37,39,71,80,87]
group interview	brainstorming; focus group; group discussion; questionnaire	[13,14,21,25,43,51,57]
interview	structured (questionnaire, role-play, twenty questions) or unstructured	[13,14,21,25,43,51,57]
retrospective interview	cognitive interview; elicitation interview	[13,14,21,25,57]
survey	interview; questionnaire	[13,14,21,43,44,47,51]
verbal protocol	co-discovery; talk-aloud; think-aloud; retrospective think-aloud	[13,14,16]
Behavioral		
constructive	collage; drawings; photographs; probes	[4,16]
experiment	A/B testing; controlled experiment; remote experiment; user test	[15,23,38,57,81,82]
observation	field or systemic observation; shadowing; applied ethnography	[14,16,25,51,57,59,82]
simulation	paper-and-pencil evaluation; Wizard of Oz experiment	[3,48,57,58,82,89]
Attitudinal and behavioral		
contextual inquiry	contextual observations and interviews	[16,25,57]
Design		
goals setting	user, usability, UX goals	[25,52,57,79]
mapping	concept map; customer journey map; service blueprint	[7,14,22,65,83]
sorting	affinity diagram; card sort	[14,25,65]
scenario	storyboard; use case; user scenario; user story	[8,22,25,52,57,66,90]
modeling	persona; work model; task model; UI model	[9,22,23,25,56,61]
lo-fi prototyping	paper prototype; PowerPoint prototype; sketch	[3,22,48,57,58]
me-fi prototyping	video prototype; wireframe; Wizard of Oz prototype	[3,22,48,50,57,58]
hi-fi prototyping	coded prototype	[3,22,58]
visual design	visual techniques	[85]
guidelines and standards	design principles; standards; style guide	[57,69]

4.3   Classification of UX Methods

To classify the UX methods, we adopted a bottom-up approach. First, we distinguished between methods and techniques by adopting the following hierarchical arrangement between approach, method and technique: *The organizational key is that techniques carry out a method which is consistent with an approach* [2]. For example, heuristic evaluation is a technique to carry out inspection (method), brainstorming is a technique to carry out a group interview (method).

Second, we grouped the UX techniques allowing to carry out the same UX method (Table 3). For example, heuristic evaluation and expert review are techniques to carry out inspection, brainstorming and focus group are techniques to carry out a group interview.

Third, we defined four levels of representation of the users and plotted the number of reference documenting a technique along two axes: the continuum from level 1 to level 4 and the methods (Fig. 4). The resulting levels of representation of the users correspond to four classes of UX methods:

1. Methods without users;
2. Attitudinal methods which focus on how users feel;
3. Behavioral methods which focus on what users do;
4. Attitudinal and behavioral methods which combine classes 2 and 3.

4.4   Classification of UX Artifacts

To classify the UX artifacts, we proceeded in a similar way by grouping UX artifacts which result from the same UX method. for example, affinity diagrams and card sorts result from sorting, and personas and work models result from modeling. We defined ten levels of representation of the system and grouped the artifacts by levels (Table 3). The levels of representation of the system are:

1. Goal setting: linguistic description of the design and UX goals;
2. Mapping: arrangement of concepts and ideas on a diagram;
3. Sorting: arrangement of data by their common features;
4. Scenario: written outline of the UX giving details on individual user actions;
5. Modeling: structured representation of users, tasks, work and environments;
6. Low-fidelity (lo-fi) prototyping;
7. Medium-fidelity (me-fi) prototyping;
8. High-fidelity (hi-fi) prototyping;
9. Visual design: visual techniques (e.g. color, shape, orientation, etc.);
10. Guidelines and standards: combination of previous levels.

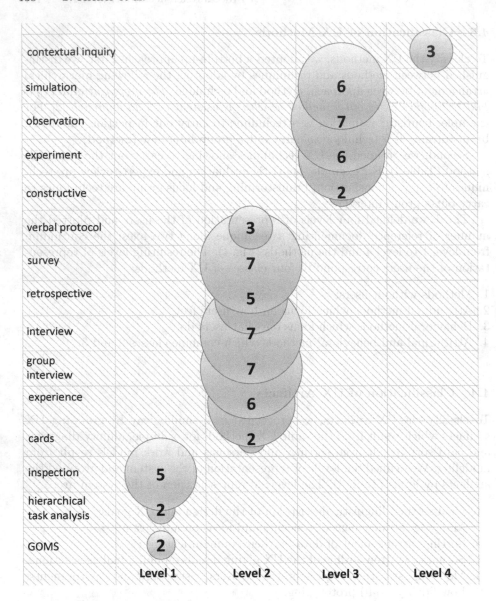

Fig. 4. Bubble chart expressing to what extent references examined in this paper cover each UX method (the size of each circle is directly proportional to the amount of references found per UX method: the larger the circle, the more references).

5   Results

5.1   Primary UX Lifecycle Processes Within a UX Lifecycle

Figure 5 depicts the primary UX lifecycle processes within a UX lifecycle specified with ISO24744 [28]. The proposed UX lifecycle involves four primary pro-

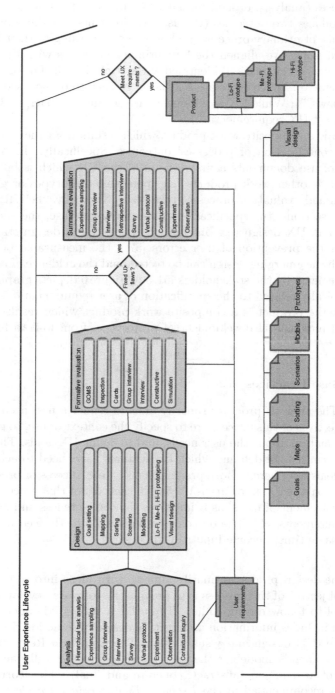

Fig. 5. Primary UX lifecycle processes specified according to ISO 24744 [28,77].

cesses or phases (analysis, design, formative evaluation and summative evaluation) and produces two outcomes (the user requirements and the product). We chose the name of primary processes and outcomes according to their frequency in the related work. We aligned the four primary processes with the sequence (analysis, design and evaluation) identified in the related work. The proposed UX lifecycle is iterative: While there are UX flaws, go back to design (see milestone "Fixed UX flaws?"); While UX requirements are not met, go back to design (see milestone "Meet UX requirements?").

The URS is a composite work product which results from the iterative and incremental consolidation of processes' outcomes. Specifically, the analysis of the context of use documents a first version of the UR which is later revised or completed by other work products (e.g. personas, prototypes or guidelines) as the design and evaluation processes take place. Typically, the URS includes the following sections: the specification of the context of use, the specification of UX goals and UX design (e.g. interaction styles, screen design standards or strategies for the prevention of user errors [57]), the measurable benchmarks against which the emerging design can be tested, and the evidence of acceptance of the requirements by the stakeholders [51,52]. Figure 6 depicts graphically the activities typically related to the specification of user requirements.

Similarly, the product is a composite work product which results from the iterative and incremental development of prototypes, from lo-fi to hi-fi prototypes.

5.2   UX Base Practices

Analysis. The analysis process primarily aims to render a first account of the UR. The objectives of this process are to specify the context of use, to gather and analyze information about the user needs, and to define UX goals. The analysis of user needs consists of defining which key features users need to achieve their goals. User goals refer here to both pragmatic goals (e.g. success or productivity) and hedonic goals (e.g. fun or arousal) [5]. The success of this process relies on the early involvement of users, as it improves the completeness and accuracy of the user requirements [52]. The outcomes of analysis directly feed into the UR and, at the same time, become inputs of the design.

Design. The design process primarily aims to turn ideas into testable prototypes. The objective of this process is to provide the software development team with a model to follow during coding. Typically, a style guide documents this model which includes information architecture design, interaction design (IxD), user interface (UI) design and visual design. The Cameleon Reference Framework [9] recommends modeling the user interface incrementally according to three levels of abstraction (abstract, concrete and final) which correspond to similar levels recommended in the *Usability Engineering Lifecycle* (conceptual model design, screen design standards and detailed UI) [57]. Another approach consists in reasoning according to the level of fidelity (low, medium and high)

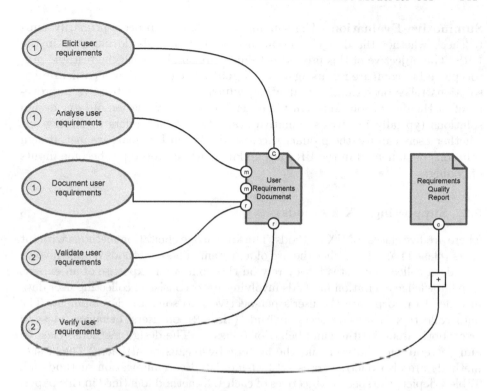

Fig. 6. Activities related to user requirements.

of prototypes [48,58,89]. The outcomes of the design process such as conceptual models or screen design standards directly feed into the URS, while the product representation (e.g. paper prototype) becomes an input of the formative evaluation process.

Formative Evaluation. The formative evaluation process primarily aims to detect UX design flaws in the product representation. As long as there are such flaws, the design process supports the production of redesign solutions to fix them. Design and formative evaluation are therefore intertwined within an iterative and incremental test-and-refine process that aims to improve the product representation. The product development team repeats this cycle until UX flaws are fixed. Once they are fixed, the redesigned solution becomes an input of the summative evaluation process. The relevant literature [9,20,25,57,68] is consistent regarding this iterative and incremental aspect of the design process. In addition, formative evaluation requires low investment in resources and effort, which efficiently supports decision-making and significantly helps reducing late design changes [3,6,57,62,82].

Summative Evaluation. The summative evaluation process primarily aims to check whether the design solution meets the UX goals documented in the URS. The objective of this process is to measure the UX with the testable prototype and to compare results against UX goals. The evaluation of earlier design solutions relies on formative evaluation, which refers to the iterative improvement of the design solutions. On the other hand, the evaluation of later design solutions typically involves summative evaluation, which refers to finding out whether users can use the product successfully. When UX goals are met, design solutions documented in the URS are updated, while prototypes become inputs of coding.

5.3   Supporting UX Methods

There are five classes of UX methods. The knowledge elicitation methods without users (class 1) aim to predict the use of a system. These methods do not involve user data collection; instead, they rely on the opinion or expertise of an expert. The knowledge elicitation methods involving users consist in collecting user data in order to incorporate the user's perspective into software development. The data collection focuses on users' attitude (class 2), on users' behavior (class 3) or on both users' attitude and behavior (class 4). The design methods (class 5) aim to represent the user and the system by means of artifacts. The design methods are also referred to as artifact-mediated communication methods [8]. Table 4 depicts the specific objectives of each UX method identified in this paper.

6   Application: Past, Present and Future Work

We use the UX process reference model in an industrial project entitled Voice Interface for Autonomous Driving based on User experienCe (VIADUCT). Launched in November 2018, the project aims to design new user experiences with the driving monitoring system (DMS) and is funded for 36 months under the reference 7982 by Service public de Wallonie (SPW), Belgium. The consortium of partners involves six institutions (two private companies, two universities and two accredited research centers) and uses an agile approach for software development. In this project, the mission of the two first authors is to support the integration and establishment of UX practice in the project organization. We use the UXPRM in the following ways.

First, we used an earlier version of the UXPRM during the project definition and planning phase (i.e. before we submitted the proposal to the project sponsor) in order to scope and define the UX methods that would be used during project development. This set of UX methods includes all the design methods referenced in this paper plus a selection of evaluation methods namely interview, survey, verbal protocol, experiment, observation and simulation (Table 4).

Second, we used the UXPRM during the project launch phase (i.e. for a period of three months starting from the kick-off) in order to communicate

Table 4. Objectives of UX methods.

Method	Objectives
Without users	
GOMS models	to predict how users will use a system
hierarchical task analysis	to capture the mental demands needed to perform a task proficiently with a system
inspection	to predict the learnability of a system or potential usability and UX problems
Attitudinal	
cards	to identify user mood and reactions about their experience with a system
experience sampling	to capture user thoughts, feelings on multiple occasions over time
group interview	to collect data from a purposely selected group of individuals about a specific topic
interview	to collect data from individuals about their attitudes, beliefs, habits, norms and values
retrospective interview	to capture the aspects of cognitive performance during user past experience with a system
survey	to collect data from a sample population about a system
verbal protocol	to make thought, cognitive processes as explicit as possible during task performance
Behavioral	
constructive	to identify unexpected uses of a system or concept
experiment	to test a hypothesis or establish a cause-and-effect relationship about a system or a sample population
observation	to capture how users perform tasks or solve problems in their natural setting
simulation	to detect UX problems with a system which has not been implemented yet
Attitudinal and behavioral	
contextual inquiry	to capture how users perform their tasks, achieve their intended goals and think about their work
Design	
goals setting	to document specific qualitative and quantitative goals that will drive UX design
mapping	to represent relationships between concepts and ideas
sorting	to cluster and label data based on relationships that make sense to users
scenarizing	to depict how users achieve their goals with a system, to describe a software feature from the user's perspective
modeling	to depict users, work organization, tasks and UIs
lo-fi prototyping	to turn design ideas into testable mock-ups
me-fi prototyping	to capture and illustrate how users will interact with a new system
hi-fi prototyping	to turn mock-ups into highly-functional and interactive prototypes
visual design	to arrange screens in such a way that they communicate ideas clearly, consistently and aesthetically
guidelines and standards	to provide developers with a model to follow during coding

about primary UX lifecycle processes, especially to advocate for the integration of analysis activities in the product development lifecycle. This allowed us to carry out little upfront analysis (i.e. refine use cases into user scenarios) by means of workshop and brainstorming, which both fall under the group discussion category.

Third, we currently use the UXPRM in order to provide the consortium with a plan of the UX activities that will take place in the next six months. In such a hybrid project environment (academic and industrial), the releases are less frequent as the agile iterations stretch from two weeks to 6–8 weeks [1]. Therefore, we will synchronize these activities with the agile retrospective, that being at the end of each iteration. The UX activities will take place in the form of workshops with the goal of delivering collectively a UX work product. The first rounds focused on modeling (e.g. personas, work models and task models) with the purpose of encouraging the reuse of such UX work products, and by doing so, increasing the return on investment of UX [6].

In the future, we will use the UXPRM to specify a companion UX process assessment model (UXPAM). Specifically, we will use the processes' attributes presented in this paper (i.e. methods, techniques, base practices, data collection and analysis, goal setting and work products) as indicators of the capability/maturity of processes. We have proposed a first version this UXPAM [73] which defines the process attributes, the measurement scales and the tools for the assessment (questionnaire and interview guide). This work-in-progress has allowed us to gain insights into the current use of UX methods in the project organization (essentially hi-fi prototyping and scenarizing), to identify the barriers in the way of UX integration (hostility towards upfront analysis and lack of understanding of the return on investment of UX) and opportunities for the integration of UX in the formal software development model (identification of "UX champion" in the consortium). Practically, we will use the UX workshops aforementioned as forums to perform the data collection for small-scale process assessment. These small-scale process assessments will allow us to populate and test the proposed UXPAM, which is essential for the development of maturity models [17] and to monitor the evolution of the UX capability/maturity.

Finally, we argue that the definition of UX activities, which is a planning process, must depend on the project scope (detailed set of deliverables or features of a project) and current capabilities of an organization to achieve UX goals. Furthermore, we argue that UX activity plans must result from the alignment between the business goals of the organization (project management), the level of UX capability/maturity of the organization (UXPAM) and the level of effort that is required to successfully implement UX methods (UXPRM). Table 5 provides an early draft of the level of effort required by the UX methods referenced in this paper as it is documented in the related work [6,51]. Building on this basis, our future work will consist of studying this alignment (Fig. 7).

Table 5. Level of effort per UX method in man-days as documented in [6,51] (CM stands for conceptual model, SDS for screen design standards; question marks indicate uncertainty in the measurement of man-days for a particular technique, blanks that the method was not documented).

Method	Techniques and effort [6]		Techniques and effort [51]	
Without users				
GOMS models				
hierarchical task analysis				
inspection			design guidelines and standards	2–5
			evaluation walk-through	3–6
			heuristic evaluation	2–3
Attitudinal				
cards				
experience sampling			diary keeping	8–15
group interview	conduct needs finding	1	context-of-use analysis	1–2
			focus group	8–15
			brainstorming	2–3
interview	questionnaire	4–8	stakeholder analysis	1
			user requirements interview	5–8
			satisfaction questionnaire	2–4
			post-experience interview	3–4
retrospective interview				
survey			survey of existing users	6–15
verbal protocol				
Behavioral				
constructive				
experiment			controlled user testing	10–16
observation			field study/user observation	5–8
			participatory evaluation	4–8
			evaluation workshop	3–6
			assisted evaluation	5–9
simulation	iterative CM evaluation	14–16	paper, video evaluation	?
	iterative SDS evaluation	14–21	WOz experiment	?
	iterative detailed UI evaluation	14–21		
Attitudinal and behavioral				
contextual inquiry	contextual task analysis	10–16	task analysis	6–15
			assessing cognitive workload	4–8
			critical incidents	6–10
Design				
goals setting	usability goals	2–4	user requirements	2–4
mapping			task/function mapping	4–6
sorting	card sorting	4–6	card sorting	2–3
			affinity diagram	2–3
scenarizing			scenario of use	3–6
			storyboarding	4–6
modeling	user profile	4	personas	1–2
	work modeling	4		
	work reengineering	10–12		
	conceptual model design	10–12		
lo-fi prototyping	CM design	4–5	parallel design	3–6
			paper prototyping	?
me-fi prototyping	SDS design	4–14	video prototyping	?
			WOz prototyping	?
hi-fi prototyping	detailed UI design	10–12	software prototyping	?
visual design				
guidelines and standards				

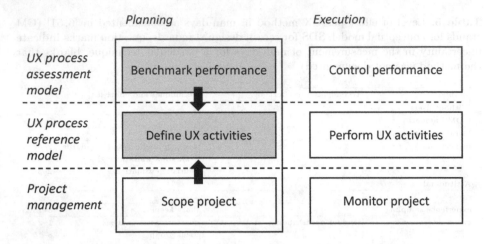

Fig. 7. Focus of our future work (grey rectangles).

7   Conclusion

We propose a process reference model for UX, which focuses on the execution of UX activities. The UXPRM includes a description of the primary UX lifecycle processes within a UX lifecycle and a set of supporting UX methods. The primary UX lifecycle processes are refined into objectives, outcomes and base practices. The supporting UX methods are refined into related techniques, specific objectives and references to the related documentation available in the literature. The contribution of the proposed UXPRM is three-fold: conceptual, as it draws an accurate picture of the UX base practices; practical, as it is intended for both researchers and practitioners and customizable for different organizational settings; methodological, as it supports researchers and practitioners to make informed decisions while selecting UX methods and techniques. This is a first step towards the strategic planning of UX activities.

Building on the promising usefulness of the proposed UXPRM for supporting UX practice, our future work consists of developing a UX processes assessment model (UXPAM), which is a measurement structure for the assessment of UX processes. We argue that UX activity planning depends on the alignment between business goals, level of UX capability/maturity and the level of effort required to successfully implement UX methods. In turn, this aims to reducing the gap between UX research and UX practice.

References

1. Anderson, E., Lim, S.Y., Joglekar, N.: Are more frequent releases always better? Dynamics of pivoting, scaling, and the minimum viable product. In: Proceedings of the 50th Hawaii International Conference on System Sciences (2017)

2. Anthony, E.M.: Approach, method and technique. Engl. Lang. Teach. **17**(2), 63–67 (1963). https://doi.org/10.1093/elt/XVII.2.63
3. Arnowitz, J., Arent, M., Berger, N.: Effective Prototyping for Software Makers. Elsevier, Amsterdam (2010)
4. Bargas-Avila, J., Hornbæk, K.: Old wine in new bottles or novel challenges? A critical analysis of empirical studies of user experience. In: Proceedings of the SIGCHI Conference on Human Factors in Computing Systems, pp. 2689–2698 (2011). https://doi.org/10.1145/1978942.1979336. http://portal.acm.org/citation.cfm?doid=1978942.1979336
5. Bevan, N.: Classifying and selecting UX and usability measures. In: Proceedings of the International Workshop on Meaningful Measures: Valid Useful User Experience Measurement, vol. 11, pp. 13–18 (2008)
6. Bias, R.G., Mayhew, D.J.: Cost-Justifying Usability: An Update for the Internet Age. Elsevier, Amsterdam (2005). https://doi.org/10.1016/B978-0-12-095811-5.X5000-7. http://dl.acm.org/citation.cfm?id=1051866
7. Braga Sangiorgi, U.: A method for prototyping graphical user interfaces by sketching on multiple devices. Ph.D. thesis, UCL-Université Catholique de Louvain (2014)
8. Brhel, M., Meth, H., Maedche, A., Werder, K.: Exploring principles of user-centered agile software development: a literature review. Inf. Softw. Technol. **61** (2015). https://doi.org/10.1016/j.infsof.2015.01.004
9. Calvary, G., Coutaz, J., Thevenin, D., Limbourg, Q., Bouillon, L., Vanderdonckt, J.: A unifying reference framework for multi-target user interfaces. Interact. Comput. **15**, 289–308 (2003)
10. Card, S.K., Newell, A., Moran, T.P.: The Psychology of Human-Computer Interaction. L. Erlbaum Associates Inc., Hillsdale (1983)
11. Carter, S., Mankoff, J.: When participants do the capturing: the role of media in diary studies. In: Proceedings of the SIGCHI Conference on Human Factors in Computing Systems, CHI 2005, pp. 899–908. ACM, New York (2005). https://doi.org/10.1145/1054972.1055098. http://doi.acm.org/10.1145/1054972.1055098
12. Chapman, L., Plewes, S.: A UX maturity model: effective introduction of UX into organizations. In: Marcus, A. (ed.) DUXU 2014. LNCS, vol. 8520, pp. 12–22. Springer, Cham (2014). https://doi.org/10.1007/978-3-319-07638-6_2
13. Cooke, N.J.: Varieties of knowledge elicitation techniques. Int. J. Hum. Comput. Stud. **41**(6), 801–849 (1994). https://doi.org/10.1006/IJHC.1994.1083. https://www.sciencedirect.com/science/article/abs/pii/S1071581984710834
14. Crandall, B., Klein, G., Klein, G.A., Hoffman, R.R.: Working Minds: A Practitioner's Guide to Cognitive Task Analysis. MIT Press, Cambridge (2006)
15. da Silva, T.S., Silveira, M.S., Maurer, F.: Usability evaluation practices within agile development. In: Proceedings of the Annual Hawaii International Conference on System Sciences, pp. 5133–5142, March 2015. https://doi.org/10.1109/HICSS.2015.607
16. Daae, J., Boks, C.: A classification of user research methods for design for sustainable behaviour. J. Clean. Prod. **106**(Complete), 680–689 (2015). https://doi.org/10.1016/j.jclepro.2014.04.056
17. De Bruin, T., Freeze, R., Kaulkarni, U., Rosemann, M.: Understanding the main phases of developing a maturity assessment model. In: Proceedings of the 16th Australian Conference on Information Systems (ACIS 2005). Australasian Chapter of the Association for Information Systems (2005)
18. Earthy, J.: Usability maturity model: human centredness scale. INUSE Proj. Deliv. D **5**, 1–34 (1998)

19. Earthy, J., Sherwood-Jones, B.: Human factors integration capability maturity model-assessment model. In: Presented at Human Interfaces in Control Rooms, pp. 320–326. IET (2000)
20. Forbrig, P., Herczeg, M.: Managing the agile process of human-centred design and software development. In: Proceedings of the 15th IFIP TC.13 International Conference on Human-Computer Interaction (INTERACT 15), pp. 223–232. ACM, New York (2015)
21. Fowler Jr., F.J.: Survey Research Methods. SAGE Publications, Thousand Oaks (2013)
22. Garcia, A., da Silva, T.S., Selbach Silveira, M.: Artifacts for agile user-centered design: a systematic mapping. In: Proceedings of the Annual Hawaii International Conference on System Sciences, pp. 5859–5868 (2017). https://doi.org/10.24251/HICSS.2017.706, http://hdl.handle.net/10125/41870
23. Ghaoui, C.: Encyclopedia of Human Computer Interaction. IGI Global, Hershey (2005)
24. HIMSS Usability Task Force: Promoting usability in health organizations: initial steps and progress toward a healthcare usability maturity model. Health Information and Management Systems Society (2011)
25. Holtzblatt, K., Wendell, J.B., Wood, S.: Rapid Contextual Design: A How-to Guide to Key Techniques for User-Centered Design. Morgan Kaufmann Publishers Inc., San Francisco (2004)
26. ISO 13407:1999: Human-centred design processes for interactive systems. Standard, International Organization for Standardization, Geneva, CH (1999)
27. ISO 21500:2012: Guidance on project management. Standard, International Organization for Standardization, Geneva, CH (2012)
28. ISO 24744:2014: Software engineering – Metamodel for development methodologies. Standard, International Organization for Standardization, Geneva, CH (2014)
29. ISO 9241-210:2019: Ergonomics of human-system interaction – Part 210: Human-centred design for interactive systems. Standard, International Organization for Standardization, Geneva, CH (2019)
30. ISO 9241-220:2019: Ergonomics of human-system interaction – Part 220: Processes for enabling, executing and assessing human-centred design within organizations. Standard, International Organization for Standardization, Geneva, CH (2019)
31. ISO/IEC 15504: Information technology – Process assessment. Standard, International Organization for Standardization, Geneva, CH (2003)
32. ISO/IEC 33004:2015: Information technology – Process assessment – Requirements for process reference, process assessment and maturity models. Standard, International Organization for Standardization, Geneva, CH (2015)
33. ISO/TR 18529:2000: Ergonomics – Ergonomics of human-system interaction – Human-centred lifecycle process descriptions. Standard, International Organization for Standardization, Geneva, CH (2000)
34. ISO/TS 18152:2010: Ergonomics of human-system interaction – Specification for the process assessment of human-system issues. Standard, International Organization for Standardization, Geneva, CH (2010)
35. John, B.E., Kieras, D.E.: The GOMS family of user interface analysis techniques: comparison and contrast. ACM Trans. Comput. Hum. Interact. 3(4), 320–351 (1996). https://doi.org/10.1145/235833.236054. http://doi.acm.org/10.1145/235833.236054
36. Jokela, T.M.: Assessment of user-centred design processes as a basis for improvement action: an experimental study in industrial settings. Ph.D. thesis, University of Oulu (2003)

37. Khan, V.J., Markopoulos, P., Eggen, B., IJsselsteijn, W., de Ruyter, B.: Recon-exp. In: Proceedings of the 10th International Conference on Human Computer Interaction with Mobile Devices and Services - MobileHCI 2008, p. 471. ACM Press, New York (2008). https://doi.org/10.1145/1409240.1409316. http://portal.acm.org/citation.cfm?doid=1409240.1409316

38. Kieffer, S.: ECOVAL: ecological validity of cues and representative design in user experience evaluations. AIS Trans. Hum. Comput. Interact. **9**(2), 149–172 (2017)

39. Kieffer, S., Batalas, N., Markopoulos, P.: Towards task analysis tool support. In: Proceedings of the 26th Australian Computer-Human Interaction Conference on Designing Futures: The Future of Design, OzCHI 2014, pp. 59–68. ACM, New York (2014). https://doi.org/10.1145/2686612.2686621. http://doi.acm.org/10.1145/2686612.2686621

40. Kieffer, S., Rukonic, L., de Meerendré, V.K., Vanderdonckt, J.: Specification of a UX process reference model towards the strategic planning of UX activities. In: 14th International Joint Conference on Computer Vision, Imaging and Computer Graphics Theory and Applications (VISIGRAPP 2019) (2019)

41. Kieffer, S., Vanderdonckt, J.: Stratus: a questionnaire for strategic usability assessment. In: Proceedings of the 31st Annual ACM Symposium on Applied Computing, pp. 205–212. ACM (2016)

42. Lacerda, T.C., von Wangenheim, C.G.: Systematic literature review of usability capability/maturity models. Comput. Stand. Interfaces **55**, 95–105 (2018)

43. Lavrakas, P.J.: Encyclopedia of Survey Research Methods. SAGE Publications, Thousand Oaks (2008)

44. Law, E.L.C., Roto, V., Hassenzahl, M., Vermeeren, A.P., Kort, J.: Understanding, scoping and defining user experience: A survey approach. In: Proceedings of the SIGCHI Conference on Human Factors in Computing Systems, CHI 2009, pp. 719–728. ACM, New York (2009). https://doi.org/10.1145/1518701.1518813. http://doi.acm.org/10.1145/1518701.1518813

45. Law, E.L.C., Vermeeren, A.P., Hassenzahl, M., Blythe, M.: Towards a UX manifesto. In: Proceedings of the 21st British HCI Group Annual Conference on People and Computers: HCI... but not as we know it-Volume 2, pp. 205–206. BCS Learning & Development Ltd. (2007)

46. Leavitt, M.O., Shneiderman, B.: Research-Based Web Design & Usability Guidelines. U.S. Department of Health and Human Services (2006)

47. Lewis, J.R.: Psychometric evaluation of the PSSUQ using data from five years of usability studies. Int. J. Hum. Comput. Interact. **14**(3–4), 463–488 (2002)

48. Lim, Y.K., Stolterman, E., Tenenberg, J.: 2008 the anatomy of prototypes: prototypes as filters, prototypes as manifestations of design ideas. ACM Trans. Comput. Hum. Interact. **15**(2) (2008). https://doi.org/10.1145/1375761.1375762

49. Losada, B., Urretavizcaya, M., Fernández-Castro, I.: A guide to agile development of interactive software with a "User Objectives"-driven methodology. Sci. Comput. Program. **78**(11), 2268–2281 (2013). https://doi.org/10.1016/j.scico.2012.07.022. http://dx.doi.org/10.1016/j.scico.2012.07.022

50. Mackay, W.E., Ratzer, A.V., Janecek, P.: Video artifacts for design: bridging the gap between abstraction and detail. In: Proceedings of the 3rd Conference on Designing Interactive Systems: Processes, Practices, Methods, and Techniques, DIS 2000, pp. 72–82. ACM, New York (2000). https://doi.org/10.1145/347642.347666. http://doi.acm.org/10.1145/347642.347666

51. Maguire, M.C.: Methods to support human-centred design. Int. J. Hum. Comput. Stud. **55**(4), 587–634 (2001). https://doi.org/10.1006/ijhc.2001.0503. http://linkinghub.elsevier.com/retrieve/pii/S1071581901905038

52. Maguire, M., Bevan, N.: User requirements analysis: a review of supporting methods. In: Hammond, J., Gross, T., Wesson, J. (eds.) Usability. ITIFIP, vol. 99, pp. 133–148. Springer, Boston, MA (2002). https://doi.org/10.1007/978-0-387-35610-5_9

53. Mahatody, T., Sagar, M., Kolski, C.: State of the art on the cognitive walkthrough method, its variants and evolutions. Int. J. Hum. Comput. **26**(8), 741–785 (2010)

54. Mallett, R., Hagen-Zanker, J., Slater, R., Duvendack, M.: The benefits and challenges of using systematic reviews in international development research. J. Dev. Eff. **4**(3), 445–455 (2012)

55. Marcus, A., Gunther, R., Sieffert, R.: Validating a standardized usability/user-experience maturity model: a progress report. In: Kurosu, M. (ed.) HCD 2009. LNCS, vol. 5619, pp. 104–109. Springer, Heidelberg (2009). https://doi.org/10.1007/978-3-642-02806-9_13

56. Markopoulos, P., Pycock, J., Wilson, S., Johnson, P.: Adept-a task based design environment. In: Proceedings of the Twenty-Fifth Hawaii International Conference on System Sciences, vol. 2, pp. 587–596. IEEE (1992)

57. Mayhew, D.J.: The Usability Engineering Lifecycle: A Practitioner's Handbook for User Interface Design. Morgan Kaufmann Publishers Inc., San Francisco (1999)

58. McCurdy, M., Connors, C., Pyrzak, G., Kanefsky, B., Vera, A.: Breaking the fidelity barrier: an examination of our current characterization of prototypes and an example of a mixed-fidelity success. In: Proceedings of the SIGCHI Conference on Human Factors in Computing Systems, CHI 2006, pp. 1233–1242. ACM, New York (2006). https://doi.org/10.1145/1124772.1124959. http://doi.acm.org/10.1145/1124772.1124959

59. Militello, L., Hutton, R.: Applied cognitive task analysis (ACTA): a practitioner's toolkit for understanding cognitive task demands. Ergonomics **41**, 1618–1641 (1998). https://doi.org/10.1080/001401398186108

60. Mostafa, D.: Maturity models in the context of integrating agile development processes and user centred design. Ph.D. thesis, University of York (2013)

61. Mulder, S., Yaar, Z.: The User is Always Right: A Practical Guide to Creating and Using Personas for the Web. New Riders, Berkeley (2006)

62. Nielsen, J.: Usability Engineering. Elsevier, Amsterdam (1993)

63. Nielsen, J.: Corporate Usability Maturity: Stages 1–4. Nielsen Norman Group, Fremont (2006)

64. Nielsen, J., Molich, R.: Heuristic evaluation of user interfaces. In: Proceedings of the SIGCHI Conference on Human Factors in Computing Systems, CHI 1990, pp. 249–256. ACM, New York (1990). https://doi.org/10.1145/97243.97281. http://doi.acm.org/10.1145/97243.97281

65. NNGroup.Com: Nielsen Norman Group (1998). https://www.nngroup.com/

66. Patton, J., Economy, P.: User Story Mapping: Discover the Whole Story, Build the Right Product. O'Reilly Media, Inc., Sebastopol (2014)

67. Paulk, M.C., Curtis, B., Chrissis, M.B., Weber, C.V.: Capability maturity model, version 1.1. IEEE Softw. **10**(4), 18–27 (1993)

68. Peres, A., da Silva, T., Silva, F., Soares, F., De Carvalho, C., De Lemos Meira, S.: AGILEUX model: towards a reference model on integrating UX in developing software using agile methodologies. In: 2014 Agile Conference, pp. 61–63. IEEE (2014)

69. Preece, J., Rogers, Y., Sharp, H.: Interaction Design: Beyond Human-Computer Interaction, 4th edn. Wiley, Hoboken (2015)

70. Raza, A., Capretz, L.F., Ahmed, F.: An open source usability maturity model (OS-UMM). Comput. Hum. Behav. **28**(4), 1109–1121 (2012). https://doi.org/10.1016/j.chb.2012.01.018. http://www.sciencedirect.com/science/article/pii/S0747563212000209

71. Rieman, J.: The diary study: a workplace-oriented research tool to guide laboratory efforts. Proceedings of the INTERACT '93 and CHI '93 Conference on Human Factors in Computing Systems (CHI 1993), pp. 321–326 (1993). https://doi.org/https://doi.org/10.1145/169059.169255. https://dl.acm.org/citation.cfm?id=169255

72. Rosenbaum, S., Rohn, J.A., Humburg, J.: A toolkit for strategic usability: Results from workshops, panels, and surveys. In: Proceedings of the SIGCHI Conference on Human Factors in Computing Systems, CHI 2000, pp. 337–344. ACM, New York (2000). https://doi.org/10.1145/332040.332454. http://doi.acm.org/10.1145/332040.332454

73. Rukonić, L., Kervyn de Meerendré, V., Kieffer, S.: Measuring UX capability and maturity in organizations. In: Marcus, A., Wang, W. (eds.) HCII 2019. LNCS, vol. 11586, pp. 346–365. Springer, Cham (2019). https://doi.org/10.1007/978-3-030-23535-2_26

74. Salah, D., Cairns, P., Paige, R.F.: A systematic literature review for Agile development processes and user centred design integration. The Agile & UCD Project View project Serious Games for Character Education View project. In: Proceedings of the 18th International Conference on Evaluation and Assessment in Software Engineering (2014). https://doi.org/10.1145/2601248.2601276. http://dx.doi.org/10.1145/2601248.2601276

75. Sauro, J., Johnson, K., Meenan, C.: From snake-oil to science: measuring UX maturity. In: Proceedings of the 2017 CHI Conference Extended Abstracts on Human Factors in Computing Systems, CHI EA 2017, pp. 1084–1091. ACM, New York (2017). https://doi.org/10.1145/3027063.3053350. http://doi.acm.org/10.1145/3027063.3053350

76. Schaffer, E.: Institutionalization of Usability: A Step-by-Step Guide. Addison-Wesley Professional, Boston (2004)

77. Sousa, K.S., Vanderdonckt, J., Henderson-Sellers, B., Gonzalez-Perez, C.: Evaluating a graphical notation for modelling software development methodologies. J. Vis. Lang. Comput. **23**(4), 195–212 (2012). https://doi.org/10.1016/j.jvlc.2012.04.001. https://doi.org/10.1016/j.jvlc.2012.04.001

78. Sward, D., Macarthur, G.: Making user experience a business strategy. In: Law, E., et al. (eds.) Proceedings of the Workshop on Towards a UX Manifesto, vol. 3, pp. 35–40 (2007)

79. Theofanos, M.F.: Common industry specification for usabilty-requirements. Technical report, U.S. National Institute of Standards and Technology (NIST) (2007)

80. Trull, T.J., Ebner-Priemer, U.: Ambulatory Assessment. Annu. Rev. Clin. Psychol. **9**(1), 151–176 (2013). https://doi.org/10.1146/annurev-clinpsy-050212-185510

81. Tsai, P.: A Survey of Empirical Usability Evaluation Methods, pp. 1–18 (1996)

82. Tullis, T., Albert, B.: Measuring the User Experience: Collecting, Analysing, and Presenting Usability Metrics. Elsevier, Amsterdam (2013). https://doi.org/http://dx.doi.org/10.1016/B978-0-12-415781-1.00007-8. http://www.sciencedirect.com/science/article/pii/B9780124157811000078

83. van den Akker, J., Branch, R.M., Gustafson, K., Nieveen, N., Plomp, T.: Design Approaches and Tools in Education and Training. Springer, Dordrecht (1999). https://doi.org/10.1007/978-94-011-4255-7

84. Tyne, S.: Corporate user-experience maturity model. In: Kurosu, M. (ed.) HCD 2009. LNCS, vol. 5619, pp. 635–639. Springer, Heidelberg (2009). https://doi.org/10.1007/978-3-642-02806-9_74

85. Vanderdonckt, J.: Visual design methods in interactive applications. In: Content and Complexity, pp. 199–216. Routledge (2014)

86. Vasmatzidis, I., Ramakrishnan, A., Hanson, C.: Introducing usability engineering into the cmm model: an empirical approach. Proc. Hum. Factors Ergon. Soc. Annu. Meet. **45**(24), 1748–1752 (2001)

87. Vermeeren, A., Kort, J.: Developing a testbed for automated user experience measurement of context aware mobile applications. In: Law, E., Hvannberg, E.T., Hassenzahl, M. (eds.) User eXperience, Towards a Unified View, p. 161 (2006)

88. Vermeeren, A.P.O.S., Law, E.L.C., Roto, V., Obrist, M., Hoonhout, J., Väänänen-Vainio-Mattila, K.: User experience evaluation methods: current state and development needs. In: Proceedings of the 6th Nordic Conference on Human-Computer Interaction: Extending Boundaries, NordiCHI 2010, pp. 521–530. ACM, New York (2010). https://doi.org/10.1145/1868914.1868973. http://doi.acm.org/10.1145/1868914.1868973

89. Walker, M., Takayama, L., Landay, J.A.: High-fidelity or low-fidelity, paper or computer? Choosing attributes when testing web prototypes. Proc. Hum. Factors Ergon. Soc. Annu. Meet. **46**(5), 661–665 (2002). https://doi.org/10.1177/154193120204600513. http://journals.sagepub.com/doi/10.1177/154193120204600513

90. Wautelet, Y., Heng, S., Kolp, M., Mirbel, I., Poelmans, S.: Building a rationale diagram for evaluating user story sets. In: Proceedings of the International Conference on Research Challenges in Information Science, 1–12 August 2016 (2016). https://doi.org/10.1109/RCIS.2016.7549299

AAT Meets Virtual Reality

Tanja Joan Eiler[1]([✉]) [iD], Armin Grünewald[1], Michael Wahl[3],
and Rainer Brück[1,2]

[1] Medical Informatics and Microsystems Engineering,
University of Siegen, 57076 Siegen, Germany
{tanja.eiler,armin.gruenewald,michael.wahl,rainer.brueck}@uni-siegen.de
[2] Life Science Faculty, University of Siegen, 57076 Siegen, Germany
[3] Digital Integrated Systems, University of Siegen, 57076 Siegen, Germany
https://www.uni-siegen.de/lwf/professuren/mim/
https://www.eti.uni-siegen.de/dis/
http://www.uni-siegen.de/lwf/

Abstract. Smoking is still one of the main causes of premature mortality and is associated with a variety of diseases. Nevertheless, a large part of the society smokes. Many of them do not seek treatment, and the path to smoking cessation is often abandoned. For this reason, we have developed a VR application that uses the AAT procedure, which has already achieved positive results in the detection and treatment of addiction disorders, as a basic procedure. We want to support the classical therapies by increasing the motivation of the affected patients through immersion, embodiment and game design elements. For this purpose, we have initially developed and evaluated a first demonstrator. Based on the results and findings, a completely revised VR application was programmed, which also eliminates the identified errors and problems of the first demonstrator. In addition, a mobile application will be developed to support the treatment. Our results show that the transfer of the AAT procedure into virtual reality, and thus into three-dimensional space, is possible and promising. We also found that three-dimensional stimuli should be preferred, since the interaction with them was more intuitive and entertaining for the participants. The benefits of game design elements in combination with the representation of interactions in the form of a hand with gripping animations also proved to be of great value, as this increased immersion, embodiment, and therefore motivation.

Keywords: Addiction · Approach Avoidance Task · AAT · Approach bias · Cognitive bias · Cognitive Bias Modification · CBM · Dual Process Model · Embodiment · Game design · Immersion · Presence · Serious gaming · Smoking · Therapy · Virtual reality · VR

1 Introduction

Tobacco consumption is one of the most common causes of premature mortality, as it often results in cardiovascular diseases or even cancer. About 29% of the adult and 7.2% of the adolescent German population smokes regularly [16].

© Springer Nature Switzerland AG 2020
A. P. Cláudio et al. (Eds.): VISIGRAPP 2019, CCIS 1182, pp. 153–176, 2020.
https://doi.org/10.1007/978-3-030-41590-7_7

This leads not only to the diseases mentioned above, but also to enormous social costs associated with them. It is estimated that a total macroeconomic loss of 79.1 billion euros can be expected [20]. In addition, smokers also endanger people around them, especially children [16]. In 2005, it was estimated that the number of deaths caused by passive smoking in Germany amounts to 3.300 per year [39].

This brings tobacco and nicotine-containing products into the top list of the most consumed addictive substances [6]. But although there are numerous methods of therapy and intervention, only one out of four smokers manages to stop smoking for more than six months [14,33]. For this reason, we are trying to develop an enhanced method, using digital medicine [24], to improve the success of smoking cessation programs. However, we would like to emphasize that we do not want to replace the traditional therapy methods, but to enhance them.

We chose the Approach Avoidance Task (AAT, [59,69]) as the foundation, as it has already been shown that it can be used to measure and modify the approach bias, which in turn can be used to treat dependency diseases [38]. The approach bias is a form of cognitive bias, which is responsible for selective perception and subsequent automatic approach to (addictive) stimuli, as it influences the emotions and motivation of the viewer [52]. This form of treatment is known as cognitive bias modification (CBM). However, the long-term success of this method is rather mediocre, as the drop-out rate is quite high [7,51,61], which is another motivation for us to improve CBM with the help of new media.

As this paper is an extension of our publication *Fighting substance dependency combining AAT therapy and virtual reality with game design elements* [21], in addition to the first prototype, which was presented extensively in the aforementioned publication, the redesigned demonstrator and the accompanying mobile application will be presented. In addition, the sections about the basics, and related works were expanded.

This leads to the following structure of the article: First, the psychological and technical foundations will be explained. Afterwards, the related works will be presented. Then the requirements as well as the hardware specifications are described. Subsequently, the first demonstrator is described in detail, as well as the evaluation carried out with it and its results. Finally, the subsequent work will be presented in the form of a revised demonstrator and a mobile application.

2   Psychological Foundations

This section explains the psychological foundations necessary to understand the work presented here. First, the Dual Process Model of Addiction is introduced, which can be used to explain dependencies. Then, the therapy method Cognitive Bias Modification (CBM), on which the Approach Avoidance Task (AAT) is based, will be presented. The AAT is also discussed in detail later on (see Sect. 2.4), as our research uses it as the basic method to treat dependencies.

2.1 Dual Process Model

Dual process models [15,35,76] assume that two different cognitive processes are responsible for our behavior, namely the reflective and impulsive processes. While the former are consciously processed, the latter are inaccessible to the consciousness and occur automatically. The impulsive processes include various sub-types of the cognitive biases, e.g. attentional and approach biases which are responsible for the selective processing of stimuli in the environment, and thereby influence emotions and motivation [52].

The approach bias is of particular importance for our studies as it can lead to an automatic approach to addictive stimuli. This can be explained by the fact that, in the case of addiction, the two cognitive processes mentioned above become unbalanced and the impulsive processes gain dominance. Thus, counteraction by the reflective processes (e.g. knowledge of the consequences of addiction) is hardly possible any more. Instead, there is an automatic approach to addictive stimuli and subsequent consumption [15,35,50,75].

2.2 Cognitive Bias Modification

One possible therapy method for the treatment of substance dependencies is the cognitive bias modification (CBM) approach. Here, systematic practice and extended exposure to task contingencies are used to directly modify the bias in order to redirect the selective processing of stimuli into regulated and desired paths [12,42,52]. Unfortunately, long-term successes are rather modest, as the drop-out rate is very high compared to other forms of treatment [7,51,61].

2.3 Hierarchical Model of Approach Avoidance Motivation

This model assumes that every individual evaluates almost all stimuli [5] and approaches those that are evaluated as positive, to hold and keep them. Stimuli rated as negative, on the other hand (e.g. pain, phobias), are avoided, for example by moving away or escaping from them [48]. These movements are not only physical, but also psychological, making them a fundamental component for understanding and generating motivation and self-regulation [25]. Studies have shown that these movements do not have to be executed with the whole body, but can also be performed by a mere pushing or pulling movement of the arms [13,53].

2.4 Approach Avoidance Task

The Approach Avoidance Task (AAT) [69] combines all the models and procedures described above. It was first evaluated for therapeutic purposes by Rinck and Becker [59]. Their studies showed that the AAT can be used to detect phobias without having to explicitly ask those affected. Subsequent research (for an overview see [38]) also showed that the cognitive bias can not only be measured with the AAT, but also be modified. Recent randomized-controlled studies

[19,77] have shown that the AAT-based version of CBM reduces relapse rates up to 13%, compared with placebo groups or non-training groups [12].

The AAT training works as follows: The participants sit in front of a computer screen where they see two-dimensional images. They are indirectly instructed to pay attention to a certain distinguishing feature (e.g. tilting, image format), and to react to them with the help of a joystick. Therefore, the joystick should be pressed away for addiction-related images in the PUSH format (e.g. tilted to the right), and pulled for neutral or positive images in the PULL format (e.g. tilted to the left). This movement makes the figures smaller or respectively larger (see Fig. 1). The movement should be done as fast and accurate as possible, by which the automatic, impulsive processes should be addressed and unconsciously modified. After sufficient training, according to the hierarchical model of approach avoidance motivation, neutral/positive stimuli should be associated positively and addiction-related stimuli negatively, so that eventually there should be no more automatic approach to addiction-related stimuli.

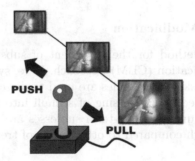

Fig. 1. Concept of the Approach Avoidance Task [21].

While in the AAT training condition all addiction related stimuli are pushed away, in the bias measurement 50% are pushed away and the other 50% are pulled. The reaction times (RT) of the participants are measured (in ms) and then evaluated for each stimulus with the following formula:

$$Bias = MedianRT[PUSH] - MedianRT[PULL] \qquad (1)$$

[21] A positive value stands for an approach bias, which goes along with an automatic approach. A negative value, on the other hand, would speak for an avoidance bias, which means an avoiding behavior towards the stimulus. As a result of this, people with addiction disorders should have a higher (approach) bias towards addiction-related stimuli than people who are not affected.

3   Technical Background

This section describes the technical basics which were used for the development of the virtual reality and mobile demonstrators.

3.1   Virtual Reality

Virtual reality (VR) is a computer-generated three-dimensional alternative reality in which users move, and in and with which they can interact [65,66]. Within this virtual environment (VE), a feeling of immersion can be achieved through the use of visual, auditory, and tactile stimuli, which in turn can improve the effectiveness of interactions within the VE [64]. If the VE is sufficiently real, an effect called *presence* is created. While immersion describes the feeling of being physically in the VE, presence is a psychological phenomenon, which leads to users adapting their behavior to the VE while interacting in and with it [63]. This can help to achieve cognitive absorption, allowing users to work more concentrated and lose the sense of time, which is especially beneficial for educational programs [1]. Further positive factors are embodiment and body-ownership which describe the feeling of artificial or virtual body parts belonging to one's own physical body [40,68]. Apart from education, VR can be used in many other domains, e.g. medicine, psychology, architecture, or entertainment [18,30].

3.2   Mobile Applications

Mobile applications, or apps for short, are encapsulated programs that can be run on mobile devices, especially smartphones and tablets [2]. Due to the enormous distribution in recent years, mobile applications are almost ubiquitous now [54].

Two major software platforms are currently dominating the global market of operating systems (OS) for mobile devices: The open-source Android platform was the clear top performer in the third quarter of 2018 with 88.2% market share, followed by Apple iOS with 11.8% market share [70]. Both platforms offer an official online marketplace to distribute previously validated apps [3]. While Android apps can also be obtained and installed from other sources, iOS does not allow this. Instead, this is only possible via an unofficial modification of the OS [3].

Due to the open source licensing of the Android OS, it is used on devices of numerous manufacturers, while iOS runs exclusively on Apple devices [3]. While the latter case results in a higher homogeneity of the hardware used, the selection of devices and market coverage are correspondingly low.

3.3   Game Design Elements

Effectiveness of an application as well as concentration and motivation of the users can be improved by implementing game design and gamification elements, e.g. high-scores, avatars, achievements or a rewarding point system [10,60]. Gamification describes the use of game design elements in a non-game context [74].

These different design elements deal with the psychological predisposition of humans in order to use them for the application. There are two types of motivation: On the one hand, there is the intrinsic motivation which arises directly from the action. On the other hand, there is the extrinsic motivation which is caused by the prospect of a distant goal such as a financial reward [8].

Gaming elements such as a rewarding score system or achievements allow the user to re-experience successes he can achieve on his own. Apart from that, high-scores or a changeable avatar can enable social interactions. Users can exchange over it or step into a competition, whereby they can be encouraged to perform uninteresting tasks with more interest and motivation [8,9].

4 Related Work

In this section, at first the potential of new technologies in the treatment of dependency diseases is introduced. Thereafter, current studies are described that use virtual reality for this purpose. On the one hand in combination with the therapy method cue exposure therapy (CET), and on the other hand those in combination with the cognitive bias modification (CBM) treatment.

The potential of new technologies in cognitive training were evaluated by Forman and colleagues [28] by treating eating-related behavior. They found out that the use of home computers can improve and maintain the effectiveness of training due to the high availability, access to the Internet, and the easy possibility to repeat training sessions as often as possible and wanted. However, compliance is not necessarily given because high motivation levels are required and there is no external accountability [4,36].

Smartphone apps also have the advantage that they are ubiquitous due to the mobility of smartphones and tablets. This allows direct intervention, especially in situations in which craving arises. Disadvantages are the screen size of mobile devices and the associated poor stimulus recognition. In addition, training with mobile apps can sometimes be accompanied by decreased focus and attention [43]. Empirical results concerning mobile apps are mixed, if available at all.

The addition of gamification elements has shown that engagement and motivation are increased and dropout rates are reduced [49]. However, after frequent repetition, training with gaming elements also becomes monotonous, resulting in decreased motivation [73]. The best and longest lasting effects have been achieved by social game elements [9].

The use of VR offers many advantages, like ecological validity and high degrees of immersion [57], the possibility to create many environments and complex contextual cues as well as individualized training [11]. VR training is associated with improved treatment outcomes, high adherence rates and generalizability, and therefore enhanced behavioral change [34,58,67]. Drawbacks of VR are the cost factor, potential emergence of cybersickness, and the required technological expertise to program and set up VR applications [11,47]. Currently, there is no existing research concerning the efficacy of therapeutic VR applications that aim at modifying cognitive biases [28].

The aim of cue exposure therapy (CET) is to reprogram the behavior of patients in connection with craving inducing environments and stimuli to allow them to rethink their actions in situations of addiction risk, instead of involuntarily giving in to their urge. Therefore, the patients are regularly exposed to addiction-related stimuli to increase their tolerance towards these stimuli to such a high level that they can achieve this goal [55]. In an experimental study Lee et al. [46] linked this therapy method to a VE by creating a combination of addiction-related environments and components in VR. Their scenario contained a bar with various three-dimensional objects, as Lee and his colleagues found out in a previous study, that three-dimensional stimuli cause higher craving than two-dimensional images [45]. Additionally, a smoking avatar and the audio track of a noisy restaurant were present in the VE. The study was composed of six sessions in which sixteen late adolescent men who consumed at least ten cigarettes a day were exposed to addictive stimuli. The results confirmed the helpfulness of VR in treatment programs, with gradually decreased cigarette consumption and craving of the participants.

Ghiţă el al. [29] conducted a systematic review of VR applications using CET for the treatment of alcohol misuse. Their research showed that the studies carried out so far have used graphics and models that do not meet current standards, which has negatively affected immersion and presence. Furthermore, the influence of immersion on the efficacy of therapeutic VR applications, and the generalization of craving responses in the real world, have not been evaluated directly. Another important factor that has been identified is the fact that none of the studies made a follow-up examination of the long-term effects of the VR treatment. Nevertheless, all studies had consistent results regarding the elicitation and reduction of addictive behavior.

Girard et al. [31] created a medieval VE in which participants had to find, grasp and destroy up to 60 hidden cigarettes. A control group collected balls instead of destroying cigarettes. After a six-week training program which included one session each week, participants smoked significantly fewer cigarettes and also dropped out of the accompanying treatment program much less. Furthermore, the abstinence rate, compared to the control group, has increased. After the study, 23% of the participants stated that they regularly remembered destroying cigarettes, while only 3% of the control group remembered collecting balls. This indicates that letting persons watch themselves destroying addiction-related stimuli and making them invests time and effort in finding more could lead to an increased belief in self-efficacy. Possibly, the perceived embodiment within the VE additionally contributed to the effectiveness of the VR training.

Kim et al. [41] developed a new AAT using a simple VE. For this study, participants viewed short video clips showing either alcoholic or non-alcoholic beverages. Following each clip, a colored dot showed up in the middle of the screen. The participants were instructed to perform a pushing movement with a joystick in the case of a red dot and an pulling movement in the case of a green dot. Results showed that craving could not only be elicited by addiction-related stimuli, but also by contextual and environmental stimuli. Additionally, it is possible to accurately measure the degree of severity of addiction-related craving.

Schroeder et al. [62] made a first attempt to transfer CBM to VR to treat eating disorders. The used hardware was a Oculus Rift DK2 as a Head-Mounted-Display (HMD) in combination with a Leap Motion infrared sensor [44] which is able to transfer real hand movements into a VE and therefore achieves a higher level of body-ownership [68]. The participants had to either reject or grab a food or a ball item, while their reaction times were measured at three points in time: At the beginning of the hand movement, at first contact with the object, and as soon as the object has been collected. To start the training, participants had to hold the dominant hand at a predefined position, center the head and hold still for 1 s. Thereafter, a three-dimensional stimulus appeared in front of them which they were supposed to either reject with a defensive hand movement or collect by grasping. While doing this, a progress bar filled up, and the last six collected stimuli were displayed at the top of the screen. The study results showed that food items were collected significantly faster than balls, especially with increasing body mass index (BMI) of the participants. In summary, VR in combination with CBM and other possible technologies, such as eye-tracking, can be a helpful tool for detecting and treating addiction disorders.

5   Requirement Analysis

The following section elaborates the requirements which are necessary for the transfer of the AAT into virtual reality and on mobile devices. These are based on the validated results of studies already conducted and should ensure a faithful conversion of the method to these new media.

5.1   General Requirements

The following requirements apply to all forms of the AAT and should therefore be applied to all platforms:

Indirect Instructions. Test persons are instructed to ignore the image content and pay attention to a certain distinguishing feature, e.g. tilting or image format. By doing so, they should learn not to be distracted in their actions by whether it is a addiction-related object or not, but to pay controlled attention to another characteristic. The changes that take place are in the millisecond range and are not necessarily consciously perceptible. However, we assume that exactly this automatic or habitual behavior can also change in everyday life, if treated correctly.

(Re-)train Automatic Reactions. As just mentioned, behavioral changes will take place in the millisecond range. To retrain these automatic reactions, movements must be 'simple' and intuitive (i.e. no chain of actions), like moving an arm forwards or backwards, or pressing one of two (arrow-)buttons.

Avoidance of Distracting Elements. Little to no distraction should be present during the AAT training, as participants have to concentrate on the task. This includes internal as well as external distracting factors, like loud noise or distractive elements within the application.

Distinguishing Feature. Addiction-related and neutral/positive stimuli must have a clearly visible distinguishing feature which should not be too dominant, as these stimuli must still be recognizable within the short time frame in which they are shown to the participants. Examples are colors, tilting of the images or the image formats (landscape or portrait).

Accurate Measurement of Reaction Times. As the changes that take place over the course of AAT training are within the millisecond range, measuring of the reaction times has to be as accurate as possible, ideally exactly to the millisecond.

Zooming. Studies [56] have shown that the visual effect of addiction-related images becoming smaller and neutral stimuli getting bigger has a great effect on the training outcome. Even without the required arm movements the results were still significant, which makes zooming an important feature that should be implemented.

5.2   VR Requirements

In the context of a pilot study, a first demonstrator should transfer the desktop AAT (DAAT) to VR, which allows a comparison of these two AAT versions. For this purpose, two virtual rooms will be implemented. While users should interact with two-dimensional images in the first room, three-dimensional objects will be used in the second room. A mock-up of this scenario, created with *eTeks Sweet Home 3D*, Version 5.6, can be seen in Fig. 2.

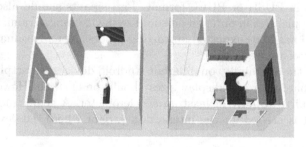

Fig. 2. Three-dimensional mock-up of the desktop room (left) and the VR enrichment room (right) [21]. (Color figure online)

As a distinguishing feature, the two-dimensional images should be tilted to the left or right. As in the studies of Machulska and colleagues [51], stimuli tilted to the left have to be pulled, and those tilted to the right have to be pushed. The desired interaction with the three-dimensional objects ought to be determined by their border color. A stimulus with a blue border has to be pulled inside a cardboard box directly in front of the participant, one with a red color pushed inside a trash can a bit further away. To train automatic reactions, stimuli which are shown one after another, must be grasped and sorted with an pulling or pushing arm movement. This should be performed as fast and accurate as possible.

While the first room completely avoids distracting features, the second room should include some game design elements. Both rooms have to include short instructions placed on a wall which show how the task should be fulfilled. The recording of the reactions times (RTs) has to be very accurate, as we want to measure and modify automatic processes and identify the cognitive bias. These involuntary processes happen very fast, usually in less than 300 ms (see P300 wave[1]). Therefore, differences resulting from AAT training are often only in the range of a few milliseconds, and such accuracy is needed to validate these differences.

For the experiments, a machine with an Intel i7-7700 CPU, an NVIDIA GTX 1070 GPU and 16 Gbyte of memory will be used to provide enough performance for implementing and running VR applications. The VR demonstrator is going to be implemented for use of the HTC Vive Head-Mounted Display (HMD), which has a resolution of 2160×1200 pixels and a refresh rate of 90 Hz [37].

5.3   App Requirements

The mobile application should have various interaction variants (pinch-to-zoom, angling device, full arm movement), which can and will be compared with each other. In addition, there should be various settings that allow to modify the training. These settings should include, among other options, the number of runs, number of images, the desired distinguishing feature (color/tilt), or the ratio of images in PUSH or PULL format. Participants should also be able to insert their own pictures in order to personalize the training. Training data must be recorded and has to be exported for further processing and evaluation, or sent encrypted to a secure web server.

In addition, the operability on different Android devices (smartphone/tablet) with different hardware and display sizes should be ensured. However, mobile devices are required to have at least an accelerometer. A geomagnetic sensor is optional, but it is needed for the interaction variant where the device must be angled.

6   VR Demonstrator Version 1

The following section describes the implementation of the first VR demonstrator which was used for a preliminary study. The study aimed to find out whether the AAT procedure can be transferred into virtual reality at all and whether bias measurements can be obtained, which can be compared with the classical desktop AAT. In addition, usability and immersion were evaluated. For the implementation of the first Windows x64 VR application, the Unreal Engine (UE) [26] was used in version 4.18. While mainly UE-Blueprints were used for the first demonstrator, some functions have been programmed in C++.

---

[1] The P300 wave is related to the decision-making process. As soon as presented stimuli are linked to a task and thus become relevant, a P300 is triggered [72].

6.1   Setup

The VR demonstrator is divided into two rooms which are connected to each other by an elevator, being equipped with an operating panel and a level indicator. Outside the window users can see an extensive grass landscape with various plants to ensure that the VE has no visible end and appears more natural. Potential oppressive feelings within the small rooms should be reduced as well to help participants feeling more comfortable during the training.

In the first room, called *desktop room* (DR), two-dimensional images which were kindly provided by Rinck and his colleagues [59], appear one after the other in the middle of the room. In the case of a smoke-related image, it is tilted to the right and has to be pushed, otherwise the image is tilted to left (see Fig. 3) and has to be pulled. While the perspective effect is sufficient for pulled images, those pushed away are artificially scaled down, since zooming is an important feature of the AAT.

Fig. 3. Example of a neutral image in PULL format [21].

After the successful completion of the DR, the ceiling lights turn green for one second, the elevator door opens, and users can enter it to reach the second room. While doing so, the ride is accompanied by a sound file that simulates a moving elevator.

The second room called *VR enrichment room* (VRER) contains some game design elements, like an environment modeled after an office space, and a particle effect, which accompanies the appearance of each stimulus. As depicted in Fig. 4, a table stands in the middle of the room. In front of it is a cardboard box, behind it a trash can. Participants are required to position themselves in front of the box to sort the stimuli according to their border color: Blue bordered stimuli should be collected inside the box, red bordered stimuli are to be thrown in the trash can (see Fig. 5). In this way, the necessary arm movements can be transferred into three-dimensional space. Contrary to the DR, stimuli in the VRER are represented by 3D models, as studies by Lee et al. [45] and Gorini et al. [32] have shown that these produce a higher craving than two-dimensional images and increase immersion.

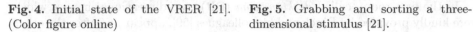

Fig. 4. Initial state of the VRER [21]. (Color figure online)

Fig. 5. Grabbing and sorting a three-dimensional stimulus [21].

While performing the training, feedback is given to the participants. Correct actions are rewarded with a positive sound effect. In case of an error, on the other hand, participants will hear a negative tone. Additionally, all ceiling lights will turn red to provide a visual feedback. Only after the correct movement has been performed, the lights regain their natural coloring, and the next stimulus can appear.

Inside the DR, two different 3D models to visually represent interactions within the VE are provided. As shown in Fig. 6, the first model represents a HTC Vive controller, while the second one is a robotic hand with gripping animations. In the VRER only the robotic hand is available. Test leaders can set the preferred model for the DR via a configuration file.

Fig. 6. Interactions are represented by a controller model (left) or a robotic hand with gripping animations (right) [21].

6.2 Experiments

This section describes the study design and the obtained results. These include observations as well as evaluations of the questionnaires and the reaction times. Furthermore, the results of the VR AAT are compared with those of the desktop AAT.

Participants and Design. Twenty participants (ten females and ten males; mean age: 29.74 years, range: 18–60; five smokers) participated in the experiment. For 75% of them this was the first contact with VR technologies. After an introduction to the framework plot, aim, and functionality of the program, participants signed a declaration of consent. Afterwards the HTC Vive HMD was put on, and the application was started.

Prior to the actual experiment, participants conducted a test run to get to learn the controls and to determine how understandable the demonstrator is only using the instructions present within the VE. Thereafter, the actual run began, during which the reaction times (RTs) were measured to calculate the cognitive bias. Participants started in the DR and had to complete both rooms. The experimental design required them to sort twenty stimuli per room. The first half of the presented stimuli had a distinguishing feature independent of the content, while in the other half, all smoking-related stimuli had to be pushed away and all neutral stimuli had to be pulled. During the first run, the controller model was used in the DR, while the robotic hand model was used in the second run.

After conducting the training, all participants were asked to fill out a non-standardized questionnaire containing fifteen questions. These questions aimed at various components of the VR application, like comprehensibility of the task, control, user-friendliness, enjoyment, or the perception of certain game design elements.

Results. Observations have shown that especially those participants who have not had any previous experience with VR have found it difficult to perform the correct arm movements or to find the right distance to the image in the DR. Besides, the distinguishing feature caused difficulties for many participants, because they got confused about which arm movement had to be done for which tilting direction. Consequently they often positioned themselves in the room in such a way that they could always see the instructions on the wall. In addition, nearly every participant forgot at least once to press the thumb-stick on the controller to make the next stimulus appear.

In the VRER even the inexperienced users had significantly fewer problems with the given task and the required movements, most likely because of the use of the robotic hands with gripping animations and the three-dimensional objects. Instead, they started experimenting with the objects during their test run. As a result, fewer mistakes were made compared to the DR (see Table 1). Beyond that, participants felt less oppressed in the VRER and had significantly more enjoyment during the task, which led to an improvement in immersion and motivation. The previously mentioned problem with pressing the controller button to make the next stimuli appear also persists in the VRER, although it occurred less frequently.

The answers given in the questionnaires show that immersion and embodiment were rated quite good (on average 7.5 and 7.4 points on a scale between 1 and 10). That also applies to user-friendliness and controls, whereby the VRER was rated higher at both aspects. 30% of the users rated the movements in the DR as unnatural and unusual, because grasping 'floating images' felt peculiar to them.

Considering the controls, 65% experienced the robotic hand model as more realistic, since the grasping of stimuli felt more intuitive, visually and tactile. 15% of them noted, that the controller model might be better for beginners, as it makes instructions easier to understand. 10% preferred the controller model, as this is more in line with reality. However, they emphasized that they would prefer hand models, if a technique would be used that would transfer own hand movements realistically into the VE, like data gloves or the Leap Motion infrared sensor. The remaining 25% did not have a preferred model to represent their interactions.

With regard to the used particle effect, a purple puff of smoke, 55% expressed positive feelings about it, as objects do not appear 'out of nowhere', and their appearance is made more interesting. 5% considered it as too distracting and intrusive. Another 5% were not disturbed by the effect, but noted that it might be inappropriate because of the smoke-associated look. Interestingly, 60% of the participating smokers stated that they did not notice the smoke effect at all.

The distribution of the recorded RTs of smokers and non-smokers, per image category and room, are shown in Figs. 7 and 8 respectively. Both groups reacted faster in the VRER, whereby smokers were significantly quicker to pull smoking-related two-dimensional images in the DR (M = 964.7, SD = 118.9) than three-dimensional objects in the VRER (M = 1252, SD = 146.7). It was also confirmed that smokers have a higher approach bias for smoking-related stimuli, whereby smokers have generally reacted faster than non-smokers.

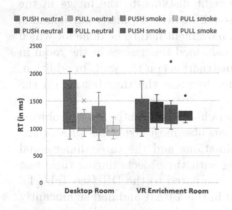
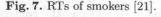
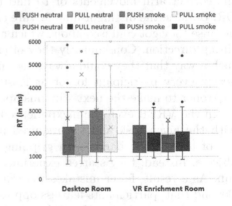

Fig. 7. RTs of smokers [21].   **Fig. 8.** RTs of non-smokers [21].

Table 1 shows the measured reaction times (RTs), the error rate, and the calculated biases compared to DAAT results, which refer to the evaluation conducted by Machulska et al. [50]. In that study participated 92 smokers and 51 non-smokers. Smoking-related two-dimensional images were tilted to the left

(PULL), non-smoking-related images to the right (PUSH). It has to be mentioned that we could not measure the RTs accurate to the millisecond. This is because the demonstrator was mainly implemented with Blueprints, a visual scripting system in the Unreal Engine. Blueprints have the disadvantage that they are frame dependent, which means that function calls occur with every frame, but not in between two frames. For our system this meant an accuracy of 16 ms.

RT distributions of the DR show that the median time elapsed before the first contact with the stimulus is 886 ms. 1416 ms are needed to finish the correct movement, whereof 530 ms were required for the execution of the arm movement itself. In the VRER, 851 ms passed until the first contact occurred, and 1319 ms until a stimulus was sorted correctly. 469 ms accounted to the arm movement. As expected, our measured RTs take longer than those measured with the DAAT, where the median RT was 610 ms up to the correct reaction, which corresponds to Fitts's law[2].

Table 1. Evaluation of RTs, error rate and bias measurements [21].

	DR	VRER	DAAT [50]
Median time until first contact	886 ms	851 ms	/
Median time until correct reaction	1416 ms	1319 ms	620 ms
Mean duration of arm movement	530 ms	469 ms	Not necessary
Error rate smokers (%)	10	0	8
Error rate non-smokers (%)	8	6	11
Ø smoking-related bias (smokers)	256	44	25
Ø smoking-related bias (non-smokers)	−312	−11	10
Ø neutral bias (smokers)	123.5	21.5	0
Ø neutral bias (non-smokers)	95	27.5	11

Although our calculated approach biases are comparable to those of the DAAT, whereby the values of the VRER come closer to those of the DATT, non-smokers show an approach bias for neutral stimuli in our results, instead of being neutral towards them. As anticipated, smokers pulled smoking-related stimuli faster than pushing them away, while for non-smokers it was the other way around, suggesting an avoidance bias. In contrast to this, non-smokers had an approach bias for smoking-related stimuli when using the DAAT.

VRER and DAAT are also comparable regarding the error rate, since smokers made significantly fewer mistakes (0% compared to 8%) than non-smokers (6% compared to 11%). On the contrary, more mistakes were made in general inside the DR, where smokers had an error rate of 10%, and non-smokers of 8%.

---

[2] Fitts's law predicts the speed required for a fast, targeted movement to a specific destination. According to the law, the time required for the start-to-finish motion can be modeled as a function of distance and target size [27].

7   VR Demonstrator Version 2

After the preliminary study presented in this article, a new demonstrator was developed, which is an extension and improvement compared to the previous demonstrator [22]. The requirements have remained essentially the same, but issues identified during the preliminary study were resolved, and additional features have been added. Major changes include the support of the Leap Motion infrared sensor [44], graphical improvements, and enhanced algorithms. For the second version of the VR application, Unity3D [71] was used as an IDE. Therefore the application is written in the C# programming language.

7.1   Improvements

Compared to the first demonstrator, significant improvements have been made. These include a graphical user interface (GUI) for the operator, which can now be used to change settings in a settings sub menu. Thereby, modifications no longer have to be made via the configuration file outside of the program. In addition to participant ID, number of stimuli, and training mode (training/bias), new setting options include starting level, control type (Leap Motion/HTC Vive Controllers), optional custom models, and the *simplified object interaction*. The latter can be used to help the subjects performing automatic actions easier, as only an impulse in the desired direction is necessary to move stimuli, and is therefore not as error-prone.

While controllers are represented as white androgynous hands with gripping animations, the use of the Leap Motion infrared sensor allows a number of different hand models (capsule hands, low poly hands, male hands with forearm (white or skin texture), female hands with forearm (white or skin texture)).

The room, in which the training takes place is, as the VRER, modeled after an office room. The distinguishing feature is still a blue or red border color, which decides whether the stimulus is to be sorted into the cardboard box, or trash can. The training is started with the thumb-stick (controller) or with a thumbs-up gesture (Leap Motion). A training just started can be seen in Fig. 9.

7.2   Preliminary Results

A first study with the VR application showed considerably better results compared to the first demonstrator. General and spatial presence greatly improved, as well as guidance through the procedure. Experienced realism, quality of interface, and sounds are still improvable, as these received a mediocre rating. Involvement got a relatively poor score, especially while using the controllers, which is why this area needs urgent revision in future versions of the application. Most importantly, however, was the realization that the Leap Motion control scheme in combination with the simplified object interaction is the most reliable type of control to obtain useful RTs for cognitive bias measurement. For that reason, this type of control will be used in the following main study.

Fig. 9. State right after starting the VR-AAT with a thumbs-up gesture. (Color figure online)

8 Mobile App

The mobile application [23] is already in an advanced state, but some optimizations and features have to be implemented before the study can start. Android Studio 3.3 and Java were used for the programming of the mobile android application.

The application offers three design variants for the transfer of the AAT procedure to mobile devices, which are described in the following, and can be seen in Fig. 10:

- *pinch-to-zoom*: This interaction variant uses the multi-touch functionality of mobile devices. In accordance with the AAT procedure, addiction-related images must be reduced in size by touching the display with two fingers and moving the fingers closer together, while neutral images must be enlarged by moving the fingers apart.
- *whole arm movement*: For this option, as with the VR demonstrator, a complete movement of the arms is required. The mobile device is held with both hands. If a addiction-related image appears, the subject must push the device away from the body, while it has to be pulled closer for neutral images.
- *tilting the device*: The device must be angled according to the image content. In the case of addiction-related images, it must be tilted backwards; in the case of neutral images, it must be tilted forwards. During movement, the images decrease or increase in size respectively.

The app already has all the minimum requirements needed for the a first study (compare Sect. 5.3). It is already possible to easily generate new language versions, or custom instructions for the sessions. Additionally, the cognitive bias is calculated and displayed on a result screen directly after a session. However, the interaction variant in which the device is moved with a whole arm movement has to be optimized, since it is still too error-prone, and the algorithms of the bias mode have to be refined.

Fig. 10. The three different interaction variants: pinch-to-zoom (left), whole arm movement (middle), and tilting device (right) [23].

9 Conclusions

Based on our results, transferring the AAT into virtual reality is promising and should be pursued further. Accuracy of the time measurement, which is an integral part of the approach bias calculation, was difficult at first due to the use of Unreal's Blueprints, but was optimized for the second VR application.

Since the VRER received better results, compared to the DR, it can be concluded that the use of game design elements offers added value for enjoyment and long-term motivation. Border colors as the distinguishing feature proved to be beneficial, since the error rate was lower compared to the tilting of the images. As expected, interacting with three-dimensional objects was perceived as more realistic, intuitive and visually appealing, resulting in increased immersion and presence. Furthermore, the VRER shows more similarities with the DAAT compared to the DR, e.g. error rate, calculated bias values, or faster RTs.

The robotic hand was the favored model for representing interactions within the VE, because is felt more natural, and increased the effect of embodiment and body-ownership, thereby enhancing immersion and presence. However, some participants mentioned that the controller model would be more suitable for beginners, since the controller is illustrated on the instructions. It was often criticized that the thumb-stick has to be pressed repeatedly to make the next stimulus appear. For that reason, this behavior has been removed from the new demonstrator. Instead, the next stimulus appears automatically after 1 s after the participant has moved far enough away from the center of the table.

The new version of the VR demonstrator which incorporates the results presented in Sect. 6.2 fixed the problems regarding the time measurement, and corrected design errors, has been developed and was presented in Sect. 7. To improve embodiment and body-ownership, using the Leap Motion infrared sensor is the preferred interaction variant. The use of data gloves has already been discussed and will be implemented in the near future. In addition, full body tracking in combination with an avatar will be another new feature, as this will, among other positive effects, allow users to move more securely within the VE.

The second study, using the VR application described in Sect. 7, is currently in progress. For this study, at least 100 daily smokers who are motivated to quit smoking will be recruited into the randomized controlled trial to evaluate the effectiveness of the VR-AAT training. While one half of the participants is training in the office room shown in Fig. 9, the other half is assigned to a placebo-control group which trains in this room as well, but the stimuli have to be sorted into boxes which are on the left and on the right side of the stimulus.

In case of positive results, a regional clinic will include the application into its portfolio in order to treat substance addiction disorders.

In addition, gamification will be added to the VR application to investigate whether this further improves the training. This includes a progress bar, a scoring system, achievements, or other elements, as there is a significant increase in user motivation when the contexts, in which rewards are received, are related to the actions performed, even if there are no stories or supporting characters [17]. Accordingly, this should also lower dropout rates.

Following the VR studies, the mobile application will be evaluated. First, the application will be further developed and optimized. Moreover, a web interface is to be implemented which can be accessed by study leaders and participants. Both the VR application and the app will send the data encrypted to the back-end of the web interface. This allows study leaders to access their participants' data independently of time and place, while the participants have access to their own data. Diagrams will be used to monitor and evaluate progress of the training.

With fulfilling the aforementioned enhancements we anticipate that our applications will lead to an improved form of therapy with a lower dropout rate.

References

1. Agarwal, R., Karahanna, E.: Time flies when you're having fun: cognitive absorption and beliefs about information technology usage. MIS Q. **24**(4), 665 (2000). https://doi.org/10.2307/3250951
2. Aichele, C., Schönberger, M. (eds.): App4U: Mehrwerte durch Apps im B2B und B2C. Springer, Wiesbaden (2014). https://doi.org/10.1007/978-3-8348-2436-3
3. Aichele, C., Schönberger, M.: App-Entwicklung – effizient und erfolgreich. Springer, Wiesbaden (2016). https://doi.org/10.1007/978-3-658-13685-7
4. Allom, V., Mullan, B., Hagger, M.: Does inhibitory control training improve health behaviour? A meta-analysis. Health Psychol. Rev. **10**(2), 168–186 (2016). https://doi.org/10.1080/17437199.2015.1051078
5. Bargh, J.A.: Advances in social cognition. In: Wyer, R.S. (ed.) The Automaticity of Everyday Life, Advances in Social Cognition, vol. 10, pp. 20–27. Erlbaum, Mahwah (1997)
6. Batra, A., Hoch, E., Mann, K., Petersen, K.U. (eds.): S3-Leitlinie Screening, Diagnose und Behandlung des schädlichen und abhängigen Tabakkonsums. Springer, Heidelberg (2015). https://doi.org/10.1007/978-3-662-47084-8
7. Beard, C., Weisberg, R.B., Primack, J.: Socially anxious primary care patients' attitudes toward cognitive bias modification (CBM): a qualitative study. Behav. Cogn. Psychother. **40**(5), 618–633 (2012). https://doi.org/10.1017/S1352465811000671
8. Blohm, I., Leimeister, J.M.: Gamification. WIRTSCHAFTSINFORMATIK **55**(4), 275–278 (2013). https://doi.org/10.1007/s11576-013-0368-0
9. Boendermaker, W.J., Boffo, M., Wiers, R.W.: Exploring elements of fun to motivate youth to do cognitive bias modification. Games Health J. **4**(6), 434–443 (2015). https://doi.org/10.1089/g4h.2015.0053

10. Boendermaker, W.J., Prins, P.J.M., Wiers, R.W.: Cognitive bias modification for adolescents with substance use problems-can serious games help? J. Behav. Therapy Exp. Psychiatry **49**(Pt A), 13–20 (2015). https://doi.org/10.1016/j.jbtep.2015.03.008

11. Bordnick, P.S., Carter, B.L., Traylor, A.C.: What virtual reality research in addictions can tell us about the future of obesity assessment and treatment. J. Diabetes Sci. Technol. **5**(2), 265–271 (2011). https://doi.org/10.1177/193229681100500210

12. Cabrera, E.A., Wiers, C.E., Lindgren, E., Miller, G., Volkow, N.D., Wang, G.J.: Neuroimaging the effectiveness of substance use disorder treatments. J. Neuroimmune Pharmacol. Off. J. Soc. Neuroimmune Pharmacol. **11**(3), 408–433 (2016). https://doi.org/10.1007/s11481-016-9680-y

13. Chen, M., Bargh, J.A.: Consequences of automatic evaluation: immediate behavioral predispositions to approach or avoid the stimulus. Personal. Soc. Psychol. Bull. **25**(2), 215–224 (2016). https://doi.org/10.1177/0146167299025002007

14. Cummings, K.M., Hyland, A.: Impact of nicotine replacement therapy on smoking behavior. Ann. Rev. Public Health **26**, 583–599 (2005). https://doi.org/10.1146/annurev.publhealth.26.021304.144501

15. Deutsch, R., Strack, F.: Reflective and impulsive determinants of addictive behavior. In: Wiers, R.W.H.J., Stacy, A.W. (eds.) Handbook of Implicit Cognition and Addiction, pp. 45–57. Sage Publications, Thousand Oaks (2006). https://doi.org/10.4135/9781412976237

16. Donath, C.: Drogen- und suchtbericht 2018. https://www.drogenbeauftragte.de/fileadmin/dateien-dba/Drogenbeauftragte/Drogen_und_Suchtbericht/pdf/DSB-2018.pdf

17. Dovis, S., van der Oord, S., Wiers, R.W., Prins, P.J.M.: Can motivation normalize working memory and task persistence in children with attention-deficit/hyperactivity disorder? The effects of money and computer-gaming. J. Abnormal Child Psychol. **40**(5), 669–681 (2012). https://doi.org/10.1007/s10802-011-9601-8

18. du Pont, P.: Building complex virtual worlds without programming. In: EUROGRAPHICS 1995 State of the Art Reports, pp. 61–70 (1995)

19. Eberl, C., Wiers, R.W., Pawelczack, S., Rinck, M., Becker, E.S., Lindenmeyer, J.: Approach bias modification in alcohol dependence: do clinical effects replicate and for whom does it work best? Dev. Cogn. Neurosc. **4**, 38–51 (2013). https://doi.org/10.1016/j.dcn.2012.11.002

20. Effertz, T.: Die volkswirtschaftlichen Kosten gefährlichen Konsums. Peter Lang, New York (2015). https://doi.org/10.3726/978-3-653-05272-5

21. Eiler, T.J., Grünewald, A., Brück, R.: Fighting substance dependency combining AAT therapy and virtual reality with game design elements. In: Proceedings of the 14th International Joint Conference on Computer Vision, Imaging and Computer Graphics Theory and Applications - Volume 2: HUCAPP, pp. 28–37. INSTICC, SciTePress (2019). https://doi.org/10.5220/0007362100280037

22. Eiler, T.J., et al.: A preliminary evaluation of transferring the approach avoidance task into virtual reality. In: Pietka, E., Badura, P., Kawa, J., Wieclawek, W. (eds.) ITIB 2019. AISC, vol. 1011, pp. 151–163. Springer, Cham (2019). https://doi.org/10.1007/978-3-030-23762-2_14

23. Eiler, T.J., et al.: One 'Stop Smoking' to Take Away, please! A preliminary evaluation of an AAT mobile app. In: Pietka, E., Badura, P., Kawa, J., Wieclawek, W. (eds.) Information Technology in Biomedicine. Advances in Intelligent Systems and Computing. Springer, Cham (accepted)

24. Elenko, E., Underwood, L., Zohar, D.: Defining digital medicine. Nat. Biotechnol. **33**(5), 456–461 (2015). https://doi.org/10.1038/nbt.3222
25. Elliot, A.J.: The hierarchical model of approach-avoidance motivation. Motiv. Emotion **30**(2), 111–116 (2006). https://doi.org/10.1007/s11031-006-9028-7
26. Epic Games: Unreal engine features (2017). https://www.unrealengine.com/en-US/features
27. Fitts, P.M.: The information capacity of the human motor system in controlling the amplitude of movement. J. Exp. Psychol. **47**(6), 381–391 (1954). https://doi.org/10.1037/h0055392
28. Forman, E.M., Goldstein, S.P., Flack, D., Evans, B.C., Manasse, S.M., Dochat, C.: Promising technological innovations in cognitive training to treat eating-related behavior. Appetite (2017). https://doi.org/10.1016/j.appet.2017.04.011
29. Ghiţă, A., Gutiérrez-Maldonado, J.: Applications of virtual reality in individuals with alcohol misuse: a systematic review. Addict. Behav. **81**, 1–11 (2018). https://doi.org/10.1016/j.addbeh.2018.01.036
30. Giraldi, G., Silva, R., Oliveira, J.: Introduction to virtual reality. LNCC Research Report, vol. 6 (2003)
31. Girard, B., Turcotte, V., Bouchard, S., Girard, B.: Crushing virtual cigarettes reduces tobacco addiction and treatment discontinuation. Cyberpsychol. Behav. Impact Internet Multimedia Virtual Real. Behav. Soc. **12**(5), 477–483 (2009). https://doi.org/10.1089/cpb.2009.0118
32. Gorini, A., Griez, E., Petrova, A., Riva, G.: Assessment of the emotional responses produced by exposure to real food, virtual food and photographs of food in patients affected by eating disorders. Ann. General Psychiatry **9**, 30 (2010). https://doi.org/10.1186/1744-859X-9-30
33. Hajek, P., Stead, L.F., West, R., Jarvis, M., Hartmann-Boyce, J., Lancaster, T.: Relapse prevention interventions for smoking cessation. Cochrane Database Syst. Rev. (8), CD003999 (2013). https://doi.org/10.1002/14651858.CD003999.pub4
34. Hoffmann, H.G.: Virtual-reality therapy. Sci. Am. **291**(2), 58–65 (2004). https://www.scientificamerican.com/article/virtual-reality-therapy
35. Hofmann, W., Friese, M., Strack, F.: Impulse and self-control from a dual-systems perspective. Perspect. Psychol. Sci. J. Assoc. Psychol. Sci. **4**(2), 162–176 (2009). https://doi.org/10.1111/j.1745-6924.2009.01116.x
36. Houben, K., Dassen, F.C.M., Jansen, A.: Taking control: working memory training in overweight individuals increases self-regulation of food intake. Appetite **105**, 567–574 (2016). https://doi.org/10.1016/j.appet.2016.06.029
37. HTC Corporation: Vive^{TM} — vive virtual reality system (2017). https://www.vive.com/us/product/vive-virtual-reality-system/
38. Kakoschke, N., Kemps, E., Tiggemann, M.: Approach bias modification training and consumption: a review of the literature. Addict. Behav. **64**, 21–28 (2017). https://doi.org/10.1016/j.addbeh.2016.08.007
39. Keil, U., Prugger, C., Heidrich, J.: Passivrauchen. Public Health Forum **24**(2), 84–87 (2016). https://doi.org/10.1515/pubhef-2016-0027
40. Kilteni, K., Groten, R., Slater, M.: The sense of embodiment in virtual reality. Presence Teleoperators Virtual Environ. **21**(4), 373–387 (2012). https://doi.org/10.1162/PRES_a_00124
41. Kim, D.Y., Lee, J.H.: Development of a virtual approach-avoidance task to assess alcohol cravings. Cyberpsychol. Behav. Soc. Netw. **18**(12), 763–766 (2015). https://doi.org/10.1089/cyber.2014.0490

42. Koster, E.H.W., Fox, E., MacLeod, C.: Introduction to the special section on cognitive bias modification in emotional disorders. J. Abnorm. Psychol. **118**(1), 1–4 (2009). https://doi.org/10.1037/a0014379
43. Lawrence, N.S., et al.: Training response inhibition to food is associated with weight loss and reduced energy intake. Appetite **95**, 17–28 (2015). https://doi.org/10.1016/j.appet.2015.06.009
44. Leap Motion: Leap motion: reach into virtual reality with your bare hands (2018). https://www.leapmotion.com
45. Lee, J.H., et al.: Experimental application of virtual reality for nicotine craving through cue exposure. Cyberpsychol. Behav. Impact Internet Multimedia Virtual Reality Behav. Soc. **6**(3), 275–280 (2003). https://doi.org/10.1089/109493103322011560
46. Lee, J., et al.: Nicotine craving and cue exposure therapy by using virtual environments. Cyberpsychol. Behav. Impact Internet Multimedia Virtual Reality Behav. Soc. **7**(6), 705–713 (2004). https://doi.org/10.1089/cpb.2004.7.705
47. de Leo, G., Diggs, L.A., Radici, E., Mastaglio, T.W.: Measuring sense of presence and user characteristics to predict effective training in an online simulated virtual environment. Simul. Healthc. J. Soc. Simul. Healthc. **9**(1), 1–6 (2014). https://doi.org/10.1097/SIH.0b013e3182a99dd9
48. Lewin, K.: Dynamic Theory of Personality - Selected Papers. McGraw-Hill Paperbacks, Read Books Ltd., Redditch (2013)
49. Lumsden, J., Edwards, E.A., Lawrence, N.S., Coyle, D., Munafò, M.R.: Gamification of cognitive assessment and cognitive training: a systematic review of applications and efficacy. JMIR Serious Games **4**(2), e11 (2016). https://doi.org/10.2196/games.5888
50. Machulska, A., Zlomuzica, A., Adolph, D., Rinck, M., Margraf, J.: A cigarette a day keeps the goodies away: smokers show automatic approach tendencies for smoking-but not for food-related stimuli. PLoS ONE **10**(2), e0116464 (2015). https://doi.org/10.1371/journal.pone.0116464
51. Machulska, A., Zlomuzica, A., Rinck, M., Assion, H.J., Margraf, J.: Approach bias modification in inpatient psychiatric smokers. J. Psychiatr. Res. **76**, 44–51 (2016). https://doi.org/10.1016/j.jpsychires.2015.11.015
52. MacLeod, C., Mathews, A.: Cognitive bias modification approaches to anxiety. Annu. Rev. Clin. Psychol. **8**, 189–217 (2012). https://doi.org/10.1146/annurev-clinpsy-032511-143052
53. Marsh, A.A., Ambady, N., Kleck, R.E.: The effects of fear and anger facial expressions on approach- and avoidance-related behaviors. Emotion (Washington, D.C.) **5**(1), 119–124 (2005). https://doi.org/10.1037/1528-3542.5.1.119
54. Maske, P.: Mobile Applikationen 1. Gabler Verlag, Wiesbaden (2012). https://doi.org/10.1007/978-3-8349-3650-9
55. Murphy, K.: Cue exposure therapy: what the future holds (2014). https://www.rehabs.com/pro-talk-articles/cue-exposure-therapy-what-the-future-holds/
56. Phaf, R.H., Mohr, S.E., Rotteveel, M., Wicherts, J.M.: Approach, avoidance, and affect: a meta-analysis of approach-avoidance tendencies in manual reaction time tasks. Front. Psychol. **5**, 378 (2014). https://doi.org/10.3389/fpsyg.2014.00378
57. Powers, M.B., Emmelkamp, P.M.G.: Virtual reality exposure therapy for anxiety disorders: a meta-analysis. J. Anxiety Disord. **22**(3), 561–569 (2008). https://doi.org/10.1016/j.janxdis.2007.04.006
58. Regenbrecht, H.T., Schubert, T.W., Friedmann, F.: Measuring the sense of presence and its relations to fear of heights in virtual environments. Int. J. Hum. Comput. Interact. **10**(3), 233–249 (1998). https://doi.org/10.1207/s15327590ijhc1003_2

59. Rinck, M., Becker, E.S.: Approach and avoidance in fear of spiders. J. Behav. Ther. Exp. Psychiatry **38**(2), 105–120 (2007). https://doi.org/10.1016/j.jbtep.2006.10.001

60. Sailer, M., Hense, J.U., Mayr, S.K., Mandl, H.: How gamification motivates: an experimental study of the effects of specific game design elements on psychological need satisfaction. Comput. Hum. Behav. **69**, 371–380 (2017). https://doi.org/10.1016/j.chb.2016.12.033

61. Schoenmakers, T.M., de Bruin, M., Lux, I.F.M., Goertz, A.G., van Kerkhof, D.H.A.T., Wiers, R.W.: Clinical effectiveness of attentional bias modification training in abstinent alcoholic patients. Drug Alcohol Depend. **109**(1–3), 30–36 (2010). https://doi.org/10.1016/j.drugalcdep.2009.11.022

62. Schroeder, P.A., Lohmann, J., Butz, M.V., Plewnia, C.: Behavioral bias for food reflected in hand movements: a preliminary study with healthy subjects. Cyberpsychol. Behav. Soc. Netw. **19**(2), 120–126 (2016). https://doi.org/10.1089/cyber.2015.0311

63. Schubert, T., Friedmann, F., Regenbrecht, H.: Embodied presence in virtual environments. In: Paton, R., Neilson, I. (eds.) Visual Representations and Interpretations, pp. 269–278. Springer, London (1999). https://doi.org/10.1007/978-1-4471-0563-3_30

64. Schultze, U.: Embodiment and presence in virtual worlds: a review. J. Inf. Technol. **25**(4), 434–449 (2010). https://doi.org/10.1057/jit.2010.25

65. Sherman, W.R., Craig, A.B.: Understanding Virtual Reality: Interface, Application, and Design. Morgan Kaufmann Series in Computer Graphics and Geometric Modeling. Morgan Kaufmann, San Francisco (2003)

66. Simpson, R.M., LaViola, J.J., Laidlaw, D.H., Forsberg, A.S., van Dam, A.: Immersive VR for scientific visualization: a progress report. IEEE Comput. Graphics Appl. **20**(6), 26–52 (2000). https://doi.org/10.1109/38.888006

67. Slater, M., Pertaub, D.P., Steed, A.: Public speaking in virtual reality: facing an audience of avatars. IEEE Comput. Graphics Appl. **19**(2), 6–9 (1999). https://doi.org/10.1109/38.749116

68. Slater, M., Perez-Marcos, D., Ehrsson, H.H., Sanchez-Vives, M.V.: Inducing illusory ownership of a virtual body. Front. Neurosci. **3**(2), 214–220 (2009). https://doi.org/10.3389/neuro.01.029.2009

69. Solarz, A.K.: Latency of instrumental responses as a function of compatibility with the meaning of eliciting verbal signs. J. Exp. Psychol. **59**(4), 239–245 (1960). https://doi.org/10.1037/h0047274

70. Statista: Marktanteile der führenden betriebssysteme am absatz von smartphones weltweit vom 1. quartal 2009 bis zum 4. quartal 2018 (2018). https://de.statista.com/statistik/daten/studie/73662/umfrage/marktanteil-der-smartphone-betriebssysteme-nach-quartalen/

71. Unity Technologies: Unity (2018). https://unity3d.com/

72. van Dinteren, R., Arns, M., Jongsma, M.L.A., Kessels, R.P.C.: P300 development across the lifespan: a systematic review and meta-analysis. PLoS ONE **9**(2), e87347 (2014). https://doi.org/10.1371/journal.pone.0087347

73. Verbeken, S., Braet, C., Goossens, L., van der Oord, S.: Executive function training with game elements for obese children: a novel treatment to enhance self-regulatory abilities for weight-control. Behav. Res. Ther. **51**(6), 290–299 (2013). https://doi.org/10.1016/j.brat.2013.02.006

74. Werbach, K.: (Re)defining gamification: a process approach. In: Spagnolli, A., Chittaro, L., Gamberini, L. (eds.) PERSUASIVE 2014. LNCS, vol. 8462, pp. 266–272. Springer, Cham (2014). https://doi.org/10.1007/978-3-319-07127-5_23

75. Wiers, C.E., et al.: Automatic approach bias towards smoking cues is present in smokers but not in ex-smokers. Psychopharmacology **229**(1), 187–197 (2013). https://doi.org/10.1007/s00213-013-3098-5

76. Wiers, R.W., Rinck, M., Dictus, M., van den Wildenberg, E.: Relatively strong automatic appetitive action-tendencies in male carriers of the OPRM1 G-allele. Genes Brain Behav. **8**(1), 101–106 (2009). https://doi.org/10.1111/j.1601-183X.2008.00454.x

77. Wiers, R.W., Eberl, C., Rinck, M., Becker, E.S., Lindenmeyer, J.: Retraining automatic action tendencies changes alcoholic patients' approach bias for alcohol and improves treatment outcome. Psychol. Sci. **22**(4), 490–497 (2011). https://doi.org/10.1177/0956797611400615

Information Visualization Theory and Applications

Orthogonal Compaction: Turn-Regularity, Complete Extensions, and Their Common Concept

Alexander M. Esser[⊠] [ID]

Fraunhofer Institute for Intelligent Analysis and Information Systems IAIS,
Sankt Augustin, Germany
`alexander.esser@iais.fraunhofer.de`

Abstract. The compaction problem in orthogonal graph drawing aims to construct efficient drawings on the orthogonal grid. The objective is to minimize the total edge length or area of a planar orthogonal grid drawing. However, any collisions, i.e. crossing edges, overlapping faces, or colliding vertices, must be avoided. The problem is *NP*-hard. Two common compaction methods are the turn-regularity approach by Bridgeman et al. [4] and the complete-extension approach by Klau and Mutzel [23]. Esser [14] has shown that both methods are equivalent and follow a common concept to avoid collisions.

We present both approaches and their common concept in detail. We introduce an algorithm to transform the turn-regularity formulation into the complete-extension formulation and vice versa in $\mathcal{O}(n)$ time, where n is the number of vertices.

Keywords: Graph drawing · Orthogonal drawing · Compaction · Turn-regularity · Complete extensions

1 Introduction

Once upon a time, an electrical engineer wanted to design the smallest possible computer chip. The area of the integrated circuit ought to be as small as possible and the total length of the wires ought to be as short as possible. However, the wires must not cross and between the electrical components, such as transistors, contacts, or logic gates, a minimum distance must be kept.

In graph drawing theory, this problem from VLSI design [24, 26] is known as the *two-dimensional compaction problem for planar orthogonal grid drawings*. One could imagine the planar orthogonal grid drawing in Fig. 1 as a circuit layout, with the vertices representing electrical components and the edges representing wires connecting these components. Minimizing the area or the total edge length of such a drawing representing a chip layout, with a certain distance between all the electrical components though, is essential for this use case.

Apart from VLSI design, compaction is important in many other contexts in information visualization. The types of drawings and domains range from UML diagrams in

© Springer Nature Switzerland AG 2020
A. P. Cláudio et al. (Eds.): VISIGRAPP 2019, CCIS 1182, pp. 179–202, 2020.
https://doi.org/10.1007/978-3-030-41590-7_8

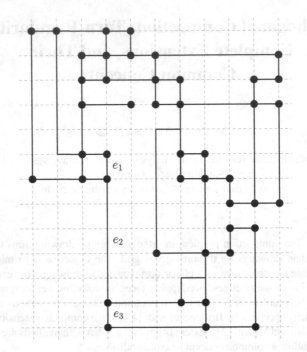

Fig. 1. An orthogonal grid drawing (adopted from [14] in a slightly modified version).

the area of software engineering, via entity-relationship diagrams for database management, through to subway maps [2,3,7,9,12,27,29].

Planar orthogonal grid drawings fulfil the following requirements: edges must not cross, all edges must consist of horizontal or vertical sections, and all vertices and bends must be located on grid points. In addition to these obligatory requirements, planar orthogonal grid drawings are often optimized with respect to some aesthetic criteria, such as a minimum number of bends, short edges or a small area. The latter two criteria are addressed by the compaction problem.

The aim of the two-dimensional compaction problem is to minimize the total edge length or the area of a planar orthogonal grid drawing. The compaction problem is one of the challenging tasks in graph drawing: On the one hand, schematic drawings are subject to size limitations in almost every application context; on the other hand, the compaction step is the final step in the so-called topology-shape-metrics scheme (see Sect. 1.2) and therefore is crucial for creating adequate drawings.

Two state-of-the-art compaction methods are the turn-regularity approach by Bridgeman et al. [4] and the complete-extension approach by Klau and Mutzel [23]. In [14] the equivalence of both approaches was formally proven: An orthogonal representation is turn-regular if and only if there exists a unique complete extension.

In this chapter, which is based on [14], we present both approaches and the proof of equivalence in more detail. We also show that the turn-regularity formulation can be transformed into the complete-extension formulation and vice versa in runtime of $\mathcal{O}(n)$, where n is the number of vertices.

This chapter is organized as follows: In the following subsections of Sect. 1 we present the topology-shape-metrics scheme and formally define the compaction problem. In Sect. 2 we present various compaction methods, especially the turn-regularity approach and the complete-extension approach. In Sect. 3 we present the proof of equivalence from [14]. In Sect. 4 we analyse the runtime for transforming one formulation into the other one. Finally, we give a summary in Sect. 5.

1.1 Preliminaries

In this subsection, we introduce basic definitions and notations. For more details on graph drawing in general, orthogonal drawings, and compaction, see e.g. [6,19,28]. This subsection is oriented on the preliminaries in [18].

Graph Drawing. In the following, we will consider an undirected, connected, planar 4-graph $G = (V, E)$, consisting of a set of n vertices V and a set of m edges E.

A graph is called *planar* if it admits a drawing in the plane without any edge crossings. We distinguish between *directed* and *undirected* graphs. In undirected graphs edges are unordered pairs of vertices, in directed graphs they are ordered pairs. A graph is a *4-graph* if all its vertices have at most four incident edges.

Every planar graph can be embedded into the plane. Such a *planar embedding* especially specifies a set of faces – multiple internal faces and one external face – as well as circular ordered lists of edges bordering each face.

An *orthogonal representation* H is an extension of a planar embedding that contains additional information about the orthogonal shape of the drawing, i.e. about bends along each edge and about the corners around each vertex – 90°, 180°, 270°, or 360° corners.

An orthogonal representation is called *simple* if it is free of bends. Then, each edge has a unique orientation – horizontal or vertical. A horizontal *segment* is a maximally set of consecutive horizontal edges. This means that there is no other consecutive horizontal edge which could be added to this set. Vertical segments are defined analogously. Every vertex belongs to exactly one horizontal and one vertical segment. In Fig. 1, for example, e_1, e_2, and e_3 form a vertical segment.

For an edge or a segment e, $\ell(e)$ denotes the vertical segment containing the leftmost vertex of e; $r(e)$ is the vertical segment with the rightmost vertex of e, $b(e)$ and $t(e)$ are the horizontal segments with the bottommost and topmost vertex of e. Note that two different edges e_1, e_2 can have the same left, right, top, or bottom segment, e.g. $\ell(e_1) = \ell(e_2)$.

When speaking of a part of an edge, in literature sometimes the term "edge segment" is used, especially for bent edges in non-simple representations. We, however, call a horizontal or vertical part of an edge an *edge section* to distinguish it from the previously defined segments.

By Γ we denote a *planar orthogonal grid drawing* of G, as described in the beginning. In addition to H, Γ contains information about the coordinates of the vertices and bends on the grid, and about the length of each edge.

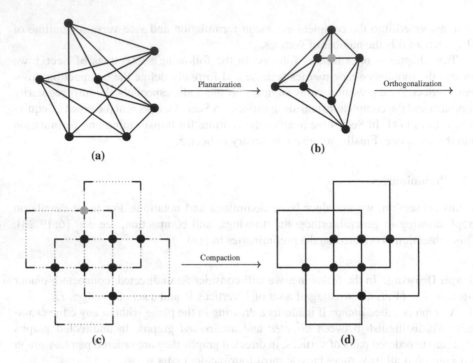

Fig. 2. Illustration of the topology-shape-metrics scheme. (Color figure online)

Minimum Cost Flows. Many compaction approaches are based on minimum cost flow techniques. Thus, we give a short introduction to networks and minimum cost flows. For more information, see e.g. [1,2].

Let $N = (V_N, E_N)$ be a directed graph. Whenever we talk about flows, we will call N a *network*, the elements in V_N *nodes*, and the elements in E_N *arcs* (in contrast to vertices and edges in undirected graphs).

Every node $n \in V_N$ has a *demand* $b(n) \in \mathbb{R}$, every arc $a \in E_N$ has a *lower bound* $\ell(a) \in \mathbb{R}^{\geq 0}$, an *upper bound* $u(a) \in \mathbb{R}^{\geq 0} \cup \{\infty\}$ and *costs* $c(a) \in \mathbb{R}^{\geq 0}$.

A *flow* is a function $x : E_N \rightarrow \mathbb{R}^{\geq 0}$ satisfying the following conditions:

$$\ell(a) \leq x(a) \leq u(a) \qquad\qquad \forall a \in E_N \qquad \text{(capacity constraints)}$$
$$f_v(x) := \sum_{a=(v,u)} x(a) - \sum_{a=(u,v)} x(a) = b(v) \quad \forall v \in V_N \quad \text{(flow conservation constraints)}$$

A *minimum cost flow* is a feasible flow x with minimum total cost $c_x = \sum_{a \in E_N} x(a)c(a)$ among all feasible flows. There are various polynomial-time algorithms to solve the minimum cost flow problem, for a comparison see e.g. [20].

1.2   Topology-Shape-Metrics Scheme

A common approach to create planar orthogonal grid drawings is the topology-shape-metrics scheme by Batini et al. [2]. The topology-shape-metrics scheme aims to create drawings with a small number of crossings, a small number of bends, and short edges. These goals, which regard topology, shape, and metrics of the drawing, are addressed in three separate phases in this order.

Planarization. In the first phase, the graph is planarized. The given 4-graph is tested for planarity, e.g. using the linear-time approach by Hopcroft and Tarjan [17]. If the graph is not planar, edge crossings are replaced by fictitious vertices. The aim is to minimize the number of edge crossings. Determining the lowest possible number of edge crossings is known as the *crossing number problem*. Garey and Johnson [15] have shown that this problem is *NP*-hard. In practice, often branch-and-cut techniques are applied, see e.g. [5]. The result of the first phase is a planar embedding.

Orthogonalization. In the second phase, an orthogonal shape of the drawing is determined. This means that the angles along each edge and the corners between edges are fixed. This shape, however, contains no information about the coordinates of the vertices and the length of the edges yet; it is dimensionless.

Here the aim is to minimize the number of bends. In general, it is *NP*-hard to find a planar orthogonal drawing of a given graph with a minimum number of bends [16]. However, with the given embedding from the first phase, an orthogonal drawing with a minimum number of bends can be computed in polynomial time by applying minimum cost flow techniques [27].

Compaction. In the third phase, the drawing is compacted. This means that the coordinates of the vertices and bends are assigned to grid points. Here the aim is to minimize the total edge length or the area of the drawing. Again, the outcome of the preceding phases is assumed to be fixed, i.e. the given embedding and the orthogonal shape have to be maintained. The compaction problem is presented in Sect. 1.3 in detail.

Figure 2 illustrates the three phases of the topology-shape-metrics scheme: Fig. 2a shows an arbitrary 4-graph $G = (V, E)$. To illustrate this arbitrary graph, all vertices were drawn on random coordinates. Figure 2b shows a planar embedding of G after the planarization phase, i.e. an embedding without any edge crossings. As the original graph G in this example is not planar, one artificial vertex (blue) had to be inserted. Figure 2c shows the result of the orthogonalization phase. To illustrate the orthogonalization phase, the fixed parts of each edge were drawn with solid lines. The angles along each edge are fixed, while the length of the blue dotted edge sections still can be varied. Figure 2d shows the final drawing resulting from the compaction phase. The artificial vertex has been removed afterwards.

1.3   The Compaction Problem

During the compaction phase, all vertices and bends are mapped to grid points, this implicitly defines the length of each edge. Meanwhile, the shape of the drawing and all angles have to be preserved. The aim is to minimize the total edge length.

Formally: Let $G = (V, E)$ be an undirected, connected, planar 4-graph. Let H be a simple orthogonal representation of G. Let Γ be a planar orthogonal grid drawing of H.

Two-dimensional Orthogonal Compaction Problem
Given H, the *two-dimensional orthogonal compaction problem* is to find a planar orthogonal grid drawing Γ of H with minimum total edge length.

We can w.l.o.g. assume G to be connected and H to be simple. If G is not connected, we can process each connected component separately. If H is not simple, bends can beforehand be replaced by artificial vertices.

Hardness and Runtime. Patrignani [25] has proven the two-dimensional orthogonal compaction problem to be *NP*-hard.

Special cases of the compaction problem, however, are solvable in polynomial time: if all faces are rectangular [6], if all faces are turn-regular (see Sect. 2.1), or if the shape description is complete or uniquely completable (see Sect. 2.2).

Objective. In the above formulation of the compaction problem, the total edge length is minimized. We will focus on this formulation.

Variations of this problem, where the area of the drawing or the length of the longest edge is minimized, can be solved using the same approaches. Only the objective function of the integer linear program or the cost function of the minimum cost flow problem, respectively, would change, but all constraints would stay the same.

2   Compaction Methods

All compaction approaches aim to avoid "collisions", i.e. crossing edges, overlapping faces, or vertices with the same coordinates, but on different ways:

– **The rectangular approach** by Tamassia [27] is to insert artificial edges and vertices in order to make all internal faces rectangular, to reduce the problem to two one-dimensional problems, and to subsequently solve these problems using minimum cost flow techniques.
– **The turn-regularity approach** by Bridgeman et al. [4] is to insert dissecting edges in order to make all faces turn-regular, to reduce the problem to two one-dimensional problems, and to subsequently solve these problems using minimum cost flow techniques.
– **The complete-extension approach** by Klau and Mutzel [23] considers specific properties of segments in order to set up a two-dimensional integer linear program (ILP); subsequently this ILP is solved.

An experimental comparison of compaction methods can be found in [22]. For an overview of compaction heuristics see e.g. [10].

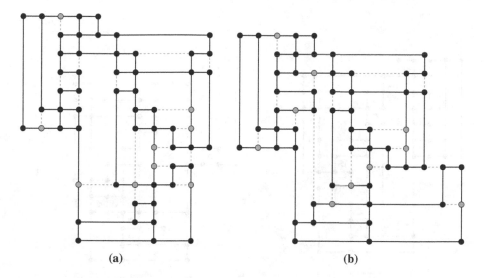

(a)   (b)

Fig. 3. Drawings for the same graph and the same orthogonal shape – with different total edge length (adopted from [14] in a slightly modified version). (Color figure online)

Dissecting Edges

Inserting artificial edges to keep a distance between vertices, so-called *dissecting edges*, is common in various compaction algorithms. However, both the orientation of these edges and the order when inserting them are important to consider.

Figure 3 shows two drawings for the same graph and the same orthogonal shape – with different total edge length. The drawings have been generated using the rectangular approach by Tamassia [27]. The dissecting edges and fictitious vertices are drawn in blue. These were inserted in different ways in both drawings. The drawing in Fig. 3a has a total edge length of 111 units, while the drawing in Fig. 3b has a total edge length of 119 units.

With the turn-regularity approach, Bridgeman et al. [4] presented a more sophisticated way to prevent collisions. They first determine all vertices which could potentially collide, vertices with so-called *kitty corners*. Only between these pairs of vertices, dissecting edges are inserted. Thereby, the turn-regularity approach practically requires much less dissecting edges than rectangular approaches [13]. This becomes important if the problem is solved as an ILP. Then, fewer dissecting edges mean fewer constraints. This avoids inserting needless place-holders to the drawing.

The Curse of One-Dimensionality

Another important difference between compaction approaches is whether they solve the compaction problem as a one-dimensional or as a two-dimensional problem.

One-dimensional heuristics transform the two-dimensional compaction problem into two one-dimensional problems [10,27]. In one dimension, either the x- or the y-dimension, the drawing is considered to be fixed, while in the other dimension, the

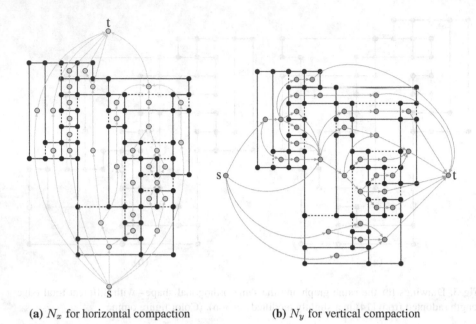

(a) N_x for horizontal compaction

(b) N_y for vertical compaction

Fig. 4. A graph (black) and the corresponding flow networks N_x (green) and N_y (red). (Color figure online)

coordinates of vertices and bends can be altered. This, however, does not lead to optimal results. A bad decision in one dimension can prevent further improvements in the other dimension.

The complete-extension approach by Klau and Mutzel [23] was one of the first approaches not splitting up the compaction problem into two one-dimensional problems but solving it as a whole – by formulating it as an ILP. Eiglsperger and Kaufmann [11] later presented a linear-time heuristic building up on the basic idea of [23].

Network Flow Techniques
For finally assigning coordinates to all vertices and a length to each edge, almost all compaction approaches use either network flow techniques or an ILP formulation based on the idea of network flows.

We present the idea of these network flow techniques using the example of the rectangular approach by Tamassia [27]: For a given orthogonal representation H, two auxiliary networks N_h, N_v are created – one horizontal and one vertical network.

The vertical network N_v is created in the following way: Every internal face in H is represented by a node in v in N_v. Additionally, N_v contains a source node s and a sink node t. N_v contains only horizontal arcs, all directed from left to right. Whenever two faces in H are bordered by a common vertical edge, the corresponding nodes in N_v are connected by a left-to-right arc. The horizontal network N_h is created analogously, containing only bottom-to-top arcs.

Figure 4 shows the horizontal (Fig. 4a, green) and the vertical (Fig. 4b, red) auxiliary network for the drawing from Fig. 3a.

The vertex coordinates and edge lengths finally are assigned by solving two minimum cost flow problems on N_h and N_v, respectively. The flow on each arc represents the length of the corresponding edge in H. To ensure that each edge has a minimum length of 1, the lower bound ℓ_a of each arc is set to 1. The upper bound u_a is set to ∞. The demand of all nodes is 0.

The flow conservation constraint ensures that two opposite sides of a rectangular face in H have the same edge length. As all dissecting edges will be removed from G afterwards, these are not taken into account when determining the costs. All other arcs have costs of 1. Thus, the cost function of the two networks is:

$$\forall \text{ arcs } a: \quad c_a = \begin{cases} 0 & \text{if the edge corresponding to a} \\ & \text{is a dissecting edge,} \\ 1 & \text{otherwise.} \end{cases} \tag{1}$$

This results in two integer linear programs for N_h and N_v, respectively:

$$\begin{aligned} \min c^T x \quad &\text{s.t.} \\ f_v(x) = 0 \quad &\forall\, v \\ x_a \geq 1 \quad &\forall\, a \\ x_a \in \mathbb{Z} \quad &\forall\, a, \end{aligned} \tag{2}$$

where a is an arc and v is a node from N_h or N_v, respectively.

Each feasible flow on these networks corresponds to feasible edge lengths for Γ. Minimizing the flow on the networks means minimizing the total edge length. Summing up the objective values of both networks gives us the total edge length.

It is important to note that the solution of this rectangular approach is in general not optimal for the compaction problem. The arrangement of dissecting edges and the reduction to two one-dimensional problems can, as described before, lead to suboptimal solutions.

Therefore, nowadays often the turn-regularity approach [4] or the complete- extension approach [23] is applied. In [14] the equivalence of both approaches was proven, more precisely: A face of an orthogonal representation is turn-regular (as defined in the first approach) if and only if the segments bounding this face are separated or can uniquely be separated (as defined in the second approach).

In the following, we describe both the turn-regularity approach and the complete-extension approach. We present basic definitions and theorems from [4] and [23], which are required for further conclusions.

2.1   Turn-Regularity Approach

The idea of the turn-regularity approach is to determine all pairs of vertices which could potentially collide. Unlike for original compaction heuristics, the faces are not required to be rectangular. The definitions and lemmata within this subsection have been adopted from [4].

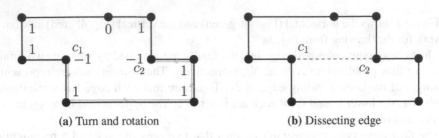

(a) Turn and rotation   (b) Dissecting edge

Fig. 5. Basic idea of the turn-regularity approach (adopted from [14] in a slightly modified version). (Color figure online)

Definition 1 (Turn). *Let f be a face in H. To every corner c in f a **turn** is assigned:*

$$turn(c) := \begin{cases} 1, & \text{if } c \text{ is a convex } 90° \text{ corner,} \\ 0, & \text{if } c \text{ is a flat } 180° \text{ corner,} \\ -1, & \text{if } c \text{ is a reflex } 180° \text{ corner.} \end{cases} \tag{3}$$

Corners enclosing $360°$ angles are treated as a pair of two reflex corners. Bridgeman et al. [4] have shown that it is sufficient to replace each $360°$ vertex by two artificial $270°$ vertices connected by an artificial edge. After the drawing has been compacted, the artificial vertices are substituted by the original vertex.

Walking along the boundary of a face and summing up the turn of each corner, yields the rotation:

Definition 2 (Rotation). *Let f be a face in H. The **rotation** of an ordered pair of corners (c_i, c_j) in f is defined as*

$$rot(c_i, c_j) := \sum_{c \in P} turn(c), \tag{4}$$

where P is a path along the boundary of f from c_i (included) to c_j (excluded) in counter-clockwise direction.

For simplifying notation, if it is clear which face is considered, we also write $rot(v_i, v_j)$ for two vertices v_i, v_j with corresponding corners c_i, c_j.

Figure 5a shows a face and the turn value assigned to each corner inside this face. The orange path along the bottom boundary illustrates $rot(c_1, c_2)$, exemplary for the two corners c_1 and c_2, summing up all turn values on the path from c_1 to c_2.

Every reflex corner either is a north-east, south-east, south-west, or north-west corner. If it is clear which face is considered, we will also speak of north-east, south-east, south-west, and north-west vertices which have a respective corner in this face. For example, in Fig. 5a, c_1 is a north-east corner and c_2 is a south-west corner.

Lemma 1. *Let f be a face in H.*

(i) For all corners c_i in f, it holds:

$$rot(c_i, c_i) = \begin{cases} 4, & \text{iff } f \text{ is an internal face,} \\ -4, & \text{iff } f \text{ is the external face.} \end{cases} \tag{5}$$

(ii) For all corners c_i, c_j in f, the following equivalence holds:

$$rot(c_i, c_j) = 2$$

$$\Leftrightarrow rot(c_j, c_i) = \begin{cases} 2, \textit{ if } f \textit{ is an int. face,} \\ -6, \textit{ if } f \textit{ is the ext. face.} \end{cases} \tag{6}$$

Definition 3 (Kitty Corners, Turn-regular). *Let f be a face in H. Two reflex corners c_i, c_j in f are named **kitty corners** if $rot(c_i, c_j) = 2$ or $rot(c_j, c_i) = 2$. A face is **turn-regular** if it contains no kitty corners. H is turn-regular if all faces in H are turn-regular.*

The reflex corners c_1, c_2 in Fig. 5a form a pair of kitty corners as the rotation $rot(c_1, c_2)$ sums up to 2. Thus, a dissecting edge between the associated vertices needs to be inserted, shown in Fig. 5b.

All pairs of kitty corners can be determined in runtime of $\mathcal{O}(n)$ [4]. If H is turn-regular, an optimal drawing can be computed in polynomial time [27]. Otherwise, all non-turn-regular faces are made turn-regular. The drawing finally is constructed by solving two one-dimensional minimum cost flow problems, which are similar to the ones in the rectangular approach.

Thereby, the turn-regularity approach defines a heuristic:

Procedure. Turn-Regularity Approach.

    Input: orthogonal representation H
    Output: planar orthogonal grid drawing Γ
1 determine all non-turn-regular faces;
2 **for** $f \in \{\text{non-turn-regular faces}\}$ **do**
3   **for** $(c_1, c_2) \in \{\text{pairs of kitty corners in } f\}$ **do**
4   insert a randomly oriented dissecting edge between c_1 and c_2;
5   **end**
6 **end**
7 solve the corresponding two minimum cost flow problems;
8 construct Γ;
9 remove all dissecting edges;
10 **return** Γ;

2.2 Complete-Extension Approach

The idea of the complete-extension approach by Klau and Mutzel [23] is to transform the compaction problem into a combinatorial problem. For this purpose, a so-called shape description is used. The definitions within this subsection have been adopted from [23] unless stated otherwise.

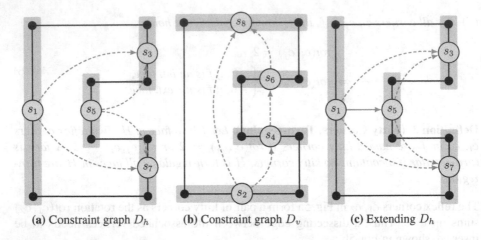

(a) Constraint graph D_h (b) Constraint graph D_v (c) Extending D_h

Fig. 6. Basic idea of the complete-extension approach. (Color figure online)

Definition 4 (Shape Description, Constraint Graph). *A **shape description** of a simple orthogonal representation H is a tuple $\sigma = \langle D_h, D_v \rangle$ of two directed so-called **constraint graphs** $D_h = (S_v, A_h)$ and $D_v = (S_h, A_v)$ with*

$$A_h := \{(\ell(e), r(e)) \mid e \; horizontal \; edge \; in \; G\}, \tag{7}$$
$$A_v := \{(b(e), t(e)) \mid e \; vertical \; edge \; in \; G\}, \tag{8}$$

and two sets of corresponding vertices S_v, S_h.

Figure 6 illustrates the basic idea of the complete-extension approach. The graph is depicted in black, the segments are marked in grey, the constraint graphs are drawn in red with dashed arcs. In the horizontal constraint graph D_h (see Fig. 6a), each vertical segment from G is represented by a segment vertex $s_1, s_3, \ldots, s_7$. These segment vertices are connected by horizontal arcs from A_h. The vertical constraint graph D_v (see Fig. 6b) is defined analogously.

Klau [21] has proven that the constraint graphs D_h, D_v are unique, can be computed in linear time, and are upward planar [21, Lemma 4.2 and Lemma 4.3]:

Lemma 2. *Given a simple orthogonal representation H with n vertices, the corresponding constraint graphs D_h, D_v are unique and can be computed in $\mathcal{O}(n)$ time.*

Lemma 3. *Given a simple orthogonal representation H, the corresponding constraint graphs D_h, D_v are upward planar.*

The arcs in $A_h \cup A_v$ determine the relative position of every pair of segments. However, this information is generally not sufficient to produce a valid orthogonal embedding. The shape description might need to be extended.

The notation $a \longrightarrow b$ denotes an immediate path from vertex a to vertex b, i.e. an edge or arc (a, b). On the path $a \xrightarrow{\star} b$, intermediate vertices are allowed.

Definition 5 (Separated Segments). *A pair of segments* $(s_i, s_j) \in S \times S$, *where* $S := S_h \cup S_v$, *is called* **separated** *if the shape description contains one of the four following paths:*

$$
\begin{array}{ll}
1. \ r(s_i) \xrightarrow{\ *\ } \ell(s_j) & 3. \ t(s_j) \xrightarrow{\ *\ } b(s_i) \\
2. \ r(s_j) \xrightarrow{\ *\ } \ell(s_i) & 4. \ t(s_i) \xrightarrow{\ *\ } b(s_j)
\end{array}
$$

A shape description is **complete** *if all pairs of segments are separated.*

For two segments $s_i, s_j \in S$ there can exist multiple paths from s_i to s_j. Imagine the two vertical segments of a simple rectangular face, for example. There exists one path along the top way and another one along the bottom way.

Klau [21] has shown that it is sufficient to consider only segments within the same face. If any two segments that share a common face are separated, the shape description is complete [21, Corollary 4.3].

Definition 6 (Complete Extension). *A* **complete extension** *of a shape description* $\sigma = \langle (S_v, A_h), (S_h, A_v) \rangle$ *is a tuple* $\tau = \langle (S_v, B_h), (S_h, B_v) \rangle$ *with the following properties:*

(P1) $A_h \subseteq B_h$ *and* $A_v \subseteq B_v$.
(P2) B_h *and* B_v *are acyclic.*
(P3) *Every non-adjacent pair of segments in* G *is separated.*

The shape description in Fig. 6 is not complete because there is no directed path from segment vertex s_1 to s_5 in D_h. If the arc (s_1, s_5) is added, as illustrated in Fig. 6c, these segments become separated. The shape description then is complete.

From every complete extension, an orthogonal drawing can be constructed which respects all constraints of this extension. Thus, the task of compacting an orthogonal grid drawing is equivalent to finding a complete extension that minimizes the total edge length. Klau and Mutzel [23] formulated this task as an ILP, which can be solved to optimality. If a shape description is complete or uniquely completable, this ILP can be constructed and solved to optimality in polynomial time [23].

The complete-extension approach is described in the following procedure:

Procedure. Complete-Extension Approach.

   Input: orthogonal representation H
   Output: planar orthogonal grid drawing Γ
1 compute the shape description σ of H;
2 compute all complete extensions τ of σ;
3 find a complete extension $\tau_{\min}$ with minimum total edge length;
4 construct Γ for $\tau_{\min}$;
5 **return** Γ;

---

3 Equivalence

In this section, we focus on the equivalence of the turn-regularity approach and the complete-extension approach and present the proof of equivalence from [14].

Note that both approaches consider different components of a drawing. While turn-regularity is a definition based on the shape of faces, completeness is a definition based on segments, more precisely: on the vertices dual to the segments. However, a segment is a set of edges so that a relation between segments and faces can be established. A segment s is said to *bound* a face f if one of the edges in s is on the boundary of f.

3.1 Auxiliary Graphs

When introducing the concept of turn-regularity, Bridgeman et al. [4] constructed auxiliary graphs G_l, G_r, H_x, and H_y.

G_l is a variation of the original graph G in which all edges are oriented leftward or upward, in G_r all edges are oriented rightward or upward.

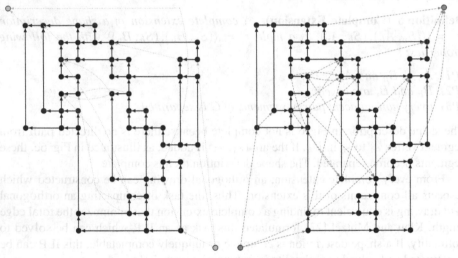

(a) G_l (black solid lines) and its saturator (pink dotted lines)

(b) G_r (black solid lines) and its saturator (purple dashed lines)

Fig. 7. Drawings of the auxiliary graphs G_l and G_r (adopted from [14] in a slightly modified version). (Color figure online)

Bridgeman et al. [4] augment G_l and G_r by so-called saturating edges. These saturating edges – or the "saturator" – are constructed based on a specific *switch property* of the edges in E. Effectively, the saturating edges simply form an acyclic directed graph with a source vertex s and a target vertex t. The saturator of G_l consists of

- additional source and target vertices s and t in the external face,
- an arc from s to every external south-east vertex,
- an arc from every external north-west vertex to t,
- arcs from every internal north-west vertex to its opposite convex vertices,
- arcs from the opposite convex vertices to any internal south-east vertex, and
- the subset of all affected vertices from V.

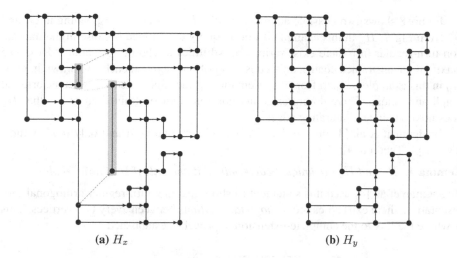

(a) H_x (b) H_y

Fig. 8. Drawings of some corresponding auxiliary graphs H_x and H_y (adopted from [14] in a slightly modified version). (Color figure online)

The saturator of G_r is defined analogously, with outgoing arcs from north-east vertices and incoming arcs towards south-west vertices.

Figure 7 shows the auxiliary graphs G_l and G_r and their saturators for the drawing from Fig. 3a.

Bridgeman et al. [4] further introduced horizontal and vertical unconstrained chains. A chain of consecutive horizontal or vertical segments in a face f is said to be *unconstrained* if both its end-vertices correspond to reflex corners. A horizontal unconstrained chain is *maximal* if no other consecutive horizontal segment in f can be added to this chain. Vertical maximal unconstrained chains are defined analogously.

G_l, G_r, and their saturators induce two more auxiliary graphs H_x, H_y. Note that H_x, H_y are graphs, not orthogonal representations. H_x and H_y are not unique; for one pair of graphs (G_l, G_r) there can exist multiple possible graphs H_x, H_y.

H_x describes the x-relation, i.e. the left-to-right relation between the segments in H. H_x contains all original edges from E and those saturating edges from G_l and G_r which are all incident to end-vertices of vertical maximal unconstrained chains. If there are multiple saturating edges incident to the same end-vertex, one of these edges is chosen and the others are omitted. Therefore H_x and H_y are not unique. The original vertical edges are kept without orientation, the original horizontal edges are all directed from left to right. The saturating edges in H_x all point from left to right (i.e. the saturating edges originating from G_l are reversed). The source vertex s, the sink vertex t, and all their incident edges are omitted.

H_y, which denotes the bottom-to-top relation between segments, is constructed in a similar way. In H_y, the saturating edges are incident to end-vertices of horizontal maximal unconstrained chains, the vertical edges are all directed upwards, the horizontal ones are undirected.

Figure 8 shows two possible auxiliary graphs H_x and H_y corresponding to G_l and G_r from Fig. 7. H_x in this example is not uniquely determined because in the internal non-turn-regular face there exist various possibilities to choose saturating edges. The maximal unconstrained chains, only consisting of one segment each, are drawn in grey. H_y in this example is, apart from the orientation of the edges, identical with the original graph and unique. There do not exist any horizontal unconstrained chains so that H_y does not contain any saturating edges.

Bridgeman et al. [4] proved that H_x and H_y are unique if and only if H is turn-regular [4, Theorem 4]:

Lemma 4. *H_x and H_y are uniquely determined if and only if H is turn-regular.*

Bridgeman et al. [4] used this statement to show that in a turn-regular orthogonal representation, there exist so-called *orthogonal relations* between every two vertices. This is where the link to the complete-extension approach is established.

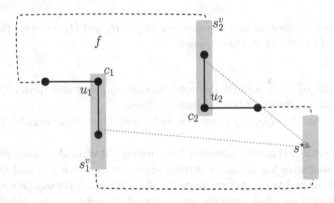

Fig. 9. Illustration of the setting in Lemma 6 and Theorem 1   (adopted from [14] in a slightly modified version).

Lemma 5. *Let $\sigma = \langle D_h, D_v \rangle$ be a shape description with $D_h = (S_v, A_h)$ and $D_v = (S_h, A_v)$. Let $S_{uc} \subseteq S_h$ be a horizontal unconstrained chain. Then, arcs in A_v between $S_h \setminus S_{uc}$ and S_{uc} either all lead towards S_{uc} or all lead away from S_{uc}.*

Take $s_i \in S_h \setminus S_{uc}$ and $s_j \in S_{uc}$. If all arcs lead towards S_{uc}, this means that all of them are of the form (s_i, s_j). If the arcs lead away from S_{uc}, all of them are of the form (s_j, s_i).

Proof. The end-vertices v_1, v_2 of a horizontal unconstrained chain both correspond to reflex corners. Let e_1, e_2 be the vertical edges incident to v_1, v_2. As the unconstrained chain is bounded by reflex corners on both sides, e_1 and e_2 are the only edges which can establish an arc in A_v between $S_h \setminus S_{uc}$ and S_{uc}. Obviously, v_1, v_2 are either both top vertices of e_1 and e_2, or v_1, v_2 are both bottom vertices of e_1 and e_2. In the first case, the corresponding arcs $(b(e_1), t(e_1)), (b(e_2), t(e_2)) \in A_v$ both lead towards S_{uc}. In the second case, $(b(e_1), t(e_1)), (b(e_2), t(e_2)) \in A_v$ both lead away from S_{uc}.   □

For vertical unconstrained chains $S_{uc} \subseteq S_v$, an analogous lemma can be proved.

3.2 Proof of Equivalence

In this subsection, we present the proof by Esser [14] stating that the turn-regularity approach and the complete-extension approach are equivalent.

Lemma 6. *Let f be a non-turn-regular face. Let (c_1, c_2) be pair of kitty corners in f with associated vertices u_1 and u_2. Neither the horizontal nor the vertical segments incident to u_1 and u_2 are separated.*

Proof. As f is non-turn-regular, there exists at least one pair of kitty corners (c_1, c_2) in f with $\mathrm{rot}(c_1, c_2) = 2$. Denote the associated vertices by u_1, u_2. Let s_1^h, s_1^v be the horizontal and vertical segment incident to u_1, and let s_2^h, s_2^v analogously be the incident segments to u_2. This setting is illustrated in Fig. 9.

Let $\mathrm{first}_p(s)$ and $\mathrm{last}_p(s)$ be the first and last vertex, respectively, of a segment s on a path p. Consider the vertical segments s_1^v, s_2^v and the path $p = u_1 \xrightarrow{*} \mathrm{last}_p(s_1^v) \xrightarrow{*} \mathrm{first}_p(s_2^h) \xrightarrow{*} u_2$. In Fig. 9, p is the path along the bottom way. If f is an internal face, it holds:

$$\mathrm{rot}(u_1, u_2) \quad = \quad 2 \quad = \quad -1 + \underbrace{\mathrm{rot}(\mathrm{last}_p(s_1^v), u_2)}_{=3} \tag{9}$$

As the rotation along the subpath $\mathrm{last}_p(s_1^v) \xrightarrow{*} u_2$ is 3, on this path, there must exist at least three convex corners and at least one more other vertical segment $s^\star$. In the constraint graph D_h, the arcs between s_1^v and $s^\star$ and between s_2^v and $s^\star$ must both either lead towards $s^\star$ or both lead away from $s^\star$.

In the example in Fig. 9, both arcs (orange dotted lines) lead towards $s^\star$.

For the sake of completeness, note that the subpath $\mathrm{last}_p(s_1^v) \xrightarrow{*} u_2$ can be even longer and contain more than three vertical segments. Then, there will not be immediate arcs between s_1^v and $s^\star$ or between s_2^v and $s^\star$ but a longer sequence of arcs.

Summarized, as both arcs either lead towards $s^\star$ or both arcs lead away from $s^\star$, path p from s_1^v to s_2^v allows none of the four connections from Definition 5.

Following from Lemma 1, it also holds:

$$\mathrm{rot}(u_2, u_1) \quad = \quad 2 \quad = \quad -1 + \underbrace{\mathrm{rot}(\mathrm{last}_p(s_2^v), u_1)}_{=3} \tag{10}$$

Thereby, the same conclusion can be shown for the other path $q = u_2 \xrightarrow{*} \mathrm{last}_q(s_2^v) \xrightarrow{*} \mathrm{first}_p(s_1^h) \xrightarrow{*} u_1$ from s_2^v to s_1^v. In Fig. 9, q is the path along the upper way.

Thus, s_1^v and s_2^v are not separated.

If f is the external face, the following two equations apply and the same conclusion can be deduced:

$$\mathrm{rot}(u_1, u_2) = \quad 2 = -1 + \mathrm{rot}(\mathrm{last}_p(s_1^v), u_2) \tag{11}$$

$$\mathrm{rot}(u_2, u_1) = -6 = -1 + \mathrm{rot}(\mathrm{last}_p(s_2^v), u_1) \tag{12}$$

When considering the horizontal segments s_1^h, s_2^h, it can be argued in the same way that these segments are not separated. Thus, neither the vertical segments s_1^v, s_2^v nor the horizontal segments s_1^h, s_2^h are separated. $\square$

Theorem 1. *A simple orthogonal representation H is turn-regular if and only if the corresponding shape description is complete or can uniquely be completed.*

Proof. Forward direction:

> **Idea**
> 1. Consider a turn-regular orthogonal representation.
> 2. There exist unique auxiliary graphs H_x, H_y.
> 3. H_x, H_y induce a complete extension.

Let H be turn-regular. We can apply Lemma 4 and conclude that H_x and H_y are uniquely determined. Thus, it needs to be proved that H_x and H_y induce a complete extension.

The horizontal arcs in H_x are all directed in right direction, the vertical arcs in H_y in top direction. Thus, from the vertical segments, we can deduce a set of segment vertices S_v. From the horizontal arcs in H_x, we can deduce a set of horizontal left-to-right arcs A_h connecting these segment vertices. Then, every two segment vertices in S_v are connected by a sequence of arcs from A_h if there is a corresponding path in H_x. Thus, the vertical segments and the horizontal arcs in H_x induce a constraint graph $D_h = (S_v, A_h)$. In the same way, a second constraint graph $D_v = (S_h, A_v)$ can be deduced from the horizontal segments and the vertical arcs in H_y.

Let B_h contain all arcs from A_h and all saturating edges from H_x. Let B_v contain all arcs from A_v and all saturating edges from H_y.

It remains to show that $\tau = \langle (S_v, B_h), (S_h, B_v) \rangle$ is a complete extension of $\sigma = \langle D_h, D_v \rangle$. We will prove that the three properties from Definition 6 are fulfilled.

As A_h and A_v have been augmented by additional arcs, for (P1) it obviously holds $A_h \subseteq B_h$ and $A_v \subseteq B_v$.

Regarding (P2), B_h and B_v are acyclic by construction. A_h contains only left-to-right arcs. The arcs from $B_h \setminus A_h$ do not close any cycles in D_h as they establish new left-to-right relations between previously unconnected vertices, i.e. they retain the left-to-right-order of D_h. The same argument applies to D_v and to the arcs from $B_v \setminus A_v$.

For (P3) we can argue that the saturating edges in H_x and H_y were added at the ends of maximal unconstrained chains.

In a general, not necessarily turn-regular representation, unconstrained chains are exactly those segments for which no unique orthogonal relation to other segments is given [4, Proof of Theorem 3, Lemma 5] – or which, in terms of complete extensions, potentially are not separated.

According to Lemma 5, arcs in A_h and A_v consistently lead towards an unconstrained chain or away from an unconstrained chain. If the arcs all lead towards the unconstrained chain, the saturating edges induce two new outgoing arcs. If the arcs all lead away from the unconstrained chain, the saturating edges induce two new incoming arcs. Thus, due to the saturating edges, one of the four paths from Definition 5 – leading towards the unconstrained chain and then leading away to another segment – can be established. As the saturating edges were added at the ends of all maximal unconstrained chains, all segments are separated.

In summary, we can deduce a complete extension from H_x and H_y. As H_x and H_y are uniquely determined, this deduced complete extension is unique.

Backward direction:

Idea
1. Consider a non-turn-regular face with a pair of kitty corners (c_1, c_2).
2. On any path from c_1 or c_2 or vice versa, there must exist a segment with only incoming or only outgoing arcs within the shape description.
3. The shape description is not complete.
4. The reflex corners c_1, c_2 affect both the horizontal and the vertical constraint graph.
5. There is no unique way to complete the shape description.

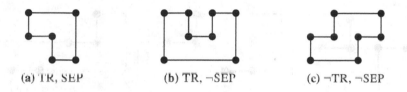

(a) TR, SEP   (b) TR, ¬SEP   (c) ¬TR, ¬SEP

Fig. 10. Possible combinations: A face can be turn-regular (TR) or not (¬TR), the bounding segments can be separated (SEP) or not (¬SEP).

Consider a non-turn-regular face with a pair of kitty corners (c_1, c_2) with associated vertices u_1 and u_2, as illustrated in Fig. 9. From Lemma 6 we know that neither the horizontal nor the vertical segments incident to u_1 and u_2 are separated.

It remains to show that there is no unique way to complete the shape description. As it holds $u_1 = r(s_1^h) = t(s_1^v)$ and $u_2 = \ell(s_2^h) = b(s_2^v)$, c_1 and c_2 affect both the horizontal and the vertical constraint graph. Thus, there are two possible ways to complete the shape description, not only one. As soon as – either in the horizontal or the vertical constraint graph – a path completing the shape description is chosen, the respective segments in the other constraint graph will be separated as well. Thus, the shape description is not complete and a complete extension cannot uniquely be chosen.   □

Note that the equivalence between both approaches is not a one-to-one equivalence. If a face is turn-regular, according to Theorem 1 there can be two cases: The segments bounding this face can either be separated already or there can be a unique way to separate them.

This distinction leads to three possible combinations, which are shown in Fig. 10. Figure 10a shows a turn-regular face; all segments are separated. Figure 10b shows a turn-regular face; the segments are not separated, however, there exists only one way to separate them. In Fig. 10c the face is not turn-regular, the segments are not separated, and there is no unique way to separate them.

4   Runtime Analysis

The proof of Theorem 1 induces an algorithm for transforming the turn-regularity formulation into the complete-extension formulation and vice versa. In this section, we present this algorithm and argue that this transformation can be done in $\mathcal{O}(n)$ time.

4.1   Turn-Regularity Formulation $\rightarrow$ Complete-Extension Formulation

Given an orthogonal representation H, a set of kitty corners in H, and a set of dissecting edges between these kitty corners, the aim is to construct the corresponding complete extension. This complete extension can be deduced, as seen before, from the auxiliary graphs H_x, H_y. We assume that the dissecting edges already have been inserted into H so that H is turn-regular.

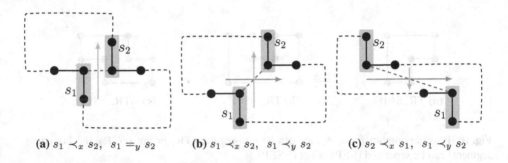

(a) $s_1 \prec_x s_2$, $s_1 =_y s_2$ (b) $s_1 \prec_x s_2$, $s_1 \prec_y s_2$ (c) $s_2 \prec_x s_1$, $s_1 \prec_y s_2$

Fig. 11. Some relative positions of segments. (Color figure online)

The transformation procedure is described below, including the runtime of each single step:

Procedure. Turn-Regularity $\rightarrow$ Complete Extensions.	
Input: turn-regular orthogonal representation H	
Output: corresponding complete extension τ	
1 construct auxiliary graphs G_l, G_r and their saturators;	$\mathcal{O}(n)$
2 construct unique auxiliary graphs H_x, H_y;	$\mathcal{O}(n)$
3 deduce the corresponding complete extension τ;	$\mathcal{O}(n)$
4 **return** τ;	

Runtime. For turn-regular orthogonal representations, the auxiliary graphs G_l, G_r and their saturators can be constructed in $\mathcal{O}(n)$ time [4,8]. The auxiliary graphs H_x and H_y, in this case, are unique (Lemma 4) and can also be constructed in $\mathcal{O}(n)$ time [4, Proof of Theorem 8].

The shape description σ can be constructed in $\mathcal{O}(n)$ time (Lemma 2). To deduce the corresponding complete extension τ, we need to add all saturating edges from H_x and H_y to σ. In a face f bounded by m_f edges, there can appear at most $\mathcal{O}(m_f)$ saturating edges. In planar graphs, it holds $\sum_f m_f \in \mathcal{O}(n)$. Thus, the above procedure for transforming the turn-regularity formulation into the complete-extension formulation can be run in total runtime of $\mathcal{O}(n)$.

Corollary 1. *For a given turn-regular orthogonal representation H, the corresponding complete extension τ can be constructed $\mathcal{O}(n)$ time.*

4.2   Complete-Extension Formulation $\rightarrow$ Turn-Regularity Formulation

When transforming the complete-extension formulation into the turn-regularity formulation, we assume that the original orthogonal representation H, the shape description σ, and one fixed complete extension τ are given, i.e. we know the relative position of every pair of segments. The aim is to express these relative positions in terms of dissecting edges between kitty corners, i.e. to construct a turn-regular orthogonal representation $\tilde{H}$ which meets exactly these relative positions.

As seen in Fig. 10, there can appear three possible cases for faces being turn-regular and the bounding segments being separated. The first two cases are unproblematic; we only need to consider the case illustrated in Fig. 10c, in which kitty corners appear and a dissecting edge needs to be inserted.

In a first step, we determine all non-separated segments in σ by simply comparing σ with its complete extension τ.

For all faces f in H, we determine all pairs of kitty corners (c_1, c_2) in f. Here we can apply the negation of Lemma 6. A pair of kitty corners can only appear if neither the incident horizontal segments nor the incident vertical segments are separated. We can use the previously determined set of non-separated segments to determine all pairs of kitty corners.

Between each pair of kitty corners, we insert a dissecting edge – in a way that exactly keeps the relative positions given in τ. As seen in Sect. 2, a dissecting edge just represents an additional constraint in the minimum cost flow problem. To allow all possible relative positions of segments, we might add constraints to both the horizontal network N_h and the vertical network N_v.

A feasible solution of this more restricted minimum cost flow problem, with additional constraints in both networks, always is a feasible solution of the less restricted minimum cost flow problem, with randomly ordered dissecting edges and only one additional constraint either in the horizontal or the vertical network.

Adding constraints to both the horizontal and the vertical network allows all possible relative positions of segments. Figure 11 shows some relative positions, the dissecting edges (dashed blue lines) and the corresponding arcs from the horizontal (green) network N_h and the vertical (red) network N_v.

The notation in Fig. 11 is inspired by the notation for orthogonal relations in [4]: $\prec_x$ denotes a distance in x-direction, e.g. $s_1 \prec_x s_2$ means that segment s_1 is left of segment s_2. Analogously, $\prec_y$ denotes some distance in y-direction and $=_x$, $=_y$ denote an equality in the respective direction. Figure 11b and c show two drawings in which constraints were added to both the horizontal and the vertical network. The red and green arcs represent exactly these additional constraints. The corresponding dissecting edge then can be imagined as a "diagonal" edge.

The transformation procedure is shown below:

Procedure. Complete Extensions → Turn-Regularity.

Input: orthogonal representation H,
shape description σ,
complete extension τ
Output: corresponding turn-regular orthogonal representation $\tilde{H}$

1	determine all non-separated pairs of segments in σ;	$\mathcal{O}(n)$
2	**for** $f \in \{\text{faces}\}$ **do**	$\mathcal{O}(n)$
3	determine all pairs of kitty corners in f;	$\mathcal{O}(m_f)$
4	**for** $(c_1, c_2) \in \{\text{pairs of kitty corners in } f\}$ **do**	$\mathcal{O}(m_f)$
5	insert a dissecting edge between c_1 and c_2;	$\mathcal{O}(1)$
6	**end**	
7	**end**	
8	**return** constructed turn-regular orthogonal representation $\tilde{H}$;	

Runtime. To determine all non-separated segments in σ, we can compare σ with the given complete extension τ. The constraint graphs D_h, D_v are upward planar (Lemma 3). Thus, we can iterative over all arcs in D_h, D_v and find all non-separated segments in $\mathcal{O}(n)$ time.

As G is a planar 4-graph, we can iterate over all faces in $\mathcal{O}(n)$ time. Determining all pairs of kitty corners in a given face f takes $\mathcal{O}(m_f)$ time, where m_f denotes the number of edges in f. In planar graphs, it holds $\sum_f m_f \in \mathcal{O}(n)$.

Regarding the dissecting edges, we know that in f there can appear at most $\mathcal{O}(m_f)$ pairs of kitty corners. Inserting a dissecting edge takes constant time. Thus, we can insert all dissecting edges in f in $\mathcal{O}(m_f)$ time.

In summary, the complete-extension formulation can be transformed into the turn-regularity formulation in $\mathcal{O}(n)$ time.

Corollary 2. *For a given orthogonal representation H, a given shape description σ, and a given complete extension τ, the corresponding turn-regular orthogonal representation $\tilde{H}$ can be computed in $\mathcal{O}(n)$ time.*

5   Summary

We presented the theorem and proof from [14] stating that two well-known compaction methods, the turn-regularity approach by Bridgeman et al. [4] and the complete-extension approach by Klau and Mutzel [23], are equivalent.

In both methods, one has to make *local* decisions, which, however, affect the *global* objective function, i.e. the total edge length or area of the drawing: In the turn-regularity approach, one has to decide how to align new artificial edges in non-turn-regular faces. In the complete-extension approach, one has to decide how to separate the segments bounding each face. The theorem means that both decisions are equivalent.

However, this equivalence is not a one-to-one equivalence. If a face is turn-regular, two cases can occur: The segments bounding this face either can be separated already or there can be a unique way to separate them. We illustrated all possible combinations.

In different use cases, for different types of graphs, the one or the other approach might be more appropriate. An orthogonal representation can, for instance, contain many non-separated segments but few kitty corners. We intend to discuss this practical perspective in future work.

The equivalence, furthermore, induces an algorithm for converting one problem formulation into the other. We presented a linear-time algorithm for transforming the turn-regularity formulation into the complete-extension formulation and vice versa.

Acknowledgements. I would like to thank Prof. Michael Jünger, University of Cologne, and his amazing team, which committed itself to Graph Drawing, enthused me with this area of research and always supported me with explanations, discussions, and suggestions: Christiane Spisla, Martin Gronemann, Sven Mallach, Francesco Mambelli, Daniel Schmidt.

My special thanks go to Joachim Köhler and all my colleagues at Fraunhofer IAIS, who support me advancing my research on Graph Drawing and bringing it together with projects from the area of document processing and table recognition.

References

1. Ahuja, R., Magnanti, T., Orlin, J.: Network Flows: Theory, Algorithms, and Applications. Prentice Hall, Englewood Cliffs (1993)
2. Batini, C., Nardelli, E., Tamassia, R.: A layout algorithm for data flow diagrams. IEEE Trans. Software Eng. **12**(4), 538–546 (1986). https://doi.org/10.1109/TSE.1986.6312901
3. Batini, C., Talamo, M., Tamassia, R.: Computer aided layout of entity relationship diagrams. J. Syst. Softw. **4**(2–3), 163–173 (1984). https://doi.org/10.1016/0164-1212(84)90006-2
4. Bridgeman, S., Di Battista, G., Didimo, W., Liotta, W., Tamassia, R., Vismara, L.: Turn-regularity and optimal area drawings of orthogonal representations. Comput. Geom. **16**(1), 53–93 (2000). https://doi.org/10.1016/S0925-7721(99)00054-1
5. Chimani, M., Mutzel, P., Bomze, I.: A new approach to exact crossing minimization. In: Halperin, D., Mehlhorn, K. (eds.) ESA 2008. LNCS, vol. 5193, pp. 284–296. Springer, Heidelberg (2008). https://doi.org/10.1007/978-3-540-87744-8_24
6. Di Battista, G., Eades, P., Tamassia, R., Tollis, I.: Graph Drawing: Algorithms for the Visualization of Graphs. Prentice Hall, Upper Saddle River (1999)
7. Di Battista, G., Garg, A., Liotta, G.: An experimental comparison of three graph drawing algorithms. In: SCG 1995: Proceedings of the 11th Annual Symposium on Computational Geometry, pp. 306–315. ACM (1995). https://doi.org/10.1145/220279.220312
8. Di Battista, G., Liotta, G.: Upward planarity checking: "faces are more than polygons". In: Whitesides, S.H. (ed.) GD 1998. LNCS, vol. 1547, pp. 72–86. Springer, Heidelberg (1998). https://doi.org/10.1007/3-540-37623-2_6
9. Eiglsperger, M.: Automatic layout of UML class diagrams: a topology-shape-metrics approach. Ph.D. thesis, University of Tübingen, Tübingen, Germany (2003)
10. Eiglsperger, M., Fekete, S.P., Klau, G.W.: Orthogonal graph drawing. In: Kaufmann, M., Wagner, D. (eds.) Drawing Graphs. LNCS, vol. 2025, pp. 121–171. Springer, Heidelberg (2001). https://doi.org/10.1007/3-540-44969-8_6

11. Eiglsperger, M., Kaufmann, M.: Fast compaction for orthogonal drawings with vertices of prescribed size. In: Mutzel, P., Jünger, M., Leipert, S. (eds.) GD 2001. LNCS, vol. 2265, pp. 124–138. Springer, Heidelberg (2002). https://doi.org/10.1007/3-540-45848-4_11

12. Eiglsperger, M., Kaufmann, M., Siebenhaller, M.: A topology-shape-metrics approach for the automatic layout of UML class diagrams. In: SOFTVIS 2003: Proceedings of the 2003 ACM Symposium on Software Visualization, pp. 189–198. ACM (2003). https://doi.org/10.1145/774833.774860

13. Esser, A.M.: Kompaktierung orthogonaler Zeichnungen. Entwicklung und Analyse eines IP-basierten Algorithmus. Master's thesis, University of Cologne, Cologne, Germany (2014)

14. Esser, A.M.: Equivalence of turn-regularity and complete extensions. In: VISIGRAPP/IVAPP 2019: Proceedings of the 14th International Joint Conference on Computer Vision, Imaging and Computer Graphics Theory and Applications, pp. 39–47. INSTICC, SciTePress (2019). https://doi.org/10.5220/0007353500390047

15. Garey, M., Johnson, D.: Crossing number is NP-complete. SIAM J. Algebr. Discrete Methods **4**(3), 312–316 (1983). https://doi.org/10.1137/0604033

16. Garg, A., Tamassia, R.: On the computational complexity of upward and rectilinear planarity testing. SIAM J. Comput. **31**(2), 601–625 (2002). https://doi.org/10.1137/s0097539794277123

17. Hopcroft, J., Tarjan, R.: Efficient planarity testing. J. ACM (JACM) **21**(4), 549–568 (1974). https://doi.org/10.1145/321850.321852

18. Jünger, M., Mutzel, P., Spisla, C.: Orthogonal compaction using additional bends. In: VISIGRAPP/IVAPP 2018: Proceedings of the 13th International Joint Conference on Computer Vision, Imaging and Computer Graphics Theory and Applications, pp. 144–155 (2018). https://doi.org/10.5220/0006713301440155

19. Kaufmann, M., Wagner, D. (eds.): Drawing Graphs: Methods and Models. LNCS, vol. 2025. Springer, Heidelberg (2001). https://doi.org/10.1007/3-540-44969-8

20. Király, Z., Kovács, P.: Efficient implementations of minimum-cost flow algorithms. arXiv preprint arXiv:1207.6381 (2012)

21. Klau, G.W.: A combinatorial approach to orthogonal placement problems. Ph.D. thesis, Saarland University, Saarbrücken, Germany (2001). https://doi.org/10.1007/978-3-642-55537-4_4

22. Klau, G.W., Klein, K., Mutzel, P.: An experimental comparison of orthogonal compaction algorithms. In: Marks, J. (ed.) GD 2000. LNCS, vol. 1984, pp. 37–51. Springer, Heidelberg (2001). https://doi.org/10.1007/3-540-44541-2_5

23. Klau, G.W., Mutzel, P.: Optimal compaction of orthogonal grid drawings (extended abstract). In: Cornuéjols, G., Burkard, R.E., Woeginger, G.J. (eds.) IPCO 1999. LNCS, vol. 1610, pp. 304–319. Springer, Heidelberg (1999). https://doi.org/10.1007/3-540-48777-8_23

24. Lengauer, T.: Combinatorial Algorithms for Integrated Circuit Layout. Wiley, Hoboken (1990). https://doi.org/10.1007/978-3-322-92106-2

25. Patrignani, M.: On the complexity of orthogonal compaction. Comput. Geom. Theory Appl. **19**(1), 47–67 (2001). https://doi.org/10.1016/S0925-7721(01)00010-4

26. Soukup, J.: Circuit layout. Proc. IEEE **69**(10), 1281–1304 (1981). https://doi.org/10.1109/proc.1981.12167

27. Tamassia, R.: On embedding a graph in the grid with the minimum number of bends. SIAM J. Comput. **16**(3), 421–444 (1987). https://doi.org/10.1137/0216030

28. Tamassia, R. (ed.): Handbook on Graph Drawing and Visualization. Chapman and Hall/CRC, Boca Raton (2013). https://doi.org/10.1201/b15385

29. Tamassia, R., Di Battista, G., Batini, C.: Automatic graph drawing and readability of diagrams. IEEE Trans. Syst. Man Cybern. **18**(1), 61–79 (1988). https://doi.org/10.1109/21.87055

A Model for the Progressive Visualization of Multidimensional Data Structure

Elio Ventocilla[1]([⊠]) and Maria Riveiro[1,2]

[1] Department of Computer and Systems Sciences, School of Informatics,
University of Skövde, Skövde, Sweden
`elio.ventocilla@his.se`
[2] Department of Computer Science and Informatics, School of Engineering,
Jönköping University, Jönköping, Sweden
`maria.riveiro@ju.se`

Abstract. This paper presents a model for the progressive visualization and exploration of the structure of large datasets. That is, an abstraction on different components and relations which provide means for constructing a visual representation of a dataset's structure, with continuous system feedback and enabled user interactions for computational steering, in spite of size. In this context, the structure of a dataset is regarded as the distance or neighborhood relationships among its data points. Size, on the other hand, is defined in terms of the number of data points. To prove the validity of the model, a proof-of-concept was developed as a Visual Analytics library for Apache Zeppelin and Apache Spark. Moreover, nine user studies where carried in order to assess the usability of the library. The results from the user studies show that the library is useful for visualizing and understanding the emerging cluster patterns, for identifying relevant features, and for estimating the number of clusters k.

Keywords: Data structure · Progressive visualization · Large data analysis · Multidimensional projection · Multidimensional data · Growing neural gas · Visual analytics · Exploratory data analysis

1 Introduction

The analysis of data is a complex task, that may include data cleaning, data transformation, data visualization, data modelling, etc. The first steps analyzing a dataset will depend on our goals and intentions, for example, whether to confirm some hypothesis or to explore and understand. The former case is normally regarded as confirmatory data analysis (CDA), whereas the latter as exploratory data analysis (EDA) [31]. EDA is "a detective work" for finding and uncovering patterns [31]; it is a process engaged by users "without heavy dependence on preconceived assumptions and models" about the data [12]. EDA forms the bases on which CDA is performed, for there is little to be confirmed unless some initial understanding of the data is gained [31].

© Springer Nature Switzerland AG 2020
A. P. Cláudio et al. (Eds.): VISIGRAPP 2019, CCIS 1182, pp. 203–226, 2020.
https://doi.org/10.1007/978-3-030-41590-7_9

A common strategy for analyzing a dataset is to build an overview of it, through, for example, a visual representation on how the data looks like in the multidimensional space [35]. Such an overview can be seen as a representation of the structure of the data and, thus, a representation of the cluster patterns or groups that naturally occur in it.

The structure of a dataset can be defined as the "geometric relationships among subsets of the data vectors in the L-space" [25], where vectors regard instances or data points, and L their dimensionality. In other words, the structure of a dataset can be given in terms of neighborhood relationships between its data points. Based on this definition, we consider that a visualization of the structure of the data is one which visually encodes such relationships.

Traditional visualization methods such as scatterplots and scatterplot matrices can produce a view of a dataset's structure without needing too much data pre-processing. However, when used in isolation, the effectiveness of these methods will be limited by the size and the dimensionality of a dataset [22,29].

Machine learning (ML) techniques can generalize the complexity of the data by employing dimensionality reduction (DR), density estimation and clustering methods. DR algorithms such as principal component analysis (PCA) [14] and the t-Distributed Stochastic Neighbor Embedding (t-SNE) [19], map the data from a multidimensional space into a two-dimensional one, allowing, thus, the use of, e.g., traditional scatter plots, even for multidimensional data. Of these two, t-SNE, has been proven to be especially powerful for visualizing the structure of multidimensional data.

Hierarchical clustering (HC) (e.g., single linkage, complete, Ward), and OPTICS [2] (density-based clustering) algorithms also provide means for visualizing the structure of multidimensional datasets. HC algorithms create trees of nested partitions where the root represents the whole dataset, each leaf a data point, and each level in between a partition or cluster. The resulting tree of an HC algorithm can be plotted using dendrograms, a tree-like structure where the joints of its branches are depicted at different heights based on a distance measure. OPTICS has a similar basis to HC algorithms since data points are brought together one by one. However, a key difference is that OPTICS computes distance values and sorts the data points in a way that can later be visualized using a reachability plot, where valleys can be interpreted as clusters and hills as outliers.

DR, HC and OPTICS construct views of neighborhood relationships for every data point in a dataset, thus requiring the computation of all distances between them. This is normally coupled to higher computational costs and time, as well as higher risks of visual clutter when datasets are large (thousands or millions of data points). In addition, the more data points that are plotted, the higher the computational burden on subsequent exploratory interactions, for instance, linked highlighting, zooming and panning.

A way to overcome the size limitation, is to 'compressed' the number of data points into a smaller set of representative units. Two examples of ML methods which provide such compression capabilities, while preserving data structure to the largest possible extent, are the Self-organizing Map (SOM) [17] and the Growing Neural Gas (GNG) [10] – K-means can also be considered as a compression algorithm if the number of k is sufficiently large, for instance, between 20 and a 100. The former two, SOM and the GNG, work in an incremental manner by design, i.e., they grow and adapt with subsets of data points. This implies that they do not need to iterate one or several times of across a whole dataset before providing a partial representation of what they have 'seen' so far. Such a feature enables the progressive construction and visualization of the data structure of large and multidimensional datasets.

Based on the framework using GNG and visual encodings with force-directed graphs (FDG) presented in [35], this paper presents a general model for the progressive visualization of the structure of multidimensional datasets, and an improved proof-of-concept prototype that expands to other incremental algorithms such as SOM.

In particular, we present a model for visualizing the structure of large datasets in an incremental manner and furthermore, we describe an instantiation of such a model, a proof-of-concept prototype. Such instantiation was developed as an off-the-shelf VA library [33] for the Apache Zeppelin[1] notebook system, while also leveraging from the Apache Spark framework[2]. The target users for both the model and the library are mainly VA researchers and VA system developers. These contributions, however, may also be found useful to a wider audience of data scientists, i.e., people "who uses code to analyze data"[3]. Finally, we describe nine users studies with domain users from different academic areas, who fitted well the data scientist profile.

Our concrete contributions are:

- A model for visualizing and exploring the structure of large datasets.
- An open source, proof-of-concept prototype in the form of an off-the-shelf VA library.
- The results of a usability evaluation through nine case studies assessing the usability of the VA library, and its contributions to GNG interpretability and insight gain.

Section 2 describes models and workflows related to progressive visualizations, as well as ML techniques for the incremental computation and visual encoding of the structure of datasets. Section 3 describes the main contribution, the model for the progressive visualization of a dataset's structure. Section 4 introduces Visual GNG, a proof-of-concept of the model in the form a VA library. Section 6 provides a summary of the case studies that showcase the use of the prototype and Visual GNG, and evaluate its usability. Finally, we discuss the results of the evaluations carried out, and the challenges associated with the use of the prototype in Sect. 7 and we finish with conclusions.

---

[1] https://zeppelin.apache.org [Accessed June 2019].
[2] https://spark.apache.org [Accessed June 2019].
[3] Kaggle "The State of Data Science & Machine Learning" 2017 https://www.kaggle. com/surveys/2017 [Accessed June 2019].

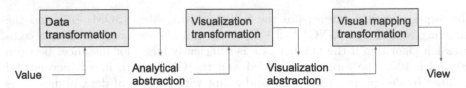

Fig. 1. The Data State Reference Model adapted from [6].

2   Background and Related Work

This section provides an overview of visualization models and data processing workflows closely related to our own. It describes their components and their purpose, and how they contribute and differentiate from our proposed model. Moreover, it gives an overview of two incremental ML techniques which provide means for the visualization of a dataset's structure, SOM and GNG, and which are suitable candidates for developing instantiations of the model.

2.1   Data State Reference Model

Chi et al. [6] presented the data state model (DSRM) as a representation of the general objects and transformations involved in the process of creating a plot. The model is composed of four data states and three transformation operators (see Fig. 1).

The first state is raw data (Value) which may come in different formats (e.g., text, images or column-separated values). Raw data is then transformed by a Data Transformation operator into an *Analytical abstraction*, i.e., into a generalization of the raw data in the form of, e.g., a set of vectors or a graph. A *Visualization Transformation* operator then takes the analytical abstraction and further abstracts the values into visualizable content (a *Visualization abstraction*). The final operator, *Visual mapping transformation*, uses the visualization abstraction in order to produce a graphical view or plot (View).

These stages define a pipeline of data transformation operations which are common to systems in the field of Information Visualization. It is, therefore, also a natural representative for VA systems or visual data analysis processes. Our proposed model can be seen as a specialization of DSRM to the process of visualizing the structure of large datasets in VA. Concretely, our model expands the Data transformation operator in order to describe the role of an incremental ML algorithm, and defines new objects through which computational status can be reported, and steering can be achieved.

2.2   Progressive and Incremental Visual Analytics

Stolper et al. [30] proposed a workflow (see Fig. 2) for developing progressive VA solutions that would provide users with early feedback in long-lasting computations (i.e., those which could take hours or days). Early and continuous feedback

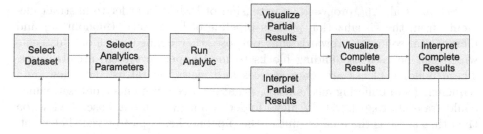

Fig. 2. The Progressive Visual Analytics Workflow adapted from [30].

is achieved via a computational loop that produces partial results: a progressive algorithm (in the *Run Analytic* stage) produces early results based on a given dataset and parameter set; the partial results are shown to the user (in the *Visualize Partial Results* stage) for he/she to analyze (represented by *Interpret Partial Results*). This cycle is repeated until the algorithm converges to a final result. Their workflow also depicts an outer cycle which is triggered by user interactions which steer the computation. The workflow provides a good overview of different analytical stages. It misses, however, details about the technical elements involved in the data transformation process; details which can facilitate the implementation of a VA system capable of enabling such analytical workflow to take place.

Fekete et al. [9] brought forward a mathematical approach for describing what they call the Progressive Analytics paradigm. Given a state S, a set of parameters P and a dataset D, a non-progressive computation can be described in terms of the following function F:

$$F(S, P, D) \rightarrow (S', R, t, m)$$

where S' defines a final state, R a result or final output, a duration time t, and memory used m. A progressive computation, on the other hand, produces partial results and states, based on a given time constraint q:

$$F_p(S_1, q, P, D_1) \rightarrow (S_2, R_1, t_1, m_1)$$

$$...$$

$$F_p(S_i, q, P, D_n) \rightarrow (S_{i+1}, R_i, t_i, m_i)$$

where i represents the number of iterations or times F_p is executed – the parameter q should be set to no longer than 10 seconds in order to keep the user within the analysis 'dialog'.

Similarly to Stolper et al., the mathematical representation lacks details about different data transformation components, as well as object states, which take part across the visualization construction pipeline. Our proposed model takes a step deeper into those details.

Schulz et al. [26], proposed an extension of DSRM in order to abstract elements from the visualization process pipeline of both static (monolithic) and incremental solutions. Towards that purpose, they defined a set of constructs which can be used as building blocks to model specific system architectures. These building blocks are: data sources and views, operators, data flow, and sequencing and buffering mechanisms. A data source represents a dataset, which could translate to, e.g., a CSV file, a folder of images or a database. A view, on the other hand, is the final output of the pipeline, i.e., a plot or visual element. Operators can be seen as functions which work on some input in order to produce an output, and whose behavior is defined by a set of parameters; operators not only transform the input, but they also report progress and/or quality metrics. Data flow represents whether data is passing from one operator to another as either a single block or in parts (batches). Finally, sequencers and buffers are elements which change the flow of data. The former splits data into batches (or finer batches), while the latter accumulates batches into larger batches.

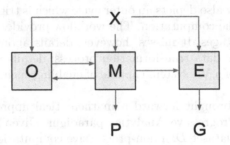

Fig. 3. Taxonomy for interactive and interpretable ML adapted from [34].

2.3 Interactive and Interpretable ML

A taxonomy was proposed in [34] in order to understand and classify how ML algorithms are made interactive and interpretable by the research community (see Fig. 3).

In it, three objects (X, P and G) and three transformations or functions (O, M and E) are represented. Object X stands for a dataset, and its role depends on the transformation it is aimed for, i.e., training, when it is used during optimization (O); prediction, when a model (M) is applied to it; and testing, when is used in the evaluation (E) of a model. P stands for a prediction, and G for goodness (of the model). The learning process of an algorithm is represented by the loop between O and M, i.e., by the continuous optimization of a model. Each of the elements of the taxonomy can be potentially enabled for user exploration, interpretation and/or interaction.

Our proposed model builds upon the logic and constructs of this taxonomy, as well as those of the previously presented models and workflows, for the purpose of visualizing datasets' structure in a progressive manner.

2.4 Unsupervised Incremental Techniques

Incremental or online learning describes a set of algorithms capable of generalizing or modeling sequential instances, or batches of instances, as they 'arrive' [3]. In that sense, an incremental learning algorithm does not require access to the whole data set in order to display some learning. This type of learning can occur both in a supervised or an unsupervised setting. We hereby describe two unsupervised incremental learning algorithms which provide means to visualize data structure progressively: SOM and GNG.

SOM. The SOM algorithm starts with an n by n set of inter-connected units, each with a model vector (or prototype) with random values, and a dimensionality L equal to that of the training dataset. The algorithm takes one input signal (or data point) at a time from the training dataset. The network of units then adapts to the signal by moving the closest unit – as given by the Euclidean distance between prototype and signal – and its topological neighbors towards the signal. As more signals are fed, the better the network will come to represent the input space by placing the units in densely populated areas.

Fig. 4. SOM applied to the Breast Cancer Wisconsin (diagnostic) dataset, visually encoded using U-Matrices. From left to right, the images represent the state of the SOM model after 1, 10, 50 and a 100 passes over the whole dataset.

The resulting network from the SOM algorithm can be plotted using a Unified Distance Matrix (U-Matrix) [32], which is a two-dimensional arrangement of color-scaled hexabins. The opacity or color of each bin can be shifted based on the distance value between model vectors. In a gray-scaled U-Matrix, the white regions can be seen as clouds of data points, whereas the dark regions as empty spaces (see Fig. 4).

GNG. GNG is described as an incremental neural network that learns topologies [10]. It constructs networks of nodes and edges as a means to describe the distribution of data points in the multidimensional space. Unlike SOM, neither the units nor the connections between units are static nor predefined (the units, however, can be limited to a specific number).

The algorithm starts with two connected neurons (units), each with a randomly initialized reference vector (prototype). Samples (signals) from the data are taken and fed, one by one, to the algorithm. For each input signal the topology adapts by moving the closest unit, and its direct neighbors, towards the signal itself. Closeness, in this case, is given by the Euclidean distance between a unit's prototype and the signal. New units are added to the topology every λ signals, in between the two neighboring units with the highest accumulated error. A unit's error corresponds to the squared distance between its prototype and previous signals. Edges (i.e., connections between units), on the other hand, are added between the two closest units to an input signal, at every iteration, and are removed when their age surpasses a threshold. Edges are aged in a linear manner at every iteration, and are only reset when they already connect the two closest units to a signal. It is also possible to remove 'obsolete' units (i.e., those which might no longer represent a populated area in the data space) by keeping a utility value as described in [11].

The topological model constructed by a GNG can be visually encoded using FDGs (see Fig. 5) as proposed in [35], where units become nodes, and connections between units edges. Other values such as units' density (i.e., number of data points they represent) and a feature's value, can be visually encoded through node size and color. Figure 5 is an example where the size of the nodes encodes the number of data points represented by each unit.

There are other incremental clustering algorithms such as Mini-batch K-Means [27], which are capable of compressing the number of data points in a dataset into a smaller number of representative units. Using this algorithm with, e.g., 100 centroids, would have a compression effect similar to that of SOM and GNG. The set of learned centroids, however, do not readily lend themselves for visually encoding datasets with a dimensionality higher than three.

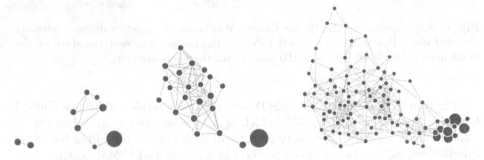

Fig. 5. GNG applied to the Breast Cancer Wisconsin (diagnostic) dataset, visually encoded with force-directed graphs. From left to right, the images represent the state of the GNG model after 1, 500, 2000 and 10000 sample signals.

3   The Model for Progressive Visualization of Structure

Figure 6 depicts our proposed model for the progressive visualization of data structure (PVSM). It is divided into three main, non-shaded, areas: Parameters, Transformations and Objects. Parameters represent sets of named values which define the behavior of different Transformation elements (e.g., the behaviour of an ML algorithm). Transformations, on the other hand, represent functions which perform some structural change on a given Object by means of abstraction; they take a set of parameters and an input object, and produce a new object. Finally, Objects are seen as data structures, e.g., vector sets, graphs or a composition of collections and values.

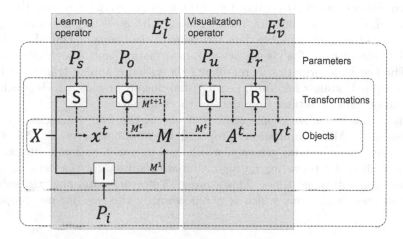

Fig. 6. The Progressive Visualization of Structure Model (PVSM).

Arrows represent flow. Incoming arrows to transformation elements represent input, while outgoing arrows represent output. Solid arrows represent a one-time input or output, while dash arrows a sequential flow of input or output.

The two shaded areas represent two types of operators or processes, one in charge of learning and the other of visualization. These operators are defined by a set of parameters, transformations and objects, which contribute to either learning (i.e., optimizing/fitting a model to a dataset) or visualizing (i.e., visually encoding or plotting data). Additionally, each has an element E representing the status of that process (e.g., number of samples, current iteration, model quality progress) at a given time t.

As with the DSRM, the transformation pipeline of our model begins with an X data object. X, however, does not stand for *raw* data, but rather training data. In that sense it is assumed here that any raw data has already been preprocessed – i.e., formatted into a vector set, scaled, null values treated, etc. – thus, made ready for the learning process.

X is the input to two transformations, a sampler element S and a model initializer I. A sampler is a function which produces samples/subsets from X at a time t, and whose behavior is defined by parameters P_s (e.g., seed and sample size). Output samples are represented by x^t. A model initializer is a function which instantiates and returns an initial, non-trained, model M^1 ($t = 1$), from a given training set X, and set of parameters P_i. Both outputs, x^t and M^1, along with a parameter set P_o, are inputs to an optimizer function O, which is responsible for producing a new, optimized model M^{t+1}. The inputs and output of an optimizer are sequential, which means that O is executed every time for each input object at a time t. Note that each newly optimized model also becomes an input to O every new iteration; it is in this manner that incremental learning is achieved. The Learning Operator may be seen as the Data Transformation stage of DSRM.

Each partial model M^t enters the *Visualization operator* realm. Here M^t becomes an input to a transformation U, along with a set of parameters P_u, in order to produce a visualizable data structure A. In terms of DSRM, M is an analytical abstraction, T a visualization transformation, and A a visual abstraction. The last transformation R represents the visual mapping transformation, whose output V stands for View, i.e., a visual output or plot, which concludes the chain of transformations from dataset to visualization.

This model is a close representative of one of the four strategies for user involvement in ML computations proposed by Mühlbacher et al. [20]. Concretely, the data subsetting strategy, which enables early system feedback, and user interaction, by providing an algorithm with subsets of the data in order to reduce computational times. This strategy, however, states that the subsets should increase with time, which is not necessarily the case for our proposed model.

4   Proof-of-Concept Prototype

In this section, we describe Visual GNG, an off-the-shelf VA library[4] for Apache Zeppelin, and a proof-of-concept with an architecture based on the proposed PVSM model. This section is divided into two parts. The first one describes the functionality of the system and its use as an exploratory data analysis tool. The second part describes the library's architecture, as seen through the PVSM model.

4.1   Functionality Overview

Apache Zeppelin is a web-based notebook system for data-driven programming. Zeppelin enables, among other things, writing and executing Scala and SQL code through Apache Spark – a framework for distributed computations in Big Data analysis. In this subsection, we explain how to deploy the library as well as its interactive capabilities.

The off-the-shelf format of the library requires that a dependency to the library is added to the Zeppelin Spark interpreter as described in [33], before it can be deployed. We assume the dependency has been added for the following steps.

To deploy Visual GNG, the user needs to – in a Zeppelin notebook – import the library, create an instance of `VisualGNG` and then call the `display`

---

[4] A version of the library can be found here: https://github.com/eliovr/visualgng.

Fig. 7. Zeppelin paragraph exemplifying the use of the Visual GNG library, and the initial state of the visual elements. Clicking on the cog button (left image) displays the list of parameters (right image) underneath it.

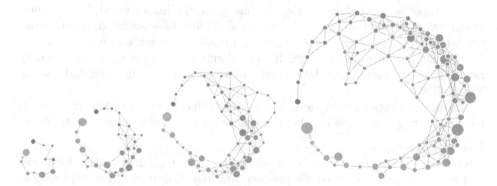

Fig. 8. Evolution of a GNG topology when applied to the E-MTAB-5367 dataset. From left to right, the images represent the state of the topology after 1, 3, 6 and a 10 thousand sample signals. The layout is given by force-directed placement. Distances between node are constraint by the Euclidean distance between units' prototypes. The color coding is based on [28], where orange represent high values, pink low, and gray values in between. Color values, in this case, are given by the `p1508_62CEL` feature values of each unit's prototype.

method (see left image in Fig. 7). Importing the library gives the user access the `VisualGNG` class. Creating an instance of it requires the user to provide the data (X) that will be used in the learning process. In the example given in Fig. 7, the data is represented by variable `data` (a Spark `DataFrame`), while the `VisualGNG` instance is represented by the variable `gng`. We make use of the E-MTAB-5367 dataset on human cells transcriptomics[5] for this example.

---

[5] https://www.ebi.ac.uk/arrayexpress/experiments/E-MTAB-5367/.

Calling the `display` method on the `gng` variable shows the following elements: two execution buttons, *Run* for running and pausing the learning process, and another *Refresh* for restarting it; a cog button which expands a drop-down element (right image in Fig. 7) with access to optimization parameters (P_o); a drop-down list *Color by* for selecting the data feature by which the fill of the nodes are colored (P_u); a status text describing the current state (E_l^t) in terms of iterations, number of edges and nodes; and, finally, two connected nodes, which is the initial state of the GNG model ($M^t, t = 1$).

Clicking on the execution button changes its text to *Pause*, and starts the learning process in the background. Updates to the FDG are made every λ number iterations, i.e., every time a new unit is added to the model. Figure 8 shows the evolution of the GNG model at different iterations steps ($M^{t=\{1..j\}}, j = 10000$) for the E-MTAB-5367 dataset – visual encodings are as suggested in [35]. It is possible to see that there are no drastic differences in the data points making the network split, or cluster nodes into well-defined groups; rather it shows a continuous change in values, with a few 'anomalies' coming out towards the center of the network. These patterns can be further investigated using a Parallel Coordinates (PC) plot (see Fig. 9).

PC is widely used for analyzing multivariate data (a book dedicated entirely to this topic is, for instance, [15]). It has been applied in multiple and very disparate application areas for the analysis of multidimensional data; state-of-the-art reports that review their usage and related research are, e.g., [7,13,16]. Spence [29] recommends using PC for visualizing and inspecting feature values, rather than for visualizing data points as units, which is its intended use in Visual GNG.

In another Zeppelin paragraph, the user can call the `parallelCoordinates()` method of the `gng` variable in other to observe the values of the units' prototypes.

User Interactions

Interactive features are described in two parts: those regarding system feedback and user control on the learning process, and those regarding user exploration for interpretation.

Feedback and Control. System feedback and user control features are described based on Mühlbacher et al.'s [20] types of user involvements (TUI): execution feedback, result feedback, execution control, and result control.

Execution feedback regards information the system gives the user about an ongoing computation. Visual GNG provides two types of execution feedback: aliveness (i.e., whether computations are under progress or something failed) and absolute progress (i.e., "information about the current execution phase"). Aliveness is given by the text in the execution button (*Run, Pause* and *Done*). Absolute progress is given in the status text, through which the user is informed about the number of iterations executed so far.

Result feedback regards information the system gives to the user about intermediate results of an ongoing computation. Visual GNG provides feedback about structure-preserving intermediate results, i.e., partial results that are "structurally equivalent" to the final result. Such feedback is given by updating the FDG when units and connections are added to, or removed from, the model. These updates are reflected by reheating the force-directed placement so that the FDG layout is recomputed, with model updates taken account.

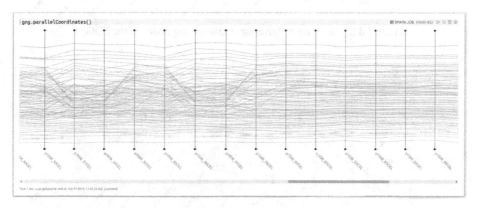

Fig. 9. Parallel coordinates plot of GNG prototypes corresponding to the units of the last iteration in Fig. 8. Stroke color is based on the prototypes' values for attribute p1508_62CEL.

Execution control involves user interactions that have an impact on an ongoing computation. Visual GNG enables this type of control through the execution buttons. By clicking on them, the user can start the optimization, pause, resume or restart it. It is also possible for the user to execute different instances of VisualGNG in different paragraphs within the same notebook, which can prove useful when comparing results for different parameter settings.

Result control defines user interactions that have an impact on the final result. This type of user involvement is often referred to as *steering* [21]. In Visual GNG the user can steer the learning process by changing learning parameters in the *Advanced* drop-down element. Changes to the learning parameters can be done before or during the learning process.

Interaction for Interpretation. Other types of interactive features aim at helping the user explore and interpret the fitted – or partially fitted – topology. Three types of interactive features were implemented for this purpose: linked highlighting, filtering and K-Means clustering.

Linked highlighting (see Fig. 10a) takes place when the user hovers over a node in the FDG. By doing so, its corresponding prototype in the PC plot is highlighted (the width of the line is increased while the opacity of other lines is reduced). Hovering out returns the prototype line to its previous visual encoding. Clicking on

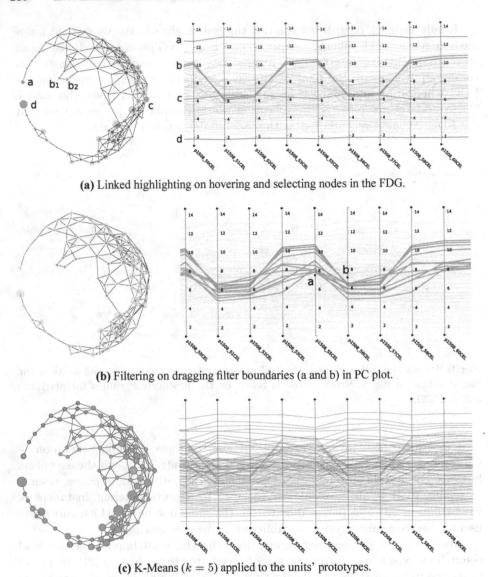

(a) Linked highlighting on hovering and selecting nodes in the FDG.

(b) Filtering on dragging filter boundaries (a and b) in PC plot.

(c) K-Means ($k = 5$) applied to the units' prototypes.

Fig. 10. Linked views and interactions between the FDG and the PC plots.

the node will make the line to remain highlighted even when hovering out, and will also demote all other non-selected nodes in the FDG by decreasing their opacity. This interactive feature aims at facilitating the interpretation of what a prototype represents (i.e., the feature values of the data points it represents), and also how prototypes differentiate from each other. More specifically, it facilitates answering the question of *what* is the general profile of the data points in a given region of the topology, and, *why* do they belong to that region and not to another.

Filtering, on the other hand, is triggered by dragging up and down feature boundaries in the PC plot (Fig. 10b). Dragging feature boundaries will cause prototype lines with values outside the boundaries, as well as their corresponding nodes in the FDG, to be demoted with gray color and lower opacity. Filtering allows users to see which areas in the GNG network have prototypes with a given profile or pattern.

K-Means clustering can be called through the kmeans(k) method of the gng variable. By providing a number of k, the system will run K-Means over the prototypes in the GNG model. In other words, it will group prototypes into k clusters, as given by K-Means. The results are then depicted in the FDG by coloring the stroke of the nodes based on their assigned cluster. Prototypes in PC are also colored based on the corresponding cluster (Fig. 10c). Calling kmeans again with another k value will recolor nodes and lines based on the new cluster assignments. This allows fast testing of different k values with visual feedback on the cluster distributions, regardless of the size and dimensionality of the data.

5 Architecture

This subsection describes the architecture of the library from a PVSM point of view. The inner architecture of the Visual GNG library is based on our proposed model, but with slight adaptations in order for it to co-exists within the Zeppelin ecosystem, as well as to follow some of Spark's semantics.

Zeppelin regards Spark as an interpreter, which can be translated as a code execution engine. Being that Zeppelin provides a web-based interface, it can be split into two modules, frontend and backend, with Spark being part of the latter. Visual GNG is a Scala plug-in library for the Spark interpreter – hence being part of the backend – but with also elements in the frontend in order to provide interactive visualizations. Concretely, all PVSM elements live in the backend with the exception of R, the visual mapping transformation, and its directly related objects and parameters. Backend elements are defined in Scala, with access to Spark libraries, while frontend elements are defined in Javascript, with access to D3.js (a data-driven visualization framework).

Learning Operator. Visual GNG's learning operator elements are as follows. A sampler S is represented by the function takeSample of an RDD[6] collection, which takes two parameters P_s, seed and sample size. The transformation elements O and I are packaged into a GNG class. Within it, O is represented by a function fit (as it is the standard for Spark's ML algorithms), which takes a data sample x^t, a GNG model M (represented by an object of the class GNGModel) and a parameter set P_o (which includes GNG's parameters as defined by Fritzke [10]). The initial model M^1 is instantiated by a function I called modelInitializer, which takes the dataset X (an RDD) as input. These elements are chained together within a learning operator method called learner,

---

[6] Resilient Distributed Dataset [36].

which keeps a learning status object E_l^t updated. Listing 1.1 shows these elements in psudo-code.

Visualization Operator. Visual GNG's visualization operator bridges backend and frontend. On the backend side resides the visualization transformation element U, while the visual mapping transformation R resides in the frontend. There are two instances of U in Visual GNG, one for the FDG plot and another for the PC plot; and the same for R, one instance for each of these.

Listing 1.1. Learning Operator.

```
def learner() {
    E_l = learnerState(iteration=0, training=True)
    P_s = sampleParams(seed=20, size=100)
    P_o = optimizerParams(maxN=100, maxA=50, ..)
    M, O = GNG(X, P_o)

    while(E_l.training) {
        x = X.takeSample(p_s)
        M = O.fit(x, M)
        P_s.seed = P_s.seed + 1
        E_l.iteration = E_l.iteration + 1
        forceDirectedGraph(M) // Visualization Operator.
    }
}
```

Listing 1.2. Visualization Operator (FDG).

```
// Backend.
def forceDirectedGraph(M) {
    P_u = params(minRadius=15, maxRadius=50, ..)
    A = visualTransformation(M, P_u)
    A_s = serializer(A)
    fdgFrontend(A_s)
}
// Frontend.
def fdgFrontend(A_s) {
    P_r = params(width=600, height=600, ..)
    visualMapping(A_s, P_r) // Creates view.
}
```

Both instances of U take the GNG model M and procure transformations that produce a visualizable object A. Examples of these transformations are, in the case of FDG, translating units into nodes with size and color, and links into edges with length and width. The U transformation elements also take the role of serializers, i.e., they convert Scala objects into JSON strings, so they can be pushed to the frontend and be read by the R transformation elements.

As previously mentioned, the logic of the R transformation instances, FDG and PC, is written in Javascript. These logics originally exist as resources in the backend, but are pushed to the frontend where they await for A^t objects in JSON string format. Whenever U pushes a new version of A (i.e., an updated, visualizable GNG model object) both FDG and PC proceed to make the corresponding updates to their views. Listing 1.2 shows the elements of the FDG Visualization Operator in pseudo-code.

System Feedback and User Interaction. So far, we have described how the library's architecture has been defined in terms of PVSM, for the purpose of providing progressive visual feedback of a dataset's structure (i.e., the GNG model). Visual GNG, however, as previously described, also provides *execution feedback* as well as elements for *user control.* PVSM does not model a user's role in the pipeline, but it does model those objects involved in the process. Execution feedback is given by the status object E_l^t of the Learning Operator, which in the case of Visual GNG represents the number of iterations, completion percentage, number of nodes and number of links. Status feedback from the Visualization Operator (E_v^t) was not implemented, but an example of it could be a 'rendering' status message. Control, on the other hand, is enabled by letting users change values in parameter sets and objects. In the case of Visual GNG, execution control is enabled through the status object E_l^t, for pausing and resuming execution, and the parameter values of P_o, with which it is possible to steer the behavior of the GNG algorithm (O).

6 Evaluation

In order to evaluate the proof-of-concept prototype, we carried out nine case studies in different application areas. The analysts that participated in the empirical evaluations fell under the category of data scientists: two engineers, three bioinformaticians, three from the area of bioinformatics, one from biomedicine, two from engineering, one from game research, and two from general data science and machine learning. Regarding their education background, one of the participants was a master student, one a doctoral student, two were post-docs and five principal investigators (PI).

The case studies were structured in four parts: (1) introductory questions regarding user domain, expertise, type of data and analysis tools employed in their daily work; (2) familiarization with the prototype using two examples corresponding to different datasets; (3) analysis of the domain data with the prototype using the thinking-aloud method; (4) and, finally, wrap up questions on user experience during the analysis, i.e., stage (3). These stages were carried out individually and were followed by a semi-structured interview. The interview guide was developed according to the recommendations provided by [23]. Each case study was carried out at each participant's workplace and lasted approximately, one hour. Each participant's workstations was prepared with a Zeppelin server to carry out the experiments. Three notebooks were created for each study: two for the examples used in part (2) and the other for the domain data needed in part (3). During the examples, we used two datasets from the UCI repository [4]: the Breast Cancer Wisconsin (Diagnostic) dataset and the Wine Quality dataset.

To overcome knowledge gaps and inexperience issues related to Zeppelin, the Scala programming language, and Spark, we chose to use notebooks, in parts (2) and (3), with a predefined layout of seven paragraphs. Each paragraph had a code for a specific task: (a) load data, (b) view data, (c) deploy Visual GNG, (d) display PC, (e) run K-means, (f) compute predictions, and (g) see predictions. Like this, we tried to focus, to the largest possible extent, to the usability of the prototype. Hence, users did not need to write code during the sessions, and focused on running the paragraphs instead. However, we intervened in a few cases to help participants removing features from the training dataset, or, e.g., setting a given feature as a label.

The responses gathered during **part (1)** showed that their experience analyzing data varied from 2 to 10 years. All of the participants worked with numerical and structured data in CSV format. Participants from the area of Bioinformatics said to have experience analyzing temporal and categorical data as well, and that their raw data usually came in a platform-specific format, e.g., .CEL files. For the analysis of the data, users reported to employ tools such as R Studio, MATLAB and Excel; two participants stated that they also worked with Python and two others with SPSS. The responses given when they were asked about what they normally looked for in the data were, five participants responded informative features for classification or clustering or classification, five said that they look for groups or clusters, two for outliers, one for anomalies, and four they try to find patterns in time series.

Participants were provided with information regarding the different technologies involved, i.e., Zeppelin, Scala, Spark and GNG, at the beginning of **part (2)**. Thereafter, users were briefed regarding the use of the library using the Wine Quality notebook example. The briefing contained information related to the purpose of each paragraph, how to execute them, how to run Visual GNG, the meaning of the visual encodings, and the supported user interactions. Once the Wine Quality example was finished, they use the prototype to analyze the Breast Cancer notebook by themselves.

Part (3) was the main part of each empirical study. The participants explored their own data with the prototype. Subjects with bioinformatics background employed two publicly available datasets: E-MTAB-5367 on "Transcriptomics during the differentiation of human pluripotent stem cells to hepatocyte-like cells", and GSE52404 on "The effects of chronic cadmium exposure on gene expression in MCF7 breast cancer cells" [7]. The biomedicine participant used data from samples taken from patients with two different conditions. One of the engineers employed machinery maintenance data, while the other one analysed data associated with clutches. The game researcher explored preprocessed data from heart and video recordings of subjects playing games, which aimed at inducing

---

[7] https://www.ncbi.nlm.nih.gov/geo/query/acc.cgi?acc=GSE52404 [Accessed June 2019].

boredom and stress in the analysis. The two data scientist used data from measurements in steel production and results from a topic modeling technique used
over telecommunications related data, respectively. The datasets were given in
.csv format and had mainly numerical values. However, some datasets included
categorical features which were, according to the participants, not relevant for
clustering and, thus, were ignored. The datasets analyzed varied in size, e.g., 46,
16,436, 33,297 up to 70,523 instances, and in the number of dimensions, e.g., 7,
58, 758 and above 2,000 features.

During the analysis, we asked participants to describe their analysis thought
process in a think-aloud manner. Some were reminded to do so, whenever they
forgot, with sporadic questions based on their interactions.

Most participants hovered and clicked on the nodes in order to see their
detailed values in the PC plot. They also selected different features from the
drop-down list to see their values color-coded. Fewer participants re-ran the
algorithm using different parameters (e.g., fewer nodes). We took note, based
on our observations and participants' comments, of three common activities:
the identification of clusters, the identification of outliers, and the assessment of
features' relevance.

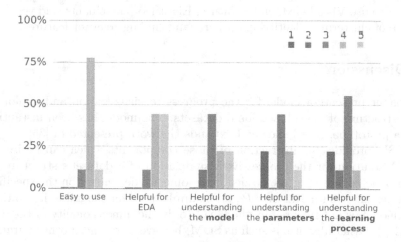

Fig. 11. Answers from nine participants on a 1 to 5 Likert scale about Visual GNG
in terms of usability (*Easy to use, Helpful for EDA*) and interpretability (*helpful for
understanding the model/the parameters/the learning process*). One (1) represents a
negative user perception (not easy, not helpful) and five (5) a positive (very easy, very
helpful). Image taken from [35].

In the final **part (4)**, we carried a structured interview with questions about
the usability of the prototype (*is the library easy to use? is it helpful for EDA?*),
the interpretability of the algorithm (*is it helpful for understanding the model,
the parameters, or the learning process?*), and whether any new insight was
drawn from the analysis (*were you able to gain new understanding of the data?*).

A 1 to 5 Likert scale was used for the first two categories of questions, and a yes or no answer for the latter. Comments in all questions were encouraged.

The results for the first two categories of questions are shown in Fig. 11. Eight (8) participants had a positive stand in regards to the two usability questions, giving ratings of 4 and 5, and one (1) participant gave a neutral rating (i.e., 3). Answers to interpretability questions were more on the neutral and negative side: five (5) participants rated 3 or less to how helpful it was for understanding the model, six (6) provided the same score to how helpful it was for understanding the parameters, and eight (8) – also a rating of 3 or less – for understanding the learning process.

To the insight-related question, four (4) participants answered yes and five (5) no. Forms of insight, expressed by the former set of participants, were: new relevant features which could be used for classification or prediction in further analysis; the clusters naturally formed in the dataset; and awareness on the variation of a set of data points (genes). Explanations provided by the latter set of participants were as follows: three (3) stated to need more time for analysis; one (1) stated to have confirmed previous insight (not new); and one (1) stated that the data chosen was not appropriate for the clustering task.

Finally, as to conclude the interview, we asked participants if they would consider to use Visual GNG in the future. Eight (8) participants said yes for the purpose of clustering, identifying outliers, and finding relevant features.

7   Discussion

This paper presents a model for the progressive visualization and exploration of the structure of multidimensional datasets. The model has been instantiated using a prototype, that builds and expands the work presented in [35].

PVSM provides an abstraction of those components which can be involved in a VA solution for the progressive visualization of a dataset's structure. Not all components, however, are needed, and some are dependent on the specifics of the implementation. Parameter P_i, for example, used for the initial instantiation of the model, is not required by the GNG (only the dimensionality of the datset is needed). Other algorithms such as SOM, however, does need other parameter specifications such as the number of units and the topological structure (hexagonal or rectangular). In this sense, the constructs of the model are there to guide rather than to constrain systems to a specific architecture.

A VA solution can enable user involvement (before and during the learning process) through any of the *Parameters* defined in the PVSM model by, e.g., changing the number of samples or seed value in P_s; setting the strategy (random or sampled) for initializing the values in the prototypes in P_i; updating the parameter values (e.g., maximum number of nodes, learning step, etc.) of the optimization through P_o; defining visual cues (e.g., maximum and minimum node radius and link distance) in P_u; and defining certain features and behaviours (e.g., width, height, or pulling and repulsive forces in the case of FDG) of a plot with P_r.

It is also possible to further extend user involvement through the Objects. A common example is that of updating the model in order to embed users' domain knowledge. Such involvement, however, is seldom enabled through the model M itself (which often is in the form of one or more data collections such as vectors and matrices), for it would require the user to have a certain degree of proficiency in programming and ML. The common strategy in VA is to allow users to modify the view V – by, e.g., adding or removing links between nodes in the FDG – and then synchronize such changes onto to the model M.

The incremental nature of the model may constrain its applicability to online ML algorithms only. Such constraint, however, is not necessarily a very restrictive one. Mühlbacher et al. [20] state that subsetting strategies such as this one, apply to most "unsupervised statistical learning" algorithms. These may include clustering methods such as K-Means, dimensionality reduction techniques such as PCA, or statistical moments such as mean and variance. What may bring challenges, however, are the types of visual outputs which can be constructed in a progressive manner, without disrupting a user's perception of progress.

The results from the case studies show that Visual GNG helps users in the exploratory analysis of large and multidimensional datasets. Participants answered positively (with a rating of four or five) to most questions, agreeing that the prototype was helpful for EDA, was easy to use, and that they would use it again. Neutral or negative answers (with a rating of three or less) were given to questions regarding the understanding of the model, the training parameters, and the learning process. In this regard, one of the participants commented that he would need a deeper understanding of the algorithm itself in order to understand the parameters it uses, and the model it produces. This is a reasonable argument, and one which Visual GNG alone cannot hope counter. The visual elements of the prototype were not aimed at encoding information about the nine iterative steps of the GNG algorithm.

Two participants of the case studies mentioned that understanding the learning process might not always be of interest. One commented that the focus during the analysis task was on "what's happening to my data", and not on process-related questions. When, then, is it useful to provide details on how the learning process works? The answer to this question probably depends on the expertise of the target audience and, at the same time, on the level of trust they have on the algorithm. If a user trusts the output of an algorithm, there is arguably little reason to dig into the details of the optimization process. These (interpretability and trust) are currently topics of discussion found in the research community, e.g., [1,8,18,24].

There are two considerations worth noting about GNG. The first regards categorical data. GNG uses Euclidean distance as a similarity measure (in order to determine the winner unit), and also as means for adaptation (when a unit moves closer to the signal). Another similarity measure could be used, such as Jaccard, but the adaptation problem remains. One suggestion for overcoming this limitation is the use of one-hot encoding, where each value of a categorical feature is turned into its own feature, with 1 or 0 values. The second consideration is GNG's sensitivity to its parameters. Changes in the maximum age value that units' connections can take, for example, can produce a significantly different topology. Tuning the GNG's parameters, as in the case of other algorithms, is an open challenge.

8   Conclusions

Tasks carried out during EDA are normally interactive, where analysts employ various support methods and tools to fuse, transform, visualize and communicate findings. A common strategy to start analyzing a dataset is to use unsupervised ML and visualization methods to reveal its structure. In order to support these tasks, this paper presents, first, a model for the progressive visualization of multidimensional datasets' structure and second, a proof-of-concept prototype that instantiates such model.

The proposed PSV model outlines technical details in terms of objects, parameters and transformations. Its constructs and logic builds upon those from previous work, i,e., the DSR model [5], the Progressive Visual Analytics workflow [30], the Progressive Analytics paradigm [9], Schulz et al.'s DSR extensions [26], and the taxonomy for interpretable and interactive ML [34]. The Visual GNG prototype stands as a proof to the PVS model's abstraction power and completeness, while the case studies sustain the usability of such VA solutions.

The qualitative feedback from expert data scientists indicates that Visual GNG is easy to use and useful for EDA, with eight out of nine participants showed interest in using the library again for the purpose of finding groups and relevant features. These are subjective results with limited external validity. Carrying studies on general (non-domain specific) EDA, with considerably more external validity, remains a challenge in terms of tasks and metrics, and will be considered in our future work.

In the future, we also aim at studying ways for enabling datasets' structure progressive visualization of with other techniques such as HC and dendrograms, and OPTICS and reachability plots. These can provide complementary perspectives of a dataset's structure during EDA.

References

1. Andrienko, N., et al.: Viewing visual analytics as model building. In: Computer Graphics Forum. Wiley Online Library (2018)
2. Ankerst, M., Breunig, M.M., Kriegel, H.P., Sander, J.: Optics: ordering points to identify the clustering structure. In: Proceedings of the 1999 ACM SIGMOD International Conference on Management of Data, pp. 49–60. ACM (1999)
3. Barber, D.: Bayesian Reasoning and Machine Learning. Bayesian Reasoning and Machine Learning. Cambridge University Press, Cambridge (2012)
4. Blake, C., Merz, C.: UCI repository of machine learning databases. University of California, Department of Information and Computer Science, Irvine, CA (1998). https://archive.ics.uci.edu/ml/datasets.html

5. Chi, E.H.: A taxonomy of visualization techniques using the data state reference model. In: IEEE Symposium on Information Visualization (InfoVis), pp. 69–75. IEEE (2000)
6. Chi, E.H.h., Riedl, J.T.: An operator interaction framework for visualization systems. In: Proceedings IEEE Symposium on Information Visualization, pp. 63–70. IEEE (1998)
7. Dasgupta, A., Chen, M., Kosara, R.: Conceptualizing visual uncertainty in parallel coordinates. In: Computer Graphics Forum, vol. 31, pp. 1015–1024. Wiley Online Library (2012)
8. Endert, A., et al.: The state of the art in integrating machine learning into visual analytics. In: Computer Graphics Forum, vol. 36, pp. 458–486. Wiley Online Library (2017)
9. Fekete, J., Primet, R.: Progressive analytics: a computation paradigm for exploratory data analysis. vol. abs/1607.05162 (2016). http://arxiv.org/abs/1607.05162
10. Fritzke, B.: A growing neural gas network learns topologies. In: Advances in Neural Information Processing Systems, vol. 7, pp. 625–632. MIT Press (1995)
11. Fritzke, B.: A self-organizing network that can follow non-stationary distributions. In: Geistner, W., Germond, A., Hasler, M., Nicoud, J.-D. (eds.) ICANN 1997. LNCS, vol. 1327, pp. 613–618. Springer, Heidelberg (1997). https://doi.org/10.1007/BFb0020222
12. Goebel, M., Gruenwald, L.: A survey of data mining and knowledge discovery software tools. SIGKDD Explor. Newsl. 1(1), 20–33 (1999)
13. Heinrich, J., Weiskopf, D.: State-of-the-art of parallel coordinates. In: Eurographics (STARs), pp. 95–116 (2013)
14. Hotelling, H.: Analysis of a complex of statistical variables into principal components. J. Educ. Psychol. 24(6), 417 (1933)
15. Inselberg, A.: Parallel coordinates. In: Liu, L., Özsu, M.T. (eds.) Encyclopedia of Database Systems, pp. 2018–2024. Springer, Boston (2009). https://doi.org/10.1007/978-0-387-39940-9
16. Johansson, J., Forsell, C.: Evaluation of parallel coordinates: overview, categorization and guidelines for future research. IEEE Trans. Visual Comput. Graphics 22(1), 579–588 (2016)
17. Kohonen, T.: Self-Organizing Maps. Springer Series in Information Sciences, 2nd edn. Springer, Heidelberg (1997). https://doi.org/10.1007/978-3-642-97966-8
18. Lu, Y., Garcia, R., Hansen, B., Gleicher, M., Maciejewski, R.: The state-of-the-art in predictive visual analytics. In: Computer Graphics Forum, vol. 36, pp. 539–562. Wiley Online Library (2017)
19. van der Maaten, L., Hinton, G.J.: Visualizing data using t-SNE. Mach. Learn. Res. 9, 2579–2605 (2008)
20. Mühlbacher, T., Piringer, H., Gratzl, S., Sedlmair, M., Streit, M.: Opening the black box: strategies for increased user involvement in existing algorithm implementations. IEEE Trans. Visual Comput. Graphics 20(12), 1643–1652 (2014)
21. Mulder, J.D., van Wijk, J.J., van Liere, R.: A survey of computational steering environments. Future Gener. Comput. Syst. 15(1), 119–129 (1999)
22. Munzner, T.: Visualization Analysis and Design. CRC Press, Boca Raton (2014)
23. Patton, M.Q.: Qualitative Research. Wiley Online Library (2005)
24. Poursabzi-Sangdeh, F., Goldstein, D.G., Hofman, J.M., Vaughan, J.W., Wallach, H.: Manipulating and Measuring Model Interpretability, pp. 1–20 (2018). http://arxiv.org/abs/1802.07810

25. Sammon, J.W.: A nonlinear mapping for data structure analysis. IEEE Trans. Comput. **C−18**(5), 401–409 (1969)
26. Schulz, H.J., Angelini, M., Santucci, G., Schumann, H.: An enhanced visualization process model for incremental visualization. IEEE Trans. Visual Comput. Graphics **22**(7), 1830–1842 (2016)
27. Sculley, D.: Web-scale k-means clustering. In: Proceedings of the 19th International Conference on World Wide Web, pp. 1177–1178. ACM (2010)
28. Spence, I., Efendov, A.: Target detection in scientific visualization. J. Exp. Psychol. Appl. **7**, 13–26 (2001). American Psychological Association
29. Spence, R.: Information Visualization: Design for Interaction, 2nd edn. Prentice-Hall, Inc., Upper Saddle River (2007)
30. Stolper, C.D., Perer, A., Gotz, D.: Progressive visual analytics: user-driven visual exploration of in-progress analytics. IEEE Trans. Visual Comput. Graphics **20**(12), 1653–1662 (2014)
31. Tukey, J.W.: Exploratory Data Analysis, vol. 2. Addison-Wesley Publishing Company, Reading (1977)
32. Ultsch, A.: Kohonen's self organizing feature maps for exploratory data analysis. In: Proceedings of INNC 1990, pp. 305–308 (1990)
33. Ventocilla, E., Riveiro, M.: Visual analytics solutions as 'off-the-shelf' libraries. In: 21st International Conference Information Visualisation (IV) (2017)
34. Ventocilla, E., et al.: Towards a taxonomy for interpretable and interactive machine learning. In: IJCAI/ECAI, Proceedings of the 2nd Workshop on Explainable Artificial Intelligence (XAI 2018), pp. 151–157 (2018)
35. Ventocilla, E., Riveiro, M.: Visual growing neural gas for exploratory data analysis. In: 14th International Joint Conference on Computer Vision, Imaging and Computer Graphics Theory and Applications (IVAPP), vol. 3, pp. 58–71. SciTePress (2019)
36. Zaharia, M., et al.: Resilient distributed datasets: a fault-tolerant abstraction for in-memory cluster computing. In: Proceedings of the 9th USENIX Conference on Networked Systems Design and Implementation, NSDI 2012, p. 2. USENIX Association (2012)

Visualization of Tree-Structured Data Using Web Service Composition

Willy Scheibel[1]([⊠]), Judith Hartmann[2], Daniel Limberger[1], and Jürgen Döllner[1]

[1] Hasso Plattner Institute, Faculty of Digital Engineering,
University of Potsdam, Potsdam, Germany
{willy.scheibel,daniel.limberger,juergen.doellner}@hpi.de
[2] Futurice, Berlin, Germany
judith.hartmann@futurice.com

Abstract. This article reiterates on the recently presented *hierarchy visualization service* HiViSer and its API [51]. It illustrates its decomposition into modular services for data processing and visualization of tree-structured data. The decomposition is aligned to the common structure of visualization pipelines [48] and, in this way, facilitates attribution of the services' capabilities. Suitable base resource types are proposed and their structure and relations as well as a subtyping concept for specifics in hierarchy visualization implementations are detailed. Moreover, state-of-the-art quality standards and techniques for self-documentation and discovery of components are incorporated. As a result, a *blueprint for Web service design, architecture, modularization, and composition* is presented, targeting fundamental visualization tasks of tree-structured data, i.e., gathering, processing, rendering, and provisioning. Finally, the applicability of the service components and the API is evaluated in the context of exemplary applications.

Keywords: Visualization as a Service · Hierarchy visualization · Tree visualization · RESTful API design · Web-based visualization

1 Introduction

Information visualization has become the prevalent way for interacting with the growing complexity, volume, and ubiquity of data. It provides the means for "bridging the two quite distinct fields of data science and human-computer interaction to help scientists and domain experts handle, explore and make sense of their data."[1] It allows users to absorb information, detect relationships and patterns, identify trends, and interact and manipulate data directly. Visualization facilitates accurate communication by aligning the mental image of and cultivating effective images to its consumers. At the same time, "web services [...] are

---

[1] L. Micallef: "AI Seminar: Towards an AI-Human Symbiosis Using Information Visualization", 2018. https://ai.ku.dk/events/ai-seminar-luana-micallef.

© Springer Nature Switzerland AG 2020
A. P. Cláudio et al. (Eds.): VISIGRAPP 2019, CCIS 1182, pp. 227–252, 2020.
https://doi.org/10.1007/978-3-030-41590-7_10

Fig. 1. Variations of hierarchy visualization techniques. From left to right and top to bottom: (1) 2.5D icicle plot, (2) 2.5D sunburst view, (3) 2.5D treemap, (4) 2.5D Voronoi treemap, (5) multiple 2.5D treemap arranged in a landscape, (5) 2.5D nested rings visualization, and (6) 2.5D treemap using sketchiness as a visual variable. Image used from "Design and Implementation of Web-Based Hierarchy Visualization Services" [51].

becoming the backbone of Web, cloud, mobile, and machine learning applications" [66] in which visualization is playing a vital part. Hence, visualization services – designed, implemented, and operated based on a Software-as-a-Service, by means of a software delivery model – are becoming an indivisible part of said web services, providing interactive, high-quality visualization techniques across all domains [10]. To this end, we focus on *hierarchy visualization*: the visualization of hierarchically-structured – strictly speaking tree-structured – multivariate data. Within the past three decades, over 300 hierarchy visualization techniques and variations have been proposed [54]. One reason for this diversity seems to be that tree-structured data is omnipresent in almost all application domains, e.g., demographics [29], business intelligence [47], health [14], and software development [69], but no single technique suits all tasks and applications (Fig. 1).

This article reiterates on the design of the recently presented hierarchy visualization service HiViSer [51]. For its API specification, the *Representational State Transfer* (REST) paradigm is employed. Structure-wise, a three-tier architecture is be implemented, comprising "a client layer which provides the user interface; a stateful web service middleware layer which provides a published interface to the visualization system; and finally, a visualization component layer which provides the core functionality of visualization techniques" [72]. To avoid a rather monolithic service design and specification, we exert a separation of concerns analogously to the well-known visualization pipeline [48] into data analysis, filtering, mapping, and rendering components. With this decomposition we anticipate various benefits, namely:

– elevated unambiguity of component interfaces and capabilities,
– simplified extensibility by means of integration of new components,
– increased re-usability of components through service composition,

- high degree of data privacy and protection [32], and
- reduced complexity and increased maintainability of components.

Supplementary, this article elaborates on relevant design decisions and incorporates technical details, eventually providing blueprints that can aid implementation of future visualization components. Finally, we ensure the visualization service's API quality by adopting the Richardson Maturity Model [70], i.e., introduce the strict use of resources, remove unnecessary variation, and improve discoverability of all resources maintained by the service.

The remainder of this article is structured as follows. Section 2 provides a rough overview of related work concerned with the fundamental building blocks required for the implementation of a modern web service. The targeted scope, prevalent use cases, and applications for a hierarchy visualization service are identified and requirements for its implementation are derived and itemized in Sect. 3. Subsequently, a concept for service orientation and service composition is derived and detailed in Sect. 4. On this basis, the actual routes specification for a web-based hierarchy visualization API is presented in Sect. 5 and evaluated in the context of exemplary applications in Sect. 6. Finally, this article is concluded by a summary of its findings, an estimate of its impact, as well as an outlook into future work in Sect. 7.

2   Related Work

This section references the fundamental building blocks and technical core ingredients that should be considered when designing and implementing modern services. Even though, these ingredients are most often interchangeable, at any time, there is only a small set of reliable, "industry-forged" options, so called de-facto standards. We do not cover solutions for deployment and storage since they should not interfere with the general design and implementation of any service: we expect that any given service built on below mentioned ingredients can most likely be ported to any technology stack and environment. Therefore, we focus on a brief discussion of the following implementation aspects: (1) image provisioning, (2) rendering frameworks, (3) visualization grammars and (4) Web API idioms and specification approaches.

As we follow a three-tier architecture for our service design [72], we propose data management and handling on a server, whereas a visualization client provides the graphical output and handles interaction. Image synthesis itself, however, can be executed either client-side or server-side, having the latter visualization pipeline fully executed on the server, i.e., all data processing, visualization mapping, and image synthesis.

Implementations of visualization services are usually not restricted by an environment and, thus, a broad range of implementation approaches are possible. However, since the hardware of the server is controllable and one server should serve visualization for multiple clients, there is a need for efficient data processing and image synthesis. The expectations towards rendering services are further heightened as they are the only hardware reliable enough for sophisticated rendering. This leaves visualization server implementations most likely to rely on hardware-accelerated image synthesis [50] using graphics APIs such as OpenGL or Vulkan (or other graphics library) for absolute control on the visual output [33], or specialized rendering engines such as game engines as well as real-time visualization engines. In highly specialized cases, intermediate resources of the visualization service such as geometry artifacts or exchange formats might be highly specialized for subsequent API or engine-dependent image synthesis (e.g., using GPU-friendly attributed vertex clouds [50]). However, such specializations should be well-considered and should not lead to limitations of the visualization services capabilities. Employing service composition, in contrast, allows for arbitrary customizability and specialization of resources for a multitude of clients and visualization consumers.

On the client side, visualizations are usually displayed by Web browsers. With our service design, we strive to support a variety of different output formats not only for convenience, but also for the various application scenarios (interactive, non-interactive, etc.). Browsers can display static images with built-in techniques, e.g., the img tag supporting raster graphic as well as SVGs. SVGs can also be used for interactive visualizations, e.g., when manipulating the DOM with the D3 library [9]. Declarative 3D is an approach to apply the properties of SVGs to 3D visualizations. So far, declarative 3D approaches like X3DOM [5] or XML3D [62] are not supported natively by browsers and are implemented by polyfill layers [62]. It remains to be seen if these approaches can attract more popularity. Lately, the HTML5 canvas is the element of choice for the creation of hardware-accelerated graphics at run-time. It allows for visual display of 2D or 3D graphics using the Web Graphics Library (WebGL). Before the canvas was introduced, interactive visualizations were usually displayed using external plugins (Java applets, Java FX, or the Macromedia/Adobe Flash) – in a way, similar to polyfill based approaches. The use of such plugins cannot be recommended anymore, since they are widely considered to be legacy.

Taking advantage of aforementioned techniques, libraries facilitate the creation of visualization in the browser. These libraries can be distinguished into two different approaches: Imperative libraries, also called visualization toolkits, that require a developer to program the resulting visualization, and declarative libraries that allow the user to create visualizations by providing data and configuration on how to display the data. Examples of visualization toolkits are Prefuse [27], its successor Flare[2] as well as Protovis [8] and its successor D3[3]. Examples of declarative libraries are charting libraries like the SVG-based Google Visualization API[4] or the canvas-based Chart.js library[5]. For the implementation of several of our service components, we preferred more low-level, use-case

---

[2] http://flare.prefuse.org/.
[3] https://d3js.org/.
[4] https://developers.google.com/chart/.
[5] https://chartjs.org/.

agnostic rendering frameworks such as webgl-operate[6]. This has the benefit of simplifying almost always required, application-specific customizations in terms of specialized target device, offscreen rendering, advanced interaction and rendering, 3D labeling, etc.

To ease the development of visualization techniques, there are multiple projects that allow to create visualizations by means of configuration. This is usually done using *visualization grammars* or domain specific languages. Examples for general visualization grammars are Vega Lite [49] and ATOM [45]. A grammar specialized for hierarchy visualization is the HiVE notation [60]. Providing a full server-client setup, the *Shiny*[7] project allows to deploy interactive visualization clients written in the R statistics language. For service composition we incorporated operations on data similar to HiVE operators, e.g., for preprocessing data or creating topology. Furthermore, multiple components (e.g., full visualization pipeline) can be invoked within a single query/request using nested objects.

The specification of possible interactions with a software component is called an *Application Programming Interface* (API). Thereby, a Web API is an API that is accessed using HTTP and related Web transmission technologies. Design approaches for Web APIs are classified as either action-based or resource-based APIs. The action-based approaches remote procedure call (RPC) protocols to trigger an action on the server. Formerly, these protocols encode calls with XML, e.g., the service-oriented architecture protocol (SOAP) [22] or XML-RPC[8]. More recent ones started to use JSON as well, e.g., JSON-RPC[9]. Although not as widespread, other notations are used as well by projects like gRPC[10] for example. The resource-based approaches mainly uses the architectural style of the *Representational State Transfer* (REST) for distributed hypermedia systems [18].

The specification of an API is usually provided using an *interface description languages* (IDL), describing the interfaces using a programming-language-agnostic syntax and semantics. Thereby, SOAP interfaces are commonly described in an XML-based IDL, the *Web Service Description Language* (WSDL) [15]. The RESTful equivalent to WSDL is the *Web Application Description Language* (WADL) [23]. A more recent RESTful API description language is *OpenAPI*[11], which comes with tooling that supports multiple programming languages for client libraries and IDE integration[12].

For hierarchy visualization, the HiViSer API [51] provides an entry point for hierarchy visualization as a service. However, the availability of APIs for server-side data processing for visualization and images is prevalent in other domains as well. For example, the OGC 3D Portrayal Service is a specification

---

[6] https://webgl-operate.org.
[7] https://shiny.rstudio.com/.
[8] http://xmlrpc.scripting.com/spec.html.
[9] https://jsonrpc.org/specification.
[10] https://grpc.io/.
[11] https://www.openapis.org/.
[12] http://openapi.tools/.

for web-based 3D geo-data portayal [24]. Likewise, processing and provisioning services are available for 3D point clouds [16] and image abstraction [46]. For this work, however, we predominantly implemented and tested service components that can be used for visualization of general business intelligence applications and visual software analytics applications for software system data and software engineering data.

3   Requirements of Hierarchy Visualization Techniques

The design and implementation for hierarchy visualization as a service should be aligned to prevalent usage scenarios. In order to increase flexibility, the services should support a broad range of prevalent visualization techniques and usage scenarios. These functional requirements are derived from available data processing and hierarchy visualization techniques (algorithms) as well as prevalent use cases (usage and integration scenarios of those algorithms). For the sake of brevity, this article includes a shortened analysis of techniques and use cases. However, we chose the subset to allow for generalization of the findings.

3.1   Scope of Processing and Hierarchy Visualization Techniques

We want the services to support a wide range of visualization techniques that operates on hierarchically-structured and tree-structured data. As data processing is highly domain-agnostic, usable algorithms are prevalent throughout the field of visualization [19]. The selection of the visualization techniques includes both *implicit* and *explicit* hierarchy visualization techniques [58]. In particular, we want to support the family of implicit hierarchy visualization techniques with their whole design space [57]. On the other hand, explicit hierarchy visualization should be supported as well [20]. Summarizing, the service design should allow for the techniques listed on treevis.net [54], i.e., over 300 tree visualization techniques.

Regarding the current state of this body of work, we want to limit the scope to visualization techniques of hierarchically-structured and tree-structured data. It is arguable if charts or unit visualizations [45] are a subset of hierarchy visualization techniques. However, for those categories, visualization systems are more wide-spread and visualization services are already available using sophisticated APIs. Further, we want to limit the supported techniques to exclude processing and depiction for additional relations among the data items (also known as *compound hierarchies*). Additionally, geo-referenced data attributes are not considered with their specific semantics but as plain data.

3.2   Use Cases of Hierarchy Visualizations

As exemplary use cases for hierarchy visualizations, we want to assess (1) the visual analysis of source code and software systems, (2) federal budget management, reporting, and controlling, (3) visual analysis on mobility data, and (4) visualization as navigational tool.

Visual Analysis of Source Code and Software. The process of creating software results in multiple sets of hierarchically-structured data. Examples are the structure of the source code modules, the team structure of developers and the association of commits to features and releases. Visualizations of these structures can facilitate conversation with stakeholders by depicting structure, behavior and evolution of the software system. As this use case is a whole field of research, we want to highlight some of the techniques. The software map [7] uses treemap subdivision to derive its layout and has a range of extensions regarding visual variables [34] and metaphors [37]. An alternative approach is the CodeCity [71] that uses quantized node weights and a recursively packing layout algorithm. A more street-focused approach is the Software City [63] that iterate on computed layouts to handle data changes. All these approaches are 2.5D visualization techniques and are provisioned using 3D scenes. As a rather exotic visualization provisioning technique, we want to highlight the use of Minecraft worlds [2].

Federal Budget Management, Reporting, and Controlling. Government spending and income can be assigned to categories and subcategories – resulting in hierarchically-structured data. Visualizing this data allows to gain a quick overview of the categories and their impact on the federal budget and facilitates a comparison throughout the years. As an example, Auber et al. visualized the government spending of the USA using GospelMaps, squarified treemaps, and icicle plots [1]. Further, a research fund dataset was visualized using treemaps with a cascading layout postprocessing [41].

Mobility Data Visual Analysis. Slingsby et al. [59] use treemaps for a visual analysis of GPS data from delivery vehicles in central London. The data contains the vehicle position, the vehicle speed, the vehicle type and the collection time of the data. For every vehicle data is collected multiple times per minute. A visual analysis of this data can be used by the courier company to optimize the vehicle allocation, scheduling and routing. It may also be used by transport authorities to assess patterns of traffics to help set up policies to reduce congestion. Since the data includes no inherent hierarchy, Slingsby et al. use categorical values to superimpose a hierarchy; we call them variable-constrained subsets. The subsets are used in different treemap visualizations to depict different aspects of the dataset. Additionally, they derive a domain-specific language HiVE to describe hierarchy visualizations [60].

Hierarchy Visualization as Navigational Tool. By providing an overview about the structure of a data hierarchy, visualization can be used as a tool to facilitate navigation through this data hierarchy. Lui et al. [41] use the nested treemaps to navigate through 30 000 proposals for fellowships and education projects on the website of the National Science Foundation. Likewise, Guerra-Gómez et al. [21] use proportional photo treemaps as a tool to navigate through the online photo service Flickr.

3.3 Derived Functional Requirements

From the scope of targeted visualization techniques and use cases, a comprehensive list of requirements can be derived. Thereby, the main feature is the operation on tree-structured or hierarchically-structured data. For flexible use from a client perspective, the required results from the processing are (1) the processed data itself, (2) mapped data by means of geometries, scenes, and 3D models, and (3) synthesized images and videos. Following a generalized pipeline for hierarchy visualization (Fig. 2), specific requirements are assigned to (a) datasets, (b) data preprocessing, (c) topology preprocessing, (d) layouting, (e) mapping and visual variables, (f) image synthesis, and, subsequent, (g) provisioning. A per-feature exposed parameterization allows for fine-grained control on the results.

Requirements on Datasets. The services should handle data from different application domains. This includes (f1) spatio-temporal, (f2) multi-variate, (f3) multi-modal, (f4) multi-run, and (f5) multi-model data [31]. Thus, a user should be able to provide heterogeneous data and the service should provide support to integrate the data. As part of dataset management, we want to support analysis on pre-existing data, e.g. train machine learning models [6], as well.

Requirements on Data Preprocessing. The services should support basic data preprocessing operations such as resampling, normalization, quantization [71], and filtering of outlier. More specific to hierarchy visualization, up-propagation and down-propagation of data is required, too. One particular metric for visualization that is computed as part of preprocessing is the degree-of-interest of nodes [68].

Requirements on Topology Preprocessing. Services should support basic operations on the topology – the relations between nodes – to specify the tree-structure for subsequent visualization. This includes subtree selection, node filtering, and subtree creation, e.g., through hierarchy operations [60]. More idiomatic, this step includes the extraction of a coarse-grained view on a hierarchy by means of aggregated views [17].

Requirements on Layouting. For tree layouting, all layout algorithms that operate on a tree data structure should be supported, i.e., all implicit and explicit tree layout algorithms. For explicit tree layouts, services should support node-link diagrams [44] and point-based depictions [56]. For implicit tree layouts, this includes rectangular treemap layout algorithms [69], as well as generalized algorithms [3]. Thereby, algorithms should not be limited in layout complexity and rather should allow for use of GosperMaps [1] and mixed polygonal layouts as well [25]. There are algorithms that use a mixed representation of implicit and explicit relations between nodes, but should be supported nevertheless. Examples include general rooted tree drawings [55]. As additional feature, the services should support for layout evolution, i.e., reusing previous layouts to compute the current one. This should work for implicit tree layouts [53], mixed representations [63], and general node-link diagrams [13].

Requirements on Visualization Mapping. Supported visual variables are target to the used visualization technique. Services should support to expose them to a client. As starting point, the standard set of visual variables has to be supported [12]. More advanced – but still considered – visual variables are sketchiness [73] and natural metaphors [74]. Next to specific visual variables, the services should support to encode two [67] and multiple points in time [63]. Additionally, labeling and glyphs should be supported, too [61].

Requirements on Geometries. The visualization service should support the geometries of implicit and explicit tree visualizations [57]. Existing techniques mainly use different types of primitive geometries (e.g., points, lines, rectangles, cuboids, polygons, or general triangle meshes). However, there are visualization techniques that use highly-specific types of scenes [2]. The type of geometry usually requires the use of specific renderers to synthesize the visualization images.

Requirements on Rendering. As renderer are highly specific, a fine-grained parameterization should be supported. Typical parameters include camera specification, enabling of rendering features (e.g., shadows, lighting, use of materials), visible semantic layers [24], or the rendering quality (e.g., number of intermediate frames for progressive rendering approaches [33]).

Requirements on Provisioning. For provisioning of results to a client, the service should be flexible as well. This means, that the services should support to provide final visualization images, but intermediate results as well (i.e., derived data [40], layouts, geometries, and models [38]). To this end, sophisticated renderer should not solely create the final image, but intermediate buffer with geometric semantics – the g-buffers – as well [36]. Typically, the visualization images and additional g-buffer can be thought of as intermediate results and can be used as input for postprocessing [46] or video creation.

4   Concept for Service Orientation and Composition

Based on the requirement analysis, we propose a service composition of multiple services to allow for the use cases and integration scenarios. Thereby, our focus is on two services that are specific to tree-structured data and its visualization. Further, an integration into a larger service landscape is outlined. The service landscape is aligned to a visualization pipeline model for tree visualization (Fig. 2). The internal data model of the services is separated into concepts on the underlying visualization implementations (capabilities) and concepts for user data (operational data).

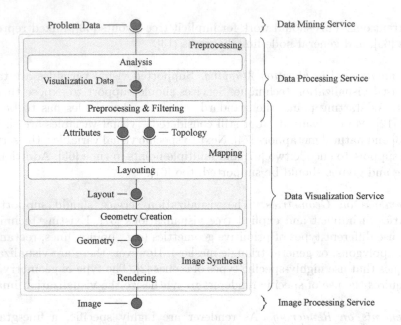

Fig. 2. The proposed tree visualization pipeline and a proposed service association. The specifics for tree visualization techniques are coined with the explicit model of attribute data and tree topology as separate inputs for the explicit layouting phase prior to geometry creation. In order to support a wide range of use cases and system integrations, having access to the final result (the image) as well as all intermediate results (attribute data, tree topologies, layouts, and geometries) is required as well.

4.1   Service Composition and Integration

Following the visualization pipeline model and the enclosing visualization process, the three technical phases to derive the visualization image are *preprocessing and filtering of data*, *mapping the preprocessed data*, and *rendering the mapped data* [48]. We propose a service landscape focusing on this pipeline model. Thereby, we introduce a separation between the data management and processing service and the data visualization service (Fig. 2). A thorough service integration has a service for problem data retrieval (data mining) as well. Services extending on the visualization images are postprocessing services or video services. Alternatively, external services may use intermediate results as subject of analysis (e.g., using layouts to derive layout metrics [69]). Although we propose a separation of data processing service and visualization service, we can imagine that an actual implementation can provide both functionalities and therefore exposes interfaces for both (e.g., HiViSer is designed this way) (Fig. 3).

4.2   Derived Operational Data Model

The operational data model represents the user data and derived data throughout visualization. For the data processing service, we propose the concepts

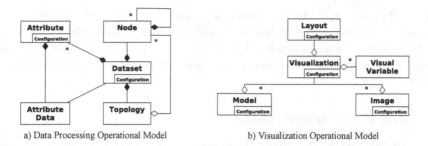

a) Data Processing Operational Model   b) Visualization Operational Model

Fig. 3. The operational models for (a) the data processing service and (b) the visualization service.

Dataset, Topology, Attribute, and *Attribute Data.* For the visualization service, we propose the concepts *Visualization, Layout, Model,* and *Image.* For the concepts Dataset, Attribute, Layout, Visualization, Model, and Image, we include explicit configurations, each. The proposed concepts are the communication artifacts between services and clients and can be actively managed by the user, i.e., create, read, update, and delete (CRUD approach).

Dataset and Topology. A Dataset represents a set of nodes, their tree-structured Topology (their parent-child relationship by means of edges), and a number of Attributes with values – their Attribute Data. Although depending the Topology Format, we suggest that a topology contains information on all nodes, their ids (identifier), and parent node. Additionally, node meta information as node labels can be added, too. As the topology is a specific view on the Dataset and changes to the Dataset are promoted through use of Dataset Processing Techniques, a Topology cannot be created, updated, or deleted independently. When not provided directly, e.g., through a data mining service or through direct upload, a dataset can be derived from another dataset by use of a Dataset Processing Technique. This is encoded using a Dataset Configuration that can be re-used for multiple Datasets.

Attribute and Attribute Data. A measure that is associated with the nodes of a Dataset is represented by an Attribute. The Attribute Data allows to query a value for each node and Attribute. An Attribute Configuration allows to derive an Attribute from another Attribute, including implementation-specific computation and derivation of the per-node values.

Visualization and Layout. A Visualization represents the concept of a configured mapping from data to a Layout and Visual Variables. Thereby, a Layout represents a spatial position and possibly an extent for each node of a Topology. A Visualization Configuration allows to re-use previously defined configurations across datasets. Likewise, a Layout Configuration allows to re-apply layout techniques to other datasets.

Model and Image. A Model represents the required graphical primitives, geometries, or the virtual scene to render depictions of a Visualization. An Image thereby represents a server-rendered image of a Visualization. The Model Configuration and Image Configuration concepts allow to re-use a configuration across datasets.

4.3   Derived Capabilities Model

The capabilities model represents the exposed implementation specifics of a service. For the data processing service, we propose the concepts *Dataset Type, Dataset Processing Technique, Topology Format, Data Processing Technique,* and *Data Format.* For the visualization service, we propose the concepts *Visualization Technique, Layout Technique, Layout Format, Visual Variable Mapping, Model Format, Rendering Technique,* and *Image Format.* The proposed concepts expose the available operations, data formats, and parameterization to other services and clients. These concepts are not meant to be actively managed by a user, i.e., they are designated to be read-only. As the content of these concepts is directly derived from implementation specifics, only a change the implementation should change the contents.

Dataset Capabilities. The Dataset Type encapsulates a set of implementation-specific properties of a Dataset. This includes associated Dataset Processing Techniques, Topology Formats, Data Processing Techniques, and Data Formats. A Dataset Processing Technique represents an implementation of a mapping of one Dataset to another, derived, Dataset. Likewise, a Data Processing Technique represents an implementation to derive Attributes and their data from other Attributes. For efficient transmission of results and intermediate results, a service can provide data using different transmission formats. The proposed services support this by use of Topology Formats for Topologies and Data Formats for Attribute Data.

Visualization Capabilities. A Visualization Technique encapsulates a set of implementation-specific properties of a Visualization. This includes associated Layout Techniques, Layout Formats, Visual Variable Mappings, Model Formats, Rendering Techniques, and Image Formats. Thereby, a Layout Technique is an implementation that maps a node to a spatial position and possibly an extent. A Layout Format represents a specific encoding of a Layout for provisioning. For Visual Variable Mapping, this concept encapsulates an Attribute that is mapped to a visual variable, i.e., a mapping to a spatial or otherwise visual encoding. A Rendering Technique represents an implementation that processes a Visualization and produces Images, i.e., color depictions and possibly with additional data buffers that encode additional per-pixel information. For efficient transmission of results and intermediate results, a visualization service can provide Models and Images using different encodings. To support this, a Model Format represents an implementation-specific encoding of a Model and an Image Format represents an encoding for Images, respectively.

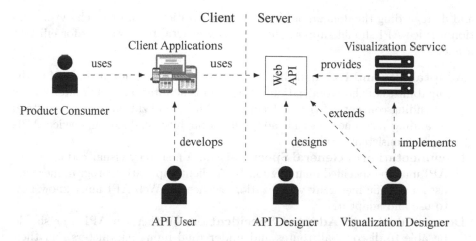

Fig. 4. There are four groups of stakeholders that interact with a hierarchy visualization service directly or indirectly. Mainly, the service is shaped by the API designer and the visualization designer. The API user develops client applications that use a hierarchy visualization service to manage tree-structured data or display depictions of the data. These client applications are used by the product consumer.

5   Hierarchy Visualization API

Derived from the service concept, we propose a RESTful API that allows to access the services. This API is designed with an extended stakeholder model in mind as well as additional non-functional requirements. Essentially, each model from the service concept is mapped to a route, and thus a REST resource. These routes are meant to be general extensions points for specific hierarchy visualization services by specialization of requests and responses, rather than adding new resources or routes to the API. Specifics of visualization implementations are expected to extend on the base resources instead of adding new ones.

5.1   Stakeholder

Stylos and Myers describe three stakeholders that interact with an API directly or indirectly: the API designer, the API user and the product consumer [64]. They imply that an API is created to provide an interface for an existing implementation. To this end, no stakeholder for implementation of a system using an existing API Design as an interface is considered. We explicitly want to design a process with such a visualization designer stakeholder and, thus, added this role to the concept of API stakeholders (Fig. 4).

5.2   API Design Requirements

In contrast to provisioning of actual features, an efficient API should be further constrained in *how* the features are provided. From an API design viewpoint –

and disregarding the domain of hierarchy visualizations –, a hierarchy visualization service API should support the following general requirements for efficient usage:

Adaptability and Extensibility. It must be possible to adapt the API to the capabilities of the visualization service. This adaptation is characterized by extending concepts of the abstract hierarchy visualization service – instead of adding new ones – so all adapted hierarchy visualization service APIs keep a consistent structure.

Documentation of General Specification. A hierarchy visualization service API must be specified unambiguously. A client application programmer who uses a specific hierarchy visualization service via Web API must know how to use and adapt it.

Documentation of Adapted Specification. Further, an API user should be able to discover all routes and understand input parameters and their meaning without looking at external documentation. A hierarchy visualization service must specify a way to include comprehensive documentation as part of the web service.

More specific to RESTful APIs, the REST standard architecture defines constraints on the API design [18]: *Client-Server Architecture, Layered System, Stateless, Cacheability, Code on Demand*, and *Uniform Interface*. In order to obtain a uniform interface, further architectural constraints are defined: *Identification of Resources, Manipulation through Representations, Self-descriptive Messages*, and *Hypermedia as the Engine of Application State*. Thus, an additional requirement is the reusability by means of re-usage of information and previously submitted or computed results on a visualization service. This facilitates the creation of visualizations with the same options for different data as the options do not have to be recreated. Further, reusability of information allows to use already processed information, instead of processing the same information twice. To ensure an efficient process, the amount of requests and the request response time should be as minimal as possible. The quality of a RESTful API can be measured by assessing its maturity with respect to the *Richardson Maturity Model* [70]. Our target is quality level 3 of the RMM, i.e., (1) using a different URL for every different resource, (2) using at least two HTTP methods semantically correct, and (3) using hypermedia in the response representation of resources.

5.3   Routes and Mapping

The design of a service composition with data processing service and visualization service as well as the distinction between an operational data model and a capabilities model results in four parts of our proposed API. There is (1) the routes for the operational data of the data processing service (main route is /datasets/, cf. Table 1), (2) the routes for the capabilities of the data processing service (main route is /datasettypes/, cf. Table 2), (3) the routes for the

Table 1. The mapping from the operational data resources of the data processing service to RESTful API routes.

Operational data	Route	Element	Collection
Datasets	/datasets/[id]	✓	✓
Topology	/datasets/<id>/topology/[id]	✓	✓
Attribute	/datasets/<id>/attributes/[id]	✓	✓
Attribute data	/datasets/<id>/attributes/<id>/data/[id]	✓	✓
Dataset configurations	/dataset-configurations/[id]	✓	✓
Attribute configurations	/attribute-configurations/[id]	✓	✓

Table 2. The mapping from the capabilities resources of the data processing service to RESTful API routes.

Capability	Route	Element	Collection
Dataset types	/datasettypes/[id]	✓	✓
Dataset processing techniques	/datasettypes/<id>/datasetprocessingtechniques/[id]	✓	✓
Topology formats	/datasettypes/<id>/topologyformats/[id]	✓	✓
Data processing techniques	/datasettypes/<id>/dataprocessingtechniques/[id]	✓	✓
Data formats	/datasettypes/<id>/dataformats/[id]	✓	✓

Table 3. The mapping from the operational data resources of the visualization service to RESTful API routes.

Operational data	Route	Element	Collection
Visualizations	/visualizations/[id]	✓	✓
Visualization configurations	/visualization-configurations/[id]	✓	✓
Layouts	/visualizations/<id>/layouts/[id]	✓	✓
Layout configurations	/visualizations/<id>/layout-configurations/[id]	✓	✓
Models	/visualizations/<id>/models/[id]	✓	✓
Model configurations	/visualizations/<id>/model-configurations/[id]	✓	✓
Images	/visualizations/<id>/images/[id]	✓	✓
Image configurations	/visualizations/<id>/image-configurations/[id]	✓	✓

Table 4. The mapping from the capabilities resources of the visualization service to RESTful API routes.

Capability	Route	Element	Collection
Visualization techniques	/visualizationtechniques/[id]	✓	✓
Layout techniques	/visualizationtechniques/<id>/layouttechniques/[id]	✓	✓
Layout formats	/visualizationtechniques/<id>/layoutformats/[id]	✓	✓
Visual variable mappings	/visualizationtechniques/<id>/visualvariables/[id]	✓	✓
Model formats	/visualizationtechniques/<id>/modelformats/[id]	✓	✓
Rendering techniques	/visualizationtechniques/<id>/renderingtechniques/[id]	✓	✓
Image formats	/visualizationtechniques/<id>/imageformats/[id]	✓	✓

operational data of the visualization service (main route is /visualizations/, cf. Table 3), and (4) the routes for the capabilities of the visualization service (main route is /visualizationtechniques/, cf. Table 4). Thereby, each route allows for query of single elements and whole collections of the associated resource.

Next to the domain-specific routes for the data processing service and the visualization service, we define two additional routes to satisfy the requirements of the documentation on the API specification and the *Hypermedia as the Engine of Application State*:

/   The root route that lists all main resources, i.e., Datasets, Dataset Types, Visualizations, and Visualization Techniques.

/api   The route that provides the API specification.

5.4   Adaptation Process

The API is designed to support various hierarchy visualization techniques and implementations. Therefore, the API designer should be able to adapt the API to the capabilities of the visualization service – at best, without structural changes to the API. This is accomplished as the API designer does not need to create new base resource types, irregardless their relation to the creation of an image, a model, or other results from the service. Instead, the base resource types have to be extended, i.e., subtypes have to be created. Typically, this adaption is done continuously within the development of the visualization service.

6   Evaluation

Our proposed service landscape and the API are evaluated by use of example applications and assessment. Thereby, we want to introduce a treemap visualization service, where we want to consider treemaps as a common example of hierarchy visualization techniques. Next, we want to take up on the former use cases and assess the use of the proposed API. Last, we describe the differences between the reiterated, proposed API and the former HIVISER API.

6.1   Treemaps as a Service

To demonstrate the service landscape, we describe a treemap visualization service. The service provides data management for tree-structured data as well as configuration and provisioning of treemap-based data visualization. Thereby, the service supports the proposed hierarchy visualization API while being restricted to treemap visualization techniques. The available processing and visualization techniques and the implementation specifics of the service are as follows:

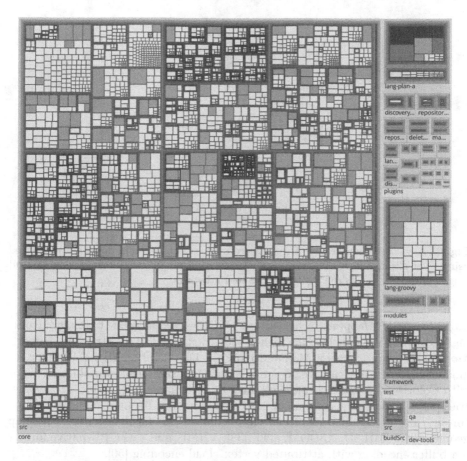

Fig. 5. A visualization result using a software dataset snapshot from ElasticSearch from 2014. The dataset is depicted using a 2D treemap, color mapping, and labeling.

Data Management and Preprocessing. Regarding source data, an end user can create own datasets via Datasets and upload data as CSV, i.e., we do not utilize a separate data mining service. The file should contain one row per leaf node and one column per attribute. The first column is expected to contain the node identifier with slashes as delimiters to encode the parent path of nodes (e.g., a file path). The parts of the node identifier are used as a nominal attribute for node labeling (for both inner and leaf nodes). The other columns of the CSV are used as further numeric attributes that are registered within the service and exposed via the API. The topology of the tree is extracted from the encoded path of the CSV file. Derived data is provided by means of attribute value normalization, transformations (e.g., quantization), and computation of a degree of interest [37]. The supported output formats are an edge list for the topology (pairs of parent identifier and child identifier for each edge using a breadth-first order) and a value buffer for attribute data (float-encoded values using breadth-first order) (Fig. 5).

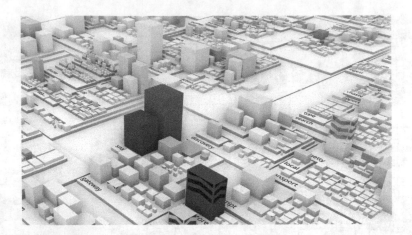

Fig. 6. A visualization result of a software system dataset. The layout was created using EvoCells layouting. The resulting visualization is a 2.5D treemap with color and height mapping, in-situ diff visualization, and labeling.

Layouting and Mapping. The service provides rectangular treemap layouting algorithms – namely Slice'n'Dice [30], Squarified [11], Strip [4], Hilbert, Moore [65], Approximation [43], and EvoCells [53] – and additional layout margins for hierarchical structure depiction. The supported visual variables are color with ColorBrewer color schemes [26], height, sketchiness [34], and in-situ geometries [39] – allowing the creation of 2.5D treemaps (Fig. 6). Labeling is supported through use of OpenLL [35]. The supported output formats for the treemap geometry are XML3D, X3DOM, glTF [38], VRML, and – as GPU-native format – a buffer encoding with attributed vertex cloud encoding [50].

Rendering. The treemap images are rendered using progressive rendering [33], allowing for both basic 2D depictions and sophisticated graphical effects for virtual 3D environments. Next to the rendered image, the g-buffers for per-pixel buffers using false-color-encoded ids and normal vectors are provided, too.

Supported Client Approaches. With this set of features and interfaces, the service allows for the following types of clients: (1) a thin client, that displays server-rendered visualization images and maps user interaction to service requests and updates to the displayed image, (2) a mixed approach where the client displays a 3D model of a visualization in the web browser by use of a polyfill layer, and (3) a rich client, that uses only preprocessed data and a topology to perform layouting, geometry creation, and rendering on the client. The latter approach provides the most flexibility with least requests to the service but imposes more load to the client.

```
 1  { "visualization": {
 2      "type": "/visualizationtechniques/2.5dtreemap",
 3      "options": {
 4        "dataset": "/datasets/Tierbestand2014",
 5        "layout": {
 6          "type": "striptreemap",
 7          "options": {
 8            "weights": { "type": "normalization",
 9              "options": { "source": "count", "min": 0, "max": "source" } },
10            "parentPadding": 0.01, "siblingMargin": 0.03 }
11        },
12        "visualVariables": [
13          { "type": "leaf-colors",
14            "options": { "attribute": "count", "gradient": "colorbrewer-3-OrRd" } },
15          { "type": "leaf-heights", "options": { "attribute": "count" } }
16        ],
17        "labeling": true
18      }
19    },
20    "options": {
21      "width": 1920, "height": 1080,
22      "eye": [ 0.0, 1.2, 0.8 ],
23      "center": [ 0.0, 0.0, 0.0 ],
24      "up": [ 0.0, 1.0, 0.0 ],
25      "quality": 100, "format": "png" } }
```

Listing 1.1. Example request JSON for a treemap image at the route /visualizations/Tierpark2014/images. This request was issued after the dataset Tierpark2014 was created and configured.

6.2   Use Case Evaluation

A use case assessment on the proposed API results in the following observations:

Visual Analysis of Source Code and Software. The API can represent software datasets with multiple revisions – multiple Datasets – and multiple attributes. Thereby, the model is flexible enough to allow different types of nodes (e.g., developers, source code modules, and commits). Further, the API allows to use different data preprocessing, layout algorithms and visual variables to allow for the targeted software visualization techniques. With the flexibility to provide implementation-defined 3D model formats, even Minecraft worlds are within the scope of the API.

Federal Budget Management, Reporting, and Controlling. As with software data, budget data is representable using the API as well. The creation of the hierarchy through use of categorical grouping is supported through dataset processing techniques. The support of polygonal layouts, e.g., for GosperMaps, is assured though an abstraction layer that places this responsibility on the implementation of the service. The evolution of layouts is supported through re-usability of precomputed layouts and specific operations that implement the evolution.

Mobility Data Visual Analysis. From an API perspective, this use case is similar to the federal budget management. However, the use of the HiVE visualization grammar is possible to be used with our API structure, e.g., by providing an implementation of a dataset processing technique that parses the HiVE grammar.

Fig. 7. A 3D printed treemap from a model that was created using a treemap service, embedded into a wooden frame.

Hierarchy Visualization as Navigational Tool. The use for a navigational tool is more broad as it adds the navigation task to the display of images. With the provisioning of both images and id-buffers, a user click into the image can be converted into the clicked node id. This node id can be used to select the original node and its metadata to extract the navigation target and, thus, allow to navigate the user further.

6.3   Comparison to HiViSer API

A comparison of the proposed API to the HiViSer API can be described as a mapping of pre-existing routes and the addition of routes to self-describe transmission formats and available processing and visualization techniques. For us, the concept of DataSources from HiViSer is conceptually located in the data mining service, and, thus, not focus of the API proposed in this article. The original concept of Datasets and Nodes is adapted and mapped to routes in a similar fashion. However, with this proposed API, the attributes are modeled as part of a dataset and Buffers are represented as Attribute Data. BufferViewOptions and BufferTransformations are remodeled to Data Processing Techniques and references to Attribute Data. The concepts Layout, Images, and Models are now part of Visualizations. The VisualVariable was moved to the self-description interface of Visualization Techniques. Labeling is currently not part of the proposed API – although we believe it can be specified by adding Labels to the data processing service and Labeling to the visualization service (similar to the HiViSer API) (Fig. 7).

7 Conclusions

We decomposed the recently presented hierarchy visualization service HiViSer analogously to the stages of the visualization pipeline model, with focus on data processing and visualization. Thereby, we described a resource-based Web API for tree-structured data management and visualization, provided instructions for the specification of its base resources, and specified routes targeting OpenAPI. We think this not only will facilitate the integration into a larger service landscape for data mining, postprocessing, video generation but also ease the implementation of future visualization APIs. Overall, we presented a proposal for Visualization-as-a-Service for tree-structured data that strives to make the development of visualization clients faster, more flexible, and less error-prone. At best, we hope this to be a valuable step towards standardization of the API and associated communication artifacts.

For future work, several directions are designated for extending the API. For example, we plan to formalize the labeling part of the API and add management for and visualization of additional relations [28]. We also imagine configurations of images, where multiple depictions are embedded within one. The latter would allow for small multiples visualizations [52]. At large, we want to evaluate if the API design is applicable to a broader class of data and corresponding information visualization techniques, to support the "[i]ntegration and analysis of heterogeneous data" and therefore tackle "one of the greatest challenges for versatile applications" [42] in information visualization.

Acknowledgments. This work was partially funded by the German Federal Ministry of Education and Research (BMBF, KMUi) within the project "BIMAP" (www.bimap-project.de) and the German Federal Ministry for Economic Affairs and Energy (BMWi, ZIM) within the projects "ScaSoMaps" and "TASAM".

References

1. Auber, D., Huet, C., Lambert, A., Renoust, B., Sallaberry, A., Saulnier, A.: GosperMap: using a Gosper curve for laying out hierarchical data. IEEE Trans. Visual. Comput. Graphics **19**(11), 1820–1832 (2013). https://doi.org/10.1109/TVCG.2013.91
2. Balogh, G., Szabolics, A., Beszédes, Á.: CodeMetropolis: eclipse over the city of source code. In: Proceedings of IEEE International Working Conference on Source Code Analysis and Manipulation, SCAM 2015, pp. 271–276 (2015). https://doi.org/10.1109/SCAM.2015.7335425
3. Baudel, T., Broeksema, B.: Capturing the design space of sequential space-filling layouts. IEEE Trans. Visual. Comput. Graphics **18**(12), 2593–2602 (2012). https://doi.org/10.1109/TVCG.2012.205
4. Bederson, B.B., Shneiderman, B., Wattenberg, M.: Ordered and quantum treemaps: making effective use of 2D space to display hierarchies. ACM Trans. Graph. **21**(4), 833–854 (2002). https://doi.org/10.1145/571647.571649

5. Behr, J., Eschler, P., Jung, Y., Zöllner, M.: X3DOM - a DOM-based HTML5/ X3D integration model. In: Proceedings of ACM International Conference on 3D Web Technology, Web3D 2009, pp. 127–135 (2009). https://doi.org/10.1145/1559764. 1559784

6. Bethge, J., Hahn, S., Döllner, J.: Improving layout quality by mixing treemap-layouts based on data-change characteristics. In: Proceedings of EG International Conference on Vision, Modeling & Visualization, VMV 2017 (2017). https://doi. org/10.2312/vmv.20171261

7. Bohnet, J., Döllner, J.: Monitoring code quality and development activity by software maps. In: Proceedings of ACM Workshop on Managing Technical Debt, MTD 2011, pp. 9–16 (2011). https://doi.org/10.1145/1985362.1985365

8. Bostock, M., Heer, J.: Protovis: a graphical toolkit for visualization. IEEE Trans. Visual. Comput. Graphics **15**(6), 1121–1128 (2009). https://doi.org/10.1109/ TVCG.2009.174

9. Bostock, M., Ogievetsky, V., Heer, J.: D3 data-driven documents. IEEE Trans. Visual. Comput. Graphics **17**(12), 2301–2309 (2011). https://doi.org/10.1109/ TVCG.2011.185

10. Bouguettaya, A., et al.: A service computing manifesto: the next 10 years. Commun. ACM **60**(4), 64–72 (2017). https://doi.org/10.1145/2983528

11. Bruls, M., Huizing, K., van Wijk, J.: Squarified treemaps. In: de Leeuw, W.C., van Liere, R. (eds.) Data Visualization 2000. Eurographics, pp. 33–42. Springer, Vienna (2000). https://doi.org/10.1007/978-3-7091-6783-0_4

12. Carpendale, M.S.T.: Considering visual variables as a basis for information visualization. Technical report, University of Calgary, Canada (2003). https://doi.org/ 10.11575/PRISM/30495. nr. 2001–693-14

13. Caudwell, A.H.: Gource: visualizing software version control history. In: Proceedings of ACM International Conference Companion on Object Oriented Programming Systems Languages and Applications Companion, OOPSLA 2010, pp. 73–74 (2010). https://doi.org/10.1145/1869542.1869554

14. Chazard, E., Puech, P., Gregoire, M., Beuscart, R.: Using treemaps to represent medical data. IOS Stud. Health Technol. Inform. **124**, 522–527 (2006). http://ebooks.iospress.nl/volumearticle/9738

15. Chinnici, R., Moreau, J.J., Ryman, A., Weerawarana, S.: Web Services Description Language (WSDL) Version 2.0 Part 1: Core Language (2007). http://www.w3.org/ TR/2007/REC-wsdl20-20070626

16. Discher, S., Richter, R., Trapp, M., Döllner, J.: Service-oriented processing and analysis of massive point clouds in geoinformation management. In: Döllner, J., Jobst, M., Schmitz, P. (eds.) Service-Oriented Mapping. LNGC, pp. 43–61. Springer, Cham (2019). https://doi.org/10.1007/978-3-319-72434-8_2

17. Elmqvist, N., Fekete, J.D.: Hierarchical aggregation for information visualization: overview, techniques, and design guidelines. IEEE Trans. Visual. Comput. Graphics **16**(3), 439–454 (2010). https://doi.org/10.1109/TVCG.2009.84

18. Fielding, R.T.: REST: architectural styles and the design of network-based software architectures. Ph.D. thesis, University of California, Irvine (2000)

19. García, S., Luengo, J., Herrera, F.: Data Preprocessing in Data Mining. Springer, Cham (2015). https://doi.org/10.1007/978-3-319-10247-4

20. Graham, M., Kennedy, J.: A survey of multiple tree visualisation. Inf. Vis. **9**(4), 235–252 (2010). https://doi.org/10.1057/ivs.2009.29

21. Guerra-Góomez, J.A., Boulanger, C., Kairam, S., Shamma, D.A.: Identifying best practices for visualizing photo statistics and galleries using treemaps. In: Proceedings of the ACM International Working Conference on Advanced Visual Interfaces, AVI 2016, pp. 60–63 (2016). https://doi.org/10.1145/2909132.2909280
22. Hadley, M., et al.: SOAP version 1.2 part 1: messaging framework, 2nd edn. Technical report, W3C (2007). http://www.w3.org/TR/2007/REC-soap12-part1-20070427/
23. Hadley, M.J.: Web application description language (WADL). Technical report, Sun Microsystems Inc. (2006)
24. Hagedorn, B., Coors, V., Thum, S.: OGC 3D portrayal service standard. Technical report, Open Geospatial Consortium (2017). http://docs.opengeospatial.org/is/15-001r4/15-001r4.html
25. Hahn, S., Döllner, J.: Hybrid-treemap layouting. In: Proceedings of the EG EuroVis 2017 - Short Papers, EuroVis 2017, pp. 79–83 (2017). https://doi.org/10.2312/eurovisshort.20171137
26. Harrower, M., Brewer, C.A.: ColorBrewer.org: an online tool for selecting colour schemes for maps. Cartogr. J. **40**(1), 27–37 (2003). https://doi.org/10.1179/000870403235002042
27. Heer, J., Card, S.K., Landay, J.A.: Prefuse: a toolkit for interactive information visualization. In: Proceedings of the ACM SIGCHI Conference on Human Factors in Computing Systems, CHI 2005, pp. 421–430 (2005). https://doi.org/10.1145/1054972.1055031
28. Holten, D.: Hierarchical edge bundles: visualization of adjacency relations in hierarchical data. IEEE Trans. Visual. Comput. Graphics **12**(5), 741–748 (2006). https://doi.org/10.1109/TVCG.2006.147
29. Jern, M., Rogstadius, J., Åström, T.: Treemaps and choropleth maps applied to regional hierarchical statistical data. In: Proceedings of the IEEE International Conference Information Visualisation, IV 2009, pp. 403–410 (2009). https://doi.org/10.1109/IV.2009.97
30. Johnson, B., Shneiderman, B.: Tree-maps: a space-filling approach to the visualization of hierarchical information structures. In: Proceedings of the IEEE Visualization, Visualization 1991, pp. 284–291 (1991). https://doi.org/10.1109/VISUAL.1991.175815
31. Kehrer, J., Hauser, H.: Visualization and visual analysis of multifaceted scientific data: a survey. IEEE Trans. Visual. Comput. Graphics **19**(3), 495–513 (2013). https://doi.org/10.1109/TVCG.2012.110
32. Koller, D., et al.: Protected interactive 3D graphics via remote rendering. ACM Trans. Graph. **23**(3), 695–703 (2004). https://doi.org/10.1145/1015706.1015782
33. Limberger, D., Döllner, J.: Real-time rendering of high-quality effects using multiframe sampling. In: Proceedings of the ACM SIGGRAPH Posters, SIGGRAPH 2016, p. 79 (2016). https://doi.org/10.1145/2945078.2945157
34. Limberger, D., Fiedler, C., Hahn, S., Trapp, M., Döllner, J.: Evaluation of sketchiness as a visual variable for 2.5D treemaps. In: Proceedings of the IEEE International Conference Information Visualisation, IV 2016, pp. 183–189 (2016). https://doi.org/10.1109/IV.2016.61
35. Limberger, D., Gropler, A., Buschmann, S., Döllner, J., Wasty, B.: OpenLL: an API for dynamic 2D and 3D labeling. In: Proceedings of the IEEE International Conference on Information Visualization, IV 2018, pp. 175–181 (2018). https://doi.org/10.1109/iV.2018.00039

36. Limberger, D., Pursche, M., Klimke, J., Döllner, J.: Progressive high-quality rendering for interactive information cartography using WebGL. In: ACM Proceedings of the International Conference on 3D Web Technology, Web3D 2017, pp. 8:1–8:4 (2017). https://doi.org/10.1145/3055624.3075951
37. Limberger, D., Scheibel, W., Hahn, S., Döllner, J.: Reducing visual complexity in software maps using importance-based aggregation of nodes. In: Proceedings of the SciTePress International Joint Conference on Computer Vision, Imaging and Computer Graphics Theory and Applications - Volume 3: IVAPP, VISIGRAPP/IVAPP 2017, pp. 176–185 (2017). https://doi.org/10.5220/0006267501760185
38. Limberger, D., Scheibel, W., Lemme, S., Döllner, J.: Dynamic 2.5D treemaps using declarative 3D on the web. In: Proceedings of the ACM International Conference on 3D Web Technology, Web3D 2016, pp. 33–36 (2016). https://doi.org/10.1145/2945292.2945313
39. Limberger, D., Trapp, M., Döllner, J.: In-situ comparison for 2.5D treemaps. In: Proceedings of the SciTePress International Joint Conference on Computer Vision, Imaging and Computer Graphics Theory and Applications - Volume 3: IVAPP, VISIGRAPP/IVAPP 2019, pp. 314–321 (2019). https://doi.org/10.5220/0007576203140321
40. Limberger, D., Wasty, B., Trümper, J., Döllner, J.: Interactive software maps for web-based source code analysis. In: Proceedings of the ACM International Conference on 3D Web Technology, Web3D 2013, pp. 91–98 (2013). https://doi.org/10.1145/2466533.2466550
41. Liu, S., Cao, N., Lv, H.: Interactive visual analysis of the NSF funding information. In: Proceedings of the IEEE Pacific Visualization Symposium, PacificVis 2008, pp. 183–190 (2008). https://doi.org/10.1109/PACIFICVIS.2008.4475475
42. Liu, S., Cui, W., Wu, Y., Liu, M.: A survey on information visualization: recent advances and challenges. Vis. Comput. 30(12), 1373–1393 (2014). https://doi.org/10.1007/s00371-013-0892-3
43. Nagamochi, H., Abe, Y.: An approximation algorithm for dissecting a rectangle into rectangles with specified areas. Discrete Appl. Math. 155(4), 523–537 (2007). https://doi.org/10.1016/j.dam.2006.08.005
44. Nguyen, Q.V., Huang, M.L.: EncCon: an approach to constructing interactive visualization of large hierarchical data. Inf. Vis. 4(1), 1–21 (2005). https://doi.org/10.1057/palgrave.ivs.9500087
45. Park, D., Drucker, S.M., Fernandez, R., Niklas, E.: ATOM: a grammar for unit visualizations. IEEE Trans. Visual. Comput. Graphics 24(12), 3032–3043 (2018). https://doi.org/10.1109/TVCG.2017.2785807
46. Richter, M., Söchting, M., Semmo, A., Döllner, J., Trapp, M.: Service-based processing and provisioning of image-abstraction techniques. In: Proceedings of the International Conference on Computer Graphics, Visualization and Computer Vision, WCSG 2018, pp. 79–106 (2018). https://doi.org/10.24132/CSRN.2018.2802.13
47. Roberts, R.C., Laramee, R.S.: Visualising business data: a survey. MDPI Inf. 9(11), 285, 1–54 (2018). https://doi.org/10.3390/info9110285
48. dos Santos, S., Brodlie, K.: Gaining understanding of multivariate and multidimensional data through visualization. Comput. Graph. 28(3), 311–325 (2004). https://doi.org/10.1016/j.cag.2004.03.013
49. Satyanarayan, A., Moritz, D., Wongsuphasawat, K., Heer, J.: Vega-Lite: a grammar of interactive graphics. IEEE Trans. Visual. Comput. Graphics 23(1), 341–350 (2017). https://doi.org/10.1109/TVCG.2016.2599030

50. Scheibel, W., Buschmann, S., Trapp, M., Döllner, J.: Attributed vertex clouds. In: GPU Zen: Advanced Rendering Techniques, Chapter: Geometry Manipulation, pp. 3–21. Bowker Identifier Services (2017)
51. Scheibel, W., Hartmann, J., Döllner, J.: Design and implementation of web-based hierarchy visualization services. In: Proceedings of the SciTePress International Joint Conference on Computer Vision, Imaging and Computer Graphics Theory and Applications - Volume 3: IVAPP, VISIGRAPP/IVAPP 2019, pp. 141–152 (2019). https://doi.org/10.5220/0007693201410152
52. Scheibel, W., Trapp, M., Döllner, J.: Interactive revision exploration using small multiples of software maps. In: Proceedings of the SciTePress International Joint Conference on Computer Vision, Imaging and Computer Graphics Theory and Applications - Volume 2: IVAPP, VISIGRAPP/IVAPP 2016, pp. 131–138 (2016).https://doi.org/10.5220/0005694401310138
53. Scheibel, W., Weyand, C., Döllner, J.: EvoCells - a treemap layout algorithm for evolving tree data. In: Proceedings of the SciTePress International Joint Conference on Computer Vision, Imaging and Computer Graphics Theory and Applications - Volume 3: IVAPP, VISIGRAPP/IVAPP 2018, pp. 273–280 (2018). https://doi.org/10.5220/0006617102730280
54. Schulz, H.J.: Treevis.net: a tree visualization reference. IEEE Comput. Graphics Appl. **31**(6), 11–15 (2011). https://doi.org/10.1109/MCG.2011.103
55. Schulz, H.J., Akbar, Z., Maurer, F.: A generative layout approach for rooted tree drawings. In: Proceedings of the IEEE Pacific Visualization Symposium, PacificVis 2013, pp. 225–232 (2013). https://doi.org/10.1109/PacificVis.2013.6596149
56. Schulz, H.J., Hadlak, S., Schumann, H.: Point-based visualization for large hierarchies. IEEE Trans. Visual. Comput. Graphics **17**(5), 598–611 (2011). https://doi.org/10.1109/TVCG.2010.89
57. Schulz, H.J., Hadlak, S., Schumann, H.: The design space of implicit hierarchy visualization: a survey. IEEE Trans. Visual. Comput. Graphics **17**(4), 393–411 (2011). https://doi.org/10.1109/TVCG.2010.79
58. Schulz, H.J., Schumann, H.: Visualizing graphs - a generalized view. In: Proceedings of the IEEE International Conference on Information Visualization, IV 2006, pp. 166–173 (2006). https://doi.org/10.1109/IV.2006.130
59. Slingsby, A., Dykes, J., Wood, J.: Using treemaps for variable selection in spatio-temporal visualisation. Inf. Vis. **7**(3), 210–224 (2008). https://doi.org/10.1057/PALGRAVE.IVS.9500185
60. Slingsby, A., Dykes, J., Wood, J.: Configuring hierarchical layouts to address research questions. IEEE Trans. Visual. Comput. Graphics **15**(6), 977–984 (2009). https://doi.org/10.1109/TVCG.2009.128
61. Soares, A.G., et al.: Visualizing multidimensional data in treemaps with adaptive glyphs. In: Proceedings of the IEEE International Conference Information Visualisation, IV 2018, pp. 58–63 (2018). https://doi.org/10.1109/iV.2018.00021
62. Sons, K., Klein, F., Rubinstein, D., Byelozyorov, S., Slusallek, P.: XML3D: interactive 3D graphics for the web. In: Proceedings of the ACM International Conference on 3D Web Technology, Web3D 2010, pp. 175–184 (2010). https://doi.org/10.1145/1836049.1836076
63. Steinbrückner, F., Lewerentz, C.: Understanding software evolution with software cities. Inf. Vis. **12**(2), 200–216 (2013). https://doi.org/10.1177/1473871612438785
64. Stylos, J., Myers, B.: Mapping the space of API design decisions. In: Proceedings of the IEEE Symposium on Visual Languages and Human-Centric Computing, VL/HCC 2007, pp. 50–60 (2007). https://doi.org/10.1109/VLHCC.2007.44

65. Tak, S., Cockburn, A.: Enhanced spatial stability with hilbert and moore treemaps. IEEE Trans. Visual. Comput. Graphics **19**(1), 141–148 (2013). https://doi.org/10. 1109/TVCG.2012.108

66. Tan, W., Fan, Y., Ghoneim, A., Hossain, M.A., Dustdar, S.: From the service-oriented architecture to the web API economy. IEEE Internet Comput. **20**(4), 64–68 (2016). https://doi.org/10.1109/MIC.2016.74

67. Tu, Y., Shen, H.: Visualizing changes of hierarchical data using treemaps. IEEE Trans. Visual. Comput. Graphics **13**(6), 1286–1293 (2008). https://doi.org/10. 1109/TVCG.2007.70529

68. Veras, R., Collins, C.: Optimizing hierarchical visualizations with the minimum description length principle. IEEE Trans. Visual. Comput. Graphics **23**(1), 631–640 (2017). https://doi.org/10.1109/TVCG.2016.2598591

69. Vernier, E.F., Telea, A.C., Comba, J.: Quantitative comparison of dynamic treemaps for software evolution visualization. In: Proceedings of the IEEE Working Conference on Software Visualization, VISSOFT 2018, pp. 96–106 (2018). https:// doi.org/10.1109/VISSOFT.2018.00018

70. Webber, J., Parastatidis, S., Robinson, I.: REST in Practice: Hypermedia and Systems Architecture, 1st edn. O'Reilly Media, Sebastopol (2010)

71. Wettel, R., Lanza, M.: Visual exploration of large-scale system evolution. In: Proceedings of the IEEE Working Conference on Reverse Engineering, WCRE 2008, pp. 219–228 (2008). https://doi.org/10.1109/WCRE.2008.55

72. Wood, J., Brodlie, K., Seo, J., Duke, D., Walton, J.: A web services architecture for visualization. In: Proceedings of the IEEE International Conference on eScience, eScience 2008, pp. 1–7 (2008). https://doi.org/10.1109/eScience.2008.51

73. Wood, J., Isenberg, P., Isenberg, T., Dykes, J., Boukhelifa, N., Slingsby, A.: Sketchy rendering for information visualization. IEEE Trans. Visual. Comput. Graphics **18**(12), 2749–2758 (2012). https://doi.org/10.1109/TVCG.2012.262

74. Würfel, H., Trapp, M., Limberger, D., Döllner, J.: Natural phenomena as metaphors for visualization of trend data in interactive software maps. In: Proceedings of the EG Computer Graphics and Visual Computing, CGVC 2015 (2015). https://doi.org/10.2312/cgvc.20151246

Breaking the Curse of Visual Analytics: Accommodating Virtual Reality in the Visualization Pipeline

Matthias Kraus[(✉)], Matthias Miller, Juri Buchmüller, Manuel Stein, Niklas Weiler, Daniel A. Keim, and Mennatallah El-Assady

University of Konstanz, Konstanz, Germany
{matthias.kraus,matthias.miller,juri.buchmueller,manuel.stein,
niklas.weiler,daniel.keim,mennatallah.el-assady}@uni-konstanz.de
https://www.vis.uni-konstanz.de/

Abstract. Previous research has exposed the discrepancy between the subject of analysis (real world) and the actual data on which the analysis is performed (data world) as a critical weak spot in visual analysis pipelines. In this paper, we demonstrate how Virtual Reality (VR) can help to verify the correspondence of both worlds in the context of Information Visualization (InfoVis) and Visual Analytics (VA). Immersion allows the analyst to dive into the data world and collate it to familiar real-world scenarios. If the data world lacks crucial dimensions, then these are also missing in created virtual environments, which may draw the analyst's attention to inconsistencies between the database and the subject of analysis. When situating VR in a generic visualization pipeline, we can confirm its basic equality compared to other mediums as well as possible benefits. To overcome the guarded stance of VR in InfoVis and VA, we present a structured analysis of arguments, exhibiting the circumstances that make VR a viable medium for visualizations. As a further contribution, we discuss how VR can aid in minimizing the gap between the data world and the real world and present a use case that demonstrates two solution approaches. Finally, we report on initial expert feedback attesting the applicability of our approach in a real-world scenario for crime scene investigation.

Keywords: Visual analytics · Virtual reality · Visualization theory

1 Introduction

Nowadays, data is collected at any time and in any context, in order to gain knowledge about real-world circumstances. Collecting data means storing

This work has received funding from the European Unions Horizon 2020 research and innovation programme under grant agreement No 740754 and from the project FLORIDA (project number 13N14253) by the German Federal Ministry of Education and Research (BMBF, Germany) and the Austrian Security Research Programme KIRAS owned by the Austrian Federal Ministry for Transport, Innovation and Technology (BMVIT).

© Springer Nature Switzerland AG 2020
A. P. Cláudio et al. (Eds.): VISIGRAPP 2019, CCIS 1182, pp. 253–284, 2020.
https://doi.org/10.1007/978-3-030-41590-7_11

digitized snapshots that reflect properties of the real world at a specific point in time. In this article as well as in previous work [42], we use the term *data world* to refer to the sum of related snapshots collected for a particular use case. To eventually gain knowledge from the collected data, visual data exploration approaches, executed on the data world, are mostly indispensable due to the constantly growing amount of data. However, the snapshots contained in the data world never completely reflect reality, but rather a small fraction of the actual real world that lacks contextual information. Crucial context information can quickly be overlooked, which may lead to false conclusions about the real world. Hence, any analysis carried out on the basis of the data world provides results that are at most applicable to the data world itself and not to the real world as initially intended.

In previous work [42], we raised awareness of this *curse of visual data exploration* (see also Sect. 3) and proposed two potential strategies which help to identify and eliminate differences between the data world and the real world. The first strategy revolves around projecting analysis results back into the real world to confirm these results. Based on the projection of analysis results into a space which is closer to the real world, the user can consider additional contextual information to reveal contradictions that might otherwise stay unnoticed. The second strategy reconstructs the real world from the data contained in the data world to verify its validity. The reconstruction is intended to consider all relevant aspects of the real world for the analysis. This enables the user to compare the reconstruction with the real world to uncover differences such as missing features. For each of the two strategies, we provided a concise proof of concept.

In this article, we build on this existing foundation and work out in detail how Virtual Reality (VR) will help us to bridge the gap between the data world and the real world in the future. In the recent past, VR applications have already gained in importance, in particular in the computer graphics sector. For the information visualization and visual analytics domain, several approaches have been proposed as well [33,55,73], but the potential of virtual reality has not yet been fully explored and established. To show that virtual reality is helpful for problems that occur within visual data exploration, we must prove that virtual reality has a place in the established information visualization pipeline [14,24,35], too (Sect. 4). Accordingly, in Sect. 4, we introduce our concept of virtual reality for information visualization and emphasize that VR is not inferior to conventional mediums (printouts, screen-based displays) when applied correctly. Instead, we argue that it is possible to obtain visualizations in VR that are as suitable as those displayed using conventional mediums. Moreover, we show that specific properties inherent to virtual reality environments (VREs) can be used to extend existing visualizations or even to establish new ones. After laying this foundation and defining the place of VR in the information visualization pipeline, we demonstrate in Sect. 5 which advantages (Sect. 5.1) VR has to minimize the effects (Sect. 5.2) or even to solve the initially described curse of visual data exploration. Ultimately, in Sect. 6, we describe an extensive real-world use case in crime scene investigation, applying both of the aforementioned strategies to break the curse. Overall, our contribution consists in (a) identifying dimensions to compare visualizations displayed on different mediums, (b) providing an overview of currently existing major drawbacks of VR for data visualization, (c) listing conceivable present and future possibilities of VR, and (d) discussing potential benefits associated with the use of VR to minimize the gap between the real world and the data world.

2   Background and Related Work

In this section, we deduce the key terms used throughout the article from related literature. For many expressions, such as VR, immersion, or presence, numerous different definitions exist. Our literature review aims at eliminating ambiguities by defining the terminology, thus paving the way for later lines of argumentation. In addition, several examples of VR visualizations are given, which show positive effects of VR in the context of InfoVis to demonstrate the potential of VR and to motivate its application.

2.1   Virtual Reality and VREs

According to the dictionary, we can define the terms 'virtual' and 'reality' as follows: Virtual is something that is "temporarily simulated or extended by computer software" (e.g., virtual memory on a hard disk) [21]; reality is "something that exists independently of ideas concerning it" [20]. To compare VR as it is used in computer science with conventional mediums, we first need to specify what exactly is meant by a VR system in this context. The first experiments with VR technology were done several decades ago [30]. Since then, a vast amount of different VR equipment has been developed, ranging from pocket-sized cardboards like the Google Cardboard [32] to room-sized installations like the CAVE [16]. Different properties of VR prototypes may be the reason why several definitions have been introduced in recent years. Although most of them are relatively similar, they differ in one crucial aspect: They describe the defining properties of VR from different perspectives. We were able to identify three general types of definitions for virtual reality.

⚙ **Hardware-centered VR.** Some VR definitions focus mainly on the hardware aspect. They usually include some kind of stereoscopic display as well as interaction controllers or data gloves. A representative definition of this category was introduced by Ellis [26], who links virtual reality with the hardware it is created with – a head-mounted display. VR displays track users' head positions and adapt themselves accordingly, enabling users to navigate through the virtual reality environment. VR is created by VR hardware.

👤 **Human-centered VR.** Latta and Oberg [45] provide an example of what we call a human-centered definition of virtual reality. They define virtual reality as an interplay of hardware and user. The hardware monitors human behavior and stimulates the human perceptual and muscle system in return.

Their model of a VR system consists of effectors and sensors. Effectors describe the hardware that the VR system uses to stimulate the human body while the sensors are responsible for detecting human actions. VR is created by the synergy of human and computer.

💡 **Concept-centered VR.** Concept-based approaches deduce virtual reality from a conceptional construct. For instance, Steuer [62] derives VR from the concept of telepresence. With regard to Steuer, someone is located in a virtual reality if he or she is telepresent in that environment. He defines telepresence as presence originating from a communication medium (such as a VR headset). Moreover, the degree of the telepresence (and therefore immersion) depends on two dimensions: vividness and interactivity. Vividness describes how well the virtual world emulates the real world, and interactivity describes how much the user can influence it. VR is created by applying a concept to the user (e.g., telepresence: mentally transporting users to a different place).

2.2 Immersion and Presence

Each of these VR definitions addresses, explicitly or implicitly, the two key aspects of VR: immersion and presence. They describe states of a user located in a virtual reality environment. Even though both terms are key elements of VREs, there is an ongoing discussion about their meaning and definition. For example, Cummings and Bailenson summarize the concept of presence as the feeling of "being there" in a mediated virtual environment. Presence is responsible for the effectiveness of VR applications as it magnifies stimuli and virtual interactions (👤 human-centered) [17]. McGloin et al. state that the term 'immersion' is synonymous with presence [49] and offer the same definition as Cummings and Bailenson.

According to Slater and Wilbur, the term 'immersion' describes the employed technological basis for the implementation of a VRE that influences its medial quality (⚙ hardware-centered) [60]. Immersion comprises the ability of a system to provide persuasive conditions that are perceived through human sensory organs. Consequently, the degree of 'immersion' of a system is an objective measure which can be used to compare the quality of different systems. Immersion depends on the ability of a system to fade out the physical reality by addressing all senses of an immersed person as good as possible. The richness of provided information eventually affects the vividness of the VRE (💡 concept-centered) as a function of screen resolution and fidelity. Besides, self-perception is a further crucial aspect of immersion, since having a virtual body in a VRE creates the perception of being part of the constituted surrounding.

Witmer et al. describe presence as the subjective experience of being transported to another place or environment (💡 concept-centered) [72]. The intensity of presence influences the sensation of an individual experiencing the VRE instead of the actual environment. Ijsselsteijn and Riva consider the experienced presence to be a complex and multi-dimensional perception shaped by various cognitive processes and the composition of sensory information (🙎 human-centered) [40]. The perceived degree of presence depends on the ability of the immersed person to have control over the mediated information. The simultaneous stimulation of multiple sensors affects cognition, perception, and emotions, which are key components of presence. The better these dimensions are addressed by the VR system, the higher the degree of presence that can be experienced in a VRE.

According to this extensive amount of recent work, optimizing presence is the primary objective when designing a VRE. Since presence directly depends on the degree of immersion, technological improvements that enhance perceived immersion (e.g., an increase in resolution) also augment presence [18]. High degrees of immersion allow users to have an improved psychological experience of "being there" [17]. The sensation of presence in a VRE can significantly vary between persons as it depends on individual factors such as attentional resources and physical conditions.

2.3   VR for Information Visualization in VA

In the past, various applications for VR have been developed. Some of them show benefits, among others, in the educational [34], biochemical [55], and physiological [5,37,69] domain. But also in the context of InfoVis, various VR applications have proven to be useful, particularly for (geo-)spatial data, volume data, and abstract multivariate data. For example, Gruchalla [33] demonstrated a beneficial effect of immersion on well-path editing tasks for oil rigs. Zhang et al. [73] were able to show that VR can help to improve the understandability of geometric models.

Attempts were also made to visualize abstract data, i.e., data that has no natural representation in the real world, in VR. For instance, Erra et al. [27] found a beneficial effect of VR on graph exploration tasks. Donalek et al. even claim that VR "leads to a demonstrably better perception of a datascape geometry, more intuitive data understanding, and better retention of the perceived relationships in the data" [23]. They presented and compared different VR systems for the immersive visualization of abstract data (e.g., scatterplots). Merino et al. [50] compared different mediums for 3D software visualizations and found that immersive 3D environments led to better recollection rates in experimental subjects than standard computer screens and physical 3D-printed representations. Wagner Filho et al. [68] investigated the performance of users in a variety of analysis tasks using a virtual desk metaphor and a traditional desktop environment, respectively. They found equal or lower error rates in perception tasks with slightly poorer time efficiency, but also a higher perceived efficiency of their proposed VRE.

2.4 Uncertainty and Validation

When analyzing data, uncertainties must always be taken into account. Uncertainties can typically occur both in the data to be analyzed (for example, when collecting data) and in the processing and displaying approaches used (for example, when aggregating data). Uncertainty factors inherent to virtual reality environments are largely caused by the latter since VREs as display technology do not interfere with the displayed data. Sources of uncertainty in VREs mostly impact a user's perception. For example, the resolution of many VR headsets is usually lower than that of current two-dimensional displays. To capture the range of uncertainties that may affect users of a VRE during a data analysis process, a generalized definition should be applied. MacEachren et al. [47] identified different components into which uncertainties can be divided. Another, more hierarchical approach is illustrated by Skeels et al. [59], who describe three different abstraction levels for uncertainty in various domains, including Information Visualization. The most basic level, *measurement precision*, refers to imprecisions in data gathering, for example in positional sensors of VR equipment. The other two levels, *completeness* and *inferences*, describe sources of uncertainty due to a loss of information in sampling or projection processes as well as in predictions of future unknown values. These three levels are completed by an additional *credibility* layer which covers the trustfulness of a data source. Specifically with regard to the visualization domain, Gershon et al. [31] apply the concept of an *imperfect world* to imperfections in data collection as well as visualizations themselves (e.g., overplotting issues).

In its original context, the concept of *measurement precision* as mentioned previously typically applies to the accuracy of the gathered data which are to be explored or analyzed. However, in VR, sensors are employed to translate a user's movements into reciprocal perspective changes to create the effect of immersion. Inaccuracies of these sensors lead to a deviation between a user's expectation with respect to the change in the field of view according to his movements and the actual visual output of the VRE system. If this deviation is too large, for example, due to high latency, negative effects on a user's immersion and engagement can occur. In addition, such sensory deviations are one of the main causes of motion sickness in VR environments. Niehorster et al. [51] conducted an in-depth analysis of sensor accuracy in a state-of-the-art VR product and concluded that, while deviations are present, current off-the-shelf technologies are sufficient for many scientific requirements.

While the accuracy of a system in terms of translating physical movements into virtual movements is critical for the usefulness of a VRE, many use cases also incorporate physical objects into the VRE. For example, physical walls or other obstacles need to be visually communicated to a user. Potential uncertainties in the mapping of physical objects and virtual representations form a second type of sensor-related uncertainty issues for VREs that must be considered. For example, Kim [41] provides a computer-vision based approach for the calibration of VREs with the physical space. Still, not all uncertainties can be countered or completely compensated. Consequently, coping mechanisms need to be applied to diminish the mentioned effects. Benford et al. [3] offer various mechanisms for this. For example, positional errors can be countered with additional position validation. If this is not sufficient, audible cues can improve users' orientation. Another proposed approach is to actively confront the user with the existing uncertainties so that the user is able to create a mental model of the uncertainties experienced and to face them in the future with the aid of experience.

In general, and independently of the visualization medium, a data validation process can be used to verify the plausibility of data values and to counter uncertainty issues, as explained by the United Nations Statistical Commission and Economic glossary [65]. A more detailed view on validation is provided by Wills and Roecker [71], who interpret validation as the ability of a model to deal with variances, missing values, and outliers in a dataset. In recent years, different techniques for dealing with outliers and anomalies have been researched, including the detection of anomalies [11], event detection in time series datasets [22], and noise detection [57]. A survey on further outlier detection techniques is provided by Hodge and Austin [36], identifying three approaches, including pure statistical distribution analysis, labeled-based classifier training on all available data points, and label-based training solely on known outliers.

2.5 Positioning of Our Work

Now that we have surveyed different types of definitions for virtual reality and immersion, we first elaborate on the generic problem we call the "curse of visual data exploration" by recapitulating previous work and bringing it into the current context. Subsequently, we position generic information visualization in a visual analytics workflow model. By deploying a theoretical, ideal VR environment as a substitute for a conventional medium (e.g., monitor screen), we can show its theoretical equality. Then, we reduce the requirements of the ideal VRE and adjust its properties to a more viable VRE. We then discuss where it still performs equally to a monitor screen (Sect. 4). Afterward, we point out benefits of VREs compared to conventional mediums by addressing each step of the visualization pipeline (Sect. 5). We will then deploy a use case with initial expert feedback to demonstrate how VR could help bridge the gap between the data world and the real world (the curse of visual data exploration).

3 The Curse of Visual Data Exploration

Visual Analytics is commonly used to gain knowledge from large datasets which have been recorded from the real world. Consequently, the resulting knowledge is considered applicable to the real world. However, the gathered data represent only a small fraction of the actual real world and lack contextual information. It is possible that crucial context information is not included in the analysis process, leading to false conclusions about the real world. As a consequence, resulting findings relate to the world represented by the collected data and may not be valid for the real world. In previous work [42], we raised awareness of this discrepancy between the data world and the real world (*The Curse of Visual Data Exploration*), which has a large impact on the validity of analysis results. In the following, we will summarize the main characteristics of the curse of visual data exploration and describe two strategies to break the curse, based on our previous work [42]. We then build on this existing foundation incorporating a thought experiment and work out in detail how Virtual Reality will help us to bridge the gap between data world and real world in the future.

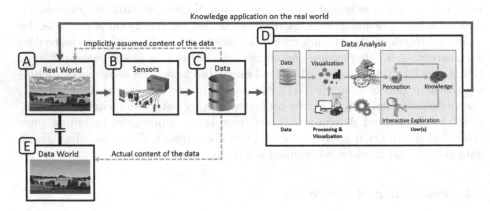

Fig. 1. The *curse of visual data exploration* displayed in an extended visual analytics model. A typical visual analytics task starts in the real world (A). The goal is to gain information about real-world properties. Information is collected using sensor technology (B) and stored in a database (C). Subsequently, the gathered data is processed in a VA pipeline. This process is illustrated in (D) as an adoption of the VA model proposed by van Wijk et al. [70]. Knowledge generated in such workflows is assumed to apply to the real world, implying that the collected data is a complete and correct copy of the real world. However, the data collected on which the analysis is based contains only a fraction of real-world aspects (the data world (E)). This can have a critical influence on the analysis process as the generated knowledge may not be complete or even invalid in the real world. Figure adopted from [42].

Visual Analytics Pipeline. Figure 1(A–D) depicts the overall visual analysis workflow. An analyst strives to gain knowledge about a particular real-world scenario (A). In order to perform an analysis, data from the real world must be collected, which can be achieved, for example, by using sensors (B). The digitized data is stored (C) to be subsequently analyzed in a VA workflow (D). Knowledge gained from the procedure is applied to the investigated real-world scenario (A).

The Curse. However, the digitized data used for the analysis do not flawlessly represent the real world. It is almost impossible to guarantee that the information used for the analysis is exclusive, complete, and correct. Thus, we claim that *data is always wrong* to some extent [42]. As shown in Fig. 1(A, E), we differentiate between the real world that is investigated and the data world, which is the entirety of the collected data based on which the actual data analysis is carried out. Insights gained during visual analysis procedures (D) therefore apply primarily to the data world. If the real world and the data world differ in essential aspects, the knowledge applicable to the data world may not be transferable to the real world as it leads to wrong assertions about properties of the real world. We call the discrepancy between the real world and the data world *the curse of (visual) data exploration.*

Breaking the Curse. We proposed two strategies to *break the curse* [42]. Both strategies aim to provide additional context information and draw the analyst's attention to potentially neglected dimensions in the analysis. The first approach consists in reconstructing the real world as far as possible from the data world so that the analyst can verify its validity. For example, when analyzing the collective behavior of birds, models of birds can be animated in a virtual environment which reflects the original scenario as far as possible, by incorporating context that was not necessarily included in the analysis itself such as the landscape or weather conditions. The second approach is to project analysis results into the real world or a less abstract reflection of the real world. This could be achieved, for instance, by projecting soccer analysis results into the original camera footage of the soccer match. By projecting the analysis results into a space that is closer to the real world, analysts can identify contradictions that would not be noticed otherwise.

Thought Experiment: Hypothetical, Ideal VR. In this paper, we investigate how and to what extent the discrepancy between the data world and the real world (see Fig. 1A, D) can be minimized by deploying VR and the first strategy (reconstructing the real world). As previously stated [42], the gap would be minimal if the data world used for the analysis perfectly resembled the real world. Therefore, we need to develop a theoretical framework to discuss this important topic at a hypothetical level where the data world perfectly resembles the real world and is perceived in a VRE. By doing so, we first assume a VRE that perfectly resembles the real world (1) and elaborate on accompanied advantages (2).

1. What if a VRE is Actually Perceived Like Reality? At least in theory, it is possible to see VR as an equivalent medium for perceiving visualizations. Consequently, we can imagine a hypothetical VR environment that perfectly reflects a real environment, for example, a so-called "Substitutional Reality" environment [58,64] that replaces each real object with a virtual representation. In our case, instead of substituting each real object with an alternative virtual object with similar properties, it is replaced by a virtual representation of itself. For instance, a room with furniture should in our case in VR be represented exactly as it is in the real world. All objects are positioned where they are in the real world. As a result, anyone entering the "duplicated world" in this example would physically walk through the real room and perceive the virtual room (which looks like the real room). Any cloned, "virtual" object could be implicitly touched and interacted with as the virtual world can be considered a "duplicate" of the real environment and each virtual object has a real counterpart.

2. What are the Advantages of Such a Possibility? As soon as we have constructed a perfect representation of the real world, we need to illustrate why we should not simply make use of a conventional medium (such as a regular monitor screen) to inspect the replication of the real world, instead of a VRE. As a first step, we show in Sect. 4 that VR has a place in the established information visualization pipeline. In Sect. 4.1 we demonstrate how the ideal VR can be embedded in the information visualization pipeline, while in Sect. 4.2 we show how perfect a reconstruction of the real world must be in order to be perceived and considered as equivalent by a user. We use the resulting findings in Sect. 5 to verify whether VR is actually preferable to a conventional screen in our scenario.

4   VR in the Visualization Pipeline

In this section, we present a hypothetical, ideal VR as part of the previously presented thought experiment and demonstrate the equality between the hereby created VRE and a conventional screen-based setup with regard to the visualization pipeline (Subsect. 4.1). Furthermore, we discuss to what extent all prerequisites and assumptions made during the thought experiment are necessary to maintain equality between the compared mediums (Subsect. 4.2).

4.1   Embedding the Ideal VR in the InfoVis Pipeline

Since we are mainly concerned with implications and effects on InfoVis and VA procedures, we examine properties of seven dimensions that constitute the main characteristics of a visualization pipeline. As illustrated in Fig. 2, we adopted the VA pipeline by van Wijk et al. [70]. We consider their model to be particularly suitable for our needs, as they provide a highly generalizable pipeline with the focus on the "user side" (knowledge generation, interaction, perception, etc.) instead of the "computer side" (parameterization, model evaluation, etc.). Therefore, we use the states and transitions they identified in the pipeline to compare two different mediums that display the same visualization. We compare the most established medium used for InfoVis and VA tasks – the monitor screen – with the ideal "VR-display" described in the previous section.

(i) Data. Input of the information visualization pipeline (Fig. 2(i)). Assuming that the data collection process is not a visualization task itself, this dimension is not directly influenced by the medium used. However, some data types are predestinated for specific visualizations, and their performance could be affected by the medium used.

(ii) Visualization. Data transformation and visual mapping steps, as they are described by Card et al. [10] in their well-established reference model for visualization, are covered by this dimension (Fig. 2(ii)). It describes how the specification is applied to the data (e.g., how the data is transformed, how visual variables are mapped). The outcome is an "Image" that can be perceived by

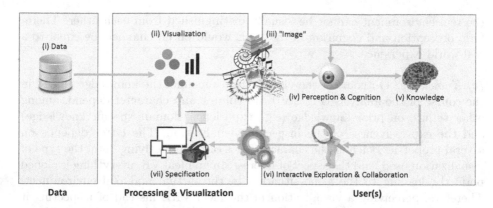

Fig. 2. Visualization pipeline adopted from van Wijk et al. [70]. Data (i) is transformed and mapped (ii) based on specifications (vii). A resulting "Image" (iii) of the visualization is perceived (iv) by the analyst who can apply prior knowledge (v) to interpret the visualization and to acquire new knowledge (v). Subsequent interactive exploration (vi), possibly through collaboration (vi), may lead to the adaption (vi) of the specification (vii).

the user. The visualization is displayed on the virtual monitor in the VRE – just exactly the same way as it is shown on the real monitor. Any data transformation, visual mapping, etc. applicable to the "real" visualization on the monitor screen would also be applicable to the VR visualization.

(iii) "Image" (Image, Animation, etc.). Outcome of the visualization step (Fig. 2(iii)). This dimension describes properties of the actually perceptible entity that is observed by the user. In the field of information visualization, this is most often a 2D image, but it could also be audio, video or haptic feedback, or any other perceivable signal. The process of transforming the visualization into a user-perceivable signal ("Image") constitutes the view transformation step in the model by Card et al. [10]. Applicable types of "Images" depend on the specification given in a respective analysis task. Vice versa, hardware specifications, such as the medium used for display, define characteristics of the "Image". Assuming that the VRE is ideal, the image presented to the user is the same as in the real world.

(iv) Perception & Cognition. Conceiving the "Image" (Fig. 2(iv)). This dimension describes how the visualization is perceived and processed by the user. This step is user-, visualization- and specification-dependent. Different users may perceive the "Image" differently, for instance, due to diverse personal notions (e.g., interpretation of "Image") or physical differences (e.g., color perception). Obviously, also the type of "Image" (e.g., line-chart visualization) and its specification (boundary conditions such as the hardware used for display), influence its perception and cognition. In an ideal VRE, limitations like low resolution, latency, or the weight of a head-mounted display are neglected. In this case, the VRE and

the real environment cannot be visually distinguished from each other. Therefore, perception and cognition in the VRE would on all channels be equal to a real-world experience.

(v) Knowledge. Outcome of the visualization pipeline: the knowledge gained in the course of the analysis (Fig. 2(v)). Its volume and character depend, among other things, on prior knowledge (e.g., experience, domain-specific knowledge) and the expressiveness of the inspected visualization. The latter depends on several properties, such as the characteristics of the underlying data, the type of visualization used, and the aspect of interest. In an ideal VR, everything is cloned perfectly, leading to the same situation in the virtual and real environment. Therefore, perceiving a visualization in the ideal VRE instead of inspecting it on a computer screen would not affect the knowledge generation process in any way.

(vi) Interactive Exploration & Collaboration. Types of interaction and their properties (Fig. 2(vi)). This dimension describes where and how users can interact with the system (i.e., adaption of specification) or with other users (collaboration). In our thought experiment, the real world is cloned, and everything in the virtual copy can be touched and interacted with. Interaction would be the same as in the real world. Collaboration could be achieved, for example, by screen sharing (of the virtual monitor screen).

(vii) Specification. Boundary conditions for the creation of the visualization (Fig. 2(vii)). The specification defines properties like the type of visualization, the dataset to be considered or the mapping of data attributes to visual variables. Moreover, this dimension describes the hardware and (physical/virtual) environmental conditions. Data processing is covered by this dimension as well: Preprocessing, such as normalization or data cleansing, as well as actual data mining steps, such as clustering, pattern detection or classification are implicitly reflected in the pipeline. Such operations can be seen as the execution of an adaption of the specification that leads to a manipulated dataset. For instance, in ML/VA workflows, the specification includes any parameterization for the training of the ML model (learning rates, activation functions, etc.). The specification can be adapted by the user through interaction. Most aspects of this dimension are independent of the medium used for the final visualization. Even though the hardware on which the visualization is displayed in the VRE is different, the outcome is the same.

4.2   Renouncing the Ideal

In this section, we will qualify statements made in Sects. 3 and 4.1 with respect to actual limitations of VREs. Due to technological flaws, the state that a VRE is perceived in exactly the same way as the real world may never be reached. However, for visual analysis, a VRE can already be considered "ideal" if it lacks nothing that would compromise the efficiency or effectiveness of the visualization or task. Therefore, the simulation of the real world does not necessarily have to reach a perfect level at which the VRE cannot be distinguished from the real environment.

Starting with the specification, it would never be possible to perfectly replicate a real-world scenario in VR without any flaws. Detail, resolution, photo-realistic rendering, real-time monitoring, and cloning of real objects into the VRE are only some examples of the many obstacles. More realistic would be an environment that is anchored at some points to the real environment, but is largely independent. The previous example in the thought experiment could, for example, be reduced to a VRE that only overlaps with the real world in certain parts – the desk and the input devices (mouse and keyboard). Anything else in the VRE could be artificial without directly affecting a particular visual analysis task. Although we would not be able to track remaining physical objects or receive haptic feedback from virtual objects in the VRE, objects would still respond to the input devices used – similar to desktop systems.

The 'workaround' used in the thought experiment to deploy hardware input devices for interaction is not the most obvious. Of course, this may be possible, as several prototypes on the market show (e.g., Logitech [46]), but there may be different interaction possibilities that work better with VR (e.g., the mouse only allows navigation on a plane). Mouse and keyboard are the two most established interaction devices most people are familiar with. Introducing new, VR-optimized devices, such as the Vive controller [38], could result in comparatively high learning effort. However, both types of input devices are possible for use in VA workflows. The main goal is to optimize the visual analysis by finding the most suitable interaction device for a particular task.

Depending on the specification, the generated "Image" would be subject to low resolution, non-photo-realistic rendering, and many other artifacts that affect later perception and cognition. The extent to which these artifacts influence the expressiveness and effectiveness of the presented visualization largely depends on the specification and the visualization, but also the task and the user. Some technological drawbacks can certainly be resolved by future technology (e.g., low resolution, latency, photo-realistic rendering). However, it is highly unlikely that one day, the human brain can be completely deceived. Nevertheless, we argue that perfect delusion is not necessary to achieve an equivalent performance in VR compared to screen-based mediums. The information displayed only needs to be conceived similarly efficient, complete, and exclusive.

5   Breaking the Curse with Virtual Reality

After having established in the previous section that VREs can theoretically perform just as well as conventional mediums in terms of conveying information, we attend to the advantages VREs can have. Figure 3 depicts two sides of VREs. The properties discussed in Sect. 4, which compensate conventional mediums, are shown in red. This section focuses on the blue side, presenting exclusive features of VREs that can have a positive impact on different dimensions. Using the previous example, one could clone the real environment and interact with it as if one perceived the real world. This would not make any difference to interacting in and with the real world directly, but it would also be possible to do things which are not possible in the real world. For instance, one could manipulate objects in the environment, install "holograms" or 3D visualizations within the room or exclude distracting parts. In the following subsection, we deal with each of the previously determined dimensions (i)–(vii). Subsequently, we discuss various dimensions that are exclusively available in VREs with regard to their ability to aid in breaking the curse of visual data exploration. This means properties of VREs that help to identify data dimensions that were neglected in the visual data analysis process but have implications for the knowledge generation process.

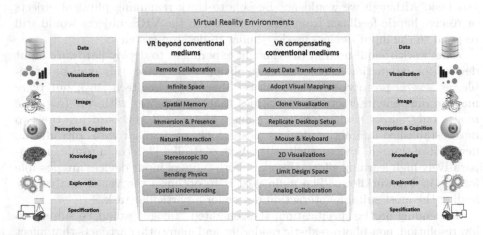

Fig. 3. We investigate if all properties in the visualization pipeline (yellow) for the display of visualizations on a conventional medium (e.g., monitor screen) can be compensated by VREs (right side, red). This is exhaustively discussed in Sect. 4. Red lines indicate to which dimension(s) the examples can be assigned. Subsequently, characteristics of VR that are not present in conventional mediums are discussed in Sect. 5 (left side, blue). Blue lines indicate which dimensions are affected by the exemplary VR characteristics listed. (Color figure online)

5.1   Implications for Dimensions

Data (i). The "Data"-dimension is not directly influenced by the medium used for display. Nevertheless, some data (e.g., spatial, 3D, volume) can be displayed more naturally in VR than in other mediums. This means that different data types allow different visualization applications and influence their effectiveness and expressiveness. As shown in the thought experiment, VR cannot only implement any conventional visualization mapping but also extend it by using, for instance, stereoscopic 3D vision.

Visualization (ii). Previous research has shown that it can be advantageous to use the immersive characteristics of VR to combine multiple views that exploit the three-dimensional aspects of such an environment [67]. The user can work with visualizations in the traditional way, while it is also possible to seamlessly combine both concepts in a single application, lowering the bar for users who are not used to VRE. Thus, the analyst is provided with a flexible system that enables him or her to investigate information as intended. For example, Huang et al. [39] combined GIS and VR in an internet application to support users in decision-making processes by deploying spatial databases and spatial data visualizations. Besides the option to map traditional visualizations of a two-dimensional setting into a VRE, it is possible to enrich the presented information with spatial views. In this way, the three-dimensionality of the visualization design space of the VRE can be utilized to integrate inherent topological information with associated spatial information. Based on these findings, VREs enable the usage of more visual mappings and a larger design space compared to monitor-based 2D applications.

"Image" (iii). If we assume that "Image" is the information of the presented visualization that is sensorily perceived by the user, then the number of possible sensory channels increases in a mediated environment. In addition to the two-dimensional visualization and potential auditory information, we can add haptic feedback, expand it with stereoscopic images, and even include the sense of smell and taste. For instance, haptic feedback can be implemented through a glove that provides advantageous touch feedback [66]. The perceived "Image" is also improved by the stereoscopic nature of identifying the exact position and depth of an object in a VR scene. The integration of three-dimensional auditory information into a VRE allows guiding through 3D visualizations and improves the intensity of the "Image" [48]. Depending on the application, using audio can be more intuitive than encoding all information visually. Porter et al. have shown how to encode spatial direction information in the sense of smell [53]. A VRE that effectively implements such techniques could subconsciously guide users in a certain direction to find the intended results or support the process of exploration. Theoretically, applying methods that stimulate the sense of taste would widen the "Image" space and substantially enrich the VR experience.

Perception & Cognition (iv). A major difference to monitor-based visualization approaches is the complete controllability of the environment, which is presented to a user in a VRE. Studies, mainly from the scientific visualization domain, such as the ones conducted by Laha et al. [44,54], indicate positive effects of immersive environments taking advantage of this aspect on the analysis efficiency of users in certain tasks. As a side effect, environmental factors of classic desktop environments such as changing lighting conditions, noise interference, or third-party interactions or distractions can be mitigated or even completely prevented. As a consequence, when using VR, a user may be able to focus more on the analytical tasks at hand.

Knowledge (v). Knowledge application and generation are indirectly influenced by the medium used. For instance, when inspecting a geo-spatial visualization on a conventional screen, the user would perceive a 2D projection of the scene, whereas he would perceive stereoscopic 3D images when inspecting the same scene in VR. Perceiving the least abstracted copy of the familiar real world (stereoscopic landscape) may help to quickly transfer and apply prior knowledge to the visualization. Vice versa, the less mental mapping is required from the visualization to the real world, the easier it is to transfer gained knowledge to the real world. The smaller distance between the visualization and the real world could ease the transfer of knowledge to the visualization, increase the understandability of visualization results, and thus augment the efficiency and effectiveness of visual analysis procedures. Previous studies (e.g., [43,50,56]) have shown that increased immersion can have positive effects on memorability as well. Spatial memory is improved by heightened spatial understanding and muscular memory. In visual analytics tasks, this could enhance efficiency (e.g., cluster identification, analytic provenance) and effectiveness (volume of knowledge extracted).

Interactive Exploration & Collaboration (vi). Traditionally, one can distinguish between remote collaboration and co-located collaboration. Using VR technology, it is possible to simulate co-located collaboration even though the collaborators are physically separated. This can be realized by projecting the avatars of all collaborators into the same VRE. This approach combines the advantages of both kinds of collaboration: The benefit of remote collaboration of not having to be present in the same room, and the improved interaction possibilities of co-located collaborations. For example, one can point at a specific position to guide other persons' attention. On a monitor, the same can be achieved by the use of a mouse pointer. However, this is limited to pointing at objects shown on the screen itself. Pointing at dashboards, persons or hardware other than the monitor is not possible with the mouse. Another advantage of remote collaboration in VR is that it is possible to see where the other person is looking at. Seeing what collaborators are currently focusing on can help to improve the common conversational grounding.

Specification (vii). In comparison to traditional mediums, one of the main advantages of VR hardware is the fully controllable stereoscopic environment. Visualizations which use a 3D structure to encode information can make good use of this as their shape is easier to identify in a stereoscopic view than in a screen-based 3D view. Compared to other mediums that offer a stereoscopic view, like a 3D print or 3D glasses, VR allows to control and influence the entire visualization space in various ways. This increases the options for displaying information and reduces distractions by removing unnecessary elements from the scene. This works well together with the large field of view that VR hardware usually offers. A large field of view allows showing more information at once since more graphical elements can be seen at the same time. Finally, head-mounted displays are easy to transport, which can be an asset in some cases. Other mediums that offer a large field of view, for example, large-scale displays, are often by far more difficult to transport or not transportable at all.

5.2 Minimizing the Gap Between the Real World and Data World Using VR

As shown in the introduction of the curse of visual data exploration, it would be optimal for the validity and applicability of analysis results if the discrepancy between the data world and the real world would not exist. In reality, however, this is hardly possible, since the digitization process of real-world properties is already subject to a loss of information. Nevertheless and as previously shown, the use of VR may minimize the gap between the data world and the real world as it allows to inspect the data world more naturally and realistically than on a conventional screen. In VR, the snippet of the real world can be inspected on a lower level of abstraction due to stereoscopic 3D perception, natural navigation, and improved spatial understanding. Because of immersion, differences between the inspected virtual environment and the familiar real world can be identified more easily. For instance, one is automatically aware of the absence of familiar properties, such as the atmosphere, wind, sound, and smell.

For the validation of analysis results, the real world could be reconstructed from the data world as close as possible. Due to an incomplete data world, the reconstruction would suffer from incompleteness as well, calling attention to missing or faulty dimensions in the data. Subsequently, analysis outcomes could be projected into the reconstruction. The analyst could then enter the resemblance of the real world and verify if displayed analysis outcomes logically fit in the displayed environment. For instance, based on experiences from the familiar real world, the user would be able to identify errors in analysis results, such as an outlier in a person's movement trajectory that describes a jump over five meters and back in two seconds. Moreover, the absence of important dimensions that possibly influence the analysis outcome could be identified and included in the next analysis iteration. For instance, in a collective behavior analysis task, tracked animals could be represented in a virtual reconstruction of the landscape in which they actually moved. By inspecting this virtual scenario, the analyst may be able to detect errors in the analysis. For example, analysts may recognize that they have not considered obstacles such as rocks, trees, or rivers that influence the animals' trajectories. Animal paths may have certain shapes due to characteristics of the environment, which are easily overlooked in a basic analysis without a more detailed visual examination. VR could also help to identify dimensions that are more difficult to detect on a screen. Being immersed in a copy of the real world allows the analyst to perceive the environment more naturally. Artifacts that contradict the familiar real-world properties, such as gravity, atmosphere, or three-dimensionality, may be identified more quickly than without immersion. For example, when analyzing collective behavior of baboons [63], often only the 2D trajectories of the animals are considered. When representing the trajectories in most detail (3D) in a virtual environment, the analyst would automatically become aware of the circumstance that the animals tend to climb tall trees overnight.

Previous studies have shown a benefit of immersion and VREs for spatial understanding [2], depth perception [1,29], and spatial memory [15]. For example, in soccer analysis, when standing immersed on the virtual soccer field, distances can be estimated more accurately due to stereoscopic perception and familiar distance estimation, since everything is presented according to reality. VR is accompanied by further benefits, as discussed in the previous section. For example, it provides a platform for natural remote collaboration, the entire space in the environment can be used for the display of visualizations, and laws of physics can be bend.

Disclaimer. We do not argue that VR is the one and only solution to minimize the gap between the data world and the real world. For many abstract data types, there is no straightforward mapping to the real world, which allows a visual validation of the real-world context. For instance, when analyzing stock exchange rates, there is no reasonable way to connect the numerical time series data to a physical real-world location. It would not make sense to map analysis results, such as parallel coordinate plots with highlighted clusters, into the virtual duplicate of a trading floor. To apply our approach, a direct connection between the data and the real world is necessary. Moreover, there are many visual analytics applications in which no sufficient data basis exists to reconstruct the real world from the data world and adequately display the analysis results in it. However, there is a wide range of domains in which VR could be used to minimize the discrepancy between the data world and the real world, such as in the emerging fields of collective behavior analysis [8], sports analysis [61], and general geo-spatial data analysis [19]. In the following section, we present an example where we deploy our solution approach for a criminal investigation use case.

6   Use Case: Crime Scene Investigation

In this section, we discuss how virtual reality is used in a visual exploration workflow deploying a use case from the field of crime data analysis. To detect differences between the data world and the real world (the curse of visual data exploration), we try to reproduce as much as possible from the real-world scenario in a virtual environment. Analysis results are, additionally, embedded in the reconstruction to be verified and analyzed by a domain expert. We first describe the developed prototype. Subsequently, we focus on the two integrated solution approaches to "break the curse" by reconstructing the real world from the data world and projecting the analysis results into the real world. Afterward, we discuss the role of VR and its benefits in this use case. Last but not least, we provide results from initial expert feedback on the expansion of the screen-based system with VR.

6.1   Prototype for 4D Crime Scene Investigation

The Austrian Security Research Programme KIRAS owned by the Austrian Federal Ministry for Transport, Innovation and Technology (BMVIT).

After a criminal act, police agencies collect all obtainable information to reconstruct the course of events in the minutest detail. Witness videos and surveillance footage are often central sources. The project FLORIDA [7], as well as the EU project VICTORIA [28], focus on facilitating the analysis of large amounts of video data for LEAs. The FLORIDA project is part of a bilateral project of the German BMBF and the Austrian Security Research Programme KIRAS, which is funded by the Austrian Ministry of Transport, Innovation and Technology (BMVIT), and is run by the FFG (Österreichische Forschungsförderungs Gesellschaft). In the scope of these projects, a tool is currently being developed that allows analysts to merge multiple video sources into a single timeline. Therefore, we visualize a 4D scene (3D space + time) from all input videos by aligning the video sources spatially and temporally. In addition, we enrich the scene with analysis results of semi-automatic object detections from machine learning algorithms. That way, large sets of videos can be inspected simultaneously. Figure 4 shows a sample crime scene (left), an exemplary frame from one of the installed surveillance cameras (center) and a point cloud reconstruction of the crime scene (right). The described use case is based on the multi-camera dataset "IOSB-4D" provided by Pollok [52]. The reconstruction and automatic detections were provided by project partners. Our contribution to the framework is the visualization of all provided information in a (VR) VA environment. In the following, we will explain the framework and its usage.

Fig. 4. Crime scene reconstruction. A crime scene (left) is monitored by multiple video cameras (red dots). Each camera provides video footage (center) that can be analyzed with machine learning algorithms (bounding boxes indicate object detections). Using multiple video sources, a 3D point cloud can be reconstructed from the 2D video streams (right). (Color figure online)

Input Data. The data base for the described analysis is a series of surveillance and witness videos that recorded the same incident from different angles. The sources can be statically installed surveillance cameras or moving hand-held devices, such as cameras or mobile phones. Figure 5 depicts the merging of all video sources into a dynamic 3D scene that is enriched with automatic detection and analysis results, which are also based on the input videos.

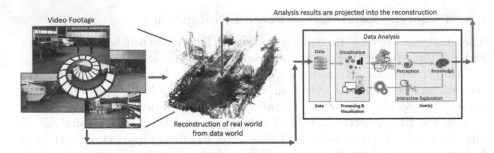

Fig. 5. Multiple video sources are positioned in a 3D scene. With feature matching algorithms, a 3D scene is generated from 2D videos. The origin of each video is registered in this 3D scene. In addition, the videos are analyzed individually, and the results from semi-automatic object detections can be visualized in the 3D scene.

4D Reconstruction. First, the videos are synchronized temporally by aligning them on a common time axis based on timestamps in the videos and striking events within the videos. In a second step, all video sources are spatially aligned using feature matching algorithms. As a result, for each timestamp on the common time axis, the origins of all video sources are positioned in a 3D scene. Next, visual features in individual video frames are manually geo-referenced to transform the 3D scene in metric space. For stereoscopic cameras and depth cameras, the videos are transformed into dynamic point clouds as depicted in Fig. 6. Additionally, a static background mesh can be created using a structure from motion approach (see Fig. 7). The credits for the reconstruction go to our project partners from Fraunhofer IOSB.

Fig. 6. Dynamic point cloud visualization. Each frame in the source video is displayed as a 3D point cloud. The 3D video stream can be played or skipped through by using a timeline slider in the interface.

Video Analysis. The input videos are processed in a semi-automatic feature detection procedure. For each video, objects and persons are detected and tracked for the entire duration of the video. Hereby, pathways of detections are created that can be positioned and visualized in the 3D scene using triangulation (see Fig. 7).

Usage of Tool. The main purpose of the developed tool is to support LEA officers in their investigations of crime scenes. The key benefit is the capability of the tool to visualize multiple source videos at once and in the correct spatial and temporal context. The user can explore the large set of witness videos by browsing through the 4D scene. But the tool is not only suited for the inspection of raw footage videos that are arranged in a 3D space. It can also be used for more advanced visual analytics procedures. For instance, results of object detection algorithms can be classified in different categories (e.g., person or car) and displayed as heatmaps (Fig. 7, bottom line). This allows the user to quickly grasp the overall distribution of the occurrence of certain objects in the entire scene. The user can refine the visualization by filtering the displayed domain or selecting some classes to display them as multiples of heatmaps. Another example is the analysis of trajectories. Movements of persons and vehicles can be tracked and analyzed. The analysis progress follows the visual analytics pipeline model depicted in Fig. 2. The analyst can select multiple time series to compare, apply operations to them (e.g., clustering, classification, event detection), display intermediate results, refine input parameters, and continue this loop until the results are satisfying. Subsequently, the outcome can be presented in the 4D scene for verification and inspection.

Figure 7 shows the interface of the 4D crime scene investigation tool. A timeline at the bottom allows for the temporal navigation within the incident. For spatial navigation, the analyst can fly through the virtual space using a keyboard. A mini-map on the top-right provides an overview of the scene and depicts all registered surveillance cameras in yellow. The original video footage can be inspected by clicking on one of the registered cameras. Detected objects are displayed in the 3D scene as snippets from the original video material, framed by green bounding boxes.

For the evaluation of evidence footage, it would be possible to dispense from visual analysis and only evaluate the results of object detection algorithms quantitatively. However, visual exploration of video sources combined with automatic detections allows domain experts to identify false classifications, outliers, unrecognized entities, and relationships between subjects that are not visible in the mere metadata from automatic procedures. The merging of all video sources into one 4D scene allows the analyst to quickly capture the entire data space without having to watch each video separately. Moreover, the spatial and temporal alignment of video sources helps to understand the spatial and temporal connection between different video sources. The presented approach combines the two presented strategies to bridge the gap between the data world and the real world [42]. The real world is reconstructed from the data world by creating a dynamic 3D scene from 2D video footage. Additionally, analysis results (machine learning object detections) are visually projected back into this reconstruction of the real world (see Fig. 4 center).

Fig. 7. Reconstruction of the real world from the data world and projection of analysis results. The generated 3D point cloud is converted to a 3D mesh of the crime scene. Object detections from all video sources can be displayed within the scene (green rectangles with snapshots from videos). (Color figure online)

6.2   Reconstruction of the Real World

There is always a discrepancy between the real world and the data world on which the analysis is based on (see Sect. 3). This gap may lead to wrong assumptions about the real-world scenario. In the present case, this may be due to dimensions not considered for analysis or incomplete video sources. For example, the scene may contain spatial or temporal black spots that were not monitored but are highly relevant for the reconstruction of the entire sequence of events.

The reconstruction of the real-world scenario from the data world provides an overview and eases the identification of errors in the data. It may also reveal dimensions that have been neglected in the analysis. If inconsistencies between the data world and the real world are found, they can be resolved, which leads to a minimization of the gap. By merging all available video sources into one 4D scene, a common frame of reference is established that maximizes the context of each individual video source. For example, when inspecting one video in the 4D scene, the analyst is aware of all sources that monitor the same area of interest and can switch between multiple videos without having to search the entire pool of footage. This connects all available sources and facilitates the overall analysis. When inspecting the virtual reconstruction of the environment, the analyst subconsciously relates perceived information to known real-world properties. Hence, not modeled dimensions can be identified that have also been neglected in the analysis of individual video sources. For example, a scene may appear unnatural if there is no sky with sun, indicating that weather conditions were not taken into account in the analysis.

6.3 Projection of Analysis Results

The basic idea of this approach is to verify the data of the analysis as well as the analysis outcomes by projecting the results of the analysis into original real-world footage. Displaying analysis results in original footage is less abstract than simply presenting the results without context. For example, when analyzing trajectories of soccer players in a soccer match, it may be more useful to visualize the extracted trajectories directly in the video from which they were extracted than to create an abstract replica of the soccer field for visualization [42]. In the current use case, the 3D reconstruction of the scene can be used to display spatially and temporally aligned automatic detections. This alignment helps to verify intermediate analysis results, such as automated machine learning outputs. Hereby, falsely classified objects can be easily identified.

Moreover, the projection of analysis results helps to associate detections of the same entity coming from different video sources. For example, if the same person is detected in multiple videos but not matched, this results in multiple individual detections for the same entity. Displaying extracted trajectories that describe the movement of entities can be visually verified. In the automatic extraction process, it often happens that affiliations of trails are mixed up when the trails of two entities cross each other. For example, if two persons go to the same location and move apart again, the trail of the first person is connected to the trail of the second person at the point of encounter and vice versa. Visually, this flaw can be identified easily, but it is hard to detect it in more abstract representations of the analysis results. Additionally, it can happen that the pathway of a single entity is split into multiple segments and is, therefore, considered to be from different entities. By visualizing the segments in the virtual environment, such segmentation can be identified and resolved by merging them into one pathway.

6.4 Advantages of Deploying Virtual Reality

Virtual reality allows the analyst to dive into the crime scene and to be fully immersed in the progress of events. The scene can be inspected "from the inside" as if the analyst had been present at the time of the incident. The course of events can be followed in a 3D video of the incident, reconstructed from all available footage. The real world is better replicated since the analyst can naturally walk through the scene. Details, such as the absence of wind or sound, may come to the analyst's mind as the reconstruction is still dissimilar to the known real world.

The reconstruction of the real world from the data world is implicitly improved by changing the way it is observed. The validation of analysis results can be enhanced by augmented depth perception and stereoscopic 3D vision. Distances may be perceived more accurately, allowing a more precise assessment of displayed trajectories. For example, if two waypoints of the trajectory of a person in the scene are far apart, being immersed can help the analyst to classify the respective segment of the trajectory as erroneous. The analyst perceives the virtual environment in familiar metric 3D space (distances are measurable in meters), allowing the user to judge properties based on personal experiences from the real world. For instance, it may become obvious that it is not possible for a human being to cover the depicted distance in the given time.

Various studies have shown a positive effect of immersion on spatial memory [6,9,25]. In our use case, too, the criminal investigator may benefit from immersion with regard to navigation and orientation capabilities, and it may help the analyst to keep track of suspicious actualities in the scene. This is also fostered by how the analyst can explore the scenario. Instead of having to use tools like mouse and keyboard, the analyst can walk naturally through the virtual environment and focus on the analysis task, not on navigation and possibly complex interaction.

The deployment of VR poses several additional opportunities. Among others, it enables remote collaboration of multiple LEA officers in the shared crime scene reconstruction. This way, a common conversational grounding is established, and the communication is eased by displaying avatars for each participating analyst [4]. The avatars can be used to perform gestures and to call attention to something by pointing at it. Moreover, the VRE offers plenty of space for the installation of infographics, control panels, and interfaces for the overall analysis.

6.5   Initial Expert Feedback

The presented tool was developed for LEAs to support them in the investigation of crime scenes. To receive initial expert feedback, we surveyed eleven criminal investigators (CI1–CI11) from the German federal police (Bundespolizei). All invited criminal investigators are employed as police inspectors (degree A10–A15) and have between 10 and 35 years of professional experience. All investigators work with video surveillance or video evidence evaluation on a daily basis. First, we demonstrated the capabilities of the tool and gave a tutorial on how to use it. Then, each expert had the opportunity to use both the standard tool on the screen and the VR extension. After that, we conducted a short interview with each participant to obtain qualitative feedback on our approach. The interview consisted of three parts. First, we addressed the crime scene reconstruction. Secondly, we put the focus on the tool's capability to display analysis results, such as object and person detections. Thirdly, we discussed the use of VR and the challenges and benefits it brings.

The results of our initial evaluation are very promising. Investigators acknowledge its potential to be deployed in the fight against crime, among others as a supporting tool for surveillance tasks, recon missions, and digital forensics. VR, in particular, was perceived as having huge potential in this specific domain. VR was even described as a quantum leap forward in criminal investigation by one of the interviewees.

Reconstruction of Crime Scene. The reconstruction of the crime scene is perceived as very helpful for investigators to re-enact the crime by providing a visual basis for imagination (CI7). Image-based reconstruction from evidence footage is not only faster, cheaper and more flexible than advanced laser scans of the scene in the retrospective forensic procedure, as it is current state-of-the-art in this domain, but also provides a snapshot of the environment from the time of the incident (CI5). Thus, situational conditions which were present at the time of the incident but changed during or shortly after the incident would not be present in a post-event recording of the environment (e.g., when the scene is scanned with lasers after the incident took place). Moreover, the detailed reconstruction of the crime scene puts all video sources into context, promotes orientation, and helps video analysts to put multiple video sources into a temporal and spatial relation (CI5, CI6). Another benefit mentioned was the versatility of the approach. Analysts do not have to physically go to the crime scene but can inspect it from anywhere (CI9). This may also be used in recon missions to gain orientation in an unknown environment before police forces are deployed. As opposed to watching video sequences from the environment, a 3D reconstruction improves spatial understanding (CI5, CI9, CI10). Currently, digital forensics makes use of 360° photo spheres to record characteristics of a crime scene. According to one criminal investigator interviewed, distances and sizes are better perceivable in a 3D scene (P7) than in image spheres or videos. Many of the LEA officers saw the need for a more photo-realistic reconstruction and expect great benefits (CI1–CI4, CI6, CI9, CI10). With advancing technological progress, we suggest that the quality of image-based reconstruction will improve significantly shortly and, therewith, minimize this constraint.

Embedding of Analysis Results. A major benefit expressed by the interviewees was that insights from all video sources are merged into one common scene. Not only are the cameras referenced locally in the scene, but also are the detections of their footage set into a temporal and spatial relation. This makes it easier for the investigators to get an overview of the scene and to re-enact the course of events (CI1–CI3, CI8). The reconstruction in combination with automatic detections from the video footage sets all analysis results into context and creates a "big picture" of the scene (CI7). The detections themselves lead the investigators to interesting and important events and characteristics in the huge pool of evidence videos (CI4, CI6). Often, videos are manually inspected, frame by frame, for hours in order to detect wanted persons or vehicles. The automated approach draws investigators' attention to areas in a video that contain movement or even certain types of detections (persons, faces/identities). Moreover, it facilitates the combined inspection of multiple video sources. For example, it allows to easily identify which cameras monitored which area or person at which time (CI4, CI6). Furthermore, the system can be used to verify automatically generated information, such as detections of persons and movement trajectories. When visualizing analysis results, the analyst can validate the information displayed by relating it to the respective context (environment and detections from other cameras) and real-world experiences (CI9). Moreover, it would be beneficial to ease the mental mapping from the data world to the real world for the display of automated detections, if they would be visualized more realistically (CI10).

For instance, detections of persons walking in the scene should be displayed as 3D models instead of image snippets moving around.

The Deployment of VR. The interviewed investigators mentioned a strong and clear benefit of VR with regard to its capability to project the analyst into the scene. According to one criminal investigator, VR allows to "dive deeper into the crime scene" as "one forgets the physical environment" (CI8). Thus, the 3D scene in VR is more comprehensible, and spatial relations are easier to understand than when inspecting the scene on a screen (CI1–CI3, CI10). Distances can be interpreted more naturally and intuitively (CI1–CI3). As further benefits and future areas of application, the experts mentioned its potential for remote collaboration and mission training in unknown areas (CI1, CI3). As for drawbacks, the investigators mentioned a high learning curve for the new medium and difficulties with the new interaction techniques with controllers (CI9, CI10). Moreover, they saw possible limitations in the high acquisition costs for the required equipment. We argue that this constraint is just a matter of time as it can be expected that prices are going down since VR devices become more and more established.

7 Discussion

We have analyzed the applicability of VR for breaking the curse of visual data exploration on a theoretical level. We first showed the theoretical equality of VR to conventional mediums by going step by step through a VA pipeline model. Thereby, we discussed on what properties a VRE has to fulfill to keep up with a monitor screen and identified characteristics of VR that can pose a benefit. Subsequently, we put the focus on VR properties that can help to break the curse of visual data exploration and demonstrated its feasibility in a use case. However, the approach from the use case cannot be applied to any analysis case. For instance, in order to reconstruct the real world from the data world, there must be an analysis scenario that has an unused visualizable context (e.g., a geo-spatial context) and enough data to actually perform a reconstruction. The strategy described is quite costly, and the potential benefit should be estimated in advance. In some cases, it may be sufficient to rely on alternative strategies to verify the accordance of the data world and the real world for the respective analysis. For example, projecting analysis results back into less abstract data can already help to detect misclassifications or discrepancies in the data (e.g., automatic object detections in videos are overplotted in the original videos). For more abstract data, such as stock market developments, the curse of visual data exploration must still be considered, but assessed formally, for instance, with a detailed influencing factor analysis. VR should only be used if the thereby introduced medium has a significant benefit compared to another medium. For instance, in geo-spatial visualizations, it may be useful to deploy VR due to its capability to project the analyst into the scene, which increases immersion and allows the analyst to inspect the scene naturally.

Currently, state-of-the-art VR technology is placed at a disadvantage because of several drawbacks. Hardware constraints – for instance low resolution or high latency – can introduce physical or psychological phenomena, such as motion sickness or excessive cognitive load. Users are more familiar with non-VR environments, which could affect their overall VR performance. Moreover, many tasks, visualizations, interaction methods, etc. were optimized for conventional desktop PC setups while interaction methods for VR have not yet been sufficiently studied. However, we live in a dynamic era of rapidly evolving technologies, and technological progress may eradicate many of the factors mentioned above shortly, e.g., by providing higher resolution, photo-realistic rendering or advanced interaction technology.

When it comes to the choice of the medium used to display information, it is all about optimizing the visual analytics procedure. According to Chen and Golan [13], this can be seen as an optimization of the cost-benefit ratio. Chen et al. [12] specifically target the cost-benefit analysis for visualizations in VREs. As pointed out in previous work [42], it is important that the analyst can verify the applicability of knowledge generated in the VA process to the investigated real-world scenario. We have demonstrated in the previous sections, how VR can aid such a verification process. However, it depends on the given analysis scenario how the cost-benefit ratio changes when switching from a display-based medium to VR. We argue that the decision to use a VRE as a design space for visualizations must be justified by improvements induced by VR. If the cost-benefit ratio is not affected, we consider it advisable to stick to other mediums.

8 Conclusion

In this paper, we identified seven dimensions which can be used to compare visualizations presented on different mediums. More precisely, we compared the mediums by means of individual characteristics that are substantial in the visual analytics pipeline. When comparing a generic VR medium with conventional mediums on a conceptional level, a theoretical equivalence of the mediums becomes apparent. Of course, this state will realistically never be fully reached. However, VREs have much potential and can be beneficial in several domains. When choosing the most suitable medium for a specific visualization task, the cost-benefit ratio is a key factor which needs to be considered. We expect technological progress to eliminate some of the current disadvantages of VR in future, which may minimize the "cost-side" and in turn strengthen the "benefit-side" of the cost-benefit equation, providing the basis for an extensive application of VR as a viable medium for presenting and exploring data. We presented a use case showing the potential benefit of VR to break the curse of visual data exploration. In a suitable analysis scenario, real-world circumstances to be analyzed can be reconstructed and enriched with intermediate analysis results. By perceiving the reflection of the real world naturally and with all available contextual information, the analyst may be able to identify discrepancies between reconstruction and reality, such as neglected dimensions that are relevant to the knowledge generation process.

References

1. Arms, L., Cook, D., Cruz-Neira, C.: The benefits of statistical visualization in an immersive environment. In: 1999 Proceedings of the IEEE Virtual Reality, pp. 88–95. IEEE (2003). https://doi.org/10.1109/vr.1999.756938
2. Bach, B., Sicat, R., Beyer, J., Cordeil, M., Pfister, H.: The hologram in my hand: how effective is interactive exploration of 3D visualizations in immersive tangible augmented reality? IEEE Trans. Visual Comput. Graphics 24(1), 457–467 (2018). https://doi.org/10.1109/TVCG.2017.2745941
3. Benford, S., et al.: Coping with uncertainty in a location-based game. IEEE Pervasive Comput. 2(3), 34–41 (2003)
4. Benford, S., Bowers, J., Fahlén, L.E., Greenhalgh, C., Snowdon, D.: User embodiment in collaborative virtual environments. In: Proceedings of the SIGCHI Conference on Human Factors in Computing Systems, CHI 1995, Denver, Colorado, USA, 7–11 May 1995. pp. 242–249 (1995). https://doi.org/10.1145/223904.223935
5. Bidarra, R., Gambon, D., Kooij, R., Nagel, D., Schutjes, M., Tziouvara, I.: Gaming at the dentist's – serious game design for pain and discomfort distraction. In: Schouten, B., Fedtke, S., Bekker, T., Schijven, M., Gekker, A. (eds.) Games for Health, pp. 207–215. Springer, Wiesbaden (2013). https://doi.org/10.1007/978-3-658-02897-8_16
6. Bliss, J.P., Tidwell, P.D.: The effectiveness of virtual reality for administering spatial navigation training to police officers. In: Proceedings of the Human Factors and Ergonomics Society Annual Meeting, vol. 39, no. 14, p. 936 (2012). https://doi.org/10.1177/154193129503901412
7. BMBF: FLORIDA - Flexibles, teilautomatisiertes Analysesystem zur Auswertung von Videomassendaten. http://www.florida-project.de/. Accessed 14 July 2019
8. Buchmuller, J., Cakmak, E., Keim, D.A., Jackle, D., Brandes, U.: MotionRugs: visualizing collective trends in space and time. IEEE Trans. Visual Comput. Graphics 25(1), 76–86 (2018). https://doi.org/10.1109/tvcg.2018.2865049
9. Cánovas, R., León, I., Roldán, M.D., Astur, R., Cimadevilla, J.M.: Virtual reality tasks disclose spatial memory alterations in fibromyalgia. Rheumatology (Oxford, England) 48(10), 1273–1278 (2009). https://doi.org/10.1093/rheumatology/kep218
10. Card, S.K., Mackinlay, J.D., Shneiderman, B.: Readings in Information Visualization: Using Vision to Think (Interactive Technologies). Morgan Kaufmann Publishers Inc., San Francisco (1999). https://doi.org/10.1234/12345678
11. Chandola, V., Banerjee, A., Kumar, V.: Anomaly detection: a survey. AMC Comput. Surv. (CSUR) 41(3), 15 (2009). https://doi.org/10.1145/1541880.1541882
12. Chen, M., Gaither, K., John, N.W., McCann, B.: Cost-benefit analysis of visualization in virtual environments. IEEE Trans. Visual Comput. Graphics 25(1), 32–42 (2018). http://arxiv.org/abs/1802.09012
13. Chen, M., Golan, A.: What may visualization processes optimize? IEEE Trans. Visual Comput. Graphics 22(12), 2619–2632 (2016). https://doi.org/10.1109/TVCG.2015.2513410
14. Chi, E.H.: A taxonomy of visualization techniques using the data state reference model. In: IEEE Symposium on Information Visualization 2000 (INFOVIS 2000), Salt Lake City, Utah, USA, 9–10 October 2000, pp. 69–75 (2000). https://doi.org/10.1109/INFVIS.2000.885092

15. Cockburn, A., McKenzie, B.: Evaluating the effectiveness of spatial memory in 2D and 3D physical and virtual environments. In: Proceedings of the SIGCHI Conference on Human Factors in Computing Systems Changing our World, Changing Ourselves - CHI 2002, p. 203 (2002). https://doi.org/10.1145/503376.503413

16. Cruz-Neira, C., Sandin, D.J., DeFanti, T.A.: Surround-screen projection-based virtual reality: the design and implementation of the CAVE. In: The Conference on Computer Graphics and Interactive Techniques (SIGGRAPH 1993), pp. 135–142. ACM (1993). https://doi.org/10.1145/166117.166134

17. Cummings, J.J., Bailenson, J.N., Fidler, M.J.: How immersive is enough? A foundation for a meta-analysis of the effect of immersive technology on measured presence. In: Proceedings of the International Society for Presence Research Conference, vol. 19, no. 2, pp. 272–309 (2012). https://doi.org/10.1080/15213269.2015.1015740

18. Bowman, D.A., McMahan, R.P.: Virtual reality: how much immersion is enough? Computer **40**(7), 36–43 (2007). https://doi.org/10.1109/MC.2007.257

19. De Smith, M.J., Goodchild, M.F., Longley, P.: Geospatial Analysis: A Comprehensive Guide to Principles, Techniques, and Software Tools. Troubador Publishing Ltd., Leicester City (2007). https://doi.org/10.1111/j.1467-9671.2008.01122.x

20. Dictionary: Reality: Dictionary. http://www.dictionary.com/browse/reality. Accessed 14 July 2019

21. Dictionary: Virtual: Dictionary. http://www.dictionary.com/browse/virtual. Accessed 14 July 2019

22. Dasgupta, D., Forrest, S.: Novelty detection in time series data using ideas from immunology. In: Proceedings of the International Conference on Intelligent Systems, pp. 82–87 (1996). http://citeseerx.ist.psu.edu/viewdoc/summary?doi=10.1.1.50.9949d

23. Djorgovski, S.G., Donalek, C., Lombeyda, S., Davidoff, S., Amori, M.: Immersive and collaborative data visualization and analytics using virtual reality. In: American Geophysical Union, Fall Meeting 2018, pp. 609–614 (2018). https://doi.org/10.1109/BigData.2014.7004282

24. Dos Santos, S., Brodlie, K.: Gaining understanding of multivariate and multidimensional data through visualization. Comput. Graph. (Pergamon) **28**(3), 311–325 (2004). https://doi.org/10.1016/j.cag.2004.03.013

25. Dünser, A., Steinbügl, K., Kaufmann, H., Glück, J.: Virtual and augmented reality as spatial ability training tools. In: Proceedings of the ACM SIGCHI New Zealand Chapter's International Conference on Computer-Human Interaction Design Centered HCI - CHINZ 2006, pp. 125–132. ACM (2006). https://doi.org/10.1145/1152760.1152776

26. Ellis, S.R.: What are virtual twins? IEEE Comput. Graphics Appl. **14**(1), 17–22 (1994). https://www.wisegeek.com/what-are-virtual-twins.htm#didyouknowout

27. Erra, U., Malandrino, D., Pepe, L.: Virtual reality interfaces for interacting with three-dimensional graphs. Int. J. Hum. Comput. Interact. **35**(1), 75–88 (2019). https://doi.org/10.1080/10447318.2018.1429061

28. EU: VICTORIA - Video Analysis for Investigation of Criminal and Terrorist Activities. https://www.victoria-project.eu/. Accessed 14 July 2019

29. Filho, J.A.W., Dal, C., Freitas, S., Rey, M.F., Freitas, C.M.D.S.: Immersive visualization of abstract information: an evaluation on dimensionally-reduced data scatterplots automatic construction of large readability corpora view project blind guardian view project immersive visualization of abstract information: an evaluation In: 2018 IEEE Conference on Virtual Reality and 3D User Interfaces (VR), pp. 483–490 (2018). https://doi.org/10.1109/VR.2018.8447558

30. Fluke, C.J., Barnes, D.G.: The ultimate display. Multimedia: from Wagner to virtual reality, pp. 506–508 (2016). https://arxiv.org/abs/1601.03459
31. Gershon, N.D.: Visualization of an imperfect world. IEEE Comput. Graphics Appl. **18**(4), 43–45 (1998). https://doi.org/10.1109/38.689662
32. Google: Google VR Cardboard. https://www.google.com/get/cardboard. Accessed 14 July 2019
33. Gruchalla, K.: Immersive well-path editing: investigating the added value of immersion. In: IEEE Virtual Reality Conference 2004, pp. 157–164 (2004). https://doi.org/10.1109/VR.2004.42
34. Gutiérrez, F., et al.: The impact of the degree of immersion upon learning performance in virtual reality simulations for medical education. J. Invest. Med. **55**, S91 (2007). https://doi.org/10.1097/00042871-200701010-00099
35. Haber, R.B., Mcnabb, D.A.: Visualization Idioms: a conceptual model for scientific visualization systems. Vis. Sci. Comput. **74**, 93 (1990)
36. Hodge, V., Austin, J.: A survey of outlier detection methodologies. Artif. Intell. Rev. **22**(2), 85–126 (2004). https://doi.org/10.1007/s10462-004-4304-y
37. He, L., Guayaquil-Sosa, A., Mcgraw, T.: Medical image atlas interaction in virtual reality. In: Proceedings of Immersive Analytics Workshop at VIS (2017)
38. HTC: Vive Controller. https://vive.com/eu/accessory/controller/. Accessed 14 July 2019
39. Huang, B., Jiang, B., Li, H.: An integration of GIS, virtual reality and the Internet for visualization, analysis and exploration of spatial data. Int. J. Geogr. Inf. Sci. **15**(5), 439–456 (2001). https://doi.org/10.1080/13658810110046574
40. Ijsselsteijn, W., Riva, G.: Being there: the experience of presence in mediated environments, p. 14. In: Emerging Communication: Studies on New Technologies and Practices in Communication. IOS Press, Netherlands (2003). citeulike-article-id: 4444927
41. Kim, W.S.: Computer vision assisted virtual reality calibration. IEEE Trans. Robot. Autom. **15**(3), 450–464 (1999)
42. Kraus, M., Weiler, N., Breitkreutz, T., Keim, D., Stein, M.: Breaking the curse of visual data exploration: improving analyses by building bridges between data world and real world. In: Proceedings of the International Joint Conference on Computer Vision, Imaging and Computer Graphics Theory and Applications - Volume 3, IVAPP, pp. 19–27. INSTICC, SciTePress (2019). https://doi.org/10.5220/0007257400190027
43. Krokos, E., Plaisant, C., Varshney, A.: Virtual memory palaces: immersion aids recall. Virtual Reality **23**(1) (2019). https://doi.org/10.1007/s10055-018-0346-3
44. Laha, B., Sensharma, K., Schiffbauer, J.D., Bowman, D.A.: Effects of immersion on visual analysis of volume data. IEEE Trans. Visual Comput. Graphics **18**(4), 597–606 (2012). https://doi.org/10.1109/TVCG.2012.42
45. Latta, J.N., Oberg, D.J.: A conceptual virtual reality model. IEEE Comput. Graphics Appl. **14**(1), 23–29 (1994). https://doi.org/10.1109/38.250915
46. Logitech: Logitech Bridge. https://github.com/Logitech/logi_bridge_sdk. Accessed 14 July 2019
47. MacEachren, A.M., et al.: Visualizing geospatial information uncertainty: what we know and what we need to know. Cartogr. Geogr. Inf. Sci. **32**(3), 139–160 (2005). https://doi.org/10.1559/1523040054738936
48. Madole, D., Begault, D.: 3-D Sound for Virtual Reality and Multimedia, vol. 19. Academic Press Professional Inc., San Diego (1995). https://doi.org/10.2307/3680997

49. McGloin, R., Farrar, K., Krcmar, M.: Video games, immersion, and cognitive aggression: does the controller matter? Media Psychol. **16**(1), 65–87 (2013). https://doi.org/10.1080/15213269.2012.752428

50. Merino, L., et al.: On the impact of the medium in the effectiveness of 3D software visualizations. In: Proceedings of the 2017 IEEE Working Conference on Software Visualization, October, vol. 2017, pp. 11–21. IEEE (2017). https://doi.org/10.1109/VISSOFT.2017.17

51. Niehorster, D.C., Li, L., Lappe, M.: The accuracy and precision of position and orientation tracking in the HTC vive virtual reality system for scientific research. i-Perception **8**(3), 2041669517708205 (2017)

52. Pollok, T.: A new multi-camera dataset with surveillance, mobile and stereo cameras for tracking, situation analysis and crime scene investigation applications. In: Proceedings of the International Conference on Video and Image Processing, pp. 171–175 (2018). https://doi.org/10.1145/3301506.3301542

53. Porter, J., Anand, T., Johnson, B., Khan, R.M., Sobel, N.: Brain mechanisms for extracting spatial information from smell. Neuron **47**(4), 581–592 (2005). https://doi.org/10.1016/j.neuron.2005.06.028

54. Prabhat, Forsberg, A.S., Katzourin, M., Wharton, K., Slater, M.: A comparative study of desktop, fishtank, and cave systems for the exploration of volume rendered confocal data sets. IEEE Trans. Visual Comput. Graphics **14**(3), 551–563 (2008). https://doi.org/10.1109/TVCG.2007.70433

55. Probst, D., Reymond, J.L.: Exploring DrugBank in virtual reality chemical space. J. Chem. Inf. Model. **58**(9), 1731–1735 (2018). https://doi.org/10.1021/acs.jcim.8b00402

56. Ragan, E.D., Sowndararajan, A., Kopper, R., Bowman, D.A.: The effects of higher levels of immersion on procedure memorization performance and implications for educational virtual environments. Presence Teleoperators Virtual Environ. **19**(6), 527–543 (2010). https://doi.org/10.1162/pres_a_00016

57. Rehm, F., Klawonn, F., Kruse, R.: A novel approach to noise clustering for outlier detection. Soft. Comput. **11**(5), 489–494 (2007). https://doi.org/10.1007/s00500-006-0112-4

58. Simeone, A.L., Velloso, E., Gellersen, H.: Substitutional reality: using the physical environment to design virtual reality experiences. In: Proceedings of the ACM Conference on Human Factors in Computing Systems, CHI, pp. 3307–3316. ACM (2015). https://doi.org/10.1145/2702123.2702389

59. Skeels, M., Lee, B., Smith, G., Robertson, G.G.: Revealing uncertainty for information visualization. Inf. Vis. **9**(1), 70–81 (2010). https://doi.org/10.1057/ivs.2009.1

60. Slater, M., Wilbur, S.: A framework for immersive virtual environments (FIVE): speculations on the role of presence in virtual environments. Presence Teleoperators Virtual Environ. **6**(6), 603–616 (1997). https://doi.org/10.1162/pres.1997.6.6.603

61. Stein, M., et al.: Director's cut: analysis and annotation of soccer matches. IEEE Comput. Graphics Appl. **36**(5), 50–60 (2016). https://doi.org/10.1109/MCG.2016.102

62. Steuer, J.: Defining virtual reality: dimensions determining telepresence. J. Commun. **42**(4), 73–93 (1992). https://doi.org/10.1111/j.1460-2466.1992.tb00812.x

63. Strandburg-Peshkin, A., Farine, D.R., Crofoot, M.C., Couzin, I.D.: Habitat and social factors shape individual decisions and emergent group structure during baboon collective movement. Elife **6**, e19505 (2017). https://doi.org/10.7554/eLife.19505

64. Suzuki, K., Wakisaka, S., Fujii, N.: Substitutional reality system: a novel. Sci. Rep. **2**, 1–9 (2012). https://doi.org/10.1038/srep00459
65. UNDP: United Nations Statistical Commission and Economic Commission for Europe glossary of terms on statistical data editing. https://webgate.ec.europa.eu/fpfis/mwikis/essvalidserv/images/3/37/UN_editing_glossary.pdf (2018). Accessed 14 July 2019
66. Van der Meijden, O.A.J., Schijven, M.P.: The value of haptic feedback in conventional and robot-assisted minimal invasive surgery and virtual reality training: a current review. Surg. Endosc. **23**(6), 1180–1190 (2009). https://doi.org/10.1007/s00464-008-0298-x
67. Verbree, E., Maren, G.V., Germs, R., Jansen, F., Kraak, M.J.: Interaction in virtual world views-linking 3D GIS with VR. Int. J. Geogr. Inf. Sci. **13**(4), 385–396 (1999). https://doi.org/10.1080/136588199241265
68. Wagner Filho, J.A., Freitas, C.M., Nedel, L.: VirtualDesk: a comfortable and efficient immersive information visualization approach. Comput. Graphics Forum **37**(3), 415–426 (2018). https://doi.org/10.1111/cgf.13430
69. Walshe, D.G., Lewis, E.J., Kim, S.I., O'Sullivan, K., Wiederhold, B.K.: Exploring the use of computer games and virtual reality in exposure therapy for fear of driving following a motor vehicle accident. CyberPsychol. Behav. **6**(3), 329–334 (2003). https://doi.org/10.1089/109493103322011641
70. van Wijk, J.J.: The value of visualization. In: IEEE Visualization Conference, pp. 79–86 (2005). https://doi.org/10.1109/VISUAL.2005.1532781
71. Wills, S., Roecker, S.: Statistics for Soil Survey. http://ncss-tech.github.io/stats_for_soil_survey/chapters/9_uncertainty/Uncert_val.html#1_introduction (2017). Accessed 14 July 2019
72. Witmer, B.G., Singer, M.J.: Measuring presence in virtual environments: a presence questionnaire. Presence Teleoperators Virtual Environ. **7**(3), 225–240 (1998). https://doi.org/10.1162/105474698565686
73. Zhang, S., et al.: An immersive virtual environment for DT-mri volume visualization applications: a case study. In: Proceedings of the Conference on Visualization, 2001, pp. 437–440. IEEE Computer Society (2001). https://doi.org/10.1109/VISUAL.2001.964545

Designing a Visual Analytics System for Medication Error Screening and Detection

Tabassum Kakar[1]([✉]), Xiao Qin[1], Cory M. Tapply[1], Oliver Spring[1],
Derek Murphy[1], Daniel Yun[1], Elke A. Rundensteiner[1], Lane Harrison[1],
Thang La[2], Sanjay K. Sahoo[2], and Suranjan De[2]

[1] Worcester Polytechnic Institute, Worcester, MA 01609, USA
{tkakar,xqin,cmtapply,obspring,dmurphy2,dyun,rundenst,ltharrison}@wpi.edu
[2] U.S. Food and Drug Administration, Silverspring, MD 20993, USA
{thang.la,sanjay.sahoo,suranjan.de}@fda.hhs.gov

Abstract. Drug safety analysts at the U.S. Food & Drug Administration analyze medication error reports submitted to the Adverse Event Reporting System (FAERS) to detect and prevent detrimental errors from happening in the future. Currently this review process is time-consuming, involving manual extraction and sense-making of the key information from each report narrative. There is a need for a visual analytics approach that leverages both computational techniques and interactive visualizations to empower analysts to quickly gain insights from reports. To assist analysts responsible for identifying medication errors in these reports, we design an interactive Medication Error Visual analytics (MEV) system. In this paper, we describe the detailed study of the Pharmacovigilance at the FDA and the iterative design process that lead to the final design of MEV technology. MEV a multi-layer treemap based visualization system, guides analysts towards the most critical medication errors by displaying interactive reports distributions over multiple data attributes such as stages, causes and types of errors. A user study with ten drug safety analysts at the FDA confirms that screening and review tasks performed with MEV are perceived as being more efficient as well as easier than when using their existing tools. Expert subjective interviews highlight opportunities for improving MEV and the utilization of visual analytics techniques in general for analyzing critical FAERS reports at scale.

Keywords: Treemaps · Visual analytics · Pharmacovigilance · Medication errors

1 Introduction

Medication errors represent major threats to public health safety [19]. A medication error involves preventable mistake that can lead to either inappropriate

© Springer Nature Switzerland AG 2020
A. P. Cláudio et al. (Eds.): VISIGRAPP 2019, CCIS 1182, pp. 285–312, 2020.
https://doi.org/10.1007/978-3-030-41590-7_12

use of the medication or patient harm while the medication is in the control of the patient, or health care professional [33]. Examples of such medication errors include but are not limited to the wrong administration or handling of drugs due to ambiguous drug names or drug labels. It is estimated that medication errors cause 1 of 854 inpatient and 1 of 131 outpatient deaths [25] and an annual cost burden of $20 billion [1]. Medication errors may also cause adverse drug events [31], which can have detrimental consequences such as patient harm, unnecessary hospitalization, time away from work and additional resource utilization [43]. Such adverse events not only impact a patient's quality of life but can even lead to death [7]. Therefore, detection and prevention of medication errors is a high priority task for health care systems worldwide [2].

To detect and prevent medication errors, developed countries have established medication vigilance systems that maintain post marketing drug surveillance programs to capture reports about potential medication errors and adverse reactions, once the drug is released to the market [35]. In the U.S., the Food and Drug Administration (FDA) maintains the Adverse Event Reporting System (*FAERS*), to which consumers, health care professionals, and drug manufacturers submit reports related to medication errors and adverse events. At the FDA, drug safety analysts in the Division of Medication Errors Prevention and Analysis (DMEPA) analyze the FAERS reports to detect any errors in the administration, prescription or distribution of drugs. If a medication error is identified then it is addressed via regulatory actions such as, revising the usage instructions, or container labels, or changing a drugname.

A drug safety analyst reviews a report based on various factors including the severity, type and cause of the error and determines if the incident corresponds to a more general problem that may potentially warrant an action. Therefore, this analysis tends to require the evaluation of many reports over a long period of time. While these reports contain useful information such as patient demographics and administered drugs in a structured format, actual details about the error tend to be discussed in more depth in the text narrative associated with the report.

Currently, the tools used by the FDA's drug safety evaluators mainly support retrieval of reports from the FAERS database associated with their assigned set of products [8]. The analyst gathers the information regarding a specific error type manually by reading through each report narrative one by one. SQL is used to collect basic statistics about a report's collection, e.g., getting a count of the total number of reports for a particular product within a specified time, or retrieving reports based on other structured information such as demographics or reporter location.

These manual approaches are tedious and time consuming particularly as the volume of reports grows. Computational techniques alone are not sufficient in supporting the drug review process, as they lack the power to enhance the human perception and cognition to interactively manipulate data [15]. Visual analytics, on the other hand, is useful in allowing analysts interactively gain insights from data by analyzing data summaries and distributions [15]. There are a few widely accepted works on leveraging automation and combining it with a human-in-the-loop approach [32] to support analysts workflows. Our goal here is to design interactive visualization and analytics techniques to address these limitations.

For this, we design MEV, a visual analytics system that supports the screening and analysis of medication error reports (Fig. 8). Based on the limitations of the FDA's current workflow, we first leverage recently developed natural language processing techniques [4,41] to extract key information from the reports' narratives crucial for the detection of the medication errors. MEV uses a multi-layer treemap visualization to display this extracted key information along with other structured information such as demographics of the patients affected by the errors. MEV's visual interactions guide analysts towards the most critical errors by empowering them to identify the associated data attributes. A timeline view allows analysts to see the overall distribution of the reports over a period of time. Finally, analysts can interactively study the reports associated with the screened errors.

This paper is an extension of our preliminary work [24] on leveraging visual analytics for the detection of medication errors. We now introduce a detailed analysis of the requirements that led to the design for MEV. In particular, we provide a description of the design process of MEV including the formative interviews with domain experts and the insights gained that helped in the design of MEV (Sect. 3). We elaborate on the importance of medication error detection via motivating use-case examples. We also present a detailed analysis of the current tools and workflow of the analysts at the FDA that lead us to gain an understanding of the analysts' pain points and limitations of the current tools. Based on these insights we solicit a set of design requirements for MEV (Sect. 3). We also include a detailed design process discussing various design alternatives and the decisions that led to the final design of MEV (Sect. 4). We enhance the evaluation of the system by including a detailed analysis of the system usability study (Sect. 6.2).

The results of our user study with ten analysts at the FDA suggest that with MEV, participants were able to perform review related tasks more quickly and perceived the tasks to be easier as compared with using their existing tools. Further, post-study qualitative interviews illustrate participants' interest regarding the use of visual analytics for medication error detection and suggestions for the improvement of the technology.

2 Related Work

In this section we focus on the existing work that study similar data types and analysis goals. The majority of our data including the key elements extracted using NLP such as drugnames, types and causes of errors are categorical, also called facets. A wide range of works have used facets as interactive filters to search and browse data [13,27,38]. FacetMap [38] focuses on using interactive visualizations for the exploration of facets in a data set. The system supports filtering, however, it does not support the analysis of relationships among facets. FacetLens [27], an extension of FacetMap, supports users to observe trends and explore relationships within faceted datasets. Although these faceted systems [27,38] are useful in data exploration, these systems are designed in a way that the interface is divided into a main view and a facet area. With such a design, only one data item (facet) can be browsed at a time. In the case of medication error screening, however, it is crucial to see the effect of selection of one item on others to support the quick identification of the data points representing more evidence towards the concerning identified errors.

Treemaps [5] have been widely used in visualization-based systems [18,28]. For example, NV [18] allows analysts and system administrators to manage and analyze vulnerabilities on their networks via histograms and treemaps. SellTrend [28] utilizes treemaps to support the analysis and exploration of transaction failures in airline travel requests. These tools do not support the extraction of name-entities from textual data. JigSaw [39], on the other hand, is a comprehensive tool for investigating textual documents by visualizing name entities and their relationships to reveal hidden criminal plots. There is a need for the support of temporal data analysis for reviewing and screening reports that is not dealt with in JigSaw.

In the medical domain, few systems have been designed to help in preventing medication errors, such as clinical information systems [21] and medication-reconciliation tools [34]. Ozturk et al. [34] designed a system to help clinicians and emergency room staff review a patient's one year long prescription history to support their medication-reconciliation efforts. Varkey et al. [40] studied the effect of interventions on decreasing drug administration related medication errors. Other tools have focused on designing user-friendly and efficient interfaces to assist in error reporting [37]. On the other hand, clinical decision support systems focus on reducing medication errors during prescription [21]. Others [43] have attempted to prevent medication errors in clinical settings by using clustering techniques to group reports based on similarity scores. However, these tools are designed with the goal of medication error prevention during the administration or prescription of the drugs.

Our work instead supports the workflow that starts after the medication errors have already occurred and have been reported to the concerned regulatory authorities such as the FDA. For example, if two drugs have similar names and FDA receives error reports about these drugs being prescribed interchangeably. Then the FDA analysts carefully examine such reports to recommend actions so that different products can be differentiated easily, in this case, changing the drugname to prevent such error from happening in the future. To the best of our knowledge, no visual analytics tool exists that directly support the exploration and analysis of medication error reports.

3   Requirement Analysis

This section provides background information on the medication error reports and the importance of the detection of these errors. We conducted formative interviews with drug safety analysts at the FDA to understand the data, tools and tasks involved in the analysis of these medication error reports.

3.1 Interviews with Domain Experts

We organized a series of one-on-one expert observations and semi-formal interview sessions with five drug safety analysts at the Division of Medication Error Prevention and Analysis (DMEPA) at the FDA. A similar approach had been taken for the design of DIVA [23], a visual analytics tool to explore adverse reactions related to drug-drug interactions. Our main goal was to gain an understanding of the current workflow of reviewing the stream of incoming error reports and in particular, to identify the challenges faced by the analysts while reviewing these reports.

Our first session was to become *familiar with the domain* and included discussions about the overall medication error detection process, the reports and the challenges of analyzing these reports. As a second step, to get precise facts and figures, we designed a *questionnaire* that included detailed questions about the workflow based on the initial discussions. Questions included, for instance, how many drugs a safety analyst gets assigned to? or, how many reports are received on a weekly basis? or, what is the most challenging aspect involved in the review of these reports?

Our second session with the same five drug safety analysts after few weeks involved field observation [26] followed by administering our carefully prepared questionnaire. We observed these analysts at the FDA in a think-aloud manner while performing their routine tasks of reports analysis. This helped us derive information on their existing practices and tools used for reviewing reports. These interviews revealed that certain information was critical to the workflow and also helped us identify the limitations of their current tools. These insights then were instrumental in being able to derive use-cases and elicit detailed requirements related to the tasks that could potentially leverage visual analytics to make the review process efficient.

The requirements were refined iteratively in a sequence of follow-up sessions through brainstorming and discussions with the same five analysts. Based on the collected requirements, we selected existing visualizations that could best support these processes and thus meet the identified requirements. That is, we showed these analysts sketches of multiple visual designs, such as parallel coordinates and node-link diagrams. The discussions about design alternatives helped us collect further design requirements, such as scalability and readability of the visualization. Based on these discussions, some of the design alternatives were rejected, while others were considered as potential candidates for the design of system. Subsequent discussions with the analysts lead to further refinements of the design and eventually one design namely, treemaps, was selected for implementation to realize the requirements which resulted in MEV.

In the later interviews, a working prototype of MEV was presented to these analysts to assess if MEV meets their needs. This then led to further suggestions to improve our visual and interaction design.

Lastly, we invited a larger group composed of ten FDA analysts, who were not involved in the design process of MEV, to participate in the user study to evaluate MEV (Sect. 6.2). This activity led to additional insights on the utility of MEV, a visual analytics tool in supporting the drug safety review process. We repeat on these final study results.

We next discuss the medication error reports, the review workflow and the challenges of current tools derived from the above interviews.

3.2   FAERS Data Description

Drug regulatory agencies worldwide maintain reporting systems as a part post-marketing drug surveillance programs to ensure public health safety [35]. In the U.S. the FDA maintains an Adverse Event Reporting Systems (FAERS) to monitor the safety of medications and therapeutic biologic products. FAERS contains voluntary reports submitted by health care professionals and consumers, and mandatory reports submitted by the drug manufacturers. These reports are composed of structured information about patient demographics, drugs being taken, therapies, and adverse reactions or medication errors. Each report also contains an unstructured textual narrative that provides a detailed description of the incidents such as medication errors or adverse reactions. These details about the incident contain richer information that can help the analysts decide if the incident is worthy of investigation and warrants an action. This information is critical for the analysis and includes the drugs or products, type and cause of error. Most of this information is categorical in nature.

Drug safety analysts review FAERS reports based on their assigned therapeutic classes of products. Each analyst receives an average number of approximately 200 reports on a weekly basis and monitors an average number of 50–100 products. These numbers may vary for different teams based on the assigned products. For instance, an analyst may maintain more than 100 products, as majority of her assigned products may rarely be causing an issue. Or, patients being familiar with an old product, might be making less mistakes than a newly approved drug which may be unfamiliar to use properly, hence generating more error reports. Additionally, for the investigation and detailed analysis of a particular error or product, thousands of reports from the past several months may be analyzed, if available.

3.3   Motivating Example of Medication Error Detection

To understand medication errors and the importance of their detection, we provide two motivating real examples. The first example includes a report received by the Institute for Safe Medication Practices (ISMP) about the medication Belbuca (Buprenorphine) [16]. Belbuca is prescribed to manage pain severe enough to require regular and long-term opioid treatment and for which alternative treatment options are inadequate. The medication is available in multiple strengths including 10 fold strengths, i.e., 75 mcg and 750 mcg (Fig. 1).

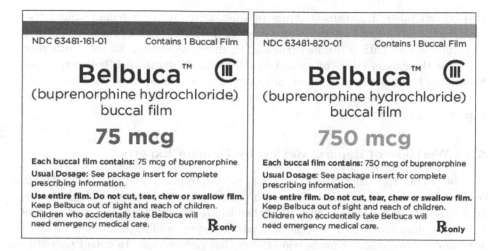

Fig. 1. Example of a potential 10-fold wrong strength error. A trailing zero (e.g., 75.0 mcg) can cause a mix-up of the two doses in prescription or dispensing of the drug.

The report describes a consumer who received a 750 mcg of Belbuca instead of 75 mcg, which is the recommended starting dose, from her pharmacy. After taking 5 doses of erroneous 750 mcg strength, the patient experienced dizziness, lightheadedness, and vomiting. Upon follow up with her doctor, she learned that he had prescribed 75 mcg, not 750 mcg. One possible reason for such error could be misreading the hand written or typed prescription with a trailing zero (i.e., 75.0 mcg). Or, selecting the wrong strength from the drug drop-down list if 75 mcg and 750 mcg strengths appear next to each other in the computer system.

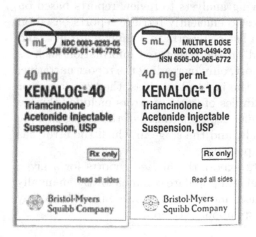

Fig. 2. Example of a look-alike carton label for the drug Kenalog resulting in an error.

Multiple instances of similar errors of mixing up 10 fold strengths of Belbuca as well as other medications such as Abilify (2 mg and 20 mg) are reported to drug safety authorities including ISMP and the FDA. A possible solution to these errors could be encouraging drug manufacturers to use strengths above or below an exact 10-fold difference, such as 749 mcg instead of 750 mcg [16].

Another example includes wrong strength error caused by a look-alike carton label. For instance, Fig. 2 represents two different strengths of Kenalog (i.e., 1 ml and 5 ml). The similarity between the two carton labels has caused errors of mistakenly using a higher concentration (5 ml) instead of 1 ml [6]. In this case, if the FDA receives multiple reports of similar errors with a product, then FDA may recommend to use a different color for the carton labels of each strength to easily distinguish between them. Seemingly benign errors like these can have adverse consequences such as death, if this were to cause patients to use the wrong drugs for treatment of critical conditions or diseases.

3.4 Workflow of Reports Review by Domain Experts

At the FDA, the drug safety analysts receive reports related to the drugs assigned to them. Every week, they receive a new set of reports which they analyze throughout the week along with other Pharmacovigilance activities such as reviewing drug labels or investigating a particular error. The goal of the analysis is to find potential severe medication errors that need action. To detect potential signals, the analysts first screen these reports based on criticality factors such as severe outcomes including hospitalization or death, particular error types or products. Then the text narrative of the screened reports are analyzed to get in-depth analysis of the errors. Once any alarming error is found, then FAERS is searched for other reports about similar errors searching backwards in time for at least last six months or a year. In some cases, the search can extend to the date when the product or error was last reviewed. Other sources are also analyzed for the investigation of a particular error.

3.5 Current Report Analysis Tools at the FDA

Currently, FDA uses FAERS Business Intelligence System (FBIS) [8] as one of the major tools to retrieve reports from FAERS database and analyze them. FBIS is designed with the goal of allowing analysts to review reports based on their assigned set of drugs, using queries to efficiently retrieve reports based on multiple dimensions. FBIS displays the list of reports in a tabular format, and provides controls to filter reports based on the structured information such as the age group or a product. For each report, only by reading the report narrative, the analysts can examine the details of the incident, such as, the type, cause or stage at which error has occurred. Examples of types of errors include taking a wrong dosage or using a wrong technique, the root causes of the errors include confusing instructions or container labels, and the stage in which the error has occurred include preparation versus dispensing.

For a detailed analysis, the analysts export the filtered reports for a given drug from FBIS into the Microsoft Excel and populate columns for the manually extracted information such as type, cause, and stage of error along with the any additional remarks added by the analysts about the report.

3.6   MEV Design Rationale

Our interviews with the analysts and their workflow observation (see Sect. 3.1) helped us identify the limitations of the current tools. These interviews also revealed that there is a need for an automatic system that first extracts the key information (type, cause and stage of error) from text narratives and then allows analysts to interactively explore and analyze this information, hence, making the report review process efficient.

This resulted in the identification of the below design requirements of a system that can support the exploration and analysis of medication error reports.

R1: *Provide an Overview of the Core Data Elements for Data Screening.* Analysts expressed a need for an overview of the key data attributes important for the analysis of medication errors. These attributes include the drugnames or products, types, causes and stages of errors. Each of these attributes consist of multiple categorical values. Such an overview should help an analyst see the distribution of reports per each attribute value to help them screen the critical errors.

R2: *Support the Exploration of Reports over Time.* Drug safety analysts review reports on a weekly basis, however, they often analyze reports accumulated over a longer period of time. Hence, a way to interactively explore reports for a specified time is needed.

R3: *Facilitate the Analysis of Demographics.* The demographics of the patient also play an important role in the analysis of medication errors. The analysts should be able to see the age, location and occupation of the reporters. This would allow them to see, for instance, if an error is more prevalent in certain groups.

R4: *Allow the Analysts to Quickly See the Related Attributes of a Selected Data Point.* Multiple errors may be reported for a given product. An error can be happening due to multiple reasons at different stages from the distribution of the drug to the administration. Therefore, analysts should be able to quickly get the gist of the associated data elements once a particular data value is selected.

R5: *Facilitate Identification of Critical Reports.* The outcome of reports is an important factor in screening and prioritization of error reports. Analysts expressed a desire of being able to quickly identify any critical data points related to serious report outcomes. These outcomes indicate if the medication errors resulted in a serious outcome such as death or hospitalization of the patient, or were non-serious.

R6: *Ready Access to the Reports Narratives Associated with Selection.* Analysts also indicated the importance of having direct access to the actual reports once a set of errors are screened. As these reports provide the details essential for decision making about the critical errors.

Besides the above mentioned requirements, the following have been found to be also equally important:

R7: *Support Smooth and Interactive Exploration of a Large Number of Reports.* Drug safety analysts review hundreds of drugs and their associated errors. Therefore, the system should be scalable to manage a large volume of reports.

R8: *Using Simple and Familiar Visual Metaphors.* Since the majority of the analysts had experience with basic visual analytics applications, it was emphasized to keep the visual designs intuitive and easy to understand. Clearly, careful design choices need to be made to consider these aspects.

4   MEV Design Process

We now discuss the design process of an interactive visual analytics system to help drug safety analysts in the screening reports by fulfilling the requirements solicited in Sect. 3. For (**R1**), a visualization was needed that display the distributions of categorical data representing our core data attributes (drugs, error types, causes and stages).

Also, visual interactions that help in assessing attribute values and their relationships with the rest of the data (**R4**) were needed. Therefore, we sketched several visual designs that meet these requirements (**R1, R4**) based on studying the state-of-the-art visualizations for multiple categorical attributes. Figure 3 through Fig. 6 depict our alternate designs for (**R1**) that were repeatedly discussed with the domain experts (safety analysts) to get their feedback. We used color hue to encode the seriousness of reports (**R5**) with brown depicting highest count of serious reports while blue depicting the lowest count. This way the brown color can help analysts quickly identify critical data points. These color encodings are kept consistent throughout all designs.

We now discuss these designs and their pros and cons highlighted by the experts.

4.1   Radial and Bar Charts

We started the design process with basic visualizations such as radial and bar charts to visualize the core data elements (**R1**) of the reports including the types, medications, stages, causes and outcomes of the errors (Fig. 3). Each layer in the radial chart (Fig. 3-left) represented the core attributes with the size of the arc mapped to the frequency of reports for each attribute value. Selecting any value on any layer updates the rest of the layers to reflect the data associated with the selection (**R4**). Similar mappings were used for the bar charts as well.

After discussions with the analysts, the radial chart was discarded due to its limited readability and scalability (**R7, R8**). The analysts expressed some concern about the complexity of the chart. On the other hand, the analysts seemed to like bar charts better than radial charts. However, their concern was that for a larger number of products (50 or more) the bars may be too tiny to read (**R7**).

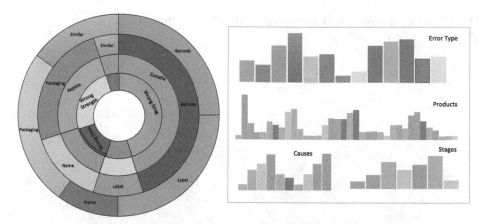

Fig. 3. Multi-layer radial chart (left) and bar-charts (right) for representing products, types, stages and causes of the errors **R1**. Color depicts the outcome of the reports.

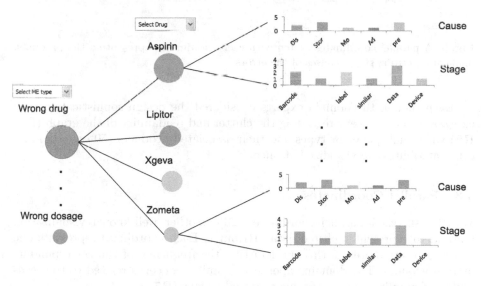

Fig. 4. A node-link diagram sketch for one medication error "Wrong drug". The second layer depicts the products associated with the error. The third layer represents the distribution of causes and stages in the form of barcharts.

4.2 Node-Link Diagrams

Inspired by the existing works [9,22] that has used node-link diagrams to visualize drug-related adverse reactions, we also considered a similar design to visualize the core attributes related to medication errors. As depicted in Fig. 4, the first layer of nodes represents the medication error type, while the second layer represents the products associated with each error type. The size of the nodes is mapped to the frequency of reports for the corresponding attribute value. The third layer uses a bar-chart to represent the distribution of stages and causes of the error (Fig. 4).

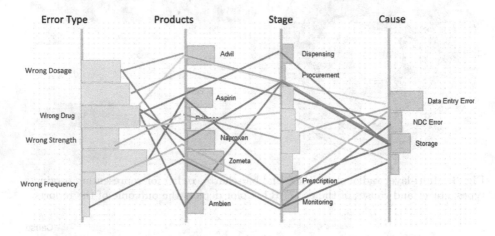

Fig. 5. A parallel coordinates sketch with each coordinate representing the products, medication errors, stage, causes of the errors.

Even though the domain experts considered the design sophisticated, they expressed some concern regarding the clutter and readability of the graph (**R7, R8**) when multiple error types and their associated products (50 or more) and other attributes are visualized at once.

4.3 Parallel Coordinates

We also sketched parallel coordinates [20], another well-known visualization method to represent multiple data attributes. Each coordinate represents one of the core attributes, with bars depicting the frequency of the corresponding attribute values. The domain experts had similar concerns related to the readability of parallel coordinates due to visual clutter (**R7**).

4.4 Treemaps

Finally inspired by Selltrend [28], a tool that uses treemaps to visualize failures in transactions related to airline travel requests, we sketched a treemap design for our medication error screening problem (Fig. 6). The multiple layers of treemaps depict the different core attributes with each layer displaying the distribution of reports related to the corresponding attribute. A rectangle within a treemap layer represents one category for the attribute represented by that particular layer.For example, a rectangle in the product treemap represents one drugname.

The size of the rectangle is mapped to the number of reports related to that specific category and the color encodes the count of the severe outcomes (**R5**) which is obtained from the structured field.

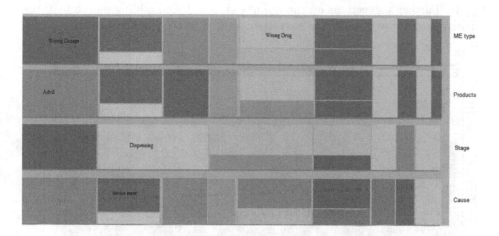

Fig. 6. A treemap based sketch that was realized into MEV.

We received positive responses from the domain experts about the treemap design. The space-filling characteristic of treemaps makes them more readable. We then iteratively refined this design through discussions with domain experts, until the final design which was developed into a prototype system (Fig. 8). Some of these refinements included the change of color, as after implementation, the shades of brown and blue colors were getting mixed and were forming new colors (green). Therefore, color saturation was identified as a better choice to encode the seriousness of reports (**R5**) instead of color hue [32].

Fig. 7. Reports count over time with color representing serious outcomes.

To fulfill (**R2, R3**), we sketched barcharts to display the distribution of reports over time (Fig. 7) as well as demographics (Fig. 8a). The analysts liked the barcharts for demographics (**R2**) and timeline (**R3**) due to their simplicity. Which is why, we do not show design alternatives for them in this paper. However, when the system was implemented, the timeline with barcharts (Fig. 7) did not look visually appealing. Hence, upon discussion with the experts, we decided to use area-curves to display the weekly reports distributions (Fig. 8c). For examining the reports associated with the screened data (**R6**), the analysts suggested to have a view with both a line-listing of the reports as well as the text narratives themselves. This resulted in the design of the reports view as depicted in Fig. 10. Further suggestions on the reports view included adding search features for searching through both the narratives as well as the line-listings.

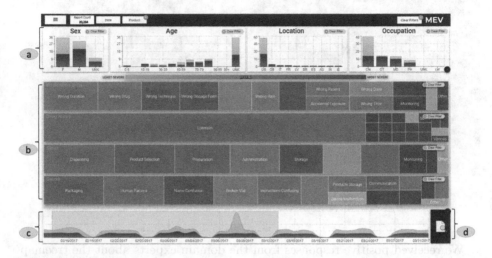

Fig. 8. The user interface of MEV (a) The demographics panel. (b) The treemap panel. (c) The timeline panel. (d) Reports icon to access the reports view to analyze the associated report narratives [24].

5   The MEV System

In this section, we describe MEV's framework and interface design in detail.

5.1   The MEV Framework

The MEV framework, depicted in Fig. 9, is designed for analysts to efficiently perform the tasks of reports review. As we introduced earlier, FAERS reports consist of structured fields as well as text narrative with details about the medical events. In the case of medication error reports, the critical information about the error types and causes are often described in the text narrative rather than explicitly being stated in the structured field. To quickly extract these valuable information from the text narrative for the analysts, we adopt rule-based name-entity recognition techniques [41] in MEV.

The extractor is equipped with domain specific lexicons [11, 33] to identify key data attributes including types, causes and stages of medication errors. In MEV, the *Natural Language Processor* (Fig. 9) also performs preprocessing such as word tokenization, stop-words removal and stemming before the core extraction process. The extracted results are then standardized by mapping them to the NCC-MERP terminology [33]. The mapping strategy is based on an *edit distance* based string matching algorithm [14].

On average, each of the standardized entity type contains approximately 15–20 categories. As illustrated in Fig. 9, both extraction results and the structured information about the demographics are stored in the *Data Store*. The *Query Executor* is responsible for handling data retrieval and transformation requests specified via online MEV visual interface. For better user experience, this module also has a caching mechanism to quickly respond to the most frequently issued requests. By using coordinated interactive visualizations, MEV assists analysts in the efficient exploration of reports as described below.

As depicted in Fig. 8, MEV consists of four core interactive displays – the treemap view, the demographics panel, the timeline panel and the reports view.

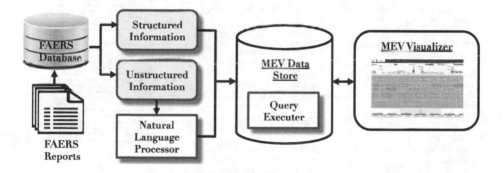

Fig. 9. The MEV framework [24].

5.2   MEV Interface Overview

The Treemap Panel. As discussed earlier, the treemap design (Fig. 8b) displays the distribution of each attribute over its possible categories.

This treemap visualization support analysts to interactively filter attribute such as drugname with even large number of categorical values (**R7**). MEV allows analysts to select multiple data values on each treemap at the same time and the other treemaps will be refreshed accordingly with the updated information corresponding to the selection. Such filtering functionality can facilitate analysts to quickly identify interesting information based on the presented data distribution (**R4**) which would take many tidy steps to achieve using their existing tools.

The Timeline Panel. The timeline panel, depicted in Fig. 8c, displays the report distribution based on volume and seriousness over a period of time using an area chart. Such design allows analyst to find spikes in the severity levels associated with the events of certain products (**R2**). It also supports interactive brushing and selection through zooming allowing analysts to quickly drill down to a particular time frame and explore specific reports. Similar to the interactivity on treemap, once a time frame is specified, other displays are then updated to reflect data corresponding to the selection (**R4**).

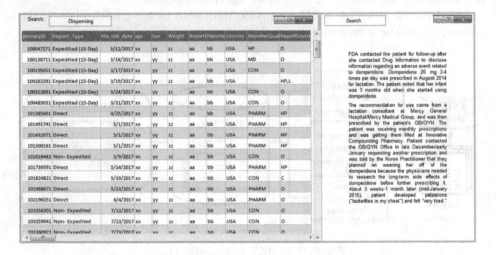

Fig. 10. Reports view with no personal information with line-listing (Left). Example of a de-identified text narrative (Right) [24].

The Demographic Panel. As pointed out by the domain analysts, the demographics information of the patients are critical for their analysis. For example, a severe outcome triggered by the administration of a certain medical product may only happen to pediatric patients as opposed to every age group. The demographics panel (Fig. 8a) can assist the analysts in quick identification and selection of problematic population defined by attributes such as age, gender or location (**R3**). This way the analysts can not only screen and prioritize reports based on selected attributes but also can immediately view the distributions of other associated attributes via linked displays (**R4**).

The Reports View. The analysts are able to browse the respective reports after they select the interesting medical products or error types as depicted in

Fig. 10. This allows them to further investigate if these reports are indicative of errors with serious consequences for patient health warranting regulatory action (**R6**). The selected reports are accessible by clicking on the reports icon shown in Fig. 8d.

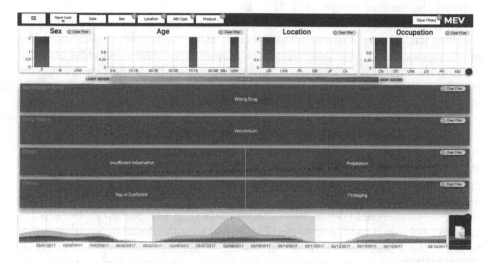

Fig. 11. MEV with selection of date range, demographics, drugname and medication error type [24].

6   Evaluation

To validate the system, we first present a use case to highlight the usage and capabilities of MEV in data analysis tasks.

6.1   Use Case

To understand the use of MEV and how it can help in screening of reports, we describe an overview of actual series of steps performed by Alex, a drug safety analyst. Alex uses MEV to review reports related to her assigned set of products. At a glance on the timeline panel, she sees an overall weekly distribution of reports count and their severity for the last month (**R2**). She instantly can see that reports are only received during the weekdays, with no report visible during the weekend. She further observes a spike in the number of reports for the week of march 3rd–7th (2017), of which around 40% of reports are severe while the rest are non-severe (Fig. 11-bottom).

Using the brush tool on the timeline panel, she selects this week's reports to analyze them (**R2**). The treemap and demographics panels are updated for the selected week (**R4**). She sees that total number of reports for this selected week are 28,123. On the demographics panel, she sees that majority of patients are female and the most prevalent age group is above 30 (**R3**). Her assigned set of products are mostly for elderly women so the distribution of the age and gender is expected. She selects females from the gender barchart and U.S. from the location barchart to investigate the reported drugs and errors. This selection reduced the number of reports to 11,174.

She notices (on the treemap) that the medication error "wrong-technique" is reported with the most severe outcomes (**R1, R5**). She is now curious about the products associated with this "wrong-technique" error. She selects this error by clicking on the rectangle labeled as "wrong-technique" in the first treemap. Now the report counts is reduced to 2,786. On the second treemap for products, she observes that the drug Lotensin has the highest count of severe outcomes. She thus selects Lotensin in the second treemap. She is now curious that at which stage are these errors happening and what is causing them.

From the third and fourth treemaps which corresponds to the cause and stage of errors respectively, she observes that the major causes of this error are "packaging" and "name confusion". She says "The errors seem to be happening in the preparation of the drug". She also notices that the number of reports that she needs to analyze in detail, has reduced to 49 (Fig. 11). She opens the reports view by clicking on the reports icon (Fig. 8d) to read the details of each narrative to see if the reports indeed have compelling evidence about these errors (**R6**). Thus, MEV supports the exploration and screening of reports and interactively guides the analysts towards concerning errors.

6.2   User Study

We conducted a user study with drug safety analysts at the Division of Medication Error and Prevention Analysis (DMEPA) at the FDA to evaluate the effectiveness of MEV in various review tasks.

Study Design. We conducted a one hour in-person study session with ten drug safety analysts at the DMEPA. These participants included nine females and one male and were within the age range of 30–50 years. These participants were pharmacists with PharmD degrees and had experience with basic visualizations such as barcharts. These analysts are responsible to conduct regular reports review to identify any concerning errors that may need regulatory action.

Tasks and Assessment Measures. Based on our initial interviews with the analysts and the study of their current practices and workflow, we designed as set of nine commonly performed tasks to evaluate the usefulness of MEV (Table 1). These tasks contained simple one-step tasks to complex multi-step tasks. Example of a one-step task (T1–T2) included finding a particular attribute value for a selected time (**R2**). While complex tasks (T3–T6) included exploring reports associated with the analysis of the distribution of multiple data attributes (**R3–R5**). On the other hand, T8 & T9 included the analysis of reports based on the extracted data entities i.e., stage and cause of error (**R1**). The composite tasks

(T3–T9) involved filtering based on the analysis of relationships between and among data attributes (**R4**). The main goal of the tasks was to find interesting data points to prioritize reports associated with critical errors (**R5**).

For each task, we measured the time taken to successfully complete the task as well as the easiness perceived by the participants. The perceived ease was recorded on a 5-point Likert scale (1 extremely difficult and 5 extremely easy). As data loading in the existing tool (FBIS) takes a long time to accomplish, so we recorded the time for completing a task **after** FAERS data was loaded into both tools for one week (from 2017). To compare the existing tools (*Control*) and MEV, the participants performed the same set of tasks with both tools.

Study Procedure. We conducted one hour in-person interview with the participants to observe them closely interacting with MEV and get their detailed feedback. The study consisted of three sessions. The first session (20 min) was dedicated to the demonstration and training. After the participants felt comfortable interacting with the system, in the second session the participants were asked to perform the set of previously designed tasks (Table 1) using both MEV and their existing tool. In the third session, a post-study questionnaire was provided to the participants at which was not timed. The first part of the questionnaire included questions related to the demographics of the participants such as age and gender. The second section contained questions about the usability [10] of MEV on a 5-point Likert scale (1 strongly disagree & 5 strongly agree). The last part included an open-ended questionnaire to solicit qualitative feedback about MEV.

Analysis Method. We used the non-parametric Mann-Whitney U Test (Wilcoxon Rank Sum Test) [30] to compare the two conditions, as the recorded time and perceived easiness for some tasks were not normally distributed (Table 2). Moreover, we calculated the 95% confidence intervals (Fig. 12) for the task completion time and perceived easiness score.

Study Results. We now discuss the participants' performance on the tasks and their response regarding the overall system usability.

Quantitative Analysis. Table 2 shows that, for the majority of the tasks except Task T1, there are significant differences between the recorded time and perceived ease score when tasks were performed using MEV and the existing tool *control*.

The differences for T1 were not significant. T1 was a simple task that involved the finding of the total number of reports for a given duration of time. One possible explanation for this outcome could be that participants knew exactly where they will find this information in their current tool as they were used to it. On the other hand, being new to MEV, they took little longer ($M = 5.11$ seconds [3.47, 6.76]) as compared to using their current tool ($M = 3.62$ [1.80, 5.44]). Participants also rated this task easier under the *control* condition than using MEV.

Table 1. List of 9 Tasks designed to evaluate the effectiveness of MEV [24]

Task #	Description
T1	How many total reports have been reported during a time period?
T2	Which medication error is reported the most for a time period?
T3	Which drug has most severe outcomes for a selected medication error?
T4	Which gender and age have most severe outcomes?
T5	Which age group is most prevalent in reports related to a selected product?
T6	What are the two most frequent medication errors reported with a select product, age group, and gender?
T7	Given the report distribution of a drug for female patients with a specified age group, what are the critical medication errors that need to be analyzed?
T8	What are the two most frequent root causes of error for a selected drug and medication error?
T9	What are the two most common reported stages of errors for a drug and a medication error?

For Task T2, there were significant differences between the performance using MEV (M =31.84 s [15.78, 47.91]) and existing tool ($M = 7.54$ s [3.57, 11.52]). T2 involved finding the most reported medication errors for a selected time period. In addition to time, we also observe that participants found it easier to perform the task using MEV ($M = 4.9$ [4.70, 5.10]) as compared to the control condition ($M = 4.0$ [3.59, 4.41]).

Similarly, significant differences can be seen between the time and perceived ease scores for the multi-step tasks (T3–T7) which involved the analysis of the distribution and severity of reports across multiple data attributes (Table 2). The composite tasks T8 and T9 involved retrieval of reports based on the stages and causes of errors with severe outcomes. The comparison for these two tasks was not possible as the causes and stages of errors were extracted using NLP for MEV and their current tools do not provide them.

Additionally, Fig. 12(Top) shows that participants have a relatively consistent performance for all tasks. That is, all participants were able to successfully complete the tasks quickly. On the other hand, under the *control* condition participants had a highly varied performance. Similarly, from Fig. 12(Bottom) we see that, the participants perceived it easier to perform tasks (T2–T9) using MEV than the existing tool. Participants rated the Task T5 which involved the analysis of the distribution of age for a selected product, as the most difficult under *control* condition. The reason is that with their existing tool it is tedious

Table 2. U-Test with significant results for both time and perceived ease for all tasks except T1 [24].

Tasks	Medviz - avg time (sec) [95% CI]	Current tools avg time (sec) [95% CI]	Significance test ($\alpha = 0.05$)	MedViz avg easiness [95% CI]	Current tools avg easiness [95% CI]	Significance test ($\alpha = 0.05$)
T1	5.11 [3.47, 6.76]	3.62 [1.80, 5.44]	(U = 29.5, p = 0.13104)	4.8 [4.54, 5.06]	4.9 [4.70, 5.10]	(U = 45, p = 0.72786)
T2	7.54 [3.57, 11.52]	31.84 [15.78, 47.91]	(U = 14, p = 0.00736)	4.9 [4.70, 5.10]	4.0 [3.59, 4.41]	(U = 14, p = 0.00736)
T3	13.74 [9.83, 17.65]	78.31 [57.80, 98.83]	(U = 0, p = 0.00018)	4.7 [4.40, 5.00]	2.5 [2.06, 2.94]	(U = 1.5, p = 0.00028)
T4	10.04 [7.39, 12.69]	82.75 [46.51, 119.00]	(U = 0, p = 0.00018)	4.4 [3.97, 4.83]	2.6 [2.00, 3.20]	(U = 7.5, p = 0.00152)
T5	12.45 [9.12, 15.77]	67.05 [46.26, 87.84]	(U = 4, p = 0.00058)	4.5 [4.06, 4.94]	1.7 [1.19, 2.21]	(U = 1, p = 0.00024)
T6	18.11 [13.75, 22.47]	83.58 [55.19, 111.96]	(U = 4, p = 0.00058)	4.4 [3.97, 4.83]	2.4 [1.73, 3.07]	(U = 7, p = 0.00132)
T7	18.52 [15.60, 21.45]	64.84 [40.38, 89.31]	(U = 11, p = 0.00362)	4.6 [4.28, 4.92]	2.2 [1.71, 2.69]	(U = 0, p = 0.00018)
T8	12.51 [8.45, 16.56]	Not supported	Not applicable	4.7 [4.40, 5.00]	Not supported	Not applicable
T9	13.07 [10.32, 15.83]	Not supported	Not applicable	4.6 [4.28, 4.92]	Not supported	Not applicable

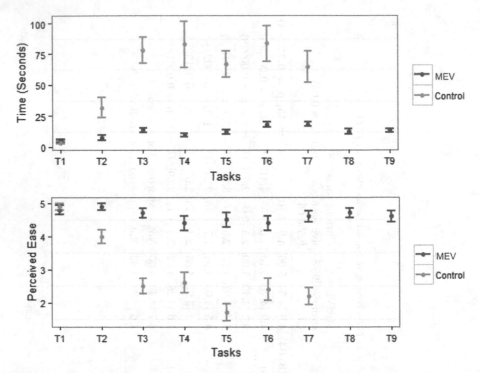

Fig. 12. 95% Confidence Interval for performing tasks using both MEV and existing tool (*Control*). (Top): Time (in sec). (Bottom): Perceived Ease Score. Task 8 & 9 are not supported by *Control* [24].

to explore the distribution of data attributes as it requires to filter each attribute value individually and then analyze the outcome of each value.

Qualitative Analysis & Overall Impression of MEV. The goal of qualitative questionnaires was to collect participants' subjective feedback about the tool and their experience using it. The analysis of the questionnaire answers suggests that the participants' experiences with the tool differed depending on their prior experience with similar interactive systems. For instance, some participants found it difficult to interact with the timeline panel to select dates, while others liked it.

Overall, the majority of participants found the goal of the MEV system in exploring and analyzing medication errors to be promising and potentially useful. Around 60% of the participants explicitly mentioned the usefulness of integrating the extracted data elements into the visualization and the intuitiveness of the tool itself. According to the participant P2: "Though the text-extraction is not perfect but it gives us a big sense of what kind of errors are being reported". Participant P5 said: "It takes some time to get used to the tool, then it is very easy and intuitive to use". Participant P10 mentioned: "Well, I think this tool

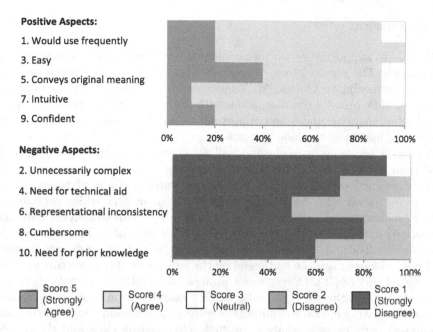

Positive Aspects:

1. Would use frequently

3. Easy

5. Conveys original meaning

7. Intuitive

9. Confident

Negative Aspects:

2. Unnecessarily complex

4. Need for technical aid

6. Representational inconsistency

8. Cumbersome

10. Need for prior knowledge

Score 5 (Strongly Agree)   Score 4 (Agree)   Score 3 (Neutral)   Score 2 (Disagree)   Score 1 (Strongly Disagree)

Fig. 13. System Usability Scale (SUS) Questionnaire Results.

makes it very easy to see what the reports are describing without going into much detail".

Participants constructive feedback about the potential improvement of the design of the system were also solicited using an open response option. For instance, four participants suggested to include an individual search option on each treemap to look up a particular error or drug. Others suggested to include a quick preview of the narratives on the treemaps view to get an instant gist of the text depicting an error.

Usability Questionnaire. At the end of the study, the participants were asked to complete the system usability scale (SUS) questionnaire [10]. Figure 13 depicts the distribution of the results of the questionnaire grouped as positive and negative questions. The majority of the participants agreed or strongly agreed with the questions related to the positive aspects of MEV, while only three participants had a neutral response. None of the participants had a negative response (scores of 1 or 2). On the other hand, for the negative questions, except for one participant who needed technical assistance with the tool, 80% of the participants disagreed or strongly disagreed (score of 1 or 2) with the questions. The overall score for the usability of the system was 85 out of 100.

7   Discussion

MEV aims to support the pharmacovigilance workflow using a visual analytics approach. The results of our user study suggest that analysts can perform review and screening tasks using MEV by exploring the relationships among data attributes. More broadly, our discussions with the drug safety analysts have highlighted additional challenges and opportunities for the medical professionals in the space of human-in-the-loop systems.

Scale is a key issue in modern visualization systems. Although the goal of MEV is to be used weekly by individual analyst for reports screening based on their assigned set of products which constitutes a count of thousands of reports. We tested MEV by loading data for the year 2017 which represented over 1.82 million reports, where the system took few seconds to load the data and transformed it to the initial overview. Other challenges of scale are associated with the visualizations themselves. If the analyst were to steer to a view with hundreds or more drugs, the rectangles on the treemap will become tiny and can cause visual clutter [36]. One solution to avoid the clutter is to only display a subset of drugs on the treemap and provide a search option to access a desired drugname. Another option to address the scale issue is to add a layer of treemap to represent drug classes. Analyst can first select a drug class and the second layer of treemap representing the drugs will be updated for the selected class. Alternatively, incorporating the domain practices into the system can also be a solution. For example, in typical cases clutter may not be a problem for MEV as the maximum number of distinct products in the reports for an analyst does not exceed 100.

During our qualitative interviews although the majority of the analysts acknowledged the usefulness of MEV in reports screening, few analysts expressed their preference for reading each and every report narrative when the number of reports was small (e.g., 10 or 20), rather than using MEV for screening. For such analysts, a feature to highlight the key information within the text narratives can be added. During our user study, we also observed that some of the extracted information was incorrect, when the participants retrieved the reports to analyze their narratives. For this system, we leveraged the MEFA [41] framework for name-entity extraction such as the cause and stage of the error. For improved accuracy, more advanced deep learning-based extraction techniques [42] could be plugged into MEV. However, name-entity recognition techniques for biomedical text itself is known to be a challenging problem and research efforts continue to improve their accuracy [3].

Our user study has a number of limitations. First, participants' familiarity with their existing tools allowed them to complete some complicated tasks in a short time. Also, few participants found some tasks irrelevant. For instance, participants who usually investigate one particular product found it irrelevant to find the reports associated with multiple products based on severity. Study participants, although a small number, are real domain experts who would be the ultimate users of MEV in daily analysis. Long term studies with these experts are needed to further assess the capabilities of MEV in their workflow.

In the future, we plan to provide direct access to external sources such as DailyMed and PubMed from within MEV. This will help the analysts to investigate these sources to reject or confirm a hypothesis about a possible medication error formed using the current views. We also plan to extend the capability of the current system by providing support for the report text analysis. Finally, we will add the capability of visual provenance [17] to allow analysts to share their thought-processes and findings with their team members during an investigation.

8   Conclusion

In this paper, we described the detailed design process of MEV—a prototype visual analytics tool for medication error detection using spontaneous reporting system. MEV uses multiple coordinated views including interactive treemaps, bar charts and timeline visualization to support the exploration and screening of spontaneous reports. MEV guides the analysts to visually identify critical errors by allowing them to evaluate the distributions of multiple facets of data across several weeks. A task-based user study with drug safety analysts at the FDA suggests that MEV is effective in supporting the analysts in their review tasks, allowing them to complete their tasks more efficiently, in comparison to their existing tool. Moreover, qualitative feedback from the participants highlights opportunities to improve the system. This is valuable for future developments of MEV specifically as well as of for the design of visual analytics tools for critical incident reports analysis in general – a ubiquitous task in domains such as aviation [29] and finance [12].

References

1. A $21 Billion Opportunity National Priorities Partnership convened by the National Quality Forum file (2010). http://www.nehi.net/bendthecurve/sup/documents/Medication_Errors_Brief.pdf. Accessed 07 Jan 2018
2. Agrawal, A.: Medication errors: prevention using information technology systems. Br. J. Clin. Pharmacol. 67(6), 681–686 (2009). https://doi.org/10.1111/j.1365-2125.2009.03427.x
3. Alshaikhdeeb, B., Ahmad, K.: Biomedical named entity recognition: a review. Int. J. Adv. Sci. Eng. Inf. Technol. 6(6), 889–895 (2016). https://doi.org/10.18517/ijaseit.6.6.1367
4. Aronson, A.R., Lang, F.M.: An overview of MetaMap: historical perspective and recent advances. J. Am. Med. Inf. Assoc. 17(3), 229–236 (2010). https://doi.org/10.1136/jamia.2009.002733
5. Asahi, T., Turo, D., Shneiderman, B.: Visual decision-making: using treemaps for the analytic hierarchy process. Craft Inf. Vis., 235–236 (2003). https://doi.org/10.1016/b978-155860915-0/50030-5
6. Authority, P.P.S.: Drug Labeling and Packaging – Looking Beyond What Meets the Eye (2017). http://patientsafety.pa.gov/ADVISORIES/documents/200709_69b.pdf. Accessed 25 June 2019
7. Bates, D.W., et al.: The costs of adverse drug events in hospitalized patients. Jama 277(4), 307–311 (1997). https://doi.org/10.1001/jama.1997.03540280045032

8. BIFACT: FAERS Business Intelligence System (FBIS). http://www.bifact.com/faers-bifact.html. Accessed 19 June 2019

9. Botsis, T., et al.: Decision support environment for medical product safety surveillance. J. Biomed. Inform. **64**, 354–362 (2016). https://doi.org/10.1016/j.jbi.2016.07.023

10. Brooke, J., et al.: SUS-a quick and dirty usability scale. Usability Eval. Ind. **189**(194), 4–7 (1996)

11. Brown, E.G., Wood, L., Wood, S.: The medical dictionary for regulatory activities (MedDRA). Drug Saf. **20**(2), 109–117 (1999). https://doi.org/10.1002/9780470059210.ch13

12. CFPB: Consumer Financial Protection Bureau. www.consumerfinance.gov/. Accessed 03 June 2019

13. Dachselt, R., Frisch, M., Weiland, M.: FacetZoom: a continuous multi-scale widget for navigating hierarchical metadata. In: Proceedings of the SIGCHI Conference on Human Factors in Computing Systems, pp. 1353–1356. ACM (2008). https://doi.org/10.1145/1357054.1357265

14. Du, M.: Approximate name matching. NADA, Numerisk Analys och Datalogi, KTH, Kungliga Tekniska Högskolan, Stockholm: un, pp. 3–15 (2005)

15. Fekete, J.-D., van Wijk, J.J., Stasko, J.T., North, C.: The value of information visualization. In: Kerren, A., Stasko, J.T., Fekete, J.-D., North, C. (eds.) Information Visualization. LNCS, vol. 4950, pp. 1–18. Springer, Heidelberg (2008). https://doi.org/10.1007/978-3-540-70956-5_1

16. Gaunt, M.J.: Preventing 10-Fold Dosage Errors. www.pharmacytimes.com/publications/issue/2017/july2017/preventing-10fold-dosage-errors. Accessed 24 June 2019

17. Groth, D.P., Streefkerk, K.: Provenance and annotation for visual exploration systems. IEEE Trans. Visual Comput. Graphics **12**(6), 1500–1510 (2006). https://doi.org/10.1109/tvcg.2006.101

18. Harrison, L., Spahn, R., Iannacone, M., Downing, E., Goodall, J.R.: NV: nessus vulnerability visualization for the web. In: Proceedings of the Ninth International Symposium on Visualization for Cyber Security, pp. 25–32. ACM (2012). https://doi.org/10.1145/2379690.2379694

19. Huckels-Baumgart, S., Manser, T.: Identifying medication error chains from critical incident reports: a new analytic approach. J. Clin. Pharmacol. **54**(10), 1188–1197 (2014). https://doi.org/10.1002/jcph.319

20. Inselberg, A.: The plane with parallel coordinates. Vis. Comput. **1**(2), 69–91 (1985). https://doi.org/10.1007/BF0189835

21. Jia, P., Zhang, L., Chen, J., Zhao, P., Zhang, M.: The effects of clinical decision support systems on medication safety: an overview. Pub. Libr. Sci. One **11**(12), e0167683 (2016). https://doi.org/10.1371/journal.pone.0167683

22. Kakar, T., et al.: DEVES: interactive signal analytics for drug safety. In: Proceedings of the 27th ACM International Conference on Information and Knowledge Management, pp. 1891–1894 (2018). https://doi.org/10.1145/3269206.3269211

23. Kakar, T., Qin, X., Rundensteiner, E.A., Harrison, L., Sahoo, S.K., De, S.: DIVA: towards validation of hypothesized drug-drug interactions via visual analysis. In: Eurographics (2019). https://doi.org/10.1111/cgf.13674

24. Kakar, T., et al.: MEV: visual analytics for medication error detection. In: 2019 International Conference on Information Visualization Theory and Applications (IVAPP), Prague. SciTePress (2019). https://doi.org/10.5220/0007366200720082

25. Kohn, L.T., Corrigan, J., Donaldson, M.S., et al.: To Err is Human: Building a Safer Health System, vol. 6. National Academy Press, Washington, DC (2000). https://doi.org/10.1016/s1051-0443(01)70072-3

26. Lam, H., Bertini, E., Isenberg, P., Plaisant, C., Carpendale, S.: Empirical studies in information visualization: seven scenarios. IEEE Trans. Visual Comput. Graphics 18(9), 1520–1536 (2011). https://doi.org/10.1109/TVCG.2011.279

27. Lee, B., Smith, G., Robertson, G.G., et al.: FacetLens: exposing trends and relationships to support sensemaking within faceted datasets. In: Proceedings of the SIGCHI Conference on Human Factors in Computing Systems, pp. 1293–1302. ACM (2009). https://doi.org/10.1145/1518701.1518896

28. Liu, Z., Stasko, J., Sullivan, T.: SellTrend: inter-attribute visual analysis of temporal transaction data. IEEE Trans. Visual Comput. Graphics 15(6), 1025–1032 (2009). https://doi.org/10.1109/TVCG.2009.180

29. Marais, K.B., Robichaud, M.R.: Analysis of trends in aviation maintenance risk: an empirical approach. Reliab. Eng. Syst. Saf. 106, 104–118 (2012). https://doi.org/10.1016/j.ress.2012.06.003

30. McKnight, P.E., Najab, J.: Mann-Whitney U test. In: The Corsini Encyclopedia of Psychology, pp. 1–1 (2010). https://doi.org/10.1002/9780470479216.corpsy0524

31. Morimoto, T., Gandhi, T., Seger, A., Hsieh, T., Bates, D.: Adverse drug events and medication errors: detection and classification methods. BMJ Qual. Saf. 13(4), 306–314 (2004)

32. Munzner, T.: Visualization Analysis and Design. AK Peters/CRC Press, Boca Raton (2014). https://doi.org/10.1201/b17511

33. NCC-MERP: National Coordinating Council for Medication Error Reporting and Prevention (1995), http://www.nccmerp.org/. Accessed 28 Feb 2018

34. Ozturk, S., Kayaalp, M., McDonald, C.J.: Visualization of patient prescription history data in emergency care. In: AMIA Annual Symposium Proceedings. vol. 2014, p. 963. American Medical Informatics Association (2014)

35. Patel, I., Balkrishnan, R.: Medication error management around the globe: an overview. Indian J. Pharm. Sci. 72(5), 539 (2010). https://doi.org/10.4103/0250-474x.78518

36. Peng, W., Ward, M.O., Rundensteiner, E.A.: Clutter reduction in multidimensional data visualization using dimension reordering. In: IEEE Symposium on Information Visualization, pp. 89–96. IEEE (2004). https://doi.org/10.1109/INFVIS.2004.15

37. Singh, R., Pace, W., Singh, A., Fox, C., Singh, G.: A visual computer interface concept for making error reporting useful at the point of care. In: Advances in Patient Safety: New Directions and Alternative Approaches (Vol. 1: Assessment). Agency for Healthcare Research and Quality (2008)

38. Smith, G., Czerwinski, M., Meyers, B.R., et al.: FacetMap: a scalable search and browse visualization. IEEE Trans. Visual Comput. Graphics 12(5), 797–804 (2006). https://doi.org/10.1109/TVCG.2006.142

39. Stasko, J., Görg, C., Liu, Z.: JigSaw: supporting investigative analysis through interactive visualization. Inf. Visual. 7(2), 118–132 (2008). https://doi.org/10.1057/palgrave.ivs.9500180

40. Varkey, P., Cunningham, J., Bisping, S.: Improving medication reconciliation in the outpatient setting. Jt. Comm. J. Qual. Patient Saf. 33(5), 286–292 (2007)

41. Wunnava, S., Qin, X., Kakar, T., Socrates, V., Wallace, A., Rundensteiner, E.A.: Towards transforming FDA adverse event narratives into actionable structured data for improved pharmacovigilance. In: Proceedings of the Symposium on Applied Computing, pp. 777–782. ACM (2017). https://doi.org/10.1145/3019612. 3022875
42. Yadav, V., Bethard, S.: A survey on recent advances in named entity recognition from deep learning models. In: Proceedings of the 27th International Conference on Computational Linguistics, pp. 2145–2158 (2018)
43. Zhou, S., Kang, H., Yao, B., Gong, Y.: Analyzing medication error reports in clinical settings: an automated pipeline approach. In: AMIA Annual Symposium Proceedings, vol. 2018, p. 1611. American Medical Informatics Association (2018)

A Layered Approach to Lightweight Toolchaining in Visual Analytics

Hans-Jörg Schulz[1]([⊠])(iD), Martin Röhlig[2], Lars Nonnemann[2](iD),
Marius Hogräfer[1](iD), Mario Aehnelt[3], Bodo Urban[2](iD), and Heidrun Schumann[2]

[1] Department of Computer Science, Aarhus University, Aarhus, Denmark
hjschulz@cs.au.dk
[2] Institute of Visual and Analytic Computing, University of Rostock,
Rostock, Germany
[3] Fraunhofer Institute for Computer Graphics Research, Rostock, Germany

Abstract. The ongoing proliferation and differentiation of Visual Analytics to various application domains and usage scenarios has also resulted in a fragmentation of the software landscape for data analysis. Highly specialized tools are available that focus on one particular analysis task in one particular application domain. The interoperability of these tools, which are often research prototypes without support or proper documentation, is hardly ever considered outside of the toolset they were originally intended to work with. To nevertheless use and reuse them in other settings and together with other tools, so as to realize novel analysis procedures by using them in concert, we propose an approach for loosely coupling individual visual analytics tools together into toolchains. Our approach differs from existing such mechanisms by being lightweight in realizing a pairwise coupling between tools without a central broker, and by being layered into different aspects of such a coupling: the usage flow, the data flow, and the control flow. We present a model of this approach and showcase its usefulness with three different usage examples, each focusing on one of the layers.

Keywords: Visual analytics · Software integration · View coordination

1 Introduction

Visual Analytics (VA) encompasses the frequent and fluent back and forth between computational analyses and interactive visual exploration. This process can get quite involved, with many specialized analysis steps from data wrangling [16], data preprocessing [18] and data exploration [6], all the way to model building [10] and investigating uncertainties [36]. Each of these specialized analysis steps comes with its own methods and tools that may or may not be used, depending on the analysis objectives known up front and on the findings made during an analysis. Such a flexible orchestration and use of VA tools at the analyst's direction is hard to capture with preconfigured analysis pipelines as

© Springer Nature Switzerland AG 2020
A. P. Cláudio et al. (Eds.): VISIGRAPP 2019, CCIS 1182, pp. 313–337, 2020.
https://doi.org/10.1007/978-3-030-41590-7_13

they are commonly employed in data science. This observation has led so far as some researchers suggesting to recoin the common phrase of "the human *in* the loop" into "the human *is* the loop" [5]. Yet in order to put the analyst at the helm of the analysis process and to be able to reconfigure and reparametrize the concerted use of multiple VA tools, a mechanism is needed to accomplish that.

We follow this line of thought and propose the concept of *layered VA toolchains* that, unlike fixed pipelines realized within an integrated VA framework, allow for a more flexible coupling of independent VA tools. Through this coupling, the otherwise autonomous tools form loose multi-tool ensembles that coordinate aspects of their joint use. This is achieved by differentiating VA tool coupling into three separate concerns:

- *Carrying out one tool after another*: The most fundamental characteristic of a concerted use of VA tools is the selection of suitable tools, as well the possible sequences in which they can be used. Capturing this first essential aspect allows guiding a user through a toolchain by automatically providing the user the right tool at the right time.
- *Funneling data from one tool into the next*: VA tools require an input to produce an output. If one tool's input is the resulting output from a preceding tool, this needs to be passed and possibly transformed between them. Capturing this aspect allows the toolchain to handle this process, automatically providing the user with the right data at the right time.
- *Coordinating functional aspects between tools*: Beyond the data, VA tools may require to set parameters or adjust the methods employed to produce their outcome. Capturing these functional characteristics allows keeping them consistent between tools, automatically providing the right parameters and presets at the right time to the user.

In this paper, which constitutes an extension of our previous work [35], we detail, discuss, and exemplify this threefold separation of concerns into layers for VA toolchaining. After looking briefly at related work for coordinating tools and views in Sect. 2, we present our conceptual model that captures these three layers in Sect. 3. We then discuss the benefits for the user that each individual layer yields and give concrete examples for them in Sect. 4 through Sect. 6.

2   Related Work

Prior research on tool coordination has resulted in a variety of frameworks such as *OBVIOUS* [7] or *VisMashup* [31]. They provide an interoperability layer on code level, which offers the necessary functionality to programmers and developers for coupling their VA tool with other tools utilizing the same framework. Coordination without code access relies usually on the exchange of data, views, or both, as they are produced by the VA tools.

Coordination among visualization tools through data exchange dates back to 1997 (cf. Visage [17]). Most approaches for data level coordination rely on a centralized mechanism. On the one hand, this can be a central database that acts as model in a model-view-controller (MVC) mechanism. Examples are *Snap-together* [22] and EdiFlow [2] that both use a relational database as underlying central data storage for a set of tools. On the other hand, coordination can employ a communication bus or service bus to broker messages among tools, as for example done by the *Metadata Mapper* [28]. In case no centralized data exchange mechanism is provided, tools use custom connectors to pass data. For closed source tools, this may rely on *screen poking* and *screen scraping* [8,14].

Coordination among visualization tools based on the views they produce can be achieved by exchanging and combining user interface (UI) components and graphical outputs from different systems, as it is proposed by approaches, such as *ManyVis* [30]. As the coordination of applications through the exchange of data or views are independent, they can be combined as necessary to achieve the required flexibility for a given analysis. One of the few examples for coordination using both, data and views, is *Webcharts* [9]. To make the interoperability of the tools available to the user, standard visualization frameworks usually display such an interlinked setup in multiple coordinated views [4,27] or fused visualizations [21]. In some cases, the toolchain itself is visually encoded in a (directed) graph layout that shows its connections [11,40]. When arranging UIs side-by-side is not sufficient, a common interface may be glued together from individual interface parts. WinCuts [39] and Façades [38] enable users to do so by cutting out pieces of multiple UIs and interactively composing them into one tailored UI. For web-based ensembles, existing UIs are combined in mashups [23].

In most application domains from climatology to biomedicine, the current practice in data analysis is to simply use independent VA tools one after the other. And as these domains also feature a diverse set of file formats and data conventions, it is often already considered a high level of interoperability when the different tools can work on the same data files, effectively easing the relay of results from one tool into the next. This current data analysis practice forms the basis of our coordination approach, which is described in the following.

3   A Layered Approach to Lightweight Toolchaining

As exemplified by the previous section, prior work on coordinating and linking visualization and analysis tools has focused mainly on the implementation and software engineering aspects of such mechanisms – i.e., the *algorithmic level* in terms of Munzner's four levels of visualization design [20, ch.4.3]. Our toolchaining approach puts its focus on the remaining three levels in the nested visualization design model that remain so far mostly undiscussed and unexplored:

- The *situational level* that aims to understand the domain situation – i.e., the application domain, the application data and questions, as well as the analytical processes that derive from them. In the context of toolchaining, the information captured on this level is about the used toolset in a domain, as well as common orders of use for domain-specific analyses. From this perspective, the situational level defines the *usage flow* among tools.

- The *abstraction level* that aims to understand the data and what is done with it in a domain-independent way. In the context of toolchaining, this captures the exchange of data between tools – i.e., which data in which data format as a result of which data transformation carried out by a VA tool. Consequently, this level defines foremost the *data flow* among tools.
- The *encoding and interaction level* that aims to understand how the data is visually displayed and how this display can be adapted. In the context of VA tools, this translates into parameter settings that govern the display, that can be changed to adapt the display, and that can be passed along for consistency of displays among tools. Hence, from the perspective of toolchains, this level defines the *control flow*.

3.1   The Principal Ideas Behind the Approach

Our approach is different from established coordination approaches in two ways: its layered structure for better separation of concerns among different notions of tool coordination, and its lightweight bottom-up realization for introducing automation to preexisting manual toolchains in an incremental and minimally invasive manner.

Our approach for toolchaining is *layered* in the sense that each of the individual levels of usage flow, data flow, and control flow can be used to affect a coordination among tools – either by themselves and with respect to the particular aspect of toolchaining they capture, or in combination with each other. For example, two tools may only be coordinated with regard to their data flow. This means that users still have to invoke one tool after the other and parametrize them manually, but that the data exchange between them is aided by a toolchaining mechanism, which can range from injecting a tool for handling format conversion to fully automated, bidirectional live data exchange. It is also possible that two tools are coordinated with regard to more than one level – e.g., their usage flow and their control flow. This could mean that after closing one tool, the next tool in the usage flow is automatically started together with a properly parametrized view that reflects any manual fine-tuning done in previously used tools, but that the data output and input still needs to be carried out manually.

Our approach for toolchaining is *lightweight* in the sense that we consider its realization to be pairwise between tools. This means, instead of having to provide an all-encompassing framework that coordinates between all possible VA tools and anticipates all possible combinations in a top-down manner, we instead use a bottom-up approach that works by coordinating only between those tools and on those levels, where this makes most sense. This approach allows us to utilize different coordination mechanisms between tools, depending on which interfaces they offer. As a result, coordination between tools in a toolchain can be introduced in a step-wise manner bridging the most tedious gaps that incur the most manual labor first, thus adding coordination and automation where it is needed most.

These ideas are directly reflected in the following coordination model that captures this vision of a layered, pairwise toolchaining.

3.2 A Model for Layered Lightweight Toolchaining

To characterize coordination among VA tools on the three levels of usage flow, data flow, and control flow, we need adequate means to describe coordination in such a faceted way. In line with common view coordination models from the literature – e.g., [43] and [3], we propose to model VA tool coordination as a graph. In this graph, VA tools constitute the nodes and directed edges capture the usage flow, data flow, and control flow between pairs of VA tools. We outline these parts of our coordination model, as well as how to establish and utilize them in the following.

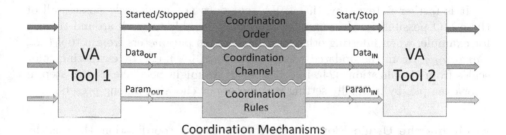

Fig. 1. Conceptual abstraction of lightweight coordination between two VA tools. The *coordination order* models any temporal dependency between two tools (i.e., their subsequent or concurrent use). The *coordination channels* capture ways to exchange data between tools, including any necessary data transformations along the way. The *coordination rules* describe automated syncing of interactive controls between tools by means of exchanging their associated parameters.

Modeling the VA Tools. VA tools form the natural basis of our coordination model, in the same way as they form the basis of the analytic toolchain. As these tools can come in any shape or form – from closed source to open source software, from simple command line tools to full-fledged VA frameworks – we follow the principle of making no prior assumptions about these tools and consider them as black boxes that are characterized by their inputs and outputs on the three levels introduced above. Figure 1 shows two tools modeled in this way, capturing the following I/O possibilities as *ports* of a VA tool:

Start(ed)/Stop(ped). Starting and stopping a tool, as well as being notified when it has started or stopped is the most fundamental of all tool capabilities. We model the invocation and termination of a tool via the port *Start/Stop* and the respective notifications via the port *Started/Stopped*.

Data (in/out). A VA tool requires input data on which to perform the analysis. The passed input may yield results that the tools passes back in return. It is common for most VA tools to have some ways of loading data, which we model as a port $Data_{IN}$ and saving results via the port $Data_{OUT}$. Note that the term data subsumes numerical data as well as image data.

Parameters (in/out). A VA tool's behavior is usually parametrizable. If parameters can be passed to the tool – e.g., as command line options when invoking it – this is modeled via the port $Param_{IN}$. If parameter settings can also be stored away for later re-use – e.g., as a config-file – this can be captured through the port $Param_{OUT}$.

Note that these ports are abstractions that we use to model the respective possibility to manage an independent VA tool, regardless of whether these are indeed provided by the tool itself or by some other entity, like the operating system. For example, invoking a tool and observing its state is on a technical level rarely done via the tool itself, but instead utilizes the operating system's process and window managers.

It is further noteworthy, that a VA tool does not necessarily possess all of these I/O possibilities. In some cases, we may be able to work around them – for example, when inferring otherwise inaccessible parameters from a tool (i.e., $Param_{OUT}$) from the updated results (i.e., $Data_{OUT}$), e.g. by extracting color scales from visualizations [24]. In other cases, we might need the user to step in – for example, by manually setting parameters or choosing among presets.

Modeling the Usage Flow. At the lowest level of coordination, there is the temporal order of VA tools – i.e., the usage flow determined by the analysis scenario and its domain. The usage flow captures the toolchain as a succession of VA tools, as these are used by the user in pursuit of a particular analysis goal. It can include subsequently used tools, as well as concurrently used tools. Tool coordination modeled and realized at this level is able to bring up the right tools at the right time, as it is foreseen by the usage flow.

In our model, this type of coordination is achieved through a bilateral connection among two tools indicating their *coordination order* – shown at the top in Fig. 1. This order basically starts one VA tool subsequently or concurrently, depending on another VA tool having been Started/Stopped:

- Subsequently: $(Tool_1.Stopped) \implies (Tool_2.Start)$
- Concurrently: $(Tool_1.Started) \implies (Tool_2.Start)$

Coordination on this level already provides as much as an automated guidance of the user along a predefined path of analysis, in the spirit of approaches such as *Stack'n'Flip* [37]. While the users still have to do everything else themselves – such as moving data back and forth between the tools, or parametrizing their visualizations to match up – the coordination order between tools allows to automatically move the VA toolchain forward as the interactive analysis progresses. Besides providing convenience for the user, this also ensures comparability between different analysis runs as VA tools are always invoked in line with the predefined usage flow. This is important in cases where carrying out analysis steps in different orders yields different results – e.g., when performing dimension reduction and clustering. But it is also useful to ensure that no VA tool is left out by mistake – e.g., forgetting to normalize the data before processing it.

Modeling the Data Flow. The next level of coordination is about getting data in and out of a VA tool. Besides starting and stopping a tool, this is arguably the most important aspect of a VA tool: without data, there is nothing to work with. The data flow captures how input data is passed into VA tools, transformed into analysis results and passed on to the next VA tool as input again. At this level, VA tool coordination automatically hands off data from tool to tool as the user proceeds with the analysis along the toolchain – i.e., the flow of data tends to follow the usage flow. Depending on where in the visualization pipeline a VA tool operates, data can refer to raw data, prepared data, focus data, geometric data, or image data [32, 41].

This coordination is achieved via *coordination channels* – shown in the middle of Fig. 1. These capture the data exchange from a VA tool's output to another's input, as well as any data transformations f_D along the way:

$$Tool_1.Data_{OUT} \xrightarrow{f_D} Tool_2.Data_{IN}$$

Coordination on this level automatically delivers the right data to the right tools at the right time, very much like data flow-oriented visualization frameworks do. While the user still has to manage what is to happen with that data in a currently used VA tool – i.e., interactively parametrizing computations and visualizations – having input data and results from previously used VA tools delivered automatically to the current VA tool takes care of any tedious conversion between different data formats, competing standards, or sometimes just different versions or interpretations of the same standard [34]. This automation also ensures that the user is not working on stale data, as the coordination channels always deliver the most current data and manually loading an old dataset can only happen on purpose, not by accident. Coordination channels can further be used to store a snapshot of the last data passed by them, so that when revisiting a previously used VA tool its last input data is still available, thus providing a coherent analysis experience forward *and* backward along the toolchain.

Note that "passing data" can be a complex problem and research on data integration and scientific workflows has established various approaches to do so [19]. For modeling the coordination on data level, it is enough to know that one of these approaches underlies a channel and realizes the data transport.

Modeling the Control Flow. The last level of coordination is about having interactively set or changed controls in one VA tool also reflected in other tools by exchanging their associated parameters. This is probably the most common understanding of coordination, where for example, filter operations, selections, or adaptations of a color scale in one tool will also affect other tools. Though, how exactly these will affect subsequently or concurrently used tools is subject to the concrete nature of the coordination. Our model captures the settings and interactive choices made within tools as expressed through their parameters – e.g., numerical filter criteria, color scales, or transformation functions that can be tuned in one tool and reused in another. At this level, VA tool coordination automates the synchronization of interactive changes made across tools, which mostly relates to the visualization or UI of these tools as these are the common means with which the user interacts.

In our model, this coordination is achieved via *coordination rules* – shown at the bottom in Fig. 1. These rules capture not only the mere passing of parameters between tools, but also the "coordination logic" behind the linking they realize. For example, in cases where this linking is supposed to have changes made in one tool being reflected likewise in another tool – e.g., a selection – the coordination rule relays the corresponding parameters as they are. Yet in cases, were this linking is supposed to have a complementing effect – e.g., the other tool not showing the selection, but everything that was not selected – the coordination rule relays the inverse of the corresponding parameters. Coordination rules are modeled as functions f_P with the identity $f_P = id$ as a default:

$$Tool_1.Param_{OUT} \overset{f_P}{\Longrightarrow} Tool_2.Param_{IN}$$

Coordination on this level can be used to provide any of the common coordination patterns, such as linking & brushing, navigational slaving, or visual linking [42] – provided that the corresponding view and data parameters are captured and accessible to be passed between tools. As it was the case for coordination channels, the user still has to manage and steer the currently used VA tool, but where applicable, this steering is picked up on and automatically mirrored or complemented in other VA tools. This includes not only tools that are used concurrently, but also those used subsequently, as interactive selections or carefully tuned color scales can be passed on as parameters to the next tool in the toolchain as well. This relieves the users from having to make the same interactive adjustments multiple times for each VA tool they are working with, and it guarantees consistent settings across tools. This is helpful when trying to compare results or for tracking certain data items across the toolchain.

3.3   Combining Tools and Flows

Taken together, we yield a 3-tiered model of lightweight coordination among VA tools comprised of the VA tools, as well as of the three coordination levels to realize different aspects of coupling among those tools. It can thus be expressed as a 4-tuple $CM = (Tools, UsageFlow, DataFlow, ControlFlow)$ consisting of the following sets:

- The set of *Tools* as it is given by the analysis setup.
- The *UsageFlow*, defined as the set of all coordination orders $(Source, [Started|Stopped], Target, [Start|Stop])$ capturing the execution sequence between a source and a target tool.
- The *DataFlow*, defined as the set of all coordination channels $(Source, Target, f_D : Data_{OUT} \mapsto Data_{IN})$ capturing data transfer and transformation (f_D) between a source and a target tool.
- The *ControlFlow*, defined as the set of all coordination rules $(Source, Target, f_P : Param_{OUT} \mapsto Param_{IN})$ capturing parameter exchange and modification (f_P) between a source and a target tool.

All three sets of *UsageFlow*, *DataFlow*, and *ControlFlow* model the pair-wise coordination between VA tools: The *UsageFlow* contains the answer to the question of which parts of the toolchain to automate tool-wise and in which order. The *DataFlow* comprises the answer to the question along which parts of the toolchain to transmit data using which data transformation. And the *ControlFlow* captures the answer to the question of which coordination logic is realized between tools of the chain. The three sets have in common that they describe the coupling between pairs of tools, so that each of these couplings can likewise be understood as sets of directed edges that taken together define a graph topology over the set of VA tools. They thus capture coordination in a bottom-up fashion by coupling two tools at a time and then combining these pairwise couplings into larger coordination mechanisms. This combination is for example done by adding reciprocity (combining the unidirectional coupling between two tools A and B, and between B and A) or by employing transitivity (coupling two tools A and B through an intermediary tool C).

The strength of this model is its decentralized, pairwise structure that requires neither a central broker or mediator, nor does it force any architectural changes on the used VA tools. Instead of trying to lift a whole toolchain up to conform to a state-of-the-art coordination framework, we coordinate directly between two VA tools. This way, we can capture the full variety of different modes, degrees, and directions of coordination between different tools, instead of boiling the coordination down to the least common denominator among all involved tools. This directly benefits:

- the VA experts, who can introduce coordination incrementally – adding one level of coordination among one pair of tools at a time – and thus adaptively expand and refine coordination as it becomes necessary,
- the software experts, who can leverage whichever features a tool already provides to connect to it,
- and the end users, who can bring in additional tools, which may or may not already have couplings to the tools from a current toolchain.

The decentralized model is mainly targeted at the VA and software experts. While the end users are also thought to benefit from the provided flexibility, overlooking and taming such a "zoo of tools" can also become quite a challenge. To aid in doing so, our approach complements the decentralized coordination mechanism with interface arrangements that provide structure on top of the tool ensemble. These interface arrangements are described in the following section.

3.4   Interface Ensembles

When handling multiple VA tools, the mere orchestration of their application windows becomes a management task by itself that requires attention, effort, and time which would be better spent on the visual analysis itself. In order to reduce this overhead, we propose the notion of *interface ensembles*. These provide an organized view on the otherwise often overwhelming mesh of any number of invisible pairwise coordination mechanisms springing to life depending on the currently used tools and the user's actions.

Interface ensembles consist first and foremost of a centralized panel for global views and interaction elements that concern the toolchain as a whole. We call this centralized panel the *Control Interface* and it serves at its core as a place in which to display information that cuts across tools and where to affect global changes and adjustments. In this function, the control interface can be utilized by all three levels of coordination:

- for the *Usage Flow*, we can use the control interface for example to display progress information and to invoke additional tools or choose between alternative analysis paths;
- for the *Data Flow*, we can use the control interface for example to archive snapshots of interesting intermediate results for later in-depth investigation along the lines of a bookmarking or multi clipboard system;
- for the *Control Flow*, we can use the control interface for example to display global information like legends and to set global parameters like color scales or which data field to use for labeling data items.

In addition to the control interface, interface ensembles employ structured ways of displaying UIs, so that they appear in a predictable manner that suits the current usage flow. Common ways of doing so are outlined in the following.

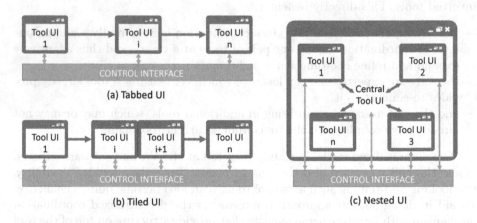

Fig. 2. Common UI layouts when dealing with multiple VA tools.

Individual VA Tool Use: The Tabbed UI. The usage flow among VA tools is a path: first a tool A is used, then a tool B, followed by a tool C, and so forth. This individual, subsequential use of VA tools as predefined by the coordination orders that connect the tools in a temporal sense, results in an exclusive use of a single UI at each point in time as shown in Fig. 2(a). What reads like an oversimplification at first is actually the most prevalent usage pattern in practice, as visual analysis is for the most part conducted as a linear series of very specific analysis steps, each carried out with a highly specialized analysis tool or view.

To the user, these tool sequences can be offered in a variety of ways. One way of displaying such sequential procedures is through a tabbed interface that opens each tool in a dedicated tab, with the tabs being ordered according to the tool sequence. This way, by clicking on the tabs, the user is always able to go back in the toolchain and to readjust some property in an earlier used tool – for example, manually moving a data item from one cluster into another one. Given that all other parameters and choices along the toolchain stay the same, these changes can be passed automatically through the appropriate channels and be processed by the appropriate rules to auto-update the current tool and its view. The tabbed interface can further be augmented by a wizard-like guidance that leads the user tab by tab along the path defined by the coordination orders.

Combined VA Tool Use: The Tiled Display. Sometimes, it makes sense to use VA tools not just one tool at a time, but to have access to subsequent tools of the toolchain at once. This can be the case, for example, when a data selection from one tool will serve as an input to the next tool and one needs to go back and forth between the two tools to try out and observe the effects of different selections. The topology is still a path topology, as shown in Fig. 2(b), but with two UIs being displayed at once to facilitate such combined use.

To the user, such setups are usually offered by tiling the display and showing the tools side by side, or by distributing them among multiple monitors. In this way, the tools are present on the screen at the same time to work with them without having to switch – i.e., sending one to the background and bringing another one to the front, as it would be the case in the tabbed interface. Synchronization features, such as linking & brushing and displaying visual links are desirable to make the back and forth between the tools more fluent.

Flexible VA Tool Use: The Nested UI. If VA tools are used more flexibly than a mere back and forth along a path of sequential tools, the resulting topology also gets more involved. A powerful example for this case is the star-shaped topology that is shown in Fig. 2(c) where all analysis steps start from a hub application or central VA tool. Such topologies support more complex usage flows that meander between multiple tools until their combined use yields an analysis result. This is often the case in comparative analyses where multiple windows and tools are needed to process, show, and relate different data subsets or different analytical procedures to each other.

To the user, the central tool is usually offered as an omnipresent overview of the data that is shown in a fashion similar to a background image. In this overview, users can select regions of interest into which to dive deeper by opening them up in other VA tools. The opened tools are shown as nested or superimposed views right in place where the selection was made. Making multiple selections opens multiple tools, effectively realizing the star-shaped topology. For this to work even with a dozen tools all scattered across the overview of the central

VA tool, the overview/background needs a map-like appearance that serves well as a context for all the other UIs and makes their spatial relation meaningful.

The following sections will give a few first impressions of how this model for lightweight tool coordination can be employed in real-world analysis scenarios. To do so, each of the three sections focuses on the coordination of one particular layer and for one particular use case, capturing and utilizing the usage flow in Sect. 4, the data flow in Sect. 5, and the control flow in Sect. 6.

4   Providing the Right Tool at the Right Time

At the first layer of VA tool coupling, we define a temporal toolchain according to existing analysis procedures in a domain. This includes selecting suitable tools for the given tasks, determining sequences in which they should be used, and guiding a user through these sequences by automatically activating the right tool at the right time. We demonstrate the benefits of this layer with an editor to specify temporal chains of VA tools and its application in ophthalmology.

4.1   An Editor for Building a Temporal Toolchain

The first step of coupling independent tools is to put them in the right order. In many cases, numerous possible orders of tools are available. For example, in a larger analysis procedure, the order between two or more tools may be predefined, e.g., due to semantic constraints, while at the same time alternative execution sequences for the remaining tools exist. Thus, the first important question is: *who defines the order of tools?*

We argue that an appropriate answer to this question is to let the domain experts create the order of tools. That is because domain experts generally have the necessary context knowledge about which individual tool needs to be used when in their analysis procedure. However, this requires a simple interface for modeling a temporal toolchain that can be understood by non-technical users. This brings us to the second important question: *how can we support domain experts in defining orders of tools?*

As with every other temporal sequence, there is some ordering implied by an analytical toolchain. We choose to represent the orders of independent tools as directed acyclic graphs that contain multiple time steps with at least a single tool and optional data inputs necessary for carrying out the tools. In addition, we provide an editor with a simple user interface in order to support domain experts in creating such graphs. This requires us to (1) allow access to all tools that are needed, (2) allow access to all data that are needed to run these tools, and (3) allow to define connections between tools and data sources in the ordering process.

The main idea of our editor is to build a metaphorical bridge between the domain expert and the internal tool coupling processes. Hence, we choose an interface design that is similar to common presentation software, where a user can configure a sequence of slides to be shown in a later talk.

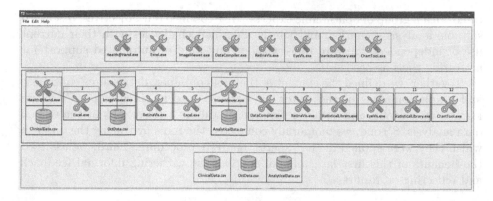

Fig. 3. Analytical process editor with temporal order of multiple tools used in an ophthalmic analysis process. The three panels in the editor's user interface represent the tool container (top), the graph container (middle), and the source container (bottom).

The display of our editor's user interface is divided into three containers that hold different types of information for the ordering process (Fig. 3):

1. The **Tool Container** holds a collection of all executable tools. Imported tools can be dragged from this container and dropped into the graph container to either create new time steps or add tools to existing time steps.
2. The **Source Container** holds a collection of data sources. These data sources can be included in the graph to supply tools at a certain time step with input or supplementary data.
3. The **Graph Container** is the main part of the editor's user interface. It provides a space for the creation and modification of directed acyclic graphs. Each tool or data source can be dragged & dropped from the other containers into the graph container to order the analytical process. The resulting temporal toolchain can then be executed or saved as a JSON file by interacting with the main menu at the top of the user interface.

All necessary tools and data sources are imported into the editor by dragging them onto the two respective containers in the user interface. In fact, all of our import and export mechanisms are based on drag & drop interactions to facilitate an intuitive user experience. This is to enable non-technical users to adapt more easily and quickly to the temporal ordering process of independent tools.

All in all, our editor provides an easy-to-use interface that abstracts the temporal coupling of tools similar to common presentation software. At the same time, the editor is highly flexible with respect to the number of imported tools or data sources as well as to the size and complexity of created toolchains.

4.2   A Temporal Toolchain in Retinal Data Analysis

We applied our editor to build a tool sequence on top of a scientific analysis procedure for retinal data in ophthalmic research. For this purpose, we collaborated with a group of ophthalmic experts. We explained our objectives, discussed suitable analysis scenarios, and jointly identified challenges with their current use of independent software tools without any kind of automated support for activating tools as needed.

Together, we defined the content of the tool container, data source container, and graph container of the editor for our use case in three steps. First, we learned which tools and data are required by taking a closer look at the retinal data analysis. Second, we temporally connected the tools by tracing the order in which they were applied in the experts' analysis procedure. Third, we assessed the benefits of this first layer of tool coupling by gathering informal feedback and reflecting on results.

Determining Tools and Data. The ophthalmic experts were interested in studying retinal changes of patients in relation to healthy controls. They started by gathering retinal 3D image data acquired via optical coherence tomography (OCT) and other clinical parameters of the two study groups from a clinical data management system. The data analysis was then carried out in three main steps:

1. Data preparation: Compilation of study groups and data quality checks.
2. Data exploration: Discovering differences between study groups and investigating relationships with clinical parameters.
3. Data presentation: Summarizing findings and reporting study results.

In this process, a set of eight diverse tools was applied. This included commercial software, e.g., device-specific OCT software and spreadsheet software, as well as statistics software and VA software, e.g., the R software environment [26] and a custom VA framework for retinal OCT data [25, 29]. The data required to carry out the tools comprised OCT scans and electronic health records for both patients and controls. All tools and datasets were imported into the tool container and the data source container of the editor, respectively.

Temporally Connecting Tools. Next, we focused on making the tools available when needed by automating their activation. We utilized our editor to model the temporal order of the tools as a directed graph according to an established analysis procedure currently in use in ophthalmic research. We started by sketching an initial tool sequence in the editor together with the experts. The visual presentation of temporal connections in the editor's user interface helped us to get an idea of when which tool was applied. Using the editor's direct drag & drop manipulation support, we were able to refine the initial sequence based on the experts' feedback and devise alternatives. Figure 3 shows the resulting sequence of applied tools in the user interface of the editor. With the editor's graph container at hand, the experts were able to create, refine, and store modeled tool sequences. They were also able to perform test runs via the editor to make sure the modeled tool sequences exactly matched their requirements.

Benefits of Temporal Tool Coupling. Our observations and the ophthalmic experts' feedback are summarized with respect to the utility of the editor and supporting a temporal coupling of tools in general. Regarding the editor, we were able to successfully model a tool sequence on top of the analysis procedure explained by the ophthalmic experts. In this regard, the editor proved to be a useful aid for this kind of work. In fact, other ophthalmic analysis procedures similar to the described use case were identified, including studying specific diseases or investigating corneal study data. While the general analysis approach and involved tasks are comparable, the required tools and the order in which they are applied may differ. In this context, our editor supports defining new temporal tool sequences and helps to revisit and adapt existing ones. The latter becomes necessary if medical devices and respective software are changed in a clinic or if additional steps are added in the analysis procedure. By serializing and storing a modeled tool sequence, it is even possible to go back to previous versions and to recreate a specific analysis result.

Regarding the temporal coupling of tools in general, the experts were relieved from having to search for each required tool and starting it independently. Although, all data funneling and parameter synchronization still had to be done by hand, the modeled tool sequence and automated tool activation already helped them to focus on the actual data processing and analysis. Especially, moving back and forth between certain analysis tasks and thus frequently switching between heterogeneous tools had previously required considerable manual effort. The basic guidance through the analysis procedure provided by this first layer of tool coupling ensured that no tool or analysis step was left out by mistake, e.g., forgetting data quality checks and applying respective repairs. In addition, it enabled comparability between different analysis runs by activating tools always in line with the experts' analysis procedure.

5   Providing the Right Data at the Right Time

At the second layer of VA tool coupling, we provide the means for VA tools to exchange data, which is probably most directly associated with the joint use of multiple tools. It should foremost make available the analysis results from one tool in a form that can be read-in by the next tool in the chain – and possibly vice-versa, depending on how the tools are used in conjunction. To provide this functionality, we introduce a library for web-based analysis tools to enable data input and output. We illustrate the usefulness of this functionality by showing how it can be used to create and augment VA tools with this functionality, and how this helped toolchaining in an astronomy use case.

5.1   A Library for Data Exchange Between VA Tools

Data exchange between multiple VA tools is not as straightforward as it may seem due to the different types of data being generated: while some tools produce numerical outputs, others produce visual outputs, as it lies in the nature of VA.

The challenge is that while numerical output can at any time be transformed into visual output, if another tool requires its input in view-form – the other way around is hardly possible. This leads to the situation where if at a later, visual-interactive step of the toolchain it becomes apparent that changes need to be made to analysis steps carried out on the numerical data – e.g., the data should have first been normalized or some aggregation parameter needs to be adjusted – the whole toolchain needs to be rolled back to that point to enact the desired change and then be carried out anew from this point onward.

This challenge is usually countered with a linking between any shown graphical object and its underlying data item – e.g., by employing centralized, MVC-like architectures – so that any interaction and manipulation in the view space can be directly reflected in the underlying data space. This approach works very well, but requires deep integration with the coordinated tools that goes far beyond the mere input and output of files. To instead realize a lightweight data exchange that follows the idea of pairwise connections between tools through passing of files, we propose a different approach: we use the visualization grammar Vega-Lite [33] to describe data, data transformations, visual mapping, and view transformations – i.e., the full visualization pipeline – all at once in one file. Consequently, VA tools that are able to read and write such grammars can adjust any aspect of a visualization, from the underlying dataset all the way to color scales and axis labels in any order.

We support the realization of "Vega-Lite enabled" VA tools by providing the open source library *ReVize* that equips web-based tools with Vega-Lite import/export functionality [15]. ReVize provides an abstraction of Vega-Lite encoded visualizations by making the contained view hierarchy and data transformation graph accessible. To do so, it uses Vega-Lite's view composition structure and decomposes it into independent layers for individual modification. ReVize's import module resolves structural dependencies (e.g., inline datasets or inherited visual encodings) that usually prevent the reuse of sub-views in Vega-Lite specifications by inferring default values from parent and child nodes. A VA tool using ReVize can thus import a Vega-Lite formatted visualization description, further process its contained data or interactively adjust the view on the data it describes, and export it as a Vega-Lite specification again.

ReVize can on one hand be used to build dedicated VA tools that are tailored to processing or adjusting individual aspects captured by the Vega-Lite format. For example, we utilized ReVize for building a data editor called *ReData* to insert/remove data transformations and a visualization editor called *ReVis* to create/adapt the visual mapping and thus the display of the data – cf. Fig. 4(left). On the other hand, ReVize can also be used to add Vega-Lite input/output functionality to pre-existing tools, thus enabling us to use them as part of a Vega-Lite based VA toolchain. We have used ReVize for instance to augment the well-known ColorBrewer web application [13] with Vega-Lite import and export functionality, so that it can be used to adjust color scales in Vega-Lite formatted visualization data. This is shown in Fig. 4(right), and all mentioned

Fig. 4. Screenshots of ReVis (left) and our Vega-Lite enabled ColorBrewer (right).

tools are furthermore also available for download from https://vis-au.github.io/toolchaining/. Together with all the tools that already feature Vega-Lite input and/or output, these VA tools form the toolset from which analysts can freely pick the most appropriate one to carry out their next analysis step. This is illustrated in the following with an analysis scenario from astronomy.

5.2   A Cross-Tool Analysis Scenario for Astronomical Data

In this use case, we worked with a dataset on extrasolar planets publicly available through the Exoplanet Data Explorer at http://exoplanets.org [12]. It contains data on more than 3,000 confirmed planets that were discovered by international research groups before 2018. For each entry, the dataset includes information on the first peer-reviewed publication for a planet, its position as well as physical attributes such as the planet's mass. The dataset is a popular reference in exoplanetary research projects.

Using this dataset, we want to investigate the planets' masses and their semi major axis, which is a positional parameter that expresses the greatest distance a planet has on its orbit around the star of its solar system. In addition to those, we are also interested in the influence of the so-called argument of periastron, i.e. the angle between the semi major axis and the direction, from which a telescope observed the planet. Looking at these values is helpful in determining characteristics of the types of measurement used to observe these planets.

For investigating this data, we rely on our Vega-Lite based toolset of ReVis, ReData, the Vega-Lite enabled ColorBrewer, and the Vega online editor at https://vega.github.io/editor/. One possible exploration path across these tools is depicted in Fig. 5.

Benefits of Coupling Tools Through Data Exchange. Besides the advantages that we have already taken into account at design time, such as the expressiveness of Vega-Lite for numerical and image data alike, there are more benefits from data exchange that emerged when working with this use case. First, by using a generic data exchange format that is not tailored to a particular application domain, we were able to bring domain-specific tools like ColorBrewer into the toolchain, even

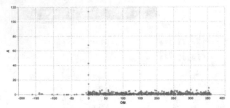
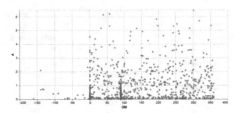

(a) ReVis is used to create a scatterplot of the exoplanets dataset, but outliers at 0° compress the view of the data vertically.

(b) Filtering these outliers using ReData reveals that many of the planets concentrate hardly discernable at 0° and 90°.

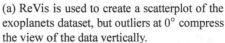
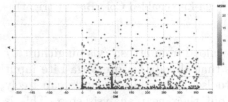
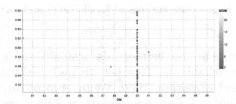

(c) Using ColorBrewer to color-code each planet's mass, does not help to better discern individual planets in these spots.

(d) Only zooming into these spots using the Vega online editor finally allows to investigate those planets individually.

Fig. 5. A possible path through a toolset of four independent tools that are coupled by Vega-Lite based data exchange.

though it is originally intended for use in cartographic applications. Second, by capturing all aspects of an intermediate result from the toolchain in one file, we could simply email these files back and forth with our collaborator from the physics and astronomy department to clarify questions or have him make adjustments to the data part of the file with his specialized tools – mostly IDL scripts. And finally, a simple form of cross-tool undo/redo functionality comes for free with this type of coupling, as snapshots of files can easily be archived every time they are passed into the next tool.

6   Providing the Right Parameters at the Right Time

At the last layer of tool coupling, we define presets for parameters or functions that are used within the analysis toolchain of our application domain. This includes the identification of tasks and their translation into modular execution steps. The Fraunhofer IGD in Rostock has already worked on multiple use case scenarios in the fields of industry, healthcare, and administrative management. One of the most popular commercial projects is the *Plant@Hand* framework [1]. This product is the fundamental structure for multiple descendants such as *Health@Hand* or *Ship@Hand* (see Fig. 6). All of these products are using presets in order to assist users in their daily task. We explain this concept with an example of a healthcare use case in *Health@Hand*.

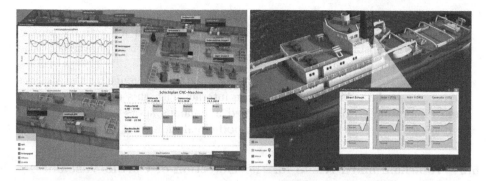

Fig. 6. Examples of visual data analysis use cases with *Plant@Hand*: factory visualization (left) and ship condition analysis (right).

6.1   A Comprehensive Application for the Visualization of Different Presets

The digitization in the health care sector has led to a growing quantity and quality of individual systems such as the clinical information system (CIS), the picture archiving and communication system (PACS), the radiological or laboratory information systems (RIS/LIS), or the electronic health record systems (EHR). Additionally, there are specific third-party tools for the analysis and visualization of real-time sensor data as well as customized information dashboards. As a result, the medical staff needs to take on the role of data scientists in order to understand the correlations between multiple tools, observe changes in the data, and assure the well-being of multiple patients at the same time.

Our *Health@Hand* interface assists this cause by providing a 3D model of an intensive care unit (ICU) in order to emphasize the changes of real-time vital data. This digital twin of the hospital ward contains different types of information about patients, rooms, staff, or inventory. We define these categories as *data sources* that each hold a set of *elements* to be handled by the *data manager*. An example of this would be the data source patient that holds elements such as pulse, heart rate variability, or respiration rate.

The visual interface is based on the nested UI model by providing a general overview on the 3D scene with relevant short information about each element. The user can interact with this UI by selecting a single short information panel to open up more detailed information. The detail information panel can contain text, images, or entire tools to further examine selected instances of a data source. The individual tools can be used independently to handle the data from typical healthcare management systems.

- The *Vital Data (VD) Dashboard* shows a patient's vital data parameters such as pulse, heart rate variability, or respiration rate.
- The *eHealth Record Browser* gives the medical staff access to the medical EHR system. This record contains context information about the patient's personal data such as name, age, and prior diagnosis.

– The *Image Viewer* uses the PACS suite to import images of medical findings.
– The *Shimmer Monitor* is used to access readings from an activity sensor. This sensor captures real-time data of the patient's vital parameters and determines the estimated activity.
– The *Cardio Anomaly (CA) Detector* uses artificial intelligence for the analysis of heart rate data. Medical experts can interact with this tool by inspecting and annotating automatically identified anomalous spots in order to point out reasons for a patient's irregular blood circulation.

6.2   Toolchain and Presets for the Detection of Cardio-Vascular Anomalies

Our *Health@Hand* system uses different sorts of parameters in the data manager to control for example (1) which information are shown in the 3D model, (2) how colors are adjusted to avoid visual clutter, (3) how screen space is allocated and organized for the parallel usage of multiple tools, and (4) which tools are linked in order to identify correlations in the visual representation.

The previously mentioned data manager can be used to switch between different data sources. However, there is also a possibility to select or deselect different parameters as every element of a data source is passed as a parameter to create the short information panels. The user can thereby interact with a side menu in order to avoid visual clutter.

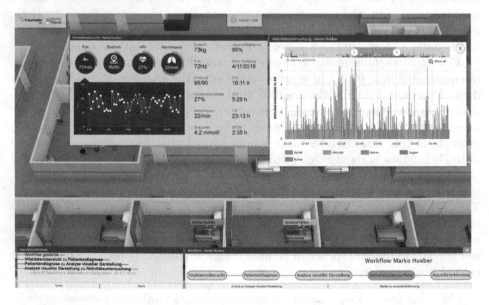

Fig. 7. Toolchain view of *Health@Hand* with the toolchain window (bottom right), the history window (bottom left) and the currently used tools (center).

In the following part, we consider the toolchain module that is used for the execution of our specific use case scenario (see Fig. 7). The module itself is a part of the data manager and consists of three different components:

The **Toolchain Window** provides information about the current progress of the user by showing an overview of the analytical toolchain graph. The graph can be explored by interacting with the buttons below in order to switch back and forth in the analytical process. The user can also find useful information about upcoming tools by hovering over the related step.

The **History Window** holds a record of all performed operations whether it is inside of a tool or by interacting with the toolchain window. These operations can then be reverted or repeated on demand.

The **Workspace Area** allows users to interact with the corresponding tools of each toolchain step. It adapts the size of each tool according to the screen space and organizes them next to each other to avoid overlap. The referred data for the scaling and positioning of each tool is thereby passed as a parameter at the initiation of the next toolchain step.

We use this module to represent the analytical process for the detection of cardio-vascular anomalies. The resulting toolchain is based on a specific temporal order of the previously mentioned tools. In the following, we describe our use case in six different steps:

1. **Situational Assessment using the Three-dimensional** *digital twin*: The medical expert observes and examines the general condition of multiple patients based on their incoming sensor data. If the vital data parameters change to an unusual state, the expert can switch to the toolchain module and select the specific patient. Thereby the patient's ID is recorded and passed through the toolchain in the following process, so that each tool opens up with the right medical data.
2. **Vital Data Overview using the** *VD Dashboard*: At the first step of the toolchain, a backlog of a patient's vital signs is examined through diagrams and gauges in the *VD Dashboard*.
3. **Evaluation of the Personal Health State using the** *eHealth Record Browser*: In the next step, the patient record entries such as *age, data of admission*, and *diagnosis* are considered in order to put the vital signs in the context of the patient's personal health state. Therefore, the workspace is reduced to show the vital data parameters next to the personal health data.
4. **Checking Diagnostic Findings using the** *Image Viewer*: The *Image Viewer* provides further context information for the medical expert to check available imaging data for related cardiovascular findings. Therefore, each tool window in the workspace is further reduced to an acceptable size.
5. **Activity Analysis using the** *Shimmer Monitor*: As both, pulse and respiration rate, correlate with the physical activity being performed, the patient's activity data (including any therapeutic stressing situations) is brought up from the *Shimmer Monitor*. At this step, Linking & Brushing is

established in order to find correlated information based on the selected time interval in the *Shimmer Monitor*. The interval is then passed as a parameter to the *VD Dashboard* and highlighted accordingly.

6. **Automated Anomaly Detection using *CA-detect*:** If the expert finds no natural cause for the current irregularities, he can use the *cardiological anomaly detection* to calculate anomaly scores based on the vital data streams of the patient. After some minor configuration time, the results are automatically opened in a D3.js line chart to show the statistical outliers.

The use case shows that there are multiple possibilities for using preset parameters in our *Health@Hand* system. We received positive feedback from different visitors at the Medica World Forum for Medicine 2018 and on many following exhibitions. The received feedback included comments such as "That's the future of clinical data exploration." (from *PAMB*) or "The integration of different data views will improve diagnosis and therapy management." (from *Poly-Projekt GmbH*). Especially, changes in the tools itself such as the Linking & Brushing between tools proved to be very helpful for the user. There have also been some minor remarks on the potential of multi-user interaction or the distribution of information across different devices, and we look forward to include these suggestions in further development.

7   Conclusion

With the proposed layered approach for lightweight toolchaining in VA, we have captured in a structured manner the stitching together of VA tools via custom scripts and copy & paste keystrokes that is the current state of affairs in coupling multiple independent VA tools in many domains. The proposed layered model that clearly separates the different aspects of working with ensembles of multiple independent tools, as well as its underlying principle of lightweightedness – i.e., their pairwise, bottom-up coordination – offer a conceptual framework in which to discuss and design toolchains and their desired and necessary degree of automation for a scenario at hand. Finally, the concrete realizations and use cases discussed for each level of coordination show the proposed approach in action, illustrating what each of them has to offer and providing arguments for the usefulness of each of them.

What these different use cases do not yet explicitly discuss, is the potential for the three levels of coordination to play together and to extend into each other's domain. For example, given an encompassing data description like a visualization grammar, adjustments to the control flow can easily be envisioned to reach well beyond mere reparametrizations by switching out not just individual values but entire parts of that description – i.e., parts of the dataset or functional building blocks in the view composition. Or given a planning environment for explicitly creating and revising the usage flow, not only tools and their order of use, but also parameters associated with particular analysis steps can be preconfigured as "temporal presets". In particular such combinations of the different coordination levels will be in the focus of future work on our layered approach.

Acknowledgements. We are indebted to the respective domain experts from ophthalmology, astronomy, and intensive care who provided their feedback and insights for Sects. 4 through 6. We also thank the anonymous reviewers for their guiding comments on earlier versions of this manuscript. Furthermore, we gratefully acknowledge funding support by the German Research Foundation through DFG project UniVA.

References

1. Aehnelt, M., Schulz, H.-J., Urban, B.: Towards a contextualized visual analysis of heterogeneous manufacturing data. In: Bebis, G., et al. (eds.) ISVC 2013. LNCS, vol. 8034, pp. 76–85. Springer, Heidelberg (2013). https://doi.org/10.1007/978-3-642-41939-3_8
2. Benzaken, V., Fekete, J.D., Hémery, P.L., Khemiri, W., Manolescu, I.: EdiFlow: data-intensive interactive workflows for visual analytics. In: Proceedings of the IEEE International Conference on Data Engineering (ICDE), pp. 780–791 (2011). https://doi.org/10.1109/ICDE.2011.5767914
3. Collins, C., Carpendale, S.: VisLink: revealing relationships amongst visualizations. IEEE TVCG **13**(6), 1192–1199 (2007). https://doi.org/10.1109/TVCG.2007.70521
4. Dörk, M., Carpendale, S., Collins, C., Williamson, C.: VisGets: coordinated visualizations for web-based information exploration and discovery. IEEE TVCG **14**(6), 1205–1212 (2008). https://doi.org/10.1109/TVCG.2008.175
5. Endert, A., Hossain, M.S., Ramakrishnan, N., North, C., Fiaux, P., Andrews, C.: The human is the loop: new directions for visual analytics. J. Intell. Inf. Syst. **43**(3), 411–435 (2014). https://doi.org/10.1007/s10844-014-0304-9
6. Fekete, J.D.: Visual analytics infrastructures: from data management to exploration. Computer **46**(7), 22–29 (2013). https://doi.org/10.1109/MC.2013.120
7. Fekete, J.D., Hémery, P.L., Baudel, T., Wood, J.: Obvious: a meta-toolkit to encapsulate information visualization toolkits - one toolkit to bind them all. In: Proceedings of the IEEE Conference on Visual Analytics Science and Technology (VAST), pp. 91–100. IEEE (2011). https://doi.org/10.1109/VAST.2011.6102446
8. Fernández-Villamor, J.I., Blasco-García, J., Iglesias, C.A., Garijo, M.: A semantic scraping model for web resources - applying linked data to web page screen scraping. In: Proceedings of International Conference on Agents and Artificial Intelligence (ICAART), pp. 451–456. SciTePress (2011). https://doi.org/10.5220/0003185704510456
9. Fisher, D., Drucker, S., Fernandez, R., Ruble, S.: Visualizations everywhere: a multiplatform infrastructure for linked visualizations. IEEE TVCG **16**(6), 1157–1163 (2010). https://doi.org/10.1109/TVCG.2010.222
10. Garg, S., Nam, J.E., Ramakrishnan, I.V., Mueller, K.: Model-driven visual analytics. In: Proceedings of the IEEE Symposium on Visual Analytics Science and Technology (VAST), pp. 19–26. IEEE (2008). https://doi.org/10.1109/VAST.2008.4677352
11. Gürdür, D., Asplund, F., El-khoury, J., Loiret, F.: Visual analytics towards tool interoperabilty: a position paper. In: Proceedings of the International Conference on Information Visualization Theory and Applications (IVAPP), pp. 139–145. SciTePress (2016). https://doi.org/10.5220/0005751401390145
12. Han, E., et al.: Exoplanet orbit database. II. updates to exoplanets.org. Publ. Astron. Soc. Pac. **126**(943), 827–837 (2014). https://doi.org/10.1086/678447

13. Harrower, M., Brewer, C.A.: ColorBrewer.org: an online tool for selecting colour schemes for maps. Cartographic J. **40**(1), 27–37 (2003). https://doi.org/10.1179/000870403235002042

14. Hartmann, B., Doorley, S., Klemmer, S.R.: Hacking, mashing, gluing: understanding opportunistic design. IEEE Pervasive Comput. **7**(3), 46–54 (2008). https://doi.org/10.1109/MPRV.2008.54

15. Hogräfer, M., Schulz, H.J.: ReVize: a library for visualization toolchaining with Vega-Lite. In: Proceedings of the Conference on Smart Tools and Applications in Graphics (STAG), pp. 129–139. Eurographics (2019). https://doi.org/10.2312/stag.20191375

16. Kandel, S., Heer, J., Plaisant, C., Kennedy, J., van Ham, F., Riche, N.H., Weaver, C., Lee, B., Brodbeck, D., Buono, P.: Research directions in data wrangling: visualizations and transformations for usable and credible data. Inf. Vis. **10**(4), 271–288 (2011). https://doi.org/10.1177/1473871611415994

17. Kolojejchick, J., Roth, S.F., Lucas, P.: Information appliances and tools in visage. IEEE Comput. Graph. Appl. **17**(4), 3–41 (1997). https://doi.org/10.1109/38.595266

18. Krüger, R., Herr, D., Haag, F., Ertl, T.: Inspector gadget: integrating data preprocessing and orchestration in the visual analysis loop. In: EuroVis Workshop on Visual Analytics (EuroVA), pp. 7–11. Eurographics Association (2015). https://doi.org/10.2312/eurova.20151096

19. Ludäscher, B., Lin, K., Bowers, S., Jaeger-Frank, E., Brodaric, B., Baru, C.: Managing scientific data: from data integration to scientific workflows. In: Sinha, A.K. (ed.) Geoinformatics: Data to Knowledge, pp. 109–129. GSA (2006). https://doi.org/10.1130/2006.2397(08)

20. Munzner, T.: Visualization Analysis & Design. CRC Press, Boca Raton (2014)

21. North, C., Conklin, N., Indukuri, K., Saini, V., Yu, Q.: Fusion: interactive coordination of diverse data, visualizations, and mining algorithms. In: Extended Abstracts of ACM SIGCHI 2003, pp. 626–627. ACM (2003). https://doi.org/10.1145/765891.765897

22. North, C., Shneiderman, B.: Snap-together visualization: a user interface for coordinating visualizations via relational schemata. In: Proceedings of the Working Conference on Advanced Visual Interfaces (AVI). pp. 128–135. ACM (2000). https://doi.org/10.1145/345513.345282

23. Pietschmann, S., Nestler, T., Daniel, F.: Application composition at the presentation layer: alternatives and open issues. In: Proceedings of International Conference on Information Integration and Web-based Applications & Services (iiWAS), pp. 461–468. ACM (2010). https://doi.org/10.1145/1967486.1967558

24. Poco, J., Mayhua, A., Heer, J.: Extracting and retargeting color mappings from bitmap images of visualizations. IEEE TVCG **24**(1), 637–646 (2017). https://doi.org/10.1109/TVCG.2017.2744320

25. Prakasam, R.K., et al.: Deviation maps for understanding thickness changes of inner retinal layers in children with type 1 diabetes mellitus. Curr. Eye Res. (2019). https://doi.org/10.1080/02713683.2019.1591463

26. R Core Team: R: A Language and Environment for Statistical Computing. R Foundation for Statistical Computing, Vienna, Austria (2019)

27. Roberts, J.C.: State of the art: coordinated & multiple views in exploratory visualization. In: Proceedings of the International Conference on Coordinated and Multiple Views in Exploratory Visualization (CMV), pp. 61–71. IEEE (2007). https://doi.org/10.1109/CMV.2007.20

28. Rogowitz, B.E., Matasci, N.: Metadata Mapper: a web service for mapping data between independent visual analysis components, guided by perceptual rules. In: Proceedings of the Conference on Visualization and Data Analysis (VDA), pp. 78650I-1–78650I-13. SPIE (2011). https://doi.org/10.1117/12.881734

29. Röhlig, M., Schmidt, C., Prakasam, R.K., Schumann, H., Stachs, O.: Visual analysis of retinal changes with optical coherence tomography. Vis. Comput. **34**(9), 1209–1224 (2018). https://doi.org/10.1007/s00371-018-1486-x

30. Rungta, A., Summa, B., Demir, D., Bremer, P.T., Pascucci, V.: ManyVis: multiple applications in an integrated visualization environment. IEEE TVCG **19**(12), 2878–2885 (2013). https://doi.org/10.1109/TVCG.2013.174

31. Santos, E., Lins, L., Ahrens, J., Freire, J., Silva, C.: VisMashup: streamlining the creation of custom visualization applications. IEEE TVCG **15**(6), 1539–1546 (2009). https://doi.org/10.1109/TVCG.2009.195

32. dos Santos, S., Brodlie, K.: Gaining understanding of multivariate and multidimensional data through visualization. Comput. Graph. **8**(3), 311–325 (2004). https://doi.org/10.1016/j.cag.2004.03.013

33. Satyanarayan, A., Moritz, D., Wongsuphasawat, K., Heer, J.: Vega-Lite: a grammar of interactive graphics. IEEE TVCG **23**(1), 341–350 (2017). https://doi.org/10.1109/TVCG.2016.2599030

34. Schulz, H.J., Nocke, T., Heitzler, M., Schumann, H.: A systematic view on data descriptors for the visual analysis of tabular data. Inf. Vis. **16**(3), 232–256 (2017). https://doi.org/10.1177/1473871616667767

35. Schulz, H.J., et al.: Lightweight coordination of multiple independent visual analytics tools. In: Proceedings of the 10th International Conference on Information Visualization Theory and Applications (IVAPP), pp. 106–117. SciTePress (2019). https://doi.org/10.5220/0007571101060117

36. Seipp, K., Gutiérrez, F., Ochoa, X., Verbert, K.: Towards a visual guide for communicating uncertainty in visual analytics. J. Comput. Lang. **50**, 1–18 (2019). https://doi.org/10.1016/j.jvlc.2018.11.004

37. Streit, M., Schulz, H.J., Lex, A., Schmalstieg, D., Schumann, H.: Model-driven design for the visual analysis of heterogeneous data. IEEE TVCG **18**(6), 998–1010 (2012). https://doi.org/10.1109/TVCG.2011.108

38. Stuerzlinger, W., Chapuis, O., Phillips, D., Roussel, N.: User interface façades: towards fully adaptable user interfaces. In: Proceedings of the ACM Symposium on User Interface Software and Technology (UIST), pp. 309–318. ACM (2006). https://doi.org/10.1145/1166253.1166301

39. Tan, D.S., Meyers, B., Czerwinski, M.: WinCuts: manipulating arbitrary window regions for more effective use of screen space. In: Extended Abstracts of CHI 2004, pp. 1525–1528. ACM (2004). https://doi.org/10.1145/985921.986106

40. Tobiasz, M., Isenberg, P., Carpendale, S.: Lark: coordinating co-located collaboration with information visualization. IEEE TVCG **15**(6), 1065–1072 (2009). https://doi.org/10.1109/TVCG.2009.162

41. Tominski, C.: Event-based visualization for user-centered visual analysis. Ph.D. thesis, University of Rostock, Germany (2006)

42. Waldner, M., Puff, W., Lex, A., Streit, M., Schmalstieg, D.: Visual links across applications. In: Proceedings of Graphics Interface (GI), pp. 129–136. Canadian Information Processing Society (2010)

43. Weaver, C.: Visualizing coordination in situ. In: Proceedings of IEEE InfoVis 2005, pp. 165–172. IEEE (2005). https://doi.org/10.1109/INFVIS.2005.1532143

Fast Approximate Light Field Volume Rendering: Using Volume Data to Improve Light Field Synthesis via Convolutional Neural Networks

Seán Bruton[✉], David Ganter, and Michael Manzke

School of Computer Science and Statistics, Trinity College Dublin,
University of Dublin, Dublin, Ireland
{sbruton,ganterd,manzkem}@tcd.ie

Abstract. Volume visualization pipelines have the potential to be improved by the use of light field display technology, allowing enhanced perceptual qualities. However, these displays will require a significant increase in pixels to be rendered at interactive rates. Volume rendering makes use of ray-tracing techniques, which makes this resolution increase challenging for modest hardware. We demonstrate in this work an approach to synthesize the majority of the viewpoints in the light field using a small set of rendered viewpoints via a convolutional neural network. We show that synthesis performance can be further improved by allowing the network access to the volume data itself. To perform this efficiently, we propose a range of approaches and evaluate them against two datasets collected for this task. These approaches all improve synthesis performance and avoid the use of expensive 3D convolutional operations. With this approach, we improve light field volume rendering times by a factor of 8 for our test case.

Keywords: Volume rendering · View synthesis · Light field · Convolutional neural networks

1 Introduction

1.1 Motivation

Many scientific and engineering domains increasingly encounter volumetric information as part of their work. In meteorology, weather system models may be represented volumetrically. 3D Medical scan data, obtained using Magnetic Resonance Imaging (MRI) and Computed Tomography (CT), is used for diagnostic purposes. Geophysical surveys can provide volumetric information about ground layers to inform geological, mining and archaeological tasks. In each of these domains, visualization of this 3D information is vital to facilitate identification of salient patterns for the respective purposes. Thus, if this visualization process is made more effective and intuitive it would have a potentially broad impact.

© Springer Nature Switzerland AG 2020
A. P. Cláudio et al. (Eds.): VISIGRAPP 2019, CCIS 1182, pp. 338–361, 2020.
https://doi.org/10.1007/978-3-030-41590-7_14

Accordingly, improving the various elements of the volume visualization pipeline are active areas of research. The problem of modelling light transport through a discretized 3D space of varying density to produce an image, known as volume rendering [7] has received particular attention. Early contributions in this area focussed on performing this volume rendering at speed using techniques such as splatting [29] and shear warping [20], before improvements in graphics hardware enabled ray casting approaches to be tractable without significant hardware resources [37]. Visualizations have been made more appealing via efficient implementations of advanced lighting effects, such as photon mapping [47] and deep shadow maps [12]. The problem of transfer function specification [27] is another aspect of volume visualization pipelines that has been focussed on. As this process can be unintuitive for users and may require specialized knowledge, providing guiding interfaces have been explored to allow easier creation of informative visualizations [18]. The display medium may also have a significant impact upon the effectiveness of volume visualizations, and as such research has been performed investigating the use of virtual reality [15] and holographic displays [9]. In this direction, we intend to enable further improvements to volume visualization pipelines by allowing future display technologies to be used at interactive rates.

Light field displays are one such technology that has the potential to improve volume visualization pipelines significantly. Representing dense volumetric data sources using a 2D display is inherently suboptimal. As multiple layers of transparent substructures may be present, the depths of these structures are not readily apparent in 2D images. Using light field displays, natural convergence and accommodation cues enable a user to differentiate multiple depths. Light field displays' potential ability to allow users to focus at a particular depth using their eyes rather than using an interface can also improve the visualization pipeline. For medical inspection purposes, it was found that the improved perception of depth allowed a physician to more easily discriminate interesting patterns in 3D medical scan data [1]. Due to the potential for these display types to improve the visualization pipeline, we contribute methods that will enable these displays to be adopted sooner for these purposes.

1.2 Problem Statement

Light fields are vector functions outputting the amount, or radiance, of light passing through all points and in all directions in space for all possible wavelengths [22]. If we assume that radiance along a single ray remains constant, and consider a single wavelength, the light field can be characterised by four dimensions. One such parametrisation allows representation by the points of intersection of the individual rays in two parallel planes. Such a representation is used by light field imaging technologies, such as the Lytro Illum, as well as by prototype light field displays [21,45]. In these settings, a light field is captured using a discrete grid of camera lenses. This can be viewed as images being produced for an aligned rectangular grid of cameras. For these cameras, the term *spatial resolution* refers to the resolution of the individual cameras, while the term *angular resolution* refers to the resolution of the rectangular grid. For the light field capture scenario, due to the fixed size of camera sensors, a trade-off must be made between these two resolutions. Similarly, to render a light field of a volumetric data source, there exists a trade-off based on the number of pixels that can be rendered given the time available to maintain real-time rates.

The volume rendering problem, as discussed, is itself demanding to be performed in real-time. The ray-casting approaches commonly used for direct rendering of the volume are expensive for large volume data sources. These approaches are more challenging when multiple dimensions of the volume are required to be represented, such as pressure and temperature in a meteorological simulation. Other methods may also need to be incorporated, such as isosurface extraction and slice representations [32] depending on the domain and use case of the visualization. Further advanced techniques, such as gradient-based transfer functions, may also add to the computational cost of rendering. Thus, volume rendering for an entire light field, where the raw number of pixels is significantly larger than for a 2D display, can be challenging to perform at interactive rates for commodity hardware.

The closely-spaced arrangement of virtual cameras for light field displays may be such that there is significant overlap in viewing frustums. Accordingly, there is potential to leverage the rendering output from one light field viewpoint to produce an image for another viewpoint. In recent years, similar problems such as video interpolation and viewpoint synthesis have been tackled effectively with convolutional neural network approaches [26, 30]. These methods train networks that can estimate the dynamics of scene components between viewpoints or video frames [48]. These approaches are targeted at the scenarios of natural imagery, rather than rendered imagery, where opaque objects dominate the images. Due to this opacity, it is possible to determine pixel correspondences between viewpoints, and so adopt optical flow or disparity based approaches to perform interpolation for novel views. For direct volume rendering, however, rays accumulate color as they pass through the model and so multiple voxels contribute to the pixel color. This violates the assumption of constant brightness that is the basis of optical flow calculations. Unique disparity values for the pixels can also not be calculated due to the contribution of multiple voxels to the final pixel value. As such, representations that work well for natural images do not translate straightforwardly to interpolation for volume rendering purposes. An alternate approach that can utilise the volume information directly could provide a novel solution.

In a learning-based approach to this image interpolation problem, there are no apparent methods that can be used to extract useful information from this volume data. If we were to perform 3D convolutional operations over the typically large volumes, the computational cost could significantly increase, precluding the use of this method as part of a real-time system. As such, in this work, we explore multiple methods that attempt to compress this volume data to allow it to be efficiently used to improve light field synthesis results.

1.3 Contributions

This work builds upon that of our previous work [5] where a number of contributions were made:

- We were first to demonstrate a neural network approach to synthesising light fields for volume visualizations.
- We proposed two compressed representation of dense real-valued volumetric data and showed that it can be used to improve image synthesis results.
- We showed that our neural network approach, when generating a light field of angular resolution 6×6, performs at real-time rates. This was a significant increase in comparison to the benchmark ray-tracing technique.
- We prepared and release two datasets for future use in tackling this problem, which we shall utilise here.

In this work, the following additional novel contributions are made:

- We propose two more volume representations that are calculated using an approximation technique for rank pooled representations. We show that these approaches perform competitively with the previous approaches.
- We identify three heuristic functions that could also be used in place of a learned representation. In our testing, we show that these approaches are competitive and hence recommendable in many cases due to their simplicity.
- We also propose two methods to allow the neural network to learn a method of compression in an end-to-end fashion. These methods are designed to add little computational cost during inference, and we show that these approaches outperform the previous techniques.
- We time each of these approaches to generate a full light field and show that we maintain a significant speed increase over the traditional ray-tracing pipeline.
- All experimental code and data for each of these approaches appears in our code repository (github.com/leaveitout/deep_light_field_interp).

We envisage that this extended range of compressed volume representations can be used for multiple similar volume visualization problems and encourage other research in this direction.

2 Related Work

2.1 Neural Network Viewpoint Synthesis

The advent of deep learning has seen the use of convolutional neural networks (CNNs) go beyond initial image classification tasks [19]. The modelling complexity inherent in CNNs is particularly applicable to many image generation tasks [26,31,49]. As mentioned, many of these techniques utilise either motion information (i.e. optical flow) or scene information in the form of disparity or depth [14,30]. Disparity and depth information have been utilised for the particular problem of viewpoint generation for light fields [14,39] where a model of the scene can be transformed for novel viewpoints. Motion information, in the form of optical flow, has also been utilised to perform image synthesis, particularly for video interpolation techniques [49]. Here, the correspondences between pixels in successive frames can be used to interpolate pixel values for generated images. As discussed in Sect. 1.2, these approaches do not directly apply to the problem of volume rendering due to the violation of unique pixel correspondences and unique depth values at pixel locations. A recent approach for video interpolation that does not rely on flow or disparity makes use of learning filters that can be used as interpolation masks between successive frames [30]. Each of these approaches, however, were developed for the target domain of natural imagery and so do not make use of volumetric information which is available in our scenario.

2.2 Learning from Volumes

Extending convolutional neural networks to higher dimensionality than images is well explored in the literature. In particular, 3D CNNs have been used to recognise spatio-temporal patterns in performing tasks such as action recognition. With the recent emergence of large-scale labelled volumetric datasets of objects [6], research has also been performed in tasks such as retrieval, recognition and generation of dense 3D object representations [46]. However, the use of 3D CNNs has been met with issues. It has been observed that the increased dimensionality of the trainable filters requires larger and more varied datasets for reliable recognition [34]. Furthermore, the increase in computational operations as a result of the extra dimensionality precludes the use of desirable deep architectures and increases training and inference time. For the task of object recognition of dense volumetric representations it was shown that the 2D CNNs trained on multiple views of the object was able to outperform 3D CNNs of similar complexity [40]. As such, alternative representations of volume data that improve learning efficiency have been explored. Point-based representations [33,35] have shown particular promise for many learning tasks [8]. In each of these use cases, the point-based representation is applied to a binary occupancy volume. However, in our case, we have dense multi-valued volumes and so these approaches do not directly translate. Other efficient representations, such as tree-based representations [17,36,44] and probing-based representations [23], similarly are applied to binary occupancy volumes and as such would require careful novel extension to be applied to multi-valued volumetric sources.

2.3 Rank Pooling

The problem of ranking query results according to a criterion has been extensively studied in the domain of information retrieval [25]. Learning-based approaches to ranking, such as Rank Support Vector Machines (RankSVM) [38], have been applied to problems such as 3D object pose estimation where features are ranked for discriminative utility [42]. Such an approach has also been applied to the problem of action recognition [10]. In this scenario, ranking spatio-temporal features according to their temporal order results in the learning of a representation that when multiplied by the feature gives its predicted rank. Such a representation can be viewed as a smaller implicit representation of the features themselves, a *rank pooled* representation, and thus can be used as an input feature to a classifier. This approach was further extended by Bilen et al. [3] to work for raw pixels rather than extracted spatio-temporal features. In this case, the rank pooling results in an output image that encodes the information of a video clip, referred to as a *dynamic image*. Beyond its original use for recognition of actions from colour and optical flow images, it has also been shown to be capable of encoding discriminative information from scene flow information for gesture recognition [43]. Our work [5] shows that such representations can be utilized beyond recognition tasks, to problems of image synthesis. Furthermore, we show that it is a viable technique for extracting useful information from dense volumetric data sources in a visualization context.

2.4   Light Field Volume Rendering

Volume rendering for light field displays has been tackled using different techniques in the literature [4,28]. Light fields were utilized by Mora et al. [28] to create volume visualizations using emissive volumetric displays. In this work, 4D light fields are used as an intermediate representation to generate a 3D representation for the volumetric display. The contribution of this work relates to this novel representation, which permits the volume appearance to mimic lighting effects present in traditional 2D volume visualizations. Other works have focussed on particular use cases of light field displays for volume visualization [1,2]. These solutions looked specifically at optimizing the traditional rendering pipeline for the use case, requiring multiple PCs equipped with graphics cards to drive the light field displays. Caching strategies have also been utilized to facilitate volume rendering for light field displays [4]. These strategies rely on idle times in interaction to fill a render cache and perform composition of these renders for display on a light field. A disadvantage of this approach is that when large viewpoint changes are made, or if changes are made to the transfer function or data source, then the cache becomes invalidated. Our work seeks to use a learning-based approach that works irrespective of viewpoint changes and does so at interactive rates for large light field display sizes.

3   Method

In this section, we explain our technique of generating novel light field viewpoints based on a rendered subset of viewpoints and compressed volume representations. We shall first describe the CNN architecture used throughout the work, and secondly, describe the devised compressed volume representations.

3.1 Viewpoint Synthesis CNN

Given our discussion in Sect. 2.1 of how CNNs are used for image generation tasks, we choose to utilise such an approach for our light field synthesis task. An early decision that we must make is what the inputs to the convolutional architecture to use. In recent work [14], a CNN encoder-decoder was able to generate a single light field viewpoint for natural images given four corner images of the light field as inputs. We shall also utilise such approach, as shown in Fig. 1. In our approach, however, we seek to synthesise the entire light field rather than a single viewpoint image.

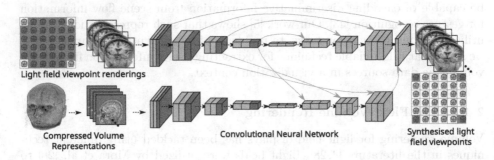

Fig. 1. An illustration (adapted from original [5]) of the convolutional encoder-decoder architecture that we use to generate novel light field viewpoint images. The network is composed of two branches. The upper branch is used to encode the rendered light field images and decode the novel viewpoint images. The lower branch encodes and decodes the compressed volume representation to add detail to the final images. In the figure, green denotes a convolutional layer, blue an average pooling layer, purple a bilinear upsampling layer and red arrows represent skip connections.

Having access to the four corner images allows coverage of how the appearance of the scene changes in the extreme viewpoints, however, it is also possible to frame the problem in an ill-posed manner. Such a framing was used by Srinivasan et al. [39], where an entire light field is synthesized from a single central viewpoint. To do so, the authors estimate the scene geometry, in the form of disparity, from the central viewpoint and utilize this to synthesize other viewpoints. However, as varying alpha values in light field renderings do not permit one-to-one mappings between viewpoints, scene estimation approaches are not ideal [5].

We adopt an approach that utilises a CNN encoder-decoder architecture to directly interpolate the novel viewpoints, in line with our previous work [5]. As an encoder-decoder architecture, it is composed of downsampling encoder blocks and upsampling decoder blocks, as shown in Fig. 1. As we are primarily interested in studying the learning ability of different volume representations, we utilise an established order of operations within these blocks that was successful in performing other image synthesis tasks [30]. As in our earlier work [5], in the

Encoder block, the order of operations is a convolution with 3×3 filter, Rectified Linear Unit (ReLU) activation $(f(x) = \max(0, x))$, followed by convolution (maintaining the same number of output channels), ReLU, and a final average pooling with 2×2 window and stride of 2. Similarly, in a Decoder block, the order of operations is a convolution with 3×3 filter, Rectified Linear Unit (ReLU) activation $(f(x) = \max(0, x))$, followed by convolution (maintaining the same number of output channels), ReLU, and a final bilinear upsampling operation which doubles the horizontal and vertical dimensions. The bilinear upsampling operation is differentiable, and thus, it allows gradients to be back-propagated through to previous operations.

The baseline architecture that we will use as a benchmark utilises only the rendered corner light field images as inputs. As in the previous work [5], we call this architecture the Light Field Synthesis Network - Direct (**LFSN-Direct**). We also utilise the same number of features as before, shown in Table 1, in the Encoder and Decoder blocks which determine the number of convolutional filters within the blocks.

Residual connections [13] have enabled deeper architectures to be utilised in recent works. Here, we utilise such a pattern to connect Encoder and Decoder blocks to frame the network as a residual function as this will encourage sharper results. This is based on the observation that, as a residual function, $\mathcal{F}$ must only learn the change to be made to input, $\mathbf{x}$, to model the output, $\mathbf{y}$, i.e.

$$\mathbf{y} = \mathcal{F}(\mathbf{x}) + \mathbf{x}. \tag{1}$$

In contexts where the output will have similar content or structure to the input, as in our case, this reduces the need to learn this shared information. In our architecture, as in our previous work [5], we use residual connections between the output of the Encoder blocks and the output of the Decoder blocks, between equivalently numbered blocks, as shown in Table 1.

There is an issue which prevents residual connections as per our desired network structure. To connect two layers of a network by a residual connection, it is necessary that the tensors at connected layers are of the same dimensions to permit addition. In our arrangement in Table 1, these dimensions do not correspond between the Encoder outputs and Decoder block inputs that we desire to connect. This mismatch is because the Decoder block requires more output features to generate the remaining light field images. We use the approach of our previous work [5], to overcome this. This approach repeats one tensor along the channel dimension to match the other tensor's dimensions, in our case the output of the convolution operation in an Encoder block and the output of the bilinear upsampling operation in a Decoder block, respectively. Using the same notation as previously, we can express this residual connection [5] as

$$\mathbf{y} = \mathcal{F}(\mathbf{x}) + [\mathbf{x}, \mathbf{x}, \mathbf{x}, \mathbf{x}]. \tag{2}$$

We call these *concatenation residual connections* and observe that they may have benefits as Encoders are encouraged to produce outputs, $\mathbf{x}$, that are more generally applicable across the generated output images in the Decoders.

Table 1. The number of features in each of the blocks of the Light Field Synthesis Network, as per our previous work [5]. The network inputs are composed of four rendered corner images, each with four channels (RGBA), concatenated on the channel dimension. The network outputs are decomposed into 32 images in a reverse fashion. These 32 images are the remaining images to be predicted for the 6×6 light field.

Block	# input features	# output features
Encoder 1	16	32
Encoder 2	32	64
Encoder 3	64	128
Encoder 4	128	256
Decoder 4	256	256
Decoder 3	256	128
Decoder 2	128	128
Decoder 1	128	128

A parallel branch is added to the network to learn from compressed volume representations in line with our previous work [5], as shown in Fig. 1. This branch (the volume branch), utilises an identical structure of Encoder and Decoder blocks as per the LFSN-Direct network. The number of input features for this branch may vary depending on the number of compressed images utilised. Concatenation residual connections are utilised to connect the main branch to this branch at the equivalent connection locations as per the main branch. The output of the final Decoder block of the volume branch is added to the output of the final Decoder block of the main branch. This addition frames the entire volume branch as a function that learns to generate fine-grained details from the volume information to improve the synthesised output images.

3.2   Volumetric Representations

We must identify possible compressed volumetric representations to use in our approach. Here, we outline techniques for performing such a compression that will allow us to leverage the volumetric information for our light field synthesis task.

Ranked Volume Images. The method of rank pooling [10] has been applied to video sequences to encode a video in a single image, known as a dynamic image [3]. Here, we generate compressed representations of volumetric data using this method. The technique utilised is equivalently used as in our previous work [5] which we shall detail for completeness. We describe the volume V, as a sequence of ordered slices $[\mathbf{v}_1, \ldots, \mathbf{v}_n]$, and let $\psi(\mathbf{v}_t) \in \mathbb{R}^d$ denote a chosen representation of a slice at index t. We use the slice information itself, i.e. $\psi(\mathbf{v}_t) = \mathbf{v}_t$. Pairwise ranked is performed by learning parameters $\mathbf{u} \in \mathbb{R}^d$, such that a ranking function $S(t|\mathbf{u}) = \mathbf{u}^T \cdot \mathbf{v}_t$ gives a score indicating the rank for the slice at t.

The optimization of the parameters $\mathbf{u}$ the scores are representative of the ordering (i.e. rank) of the slices, i.e. $q > p \implies S(q|\mathbf{u}) > S(p|\mathbf{u})$. This optimization can be formulated as a convex problem and solved using RankSVM [38]:

$$\mathbf{u}^* = \rho(\mathbf{v}_1, \ldots, \mathbf{v}_n; \psi) = \underset{\mathbf{u}}{\operatorname{argmin}} E(\mathbf{u}); \tag{3}$$

$$E(\mathbf{u}) = \frac{2}{n(n-1)} \times \sum_{q>p} \max\{0, 1 - S(q|\mathbf{u}) + S(p|\mathbf{u})\} + \frac{\lambda}{2}\|\mathbf{u}\|^2. \tag{4}$$

In Eq. 3, the optimized representation, $\mathbf{u}^*$, is a function of the slices, $\mathbf{v}_1, \ldots, \mathbf{v}_n$ and the representation type ψ, which can be as an energy minimisation of the energy term $E(\mathbf{u})$. The energy term as expressed in Eq. 4 is composed of a hinge-loss soft-counting of incorrectly scored pair, and a quadratic regularization term weighted by hyperparameter λ. The first term in Eq. 4 is the hinge-loss soft-counting of the number of incorrectly ranked pairs, and the second term is the quadratic regularization term of support vector machines, where the regularization is controlled by hyperparameter λ. Using hinge-loss soft-counting ensures unit score difference between correctly scored pairs, i.e. $S(q|\mathbf{u}) > S(p|\mathbf{u}) + 1$.

As this ranking representation, $\mathbf{u}^*$, captures discriminative information about the order of the slices, we view the function $\rho(\mathbf{v}_1, \ldots, \mathbf{v}_n; \psi)$ as a mapping from an arbitrarily sized volume to a compressed representation. This representation allows 3D volumes with 2D images and hence permits 2D CNNs to be used to learn from volume data.

It remains to decide the slice order to optimise our representation to predict. An obvious choice is the order of the slices along an axis. Alternatively, we can rank the slices according to the Shannon entropy of the slice $H = -\sum_i p_i \log p_i$, where p_i is the normalised probability of a specific cell value in a slice. As entropy provides a measurement of the complexity of slices, the compressed representation should learn to rank this aspect. We refer to the compressed representations for axis-order ranking and entropy ranking as Volume Images - Order-Ranked (VIOR) and Volume Images - Entropy-Ranked (VIER), respectively. We also refer to network architectures as **LFSN-VIOR** or **LFSN-VIER**, depending on whether we use VIOR or VIER images, respectively.

Approximate Ranked Volume Images. The previous formulation for ranking the volume slices will require learning the compressed representation before training. This requirement precludes the use of such a compression stage as part of an intermediate layer in a neural network. Furthermore, for specific use cases, it may be desirable to slice the volume along many different axes such as the axes of the cameras used for capturing the images. Accordingly, we identify a solution to these issues and validate its use case for light field synthesis of volume rendering.

In their work on Dynamic Image Networks [3], Bilen et al. show that an approximate solution, using a single gradient descent step, to Eq. 3 can be used effectively. The approximation results in a solution of the form:

$$\mathbf{u}^* \propto \rho(\mathbf{v}_1, \ldots, \mathbf{v}_n; \psi) = \sum_{t=1}^{n} \alpha_t \psi(\mathbf{v}_t), \tag{5}$$

which is a weighted combination of the slice representations $\psi(\mathbf{v}_t)$. If using an identity transformation for ψ, as we do, the weighting parameters are derived as

$$\alpha_t = 2t - n - 1, \tag{6}$$

which is linear in t [3]. This approximation allows use of volume compression as an intermediate layer within a CNN architecture. We refer to network architectures that utilise these approximate ranking representation as **LFSN-VIOR-A** or **LFSN-VIER-A**, depending on whether we use axis ordering or entropy ordering, respectively.

Vector Norm Volume Images. We also propose a more straightforward method for calculating volume images. Inspired by the Maximum Intensity Projection method for visualization of CT scan data [11], we formulate approaches based on vector norms of the vector of voxels, $\mathbf{x} = [x_1, \ldots, x_n]$, along a chosen axis, at a position, (i, j), in the slices. We use the ℓ_1, ℓ_2, and ℓ_∞ norms, and calculate them respectively as:

$$\|\mathbf{x}\|_1 = |x_1| + \ldots + |x_n|; \tag{7}$$

$$\|\mathbf{x}\|_2 = (|x_1|^2 + \ldots + |x_n|^2)^{1/2}; \tag{8}$$

$$\|\mathbf{x}\|_\infty = \max\{|x_1|, \ldots, |x_n|\}. \tag{9}$$

To generate the volume images, we calculate the norms for all positions (i, j). The max operation permits back propagation, as per the maximum pooling operation, and so each of these functions could also be used as an intermediate layer in an architecture to compress a volume input. We refer to our network architectures that utilise these norm images as **LFSN-VI-ℓ_1**, **LFSN-VI-ℓ_2**, and **LFSN-VI-ℓ_∞**, depending on the norm used to compress the input volume data. In Fig. 2, we show example volume images of each of the previous formulations.

Learned Volume Images. Thus far, our approaches for extracting useful information from a volume could be performed before the training of our Light Field Synthesis Network. As such, we have focussed on techniques that rely on learned representations, approximations and functions. However, we could obtain performance improvements by allowing the network to learn how to compress the

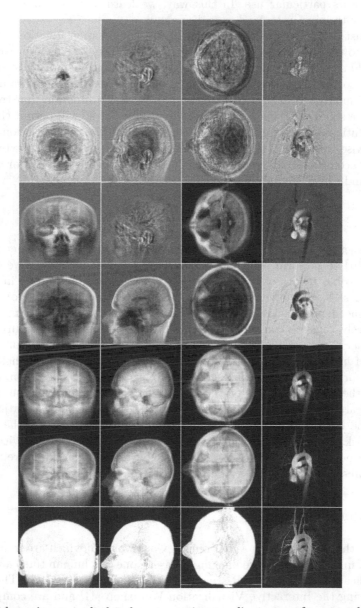

Fig. 2. Volume images calculated across entire coordinate axes for two volume data sources. Rows from the top are VIOR, VIER, VIOR-A, VIER-A, VI-ℓ_1, VI-ℓ_2, and VI-ℓ_{inf}. The VIER images (second row) appear to capture greater detail than in the VIOR images (top row). We observe that the axis-order representation (VIOR) devotes more features discriminating early elements such as the nose and ear, and less to later internal structures. Here, we encode the volume in a single image, however, we can use more than one image to encode the volume when training.

volume for its particular use. In this way, we learn a nonlinear compression function by training in an 'end-to-end' manner.

We must decide on the arrangement of this compression function. One possibility is to utilise successive 3D convolutional operations. This approach, however, would significantly increase the computational and memory cost of a network due to the large size of the volume data, precluding high frame rates. Instead, similarly to our previous approaches, we seek a function that treats the volume as a series of 2D slices. Treating the volume like so will permit the use of 2D convolutional operations, which are cheaper to utilise, computationally and memory-wise, and so will not hinder inference speed to as great an extent.

In a 2D convolutional operation, the output at a pixel location $\mathbf{c}$ for an input image f and convolutional filter g is

$$(f * g)(\mathbf{c}) = \sum_{\mathbf{a} \in \mathcal{A}} f(\mathbf{a}) \cdot g(\mathbf{c} - \mathbf{a}), \tag{10}$$

where the set $\mathcal{A}$ is the set of locations in the convolutional filter. We can choose $\mathcal{A} = \{(0,0)\}$ to perform what is referred to as 1×1 convolution. We can also choose $\mathcal{A}$ to be the set of locations in a 3×3 window about $(0,0)$ in what is referred to as 3×3 convolution. The 1×1 convolution is capable of learning inter-channel dependencies [24] and has the benefit of requiring less computation. Alternatively, the 3×3 convolution has the benefit of permitting the surrounding context, i.e. the receptive field, in a volume slice to affect the output, at the expense of higher computational and memory costs. The receptive field size is increased further when multiple operations are performed in succession. As such, we select these two variations of 2D convolution to learn from volume data.

In our architecture, we replace the compressed volume representations with the volume itself and pass it through a Volume Image Block. In the Volume Image Block, we perform convolution followed by ReLU activation twice in succession. Depending on whether 1×1 or 3×3 convolution is used, we refer to the architectures as **LFSN-VIB-1** or **LFSN-VIB-3**, respectively.

4   Implementation

We utilize the same datasets as per our previous work [5] which are from visualizations of magnetic resonance imaging datasets, one of a human torso and one of a human head referred to as MRI-Head and MRI-Torso, respectively. These were created using the Interactive Visualization Workshop [41] and are composed of 2,000 light field images of angular resolution 6×6 and spatial resolution 256×256. In the MRI-Head dataset, clipping planes were used to expose internal structures of the data. For the MRI-Torso dataset, global illumination effects based on an extracted isosurface are present, making synthesis more difficult. The transfer function was kept fixed for collection of both datasets.

As before [5], to calculate the volume images, we select an axis along which to calculate the ranked volume images (the side profile for MRI-Head and front profile for MRI-Torso). Apart from for Volume Image Blocks, where we pass the entire volume to the block, volume images are calculated for adjacent sub-volumes of sizes $16 \times 256 \times 256$ and concatenated to form the compressed representation. As noted before [5] there exists a trade-off here between volume compression and minimising the training and inference cost.

During training, the network takes as input the four corner viewpoint images of the light field and outputs the remaining 32 viewpoint images. We use the first 1,600 images for training, and the remaining 400 for testing. We train the network for 150 epochs due to overfitting being present in some instances if trained further.

In line with our previous work [5] the Adamax learning rate scheduling is used to train the network with the recommended parameters [16] and the data was transformed to lie in the $[0, 1]$ range prior to use during training. The loss function used was $\mathcal{L}_1 = \|I - I_{gt}\|_1$ where I and I_{gt} are the synthesised and ground truth light field images respectively. We use the Pytorch framework and an Nvidia Titan V graphics card for training (Fig. 3).

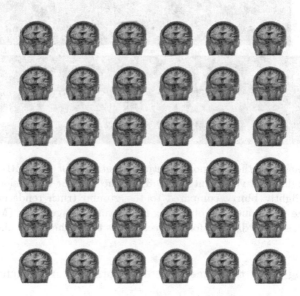

Fig. 3. A sample rendered light field for the MRI-Head dataset, originally presented in [5].

5   Evaluation

5.1   Qualitative

We inspect the outputs of the tested approaches to compare their performances qualitatively (an example set of outputs is shown in Fig. 4). We note, as before [5], that viewpoint images that are further within the grid of viewpoints in the light field perform worse. This is likely due to the synthesis problem becoming

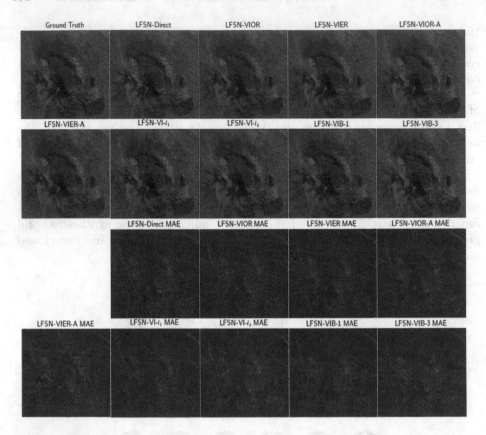

Fig. 4. Result images of the viewpoint synthesis methods for the MRI-Torso dataset, generating the light field viewpoint at (3, 3) in the angular dimension. Synthesised images appear slightly blurry compared to the ground truth rendering, to varying degrees across the techniques. By observing the Mean Average Error (MAE), we note that specular areas prove difficult for the techniques to synthesise to high accuracy.

more challenging when the camera distance form the input rendered images increases.

In the case of the MRI-Head model, all models perform well and it is difficult to distinguish any rendering artefacts. On inspection of Mean Average Error images, however, we do note that the edges of surfaces are synthesised less accurately with the LFSN-Direct approach as shown in Fig. 5.

In the case of the MRI-Torso model, the LFSN-Direct model has overall a higher level of errors throughout the image, as can be seen in Fig. 6. In line with our previous work [5], the finer details appear sharper in the volume images approaches. Throughout all approaches, however, there is a slight observable blurriness in the results. As observed previously [5], this may be due to the use of the L_1 loss for training as other authors have also noted [30].

5.2   Quantitative

We select common image metrics, Structural Similarity (SSIM) and Peak Signal to Noise (PSNR), to compare our various approaches. These metrics are calculated by comparing the synthesised test set images to the ground truth images. We calculate and report the mean value across the generated viewpoints and the corresponding standard deviations.

Table 2. Metrics calculated for the test set of images on MRI-Head dataset.

Network type	SSIM	PSNR
LFSN-Direct	0.983 ± 0.013	38.60 ± 3.14
LFSN-VIER	0.987 ± 0.006	39.75 ± 2.02
LFSN-VIOR	0.983 ± 0.004	38.34 ± 1.83
LFSN-VIER-A	0.988 ± 0.004	40.12 ± 1.35
LFSN-VIOR-A	0.086 ⊥ 0.006	39.31 ± 1.92
LFSN-VI-ℓ_1	0.989 ± 0.004	39.50 ± 1.21
LFSN-VI-ℓ_2	0.988 ± 0.004	39.78 ± 1.22
LFSN-VI-ℓ_∞	0.989 ± 0.006	39.88 ± 1.20
LFSN-VIB-1	**0.991** ± 0.003	**41.24** ± 1.22
LFSN-VIB-3	0.990 ± 0.004	41.13 ± 1.24

Table 3. Metrics calculated for the test set of images on MRI-Torso dataset.

Network type	SSIM	PSNR
LFSN-Direct	0.918 ± 0.044	35.07 ± 3.86
LFSN-VIER	0.936 ± 0.031	36.42 ± 3.36
LFSN-VIOR	0.932 ± 0.034	36.28 ± 3.65
LFSN-VIER-A	0.931 ± 0.032	36.70 ± 3.51
LFSN-VIOR-A	0.936 ± 0.030	**37.00** ± 3.46
LFSN-VI-ℓ_1	0.936 ± 0.030	36.74 ± 3.33
LFSN-VI-ℓ_2	0.936 ± 0.030	36.55 ± 3.40
LFSN-VI-ℓ_∞	0.936 ± 0.031	36.42 ± 3.35
LFSN-VIB-1	0.933 ± 0.033	36.09 ± 3.46
LFSN-VIB-3	**0.943** ± 0.030	36.95 ± 3.57

The results for the MRI-Head dataset are presented in Table 2. We observe that all volume image approaches improve upon the LFSN-Direct approach. In particular, we note that the standard deviation is significantly reduced in the volume image approaches. The learned approach, LFSN-VIB-1, performs marginally better than the other approaches overall. We note also that the function based approaches, LFSN-VI-ℓ_1, LFSN-VI-ℓ_2 and LFSN-VI-ℓ_∞, all perform comparatively well, indicating these simpler approaches may suffice in certain cases.

The results for the MRI-Torso data are presented in Table 3. We note that this dataset is more challenging for the LFSN-Direct approach, in comparison to the volume image approaches. The global illumination effects present in the volume rendering are the likely cause of these lower metrics overall. Once again, a learned approach, LFSN-VIB-3, performs best in terms of SSIM. An approximate approach, LFSN-VIOR-A, slightly outperforms this method in terms of PSNR. The function-based approaches also perform comparatively well. Based on these results, in the case where computation speed is important, these approaches can be recommended, given their simplicity to implement compared to the LFSN-VIER and LFSN-VIOR approaches.

5.3   Performance Speed

The speed at which the entire light field can be synthesised is important to consider. If the volume images approaches are worthwhile, there needs to be a significant increase in frame rate to justify its use over the ground-truth rendering approach. As in our previous work [5], we recorded the time to render all light field images in the MRI-Torso dataset with a single computer, under similar load, with an Nvidia Titan V graphics card used for both light field synthesis and render timings. The Baseline approach in Table 4 utilizes ray-tracing and

Table 4. The timings of all the different approaches to output a complete light field for the MRI-Torso dataset are shown below. We take the average over the entire test set of 400 light fields and calculate the standard deviation for the same set. As before [5], we include the time to render the four corner images for all variants, amounting to 0.08 seconds on average.

Network type	Time (s)
Baseline	0.716 ± 0.223
LFSN-Direct	0.084 ± 0.026
LFSN-VIOR	0.088 ± 0.026
LFSN-VIER	0.088 ± 0.026
LFSN-VIER-A	0.088 ± 0.026
LFSN-VIOR-A	0.088 ± 0.026
LFSN-VI-ℓ_1	0.088 ± 0.026
LFSN-VI-ℓ_2	0.088 ± 0.026
LFSN-VI-ℓ_∞	0.088 ± 0.026
LFSN-VIB-1	0.088 ± 0.026
LFSN-VIB-3	0.088 ± 0.026

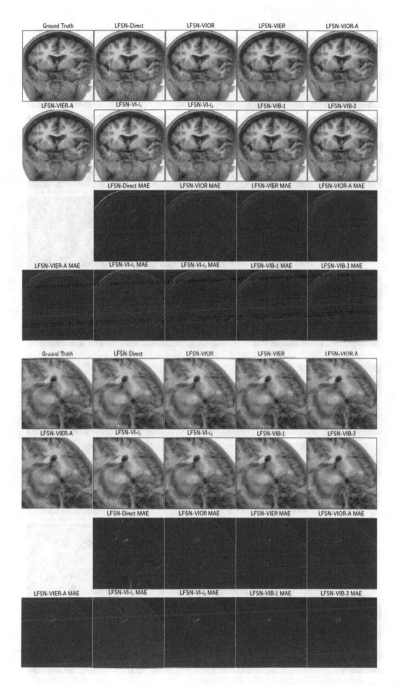

Fig. 5. Result images of the viewpoint synthesis method for the MRI-Head dataset, generating light field viewpoints at (1, 1) in the angular dimension. We observe that for the LFSN-Direct approach there are acute errors around discontinuities, such as edges, in the volume data.

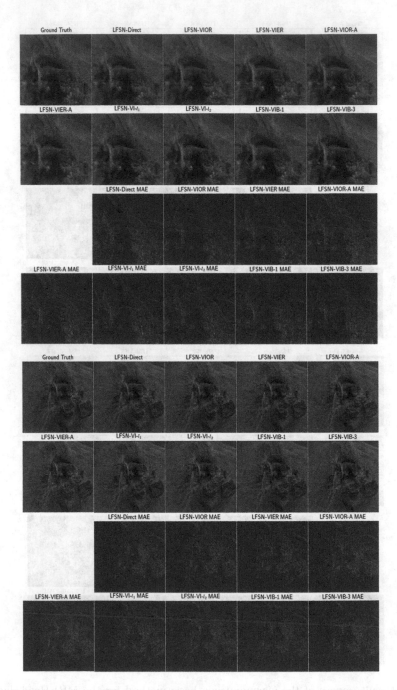

Fig. 6. Result images of the viewpoint synthesis methods for the MRI-Torso dataset, generating the light field viewpoint at (2, 2) in the angular dimension. The LFSN-Direct approach has overall a larger error observable throughout the synthesised image.

global illumination as part of the volume rendering workspace implemented in the Interactive Visualization Workshop.

In Table 4, we observe that the light field generation time is significantly reduced. All generation approaches are within 1 millisecond of each other, apart from LFSN-Direct which is 4 ms faster. The results in Table 4 show that we substantially reduce the time taken to generate a light field. We note that, when using this approach, it will be possible to fall back to the traditional rendering pipeline if such a time frame is available, e.g. if the user has stopped interacting with the visualization. Each approach improves rendering time by a factor of 8, achieving 11 frames per second. The neural network is also able to generate images at a fixed rate regardless of the complexity of the volumetric scene. As in our previous work [5], we note that the majority of time taken for the LFSN variants is spent rendering the four corner images. Possible speed improvement could be made, for example, by utilizing a neural network to perform a prior super-resolution stage on these four corner images.

6 Conclusions

In this work, we extend our previous work [5] that showed that convolutional neural networks could be used to synthesizes light field viewpoints of dense volumetric renderings. In addition to the original two ranked volume representations that allow the network access to the volume data during the synthesis task, we propose seven more methods that can be used to improve synthesis results. We have shown that some of these approaches are simpler to calculate and so can be implemented straightforwardly for similar levels of synthesis improvement. We also devise an approach which allows the network access to learn a compressed representation in an end-to-end manner and show that it improves results over the previous work. As before [5], the advantage of these techniques is the speed with which a light field can be synthesized, making real-time light field generation achievable. We release the code and data and encourage the community to build upon it to tackle related problems in light field volume visualization.

In this work, we have used an image synthesis approach for visualization of medical imaging data. Here, there may be valid concerns about utilizing a predictive method to inspect volume data, as it may incorrectly synthesize the salient property of the volume data This poses a key research question to be tackled to dispel such concerns and allow widespread adoption of such techniques in this domain. As previously stated [5] this could be addressed by improving performance to within an acceptable limit on a range of datasets, or by characterizing uncertainty in the synthesized images.

Other possible avenues for further research include improvements to the generation speed by utilizing scene estimations to reduce the number of required views. Exploring techniques for spatial super-resolution in combination with our angular super-resolution approaches could also provide speed improvements. Finally, investigating whether utilizing compressed volume representations benefits tasks such as frame extrapolation may also be worthwhile.

Acknowledgements. The authors would also like to thank the anonymous referees for their valuable comments and helpful suggestions. This research has been conducted with the financial support of Science Foundation Ireland (SFI) under Grant Number 13/IA/1895.

References

1. Agus, M., et al.: An interactive 3D medical visualization system based on a light field display. Vis. Comput. **25**(9), 883–893 (2009). https://doi.org/10.1007/s00371-009-0311-y
2. Agus, M., Gobbetti, E., Guitián, J.A.I., Marton, F., Pintore, G.: GPU accelerated direct volume rendering on an interactive light field display. Comput. Graph. Forum **27**(2), 231–240 (2008). https://doi.org/10.1111/j.1467-8659.2008.01120.x
3. Bilen, H., Fernando, B., Gavves, E., Vedaldi, A., Gould, S.: Dynamic image networks for action recognition. In: 2016 IEEE Conference on Computer Vision and Pattern Recognition (CVPR), pp. 3034–3042 (2016). https://doi.org/10.1109/CVPR.2016.331
4. Birklbauer, C., Bimber, O.: Light-field supported fast volume rendering. In: ACM SIGGRAPH 2012 Posters on - SIGGRAPH 2012, p. 1. ACM Press, Los Angeles, California (2012). https://doi.org/10.1145/2342896.2343040
5. Bruton, S., Ganter, D., Manzke, M.: Synthesising light field volumetric visualizations in real-time using a compressed volume representation. In: Proceedings of the 14th International Joint Conference on Computer Vision, Imaging and Computer Graphics Theory and Applications - Volume 3: IVAPP. pp. 96–105. SciTePress (2019). https://doi.org/10.5220/0007407200960105
6. Chang, A.X., et al.: ShapeNet: an information-rich 3D model repository. arXiv:1512.03012 [cs] (2015)
7. Drebin, R.A., Carpenter, L., Hanrahan, P.: Volume rendering. In: Proceedings of the 15th Annual Conference on Computer Graphics and Interactive Techniques, pp. 65–74. SIGGRAPH 1988, ACM, New York, NY, USA (1988). https://doi.org/10.1145/54852.378484
8. Engelmann, F., Kontogianni, T., Hermans, A., Leibe, B.: Exploring spatial context for 3D semantic segmentation of point clouds. In: 2017 IEEE International Conference on Computer Vision Workshops (ICCVW). pp. 716–724 (2017). https://doi.org/10.1109/ICCVW.2017.90
9. Favalora, G.E.: Volumetric 3D displays and application infrastructure. Computer **38**(8), 37–44 (2005). https://doi.org/10.1109/MC.2005.276
10. Fernando, B., Gavves, E.M., Oramas, J., Ghodrati, A., Tuytelaars, T.: Rank pooling for action recognition. IEEE Trans. Pattern Anal. Mach. Intell. **39**(4), 773–787 (2017). https://doi.org/10.1109/TPAMI.2016.2558148
11. Fishman, E.K., Ney, D.R., Heath, D.G., Corl, F.M., Horton, K.M., Johnson, P.T.: Volume rendering versus maximum intensity projection in CT angiography: what works best, when, and why. Radiographics: A Review Publication of the Radiological Society of North America, Inc 26(3), 905–922 (2006). https://doi.org/10.1148/rg.263055186
12. Hadwiger, M., Kratz, A., Sigg, C., Bühler, K.: GPU-accelerated deep shadow maps for direct volume rendering. In: Proceedings of the 21st ACM SIGGRAPH/EUROGRAPHICS Symposium on Graphics Hardware, pp. 49–52. GH 2006, ACM, New York, NY, USA (2006). https://doi.org/10.1145/1283900.1283908

13. He, K., Zhang, X., Ren, S., Sun, J.: Deep residual learning for image recognition. In: 2016 IEEE Conference on Computer Vision and Pattern Recognition (CVPR), pp. 770–778 (2016). https://doi.org/10.1109/CVPR.2016.90

14. Kalantari, N.K., Wang, T.C., Ramamoorthi, R.: Learning-based view synthesis for light field cameras. ACM Trans. Graph. **35**(6), 193:1–193:10 (2016). https://doi.org/10.1145/2980179.2980251

15. Kühnapfel, U., Çakmak, H.K., Maaß, H.: Endoscopic surgery training using virtual reality and deformable tissue simulation. Comput. Graph. **24**(5), 671–682 (2000). https://doi.org/10.1016/S0097-8493(00)00070-4

16. Kingma, D.P., Ba, J.: Adam: A method for stochastic optimization. arXiv:1412.6980 [cs] (2014)

17. Klokov, R., Lempitsky, V.: Escape from cells: deep kd-networks for the recognition of 3D point cloud models. In: 2017 IEEE International Conference on Computer Vision (ICCV), pp. 863–872 (2017). https://doi.org/10.1109/ICCV.2017.99

18. Kniss, J., Kindlmann, G., Hansen, C.: Multidimensional transfer functions for interactive volume rendering. IEEE Trans. Vis. Comput. Graph. **8**(3), 270–285 (2002). https://doi.org/10.1109/TVCG.2002.1021579

19. Krizhevsky, A., Sutskever, I., Hinton, G.E.: Imagenet classification with deep convolutional neural networks. In: Pereira, F., Burges, C.J.C., Bottou, L., Weinberger, K.Q. (eds.) Advances in Neural Information Processing Systems, vol. 25, pp. 1097–1105. Curran Associates, Inc., New York (2012)

20. Lacroute, P., Levoy, M.: Fast volume rendering using a shear-warp factorization of the viewing transformation. In: Proceedings of the 21st Annual Conference on Computer Graphics and Interactive Techniques, pp. 451–458. SIGGRAPH 1994, ACM, New York, NY, USA (1994). https://doi.org/10.1145/192161.192283

21. Lanman, D., Luebke, D.: Near-eye light field displays. In: ACM SIGGRAPH 2013 Emerging Technologies, p. 11:1. SIGGRAPH 2013, ACM, New York, NY, USA (2013). https://doi.org/10.1145/2503368.2503379

22. Levoy, M., Hanrahan, P.: Light field rendering. In: Proceedings of the 23rd Annual Conference on Computer Graphics and Interactive Techniques, pp. 31–42. SIGGRAPH 1996, ACM, New York, NY, USA (1996). https://doi.org/10.1145/237170.237199

23. Li, Y., Pirk, S., Su, H., Qi, C.R., Guibas, L.J.: FPNN: field probing neural networks for 3D data. In: Lee, D.D., Sugiyama, M., Luxburg, U.V., Guyon, I., Garnett, R. (eds.) Advances in Neural Information Processing Systems, vol. 29, pp. 307–315. Curran Associates, Inc., New York (2016)

24. Lin, M., Chen, Q., Yan, S.: Network in network. arXiv:1312.4400 [cs] (2013)

25. Liu, T.Y.: Learning to rank for information retrieval. Found. Trends Inf. Retr. **3**(3), 225–331 (2009). https://doi.org/10.1561/1500000016

26. Liu, Z., Yeh, R.A., Tang, X., Liu, Y., Agarwala, A.: Video frame synthesis using deep voxel flow. In: 2017 IEEE International Conference on Computer Vision (ICCV), pp. 4473–4481 (2017). https://doi.org/10.1109/ICCV.2017.478

27. Ljung, P., Krüger, J., Groller, E., Hadwiger, M., Hansen, C.D., Ynnerman, A.: State of the art in transfer functions for direct volume rendering. Comput. Graph. Forum **35**(3), 669–691 (2016). https://doi.org/10.1111/cgf.12934

28. Mora, B., Maciejewski, R., Chen, M., Ebert, D.S.: Visualization and computer graphics on isotropically emissive volumetric displays. IEEE Trans. Vis. Comput. Graph. **15**(2), 221–234 (2009). https://doi.org/10.1109/TVCG.2008.99

29. Mueller, K., Yagel, R.: Fast perspective volume rendering with splatting by utilizing a ray-driven approach. In: Proceedings of Seventh Annual IEEE Visualization 1996, pp. 65–72 (1996). https://doi.org/10.1109/VISUAL.1996.567608

30. Niklaus, S., Mai, L., Liu, F.: Video frame interpolation via adaptive separable convolution. In: 2017 IEEE International Conference on Computer Vision (ICCV), pp. 261–270 (2017). https://doi.org/10.1109/ICCV.2017.37
31. Park, E., Yang, J., Yumer, E., Ceylan, D., Berg, A.C.: Transformation-grounded image generation network for novel 3D view synthesis. In: 2017 IEEE Conference on Computer Vision and Pattern Recognition (CVPR), pp. 702–711 (2017). https://doi.org/10.1109/CVPR.2017.82
32. Philips, S., Hlawitschka, M., Scheuermann, G.: Slice-based visualization of brain fiber bundles - a lic-based approach. In: Proceedings of the 13th International Joint Conference on Computer Vision, Imaging and Computer Graphics Theory and Applications - Volume 3: IVAPP, pp. 281–288. SciTePress (2018). https://doi.org/10.5220/0006619402810288
33. Qi, C.R., Su, H., Kaichun, M., Guibas, L.J.: PointNet: deep learning on point sets for 3D classification and segmentation. In: 2017 IEEE Conference on Computer Vision and Pattern Recognition (CVPR), pp. 77–85 (2017). https://doi.org/10.1109/CVPR.2017.16
34. Qi, C.R., Su, H., Nießner, M., Dai, A., Yan, M., Guibas, L.J.: Volumetric and multi-view CNNs for object classification on 3D data. In: 2016 IEEE Conference on Computer Vision and Pattern Recognition (CVPR), pp. 5648–5656 (2016). https://doi.org/10.1109/CVPR.2016.609
35. Qi, C.R., Yi, L., Su, H., Guibas, L.J.: PointNet++: deep hierarchical feature learning on point sets in a metric space. In: Guyon, I., et al. (eds.) Advances in Neural Information Processing Systems, vol. 30, pp. 5099–5108. Curran Associates, Inc., New York (2017)
36. Riegler, G., Ulusoy, A.O., Geiger, A.: OctNet: Learning deep 3D representations at high resolutions. In: 2017 IEEE Conference on Computer Vision and Pattern Recognition (CVPR), pp. 6620–6629 (2017). https://doi.org/10.1109/CVPR.2017.701
37. Salama, C.R.: GPU-based monte-carlo volume raycasting. In: 15th Pacific Conference on Computer Graphics and Applications (PG 2007). pp. 411–414 (2007). https://doi.org/10.1109/PG.2007.27
38. Smola, A.J., Schölkopf, B.: A tutorial on support vector regression. Stat. Comput. 14(3), 199–222 (2004). https://doi.org/10.1023/B:STCO.0000035301.49549.88
39. Srinivasan, P.P., Wang, T., Sreelal, A., Ramamoorthi, R., Ng, R.: Learning to synthesize a 4D RGBD light field from a single image. In: 2017 IEEE International Conference on Computer Vision (ICCV), pp. 2262–2270 (2017). https://doi.org/10.1109/ICCV.2017.246
40. Su, H., Maji, S., Kalogerakis, E., Learned-Miller, E.: Multi-view convolutional neural networks for 3D shape recognition. In: Proceedings of the 2015 IEEE International Conference on Computer Vision (ICCV), pp. 945–953. ICCV 2015, IEEE Computer Society, Washington, DC, USA (2015). https://doi.org/10.1109/ICCV.2015.114
41. Sunden, E., et al.: Inviwo - an extensible, multi-purpose visualization framework. In: 2015 IEEE Scientific Visualization Conference (SciVis), pp. 163–164 (2015). https://doi.org/10.1109/SciVis.2015.7429514
42. Tuzel, O., Liu, M.-Y., Taguchi, Y., Raghunathan, A.: Learning to rank 3D features. In: Fleet, D., Pajdla, T., Schiele, B., Tuytelaars, T. (eds.) ECCV 2014. LNCS, vol. 8689, pp. 520–535. Springer, Cham (2014). https://doi.org/10.1007/978-3-319-10590-1_34

43. Wang, P., Li, W., Gao, Z., Zhang, Y., Tang, C., Ogunbona, P.: Scene flow to action map: a new representation for RGB-D based action recognition with convolutional neural networks. In: 2017 IEEE Conference on Computer Vision and Pattern Recognition (CVPR), pp. 416–425 (2017). https://doi.org/10.1109/CVPR.2017.52

44. Wang, P.S., Liu, Y., Guo, Y.X., Sun, C.Y., Tong, X.: O-CNN: octree-based convolutional neural networks for 3D shape analysis. ACM Trans. Graph. 36(4), 1–11 (2017). https://doi.org/10.1145/3072959.3073608

45. Wetzstein, G., Lanman, D., Hirsch, M., Raskar, R.: Tensor displays: compressive light field synthesis using multilayer displays with directional backlighting. ACM Trans. Graph. 31(4), 80:1–80:11 (2012). https://doi.org/10.1145/2185520.2185576

46. Xie, J., Dai, G., Zhu, F., Wong, E.K., Fang, Y.: Deepshape: deep-learned shape descriptor for 3D shape retrieval. IEEE Trans. Pattern Anal. Mach. Intell. 39(7), 1335–1345 (2017). https://doi.org/10.1109/TPAMI.2016.2596722

47. Zhang, Y., Dong, Z., Ma, K.: Real-time volume rendering in dynamic lighting environments using precomputed photon mapping. IEEE Trans. Vis. Comput. Graph. 19(8), 1317–1330 (2013). https://doi.org/10.1109/TVCG.2013.17

48. Zhou, T., Tucker, R., Flynn, J., Fyffe, G., Snavely, N.: Stereo magnification: learning view synthesis using multiplane images. ACM Trans. Graph. 37(4), 65:1–65:12 (2018). https://doi.org/10.1145/3197517.3201323

49. Zhou, T., Tulsiani, S., Sun, W., Malik, J., Efros, A.A.: View synthesis by appearance flow. In: Leibe, B., Matas, J., Sebe, N., Welling, M. (eds.) ECCV 2016. LNCS, vol. 9908, pp. 286–301. Springer, Cham (2016). https://doi.org/10.1007/978-3-319-46493-0_18

A Reproducibility Study for Visual MRSI Data Analytics

Muhammad Jawad, Vladimir Molchanov, and Lars Linsen[✉]

Westfälische Wilhelms-Universität Münster, Münster, Germany
{jawad,molchano,linsen}@uni-muenster.de

Abstract. Magnetic Resonance Spectroscopy Imaging (MRSI) is a spectral imaging method that measures per voxel spectral information of chemical resonance, from which metabolite concentrations can be computed. In recent work, we proposed a system that uses coordinated views between image-space visualizations and visual representations of the spectral (or feature) space. Coordinated interaction allowed us to analyze all metabolite concentrations together instead of focusing only at single metabolites at a time [8]. In this paper, we want to relate our findings to different results reported in the literature. MRSI is particularly useful for tumor classification and measuring its infiltration of healthy tissue. We compare the metabolite compositions obtained in the various tissues of our data against the compositions reported by other brain tumor studies using a visual analytics approach. It visualizes the similarities in a plot obtained using dimensionality reduction methods. We test our data against various sources to test the reproducibility of the findings.

Keywords: Reproducibility · Multidimensional data visualization · Medical visualization · Coordinated views · Spectral imaging analysis

1 Introduction

Magnetic Resonance Imaging (MRI) is a routinely applied anatomical imaging procedure for varying medical applications including the detection and monitoring of tumors, more precisely, of tumor locations, shapes, and sizes. Magnetic Resonance Spectroscopy Imaging (MRSI) allows the radiologist to extract further information on the chemical composition of the scanned tissues. It measures for each voxel the chemical resonance spectra [5], from which the metabolite concentrations can be computed, see Sect. 2.2. Hence, the medical expert can draw conclusion on how active the tumor cells, where the tumor is still growing, and which surrounding regions are already infiltrated [2]. Moreover, one can deduce from the metabolite concentration of the tumor voxel, which tumor type is present. Of particular interest, is to rate the malignancy of a tumor.

While it is common to focus on a few metabolites in medical studies, we recently presented an approach to analyze all available metabolic information

© Springer Nature Switzerland AG 2020
A. P. Cláudio et al. (Eds.): VISIGRAPP 2019, CCIS 1182, pp. 362–388, 2020.
https://doi.org/10.1007/978-3-030-41590-7_15

using an interactive visual approach [8]. Our approach is based on coordinated views of visualizations in image space and the so-called feature space formed by the set of detected metabolites. For image-space views, we use slice-based visualizations, as they avoid occlusion and it is common to only scan a few slices in MRSI. MRSI and MRI visualizations are overlaid and combined with the outcome of automatic MRI segmentation algorithms, which can then be interactively refined. For feature-space views, labeled multidimensional data visualization methods are used, where the label stems from the semi-automatic image segmentation We apply interaction in star-coordinates plots to separate the labeled classes. The findings can be related back to the individual dimensions, basically allowing for conclusions, which metabolite concentration allow for class separation. The methodology was successfully applied to brain tumor data, as reported in our previous paper [8]. Section 4 reviews the methodology and the findings, which lay the foundation for the further steps taken in this paper.

Common analysis tools for MRSI data in clinical settings visualize the spatial distribution of individual metabolite concentrations using color mapping of an image slice. Thus, only a single metabolite is investigated at a time. Therefore, a lot of information is being neglected and the interplay of metabolites concentrations cannot be analyzed. We propose a novel tool that integrates all information at hand for a comprehensive analysis of MRSI data. Our approach is based on coordinated views of visualization in image and feature space. For image space, we also use slice-based visualizations, as they avoid occlusion and it is common to only scan a few slices in MRSI. MRSI visualizations are overlaid with MR images and combined with automatic MRI segmentation results. For feature space, multidimensional data visualization methods are used. Given the image segmentation result, the multidimensional data are labeled accordingly and we apply interaction methods to separate the labeled classes. Using star-coordinates encoding, the separations can be related back to the individual dimension, basically allowing for conclusions, which metabolite concentration allow for class separation. The methodology is detailed in Sect. 4.

In Sect. 4.4, we apply our methods to a scenario for MRSI data analysis to investigate brain tumors. In 2012, WHO reported 256,213 brain cancer cases, of which 189,382 deaths have been recorded [1]. We show how our coordinated views allow for analyzing the brain tumor's chemical composition as well as the infiltration of surrounding tissue that when based on MRI data may be classified as non-tumor region.

2   Data Description

Before we explain our methodology and that of related studies, we first want to provide background information on the spectral imaging method, how to obtain metabolite concentration from the spectral image, and what are the specifics of the data we are using for our application scenario.

2.1 Data Acquisition

MRSI is an in-vivo medical imaging method, where per voxel a whole spectrum of intensities is recorded. While MRI only measures water intensities per voxel, MRSI measures intensities at different radio frequencies. Most commonly ^{1}H MRSI is used, where the signal of hydrogen protons (^{1}H) in different chemicals is measured in the form of intensity peaks. As this measurement is performed for different frequencies, different chemicals can be detected within each voxel. The intensity peaks within the frequency spectrum are quantified as parts per million (ppm).

Measuring entire spectra in MRSI comes at the expense of much lower spatial resolution when compared to MRI. Using 1.5T or 3T scanners yields voxel sizes of about 10 mm × 10 mm × 8 mm [16,19]. 7T scanners yielding higher resolutions are barely used in clinical practice due to high costs [27]. Typically, MR images are taken (using T1 or T2 relaxation) in addition to MRS images within the same session. The MR images provide anatomical information at a higher resolution.

2.2 Quantification of Metabolites

The main goal of the pre-processing step is to quantify the various chemicals, referred to as metabolites, within each voxel from the respective intensity spectrum. MRSI is not yet standardized (e.g., using a DICOM format), but there is open source and proprietary software available for pre-processing and metabolite quantification like LCModel [23], jMRUI [29], and TARQUIN [25].

Common pre-processing steps provided by the tools are eddy current compensation, offset correction, noise filtering, zero filling, residual water suppression, and phase and base line correction. These steps are executed in the measured time domain or in the frequency domain after performing a Fourier transform. Signal strength in time domain indicates the metabolites' concentration, while the area under the metabolite curve is used to compute the concentration in frequency domain. The computed values are similar when measurements with high signal-to-noise ratio are provided [31]. For the actual metabolite concentration, mathematical models are used and fitted to the measured data. Different tools use different fitting approaches.

For the pre-processing of our data, we used TARQUIN, which is short for Totally Automatic Robust Quantitation in NMR. It is an open-source GUI-based software for in-vivo spectroscopy data pre-processing and quantification. It operates in time domain and uses a non-negative least-squares method for model fitting to compute the metabolite concentrations. We incorporate TARQUIN in our data preparation step due to free availability, friendly user interface, automatic quantification, support for multi-voxel spectroscopy, being able to import the imaging data and export the quantification results in various formats [25,32].

2.3 Application Scenario

The data we use in this paper for testing against results obtained in other studies are courtesy of Miriam Bopp and Christopher Nimsky from the University

Hospital Marburg, Germany. They were acquired using an ^{1}H MRSI technology on a 3T Siemens scanner (TR/TE/flip $= 1700\,\text{ms}/135\,\text{ms}/90°$). Two MRSI series are taken, each having a $160\,\text{mm} \times 160\,\text{mm} \times 1\,\text{mm}$ field of view. In addition, an MRI volume is captured and registered with the MRSI volume. The MRI volume is $224\,\text{mm} \times 256\,\text{mm} \times 144\,\text{mm}$ with $1\,\text{mm}$ slice thickness. We clip the MRI volume to the MRSI volume. The resolution of the MRI volume is much higher such that each MRSI voxel stretches over $10\,\text{mm} \times 10\,\text{mm} \times 12\,\text{mm}$ MRI voxels. We used TARQUIN for MRSI pre-processing and metabolite quantification. The quantification process resulted in 33 metabolites that are listed in Table 1.

Table 1. Metabolites delivered by a quantification from brain MRSI data using TARQUIN (taken from [8]).

No.	Name	No.	Name	No.	Name
1	Ala	12	Lip09	23	PC
2	Asp	13	Lip13a	24	PCr
3	Cr	14	Lip13b	25	Scyllo
4	GABA	15	Lip20	26	Tau
5	GPC	16	MM09	27	TNAA
6	Glc	17	MM12	28	TCho
7	Gln	18	MM14	29	TCr
8	Glth	19	MM17	30	Glx
9	Glu	20	MM20	31	TLM09
10	Ins	21	NAA	32	TLM13
11	Lac	22	NAAG	33	TLM20

After metabolite quantification, we can characterize the available data as two slabs of MRSI voxels with registered MRI voxels. For each MRSI voxel, we have computed concentrations of 33 metabolites, leading to a 33-dimensional feature space, where each point in the feature space represents the chemical composition of one voxel. In addition, we know for each MRSI voxel, which are the matching MRI voxels (single intensity values). For our investigations the information of all metabolite concentrations of all voxels shall be considered.

3   Existing Studies and Tools

3.1   Existing Studies

We want to relate our observations to findings in other studies. We took results from three in-vitro studies of brain tumors in human brain from the literature [7, 11, 22]. The purpose is to analyze, whether we can make use of these studies to draw conclusions about the tumor type for the human brain tumors detected in our data.

The first study we consider was performed by Kinoshita and Yokota [11]. They collected samples from 60 patients. Those samples were categorized in 14 different brain tumors that are graded from benign to malignant. For the investigation of changes in the concentration for various brain metabolites in case of a tumor, a control was formed by sampling the cerebral cortex from four healthy subjects. Statistics of eight metabolite concentrations (measures in μmol/100 g) in the detected tumor types and in the cerebral cortex are provided in the table shown in Fig. 1, which is taken from the paper by Kinoshita and Yokota. The statistics list means and standard deviations from the means for each metabolite and each tumor type (plus normal tissue). For our investigations, we can only relate those metabolites to our studies that were quantified both in their and our study (cf. Table 1). More precisely, we only considered the concentration of the metabolites NAA, TCr, TCho, Tau, Ala and Ins. Moreover, Kinoshita and Yokota do not explain how missing entries (ND and NQ) are to be interpreted. Therefore, we decided to not consider the tumor types with entries ND or NQ. Thus, we were left with the six tumor types astrocytoma, anaplastic astrocytoma, glioblastoma, medulloblastoma, malignantLymphoma, and pituitary adenoma plus the cerebral cortex, which we use in this paper for our studies.

The second study we use in our analysis was conducted by Howe et al. [7]. They sampled five different human brain tumors from 42 patients. For control they chose white matter as normal brain region, which was sampled from eight healthy subjects. Metabolites concentrations (in mM) were measured from tumor and white matter regions to estimate the changes in tumor regions. Statistics of the acquired data in the form of means and standard deviations from the means are provided in Fig. 2. Again, we match the provided metabolites with the ones we quantified with TARQUIN. From the six listed metabolites, we could use five (Cho, Cr, NAA, Ala, and Lac) for the comparison to our study. The list provided by Howe et al. does not have any missing entries such that we can consider all of given tumors for our investigations.

Finally, the third study we consider in our analysis was conducted by Peeling and Sutherland [22]. They diagnosed 32 patients and categorized the respective tumors into six different human brain tumor types. Top gather control data they collected samples from the grey matter of 43 healthy subjects. They quantified 13 metabolite concentrations (in μmol/100 g) in all patients and volunteers. Statistics of the acquired data in the form of means and standard deviations from the means are provided in Fig. 3. Again, we neglected those tumor types with missing values and those metabolites that do not match with our list of quantified metabolites. From the 13 quantified metabolites, we were able to include 9 (Ala, NAA, GABA, Glu, Asp, Tau, Cr, Cho, and Ins) for our analyses. Besides normal brain tissue we used data for malignant astrocytoma, meningioma, metastatic, and oligodendroglioma tumor types. Note that the oligodendroglioma tumor type only has one missing value for the glycine metabolite. Since we do not consider glycine, we can include oligodendroglioma.

Pathology (numbers)		NAA	Total creatine	Choline-containing compounds	Glycine	Taurine	Alanine	Inositol	PEA
Cerebral cortex	(n=4)	555.6±32.9	1076.3±102.7	64.3±10.1	82.2±19.7	153.2±29.1	91.9±18.9	174.3±17.0	105.8±10.3
Neuroectodermal tumor									
Astrocytoma	(n=6)	180.8±27.7	780.9±84.1	68.6±6.5	91.3±15.5	169.0±16.7	101.0±20.3	364.7±94.2	127.1±10.6
Anaplastic astrocytoma	(n=4)	110.2±43.4	564.8±89.0	132.6±7.9	269.5±76.1	227.7±47.7	52.4±17.9	479.6±97.6	205.1±40.5
Glioblastoma	(n=12)	40.6±10.9	365.9±40.7	119.2±29.8	501.7±43.9	214.9±42.0	325.8±31.1	333.9±72.5	238.2±38.3
Ependymoma	(n=1)	ND	445.8	88.4	677.9	304.8	324.2	502.8	106.2
Medulloblastoma	(n=1)	40.3	888.4	205.8	913.5	1930.0	225.2	184.8	462.3
Non-neuroectodermal tumor									
Meningioma	(n=14)	ND	130.8±24.9	78.8±12.6	158.3±20.6	338.6±98.8	320.8±43.2	179.7±36.1	161.3±24.1
Neurinoma	(n=6)	ND	72.9±15.2	124.5±23.7	75.0±6.9	215.6±28.4	156.0±28.4	873.1±142.5	186.7±30.7
Craniopharyngioma	(n=3)	ND	86.2±14.5	15.7±9.1	85.6±23.6	128.0±31.0	138.3±32.1	104.6±66.3	68.3±12.4
Chordoma	(n=4)	ND	151.6±63.5	110.2±41.6	166.3±44.3	315.6±115.0	131.2±34.5	267.0±141.4	205.4±68.8
Malignant lymphoma	(n=2)	51.2±5.7	160.8±37.0	74.3±0.4	197.6±13.3	201.1±11.7	117.1±2.7	160.9±76.2	428.8±16.5
Pituitary adenoma	(n=2)	147.9±147.9	311.6±176.6	132.6±18.8	245.1±245.1	853.6±64.9	272.3±80.3	327.6±56.7	440.5±105.8
Metastatic tumor									
Pulmonary adenocarcinoma	(n=3)	ND	102.7±12.0	101.4±18.1	136.6±13.6	236.6±26.7	140.7±30.8	337.5±40.1	212.1±20.4
Renal cell carcinoma	(n=1)	ND	72.8	72.8	93.7	786.2	103.2	NQ	NQ
Hepatocellular carcinoma	(n=1)	ND	212.8	572.2	166.8	117.7	142.7	NQ	NQ

Values shown are means ± SEM (μmol/100 g wet wt).
ND, not detected.
NQ, not quantifiable.

Fig. 1. List of brain tumors graded from benign to malignant that are detected from 60 in-vitro samples. Metabolites profile of each tumor type is provided by mean values and standard deviations to the mean (taken from [11]).

Metabolite Concentrations (mM) Given as Mean ± SD. Calculated Relative to Tumor Water From STEAM TE 30ms Spectra

	mIG	Cho	Cr	NAA	Ala	Lac
PWM, normal parietal white matter (N = 8)	7.4 ± 0.9	1.7 ± 0.2	6.2 ± 0.3	10.1 ± 0.6	0.4 ± 0.3	0.5 ± 0.4
MNG, meningioma (N = 8)	1.4 ± 1.4	2.7 ± 1.7	1.0 ± 0.7	0.6 ± 0.6	2.0 ± 1.9	2.7 ± 2.3
AS, astrocytoma grade 2 (N = 5)	10.4 ± 2.7	2.2 ± 0.4	3.8 ± 1.0	1.2 ± 0.9	0.8 ± 0.6	1.5 ± 1.0
AA, anaplastic astrocytoma (N = 7)	8.1 ± 2.4	2.5 ± 0.6	4.2 ± 1.4	2.2 ± 1.8	1.8 ± 2.5	5.8 ± 8.5
GBM, glioblastoma (N = 10)	2.9 ± 2.4	1.5 ± 1.0	1.8 ± 1.8	0.8 ± 1.0	1.7 ± 1.9	11.7 ± 7.0
MET, metastases (N = 6)	3.5 ± 3.3	2.0 ± 1.0	2.1 ± 2.0	1.8 ± 2.1	3.0 ± 3.8	14.2 ± 7.1

Fig. 2. List of brain tumors detected from 42 in-vitro samples. Metabolites profile of each tumor type is provided by mean values and standard deviations to the mean (taken from [7]).

Metabolite Levels (μmol/100 g tissue ± SEM) in Human Brain Tissue

Metabolite	Brain (43)[a]	Malignant astrocytoma (14)	Benign astrocytoma (5)	Meningioma (6)	Metastatic tumor (5)	Schwannoma (1)	Oligodendroglioma (1)
Alanine	26 ± 2	92 ± 13	37 ± 29	127 ± 61	146 ± 61	116	75
NAA	520 ± 30	21 ± 7 [3][b]	58 ± 15	18 [1]	63 ± 48 [3]	nd	63
GABA	70 ± 2	17 [1]	nd	25 [1]	9 [1]	nd	181
Glutamate	775 ± 35	171 ± 20	129 ± 20	234 ± 62	189 ± 65	144	221
Glutamine	nq	358 ± 60	216 ± 20	231 ± 63	185 ± 75	95	450
Aspartate	125 ± 6	67 [1]	24 ± 4 [2]	62 ± 13	55 ± 35 [3]	35	21
Glycine	nq	153 ± 26 [8]	126 [1]	101 ± 30	81 ± 28	nq	nq
Taurine	125 ± 8	177 ± 34	33 ± 9 [2]	236 ± 48	116 [1]	172	237
Succinate	22 ± 1	15 ± 2	9 ± 1	9 ± 2	7 ± 2	7	16
Creatine	690 ± 40	185 ± 41	194 ± 24	42 ± 19	114 ± 60	17	375
Cholines	110 ± 5	136 ± 19	101 ± 18	101 ± 26	130 ± 23	143	241
Inositol	555 ± 20	343 ± 79 [7]	348 ± 58	179 [1]	330 ± 189 [2]	343	492
Glucose	nd	100 ± 18	93 ± 18	103 ± 20	128 ± 38	300	135

Note. nq, not quantitated; nd, not detected.
[a] Number of samples analyzed.
[b] Number of samples in which metabolite was detected.

Fig. 3. List of brain tumors detected from 43 in-vitro samples. Metabolites profile of each tumor type is provided by mean values and standard deviation to the mean (taken from [22]).

3.2   Analysis Tools

in 2014, Nunes et al. [19] provided a survey of existing methods for analyzing MRSI data. They concluded that no approach exists that analyzes all metabolic information. They report that MRSI data visualization packages provided with commercial scanners such as SyngoMR and SpectroView only provide color mapping of individual metabolic concentrations (or a ratio of metabolic concentrations) in image space. Other tools such as Java-based Magnetic Resonance User Interface (jMRUI) [29] and Spectroscopic Imaging Visualization and Computing (SIVIC) [3] that are also widely used in clinical practice provide a similarly restricted functionality.

Raschke et al. [24] proposed an approach to differentiate between tumor and non-tumor regions by showing the relation of concentrations of two selected metabolites in a scatterplot. By plotting the regression line, they identified abnormal regions by data points that are far from the line. Similarly, Rowland et al. [26] use scatterplots to inspect three different pairs of metabolites tumor analysis. These procedures only target selected metabolites, which are chosen a priori.

An approach to exploit the metabolic information better was presented by Maudsley et al. [15]. They developed the Metabolite Imaging and Data Analysis System (MIDAS) for MRSI pre-processing and visualization. Users can view histograms of metabolites, but relations between metabolites cannot be studied. Feng et al. [4] presented the Scaled Data Driven Sphere (SDDS) technique, where information of multiple metabolite concentrations is combined in a glyph-based visualization using mappings to color and size. Obviously, the amount of dimensions that can be mapped is limited. They overlay the images with the glyphs and link them to parallel coordinate plots. Users can make selections in the parallel coordinate plot and observe respective spatial regions in the image space. As a follow-up of their survey, Nunes et al. [20] presented a system that couples the existing systems ComVis [14] and MITK [33]. They use scatterplot, histogram, and parallel coordinate plot visualizations for analyzing the metabolic features space. The visualizations provide linked interaction to image-space representations such that brushing on metabolic data triggers highlighting in image space.

In our recent work, we built upon the idea of using coordinated interactions in feature and image space, but enhanced the functionality significantly [8]. We incorporate segmentation results, provide means to separate the labeled data in feature space, allow for a comparative visualization of classes, and provide a single tool that integrates all information and allows for fully coordinated interaction in both directions. Details are provided below in Sect. 4. In this paper, we propose novel ideas for a visual analytics approach to relate the findings from our study to the studies in literature provided in Sect. 3.1.

4 Visual MRSI Data Analytics

MRSI data are mainly acquired to analyze the metabolic compositions of tumors, e.g., brain tumors, which is what we use in our application scenario. The main driving question is tumor classification, i.e., one would want to find out what type of tumor it is and how its malignancy is rated.

In order to relate our data against other results from literature, we first need to analyze our data at hand. This involves a number of subtasks. First, we need to localize the tumor, i.e., we need to identify the MRSI voxels that belong to the tumor. Second, we need to investigate the chemical compositions of the tumor voxels. Since the tumor may already have infiltrated surrounding tissues, those voxels are also of interest. In particular, the infiltration is an indication of the malignancy of the tumor. So, the third subtask is to analyze voxels surrounding the tumor and check how much they relate to the chemical compositions of healthy tissue and tumor to rate, whether the surrounding voxels are already infiltrated. Having identified the compositions of tumor voxels and surrounding ones, one can, finally, check whether there are other regions/voxels with similar compositions, which may be an indicator of another tumor.

In the following, we will describe the methodology for analyzing the entire metabolic information with respect to the formulated tasks. The methodology and analyses described in this section are that from our previous paper [8]. The subsequent sections will then build upon the findings we made here.

4.1 Image-Space Visualization

In order to analyze image regions such as tumors or surrounding tissue, one would need to be able to visually inspect those regions and interactively select them. We support this using a slice-based viewer, which we implemented using the VTK [28] and ITK [9] libraries. Within the slice, we have two visualization layers. The first layer represents the MRI volume. It provides the anatomical context for the MRSI data analysis, see Fig. 4. A gray scale luminance color map is used, as this is a common standard in MRI visualization. The second layer represents the MRSI volume. Here, individual voxels can be selected interactively by brushing on the image regions. Selected voxels are highlighted by color, see Fig. 10 (left).

In addition to manual selection of MRSI voxels based on visual inspection, we also support an automatic image segmentation method. The automatic image segmentation predefines regions for quick selection and region analysis. The automatic segmentation method partitions the MR image. A large range of algorithms exist each having certain advantages and drawbacks. For the data at hand, the best results in our tests were obtained by the Multiplicative Intrinsic Component Optimization (MICO) [12] segmentation method for auto characterization of various tissues in brain MRI. MICO could handle MRI with low signal-to-noise ratio that is due to magnetic field disturbance and patient movement during the scanning process. MICO performs well in bias field estimation and in discarding intensity inhomogeneity.

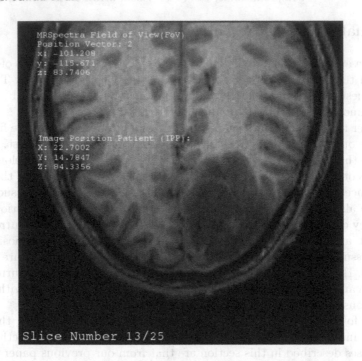

MRSpectra Field of View(FoV)
Position Vector: 2
x: -101.208
y: -115.671
z: 83.7406

Image Position Patient (IPP):
X: 22.7002
Y: 14.7847
Z: 84.3356

Slice Number 13/25

Fig. 4. Anatomical context in slice-based image-space MRI visualization (taken from [8]).

If we impose the MR image segmentation result on the MRSI voxels, we have to deal with partial volume effects, as one MRSI voxel corresponds to many MRI voxels, which may be classified differently. We propose to use a visual encoding that conveys the partial volume effect. Instead of simply color-coding the MRSI voxel by the color for the dominant class among the MRI voxels, we render square glyphs that are filled with different colors, where the color portions reflect the percentages how often the MRI voxels are assigned to the respective class. Figure 5 shows a respective slice-based visualization, where the segmentation result is shown in the MRSI layer overlaying the MRI layer. The mixture of colors indicate the uncertainty of the anatomical region segmentation at the MRSI resolution. In the image, we show the glyphs for each of the dominant class, i.e., the five columns (from left to right) show MRSI voxels, where the respective MRI voxels have primarily been classified as background, cerebrospinal fluid (CSF), tumor, gray matter, and white matter, respectively. The colors used for the respective classes are shown above the images. Results are shown for both MRSI slabs.

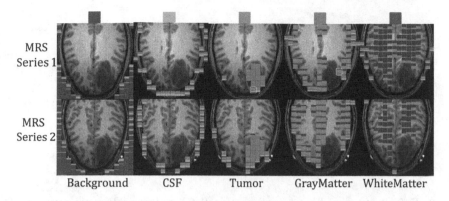

Background   CSF   Tumor   GrayMatter WhiteMatter

Fig. 5. MRI segmentation visualized at MRSI resolution using glyphs with color ratios that reflect class distribution of each voxel (taken from [8]). (Color figure online)

4.2 Feature-Space Visualization

The feature-space visualization problem is that of a multidimensional data visualization problem, where dimensionality is in the range of tens, while the number of data points is in the range of hundreds. Hence, we need an approach that scales sufficiently well in both aspects. Moreover, the tasks require us to observe patterns such as clusters of data points, i.e., sets of data points that are close to each other and different from other data points. Many multidimensional data visualization approaches exist and we refer to a recent survey for their descriptions [13]. They can be distinguished by the performed data transformations, by the visual encodings, and by their mappings to visual interfaces. Point-based visual encodings scale well in the number of data points and allow for an intuitive detection of clusters. Among them, projection-based dimensionality reduction approaches use data transformations to a visual interface supporting the handling of high dimensionality. Linear projections have the advantage over non-linear projections that the resulting plots can be easily related back to the original dimensions by the means of star coordinates. Thus, we propose to use star-coordinates plots [10] to visualize the feature space.

A projection from an n-dimensional feature space to a 2D visual space is obtained by a projection matrix of dimensions $2 \times n$, where each of the n columns represent the tip of the n star-coordinates axes. The default set-up is to locate the tips equidistantly on the unit circle, see Fig. 6. The tips can be moved to change the projection matrix, which allows for an interactive multidimensional data space analysis supporting the detection of trends, clusters, outliers, and correlations [30].

When assuming an image-space segmentation of the voxels as described in the preceding section, each voxel is assigned to a class. Hence, we are dealing with *labeled* multidimensional data. A common task in the visual analysis of labeled multidimensional data is to find a projected view with a good separation of the classes. This is also of our concern, as we want to separate tumor

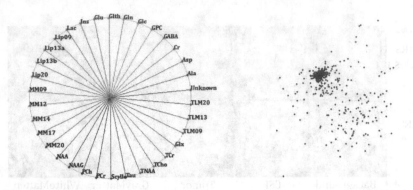

Fig. 6. Default star-coordinates configuration (left) and respective projection of 33-dimensional feature space to a 2D visual space. Each dimension represents a metabolite concentration, each point corresponds to a MRSI voxel (taken from [8]).

from healthy tissues. Molchanov et al. [17] proposed an approach for intuitive interactive class separation in linearly projected views. We adopt their idea for our purposes. The idea is to use a control point for each class, e.g., being the classes' medians. Classes can then be separated by dragging control points apart. Since the visual representation is restricted to linear projections, the position to which the control points are dragged can, in general, not be exactly obtained in a linear projection. Molchanov et al. proposed to use a least-squares approach to find the best match to the desired interaction. Using this idea, the user just needs to move the control points of the classes in an intuitive manner, where the number of classes is usually low (five in the case of brain imaging data), instead of interacting with all star-coordinates axes, which becomes tedious for larger number of dimensions (33 in our application scenario). Figure 7 illustrates this by showing the star-coordinates configuration to the left and the linear projection of the labeled multidimensional data to the right, where the control points that can be interactively moved are the ones with a black frame. Since the visualization would be too cluttered when overlaying the two views, we decided to show them in a juxtaposed manner. Further argumentation for juxtaposed views of star coordinates is provided by Molchanov and Linsen [18].

Since the points in our projected view represent MRSI voxels, while the image segmentation is performed on the higher-resolution MR image, we have partial volume effects as discussed in the previous section. The uncertainty of the labeling result shall also be conveyed in our projected view. For example, if some points are identified as outliers of a cluster, one should be able to reason if the labeling of the respective voxel is uncertain or not. If it was uncertain, the outlyingness may be due to a wrong labeling decision. To visually convey the labeling uncertainty, the probabilities of each point belonging to each class label shall be represented. Since the probabilities sum up to one, they are well represented by a pie chart, i.e., each point in the projected view is displayed by

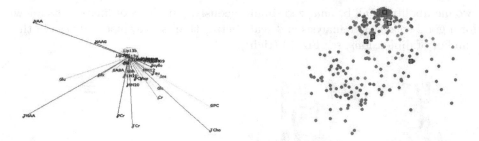

Fig. 7. Interactive visual analysis of labeled multidimensional data: star-coordinates configuration (left) and respective projected view (right) obtained by interacting with control points (black frames) of the classes induced by image segmentation (taken from [8]).

Fig. 8. Pie-chart representation of uncertainty for projected points (taken from [8]).

a pie chart. Figure 8 shows a projected view with labeling uncertainty conveyed by pie charts.

While interacting with the control points in the projected view, the star-coordinates axes are updated accordingly. Hence, when separating classes in the projected view, one can observe in the star-coordinates view, which axes are mainly responsible for the separation, i.e., which axes are the ones that allow for such a separation. The dimensions that are associated with these axes are subject to further investigations, as one of our tasks was to detect the metabolic compositions of tumors and surrounding tissues. For selected metabolites, we use different statistical plots supporting different analysis steps.

If we are interested in investigating the metabolic compositions of two classes for selected metabolites, we use juxtaposed box plots to show the statistical information of the metabolic concentrations of all voxels that were assigned to each class. The box plots convey the median, the interquartile range, as well as minimum and maximum, see Fig. 15.

If we are interested in investigating the interplay of two metabolites, we can analyze their correlation using a 2D scatterplot. Scatterplots are most effective in showing correlation and detecting outliers, see Fig. 9 (right). Correlation analysis is supported by computing and displaying a regression line. In case more than two metabolites shall be analyzed simultaneously, parallel coordinate plots allow for a good correlation analysis and scale better than scatterplot matrices in the number of dimensions, see Fig. 10 (right).

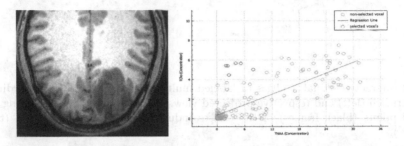

Fig. 9. Correlation analysis of the metabolites TCho and TNAA in a scatterplot (right) and detection of outliers. Voxels corresponding to the interactively selected outliers are shown in the image-space visualization (left), which conveys their relation to the tumor (taken from [8]).

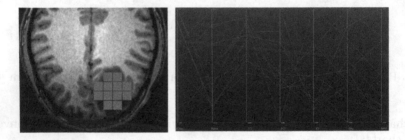

Fig. 10. (left) Interactive selection of MRSI voxels in image space and (right) linked view for correlation analysis of seven selected metabolites (TCho, TNAA, TCr, PCr, Cr, Glu, and NAA) using parallel coordinate plot for the selected voxels (taken from [8]).

4.3   Coordinated Interaction in Image and Feature Space

The interactive visual analysis becomes effective when using coordinated interaction of the image- and feature-space views. In the image space, one can brush the MRSI voxels in the slice-based visualization and group them accordingly. Hence, interactive labeling is supported. Of course, one can also use the labeling that is implied by the MRI segmentation (including the uncertainty visualization). The resulting labeling is transferred to the projected feature-space visualization by using the same colors in both visualizations, see image-space visualization in Fig. 5 and respective projected feature-space visualization in Figs. 7 and 8. This coordinated interaction allows for investigating whether image segments form coherent clusters in feature space, which dimensions of the feature space allow for a separation of the clusters, and whether there are some outliers. Of course, the same holds true when using the statistical plots as a coordinated feature-space view. Figure 9 shows a brushing in image space and investigation of the selection in a scatterplot visualization of two (previously selected) dimensions, while Fig. 10 shows a brushing in image space and investigation of the selection in a parallel coordinate plot of seven (previously selected) dimensions.

The coordinated interaction of image and feature space is bidirectional. Thus, the user may also brush in the feature space, e. g., by selecting a group of points in the feature space that form a cluster, and observe the spatial distribution of the selection in the image space. Again, one can use any of the feature-space visualizations or a combination thereof. For example, in Fig. 11, the four voxels to the left are chosen (voxels having low concentration of TNAA relative to TCho and TCr) in the bar chart and the respective highlighting in the image space convey that these voxels belong to the tumor region.

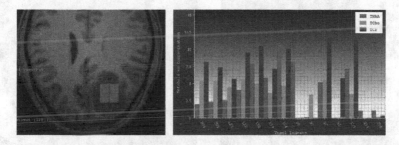

Fig. 11. Concentration of three selected metabolites shown side by side in the bar chart (right). Four voxels with low concentration of TNAA relative to TCho and TCr are selected by brushing. The coordinated image-space view (left) highlights the selected voxels, which form the core of the tumor (taken from [8]).

We also support a heatmap visualization as commonly used when observing the spatial distribution of individual metabolite concentrations. The user may select a single metabolite or a combination of metabolites. Figure 12 shows heatmaps of TCho and TNAA concentrations and In Fig. 13, a heatmap of the logarithm of the ratio of TCho/TNAA is shown.

The full potential of our system is reached by using coordinated interaction in both ways simultaneously. For example, one can select the tumor voxels in the image space, can investigate the respective set of points in feature space possibly forming a cluster, detect further points that fall into the cluster, select those further points, and observe their spatial distribution. These newly selected voxels may be voxels surrounding the tumor, in case the tumor has already infiltrated surrounding regions, or may be voxels that form a region elsewhere, in case there is a second tumor.

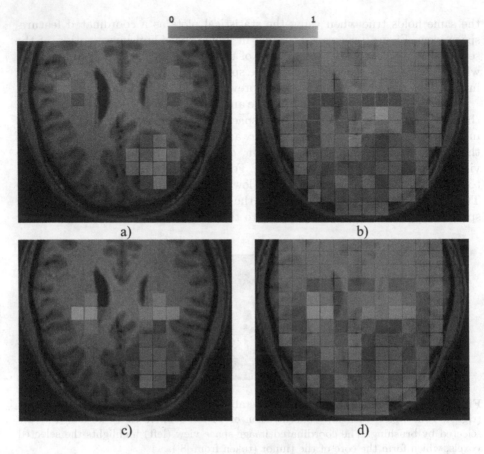

Fig. 12. Heatmaps of TCho (top) and TNAA concentration (bottom) for all brain voxels (right) and selected regions of interest (left) (taken from [8]).

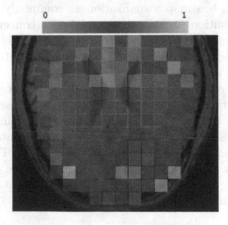

Fig. 13. Heatmap of ratio of TCho/TNAA concentrations (in logarithmic scaling) (taken from [8]).

4.4   Application Scenario, Results, and Discussion

In our application scenario, we applied the developed methods to the data acquired from a 26-year old male patient with a brain tumor. MRSI and MRI head scans were obtained as described in Sect. 2.3. We preprocessed the data as described in Sect. 2.2. Then, we first investigated the projected feature space using the default star-coordinates layout as in Fig. 6. We observed no obvious clusters in the projected space. Thus, we next applied an automatic segmentation of the MR image and imposed the segmentation onto the MRSI voxels using our uncertainty-aware visualization in Fig. 5. The extracted segments represent grey matter, white matter, CSF, the tumor, and background. This segmentation implies a labeling as shown in Fig. 7, where the colors match with the ones in Fig. 5. We applied the interactive technique to separate the classes in feature space using the interaction with the classes' control points. In particular, we applied it to separate the tumor class (red) from the other classes. We observe that certain dimensions of our 33-dimensional feature space are affected strongly by this interactive optimization. Hence, the respective metabolites may be the ones that distinguish tumor from the other segments. In Fig. 7, we see that the axes labeled NAA, Glu, TNAA, PCr, TCr, TCho, and GPC are longest. These metabolites are candidates for further investigations. In Fig. 9, we select TNAA and TCho concentrations, which had the longest axes in Fig. 7 and plot all voxels in a scatterplot. We select a group of outliers (red) with high TCho and low TNAA concentrations and observe that these form the core of the tumor region. Please note that by just looking at one of the two metabolites the voxels would have not been outliers. In Fig. 12, we use heatmaps to investigate the spatial distributions of TCho and TNAA concentrations, respectively. When looking at the entire brain region (right image column), it is hard to detect structures. However, when selecting regions of interest such as a white matter region and the tumor region, we can spot the differences (left image column). The two columns apply the red-to-green color map to the minimum-to-maximum range of selected voxels only. In Fig. 13, a heatmap of the logarithm of the ratio of the two metabolites is shown.

When investigating the projected feature space in Fig. 7, we observe that the classes are actually not well separated. We further investigate the pie chart visualization of the labeling result, see Fig. 8 to observe quite a few voxels that are rather uncertain with respect to the automatic labeling, which may be due to the partial volume effect. Also, when selecting all voxels that contain parts of the tumor as in Fig. 10, the distribution of values in the parallel coordinate plot appear rather diverse. Thus, we decided to manually select regions of low-uncertainty voxels and create new labels, see Fig. 14. Figure 15 shows the juxtaposed box plots of the four selected voxel groups. We observe that the tumor region (red) is quite different from the other regions in NAA and TNAA concentrations, where TNAA is the sum of NAA and NAAG. We also observe that the voxels surrounding the tumor (orange) behave like gray/white matter voxels rather than tumor voxels for these metabolites, which makes us believe that these voxels are not yet infiltrated by the tumor. However, this may not be true for

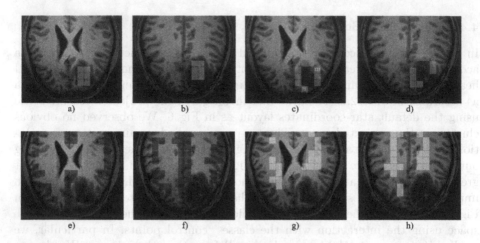

Fig. 14. Interactive labeling of voxels within both MRSI slabs representing tumor (red), voxels in the vicinity of the tumor (orange), grey matter (blue), and white matter (green) (taken from [8]). (Color figure online)

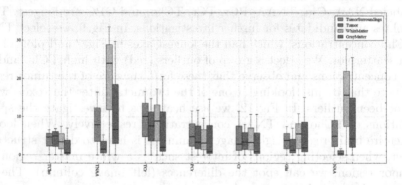

Fig. 15. Juxtaposed box plots to compare class statistics for selected classes (voxels labeled as white matter, grey matter, tumor, and surrounding tumor) and selected metabolites (taken from [8]). (Color figure online)

individual voxels of that group. Since the tumor size and shape of its boundary is important for diagnosis and treatment, we selected individual voxels at the border of the tumor and investigated the metabolite concentrations and individually compared their metabolite concentrations to those of tumor as well as white and grey matter. Figure 16 shows such a comparison for the voxels labeled 73 and 123 in Fig. 14. These two voxels showed high uncertainties in the automatic segmentation result. We can observe in Fig. 16 that voxel 73 matches well the tumor class, while voxel 123 does not. Hence, we conclude that the tumor may already have infiltrated the area at voxel 73, while it may have not yet done so for voxel 123.

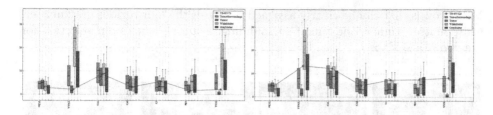

Fig. 16. Metabolite concentration of voxels 73 (left) and 123 (right) in vicinity of the tumor selected in Fig. 14 in comparison to the concentrations of the tumor, white matter, and grey matter classes (taken from [8]).

We invited two MRSI experts with many years of experience of acquiring and analyzing MRSI data to our lab to show them our tool on a large multi-touch display. The visual encodings were mostly intuitive to them and they were quickly able to suggest interactive analysis steps themselves. Only the projected view needed some explanations, but the intuitive interaction with the control points made them adopt the concept quickly. They also quickly brought in their expertise into the analytical work flow by excluding some metabolites that they knew would not be important for the given tasks such as lipids and macromolecules and by interpreting correctly combinations such as TNAA being a combination of NAA and NAAG. In the session, we jointly looked into the metabolic composition of voxels in the vicinity of the tumors as documented above. To test the reproducibility of the analysis, it would be desirable to test our tool with a large number of experts on a large number of data sets. We hope that we can conduct such a study in future work, but acknowledge that it will be challenging to recruit a large number of experts.

5   Reproducibility Study

We want to investigate how the findings in Sect. 4 relate to the findings in literature that were reported in Sect. 3. The goal is to investigate whether the tumor voxels we detected as well as the voxels surrounding the tumor match the chemical compositions of tumors detected in other studies. In particular, we detailed in Sect. 3.1 that in the three listed studies multiple tumor types were investigated in comparison to normal tissue. Thus, we can compute similarities of the metabolite concentrations computed from our tumor regions with the metabolite concentrations reported for various tumor types in literature. Since the investigated tumor types and metabolites in the three studies in Sect. 3.1 do not match, we cannot combine the three studies. Instead, we will compare our data against each of the three studies individually to investigate how reproducible the findings are. In addition to the tumor regions we detected in our studies, we also define gray-matter and white-matter region. Then, we can also check how reproducible the results for normal regions are, i.e., regions without tumors. Following this discussion and following our analysis in Sect. 4, we select

MRSI voxels that can be clearly assigned to one of the four regions gray matter, white matter, tumor, and tumor-surrounding. Figure 17 shows our selection for the two MRSI slabs.

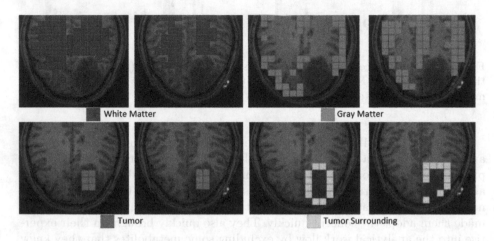

Fig. 17. Selection of voxels belonging to the regions of gray matter, white matter, tumor, and tumor surrounding. Selections are made for each region in both MRSI slabs.

5.1 Methodology

Normalization. A first observation that can be made from the tables in Sect. 3.1 is that the absolute values of metabolite concentration differ quite substantially from study to study. This is most obvious when looking at normal tissue. Thus, the measurement set-up plays an important role here. In order to compensate for this systematic scaling artifact, we apply a normalization of the data given in the three studies from literature and for our data. Since the different studies contain different types of tumor, we can only reliably obtain a normalization factor from the normal tissue. Thus, we scale all values of each study by mean and standard deviation of the normal tissue. We apply a whitening step fr normalization, where the mean is subtracted from all values and we, subsequently, divide by the standard deviation. Thus, the metabolite concentrations for normal tissue is scaled such that the range of mean ± standard deviation maps to the interval [−1,1], while the metabolite concentrations for the various tumor types are translated and scaled by the same factor. We can apply the same procedure to our study and the three studies from literature. One aspect needs to be considered: In the first and third study normal tissue represents the cerebral cortex (gray matter), while in the second study normal tissue represents white matter. Thus, when comparing our data to the data from the first or third study, we take all our gray-matter voxels and compute the respective mean, which is then used for normalizing all regions. Instead, when comparing our data to the data from the second study, we take all our white-matter voxels and compute the respective mean, which is then used for normalizing all regions.

Data Transformation. In Sect. 4, we have been comparing the various voxels of our data sets with respect to their metabolite concentrations. In this section, we want to compare our voxels against the metabolite concentrations that have been reported in literature for different tumor types (and normal tissue). Thus, we are again interested in similarities of multi-dimensional feature vectors. However, we do not anymore compare how the multi-dimensional feature vectors of our voxels relate to each other, but how these vectors compare to the vectors of different tumor types. Thus, our feature space has undergone a data transformation.

Feature-Space Visualization. To visualize our feature space, we have to accommodate for the changed analysis task. We still want to visualize similarities in a multi-dimensional space, which is best performed by using some dimensionality projection method such as the star coordinates projection we used in Sect. 4. However, now the goal is to perceive how much the individual multi-dimensional data items concur with the multi-dimensional data values reported for different tumor types. Thus, we want to observe how similar our voxels are to the tumor type values. A direct implementation of this idea leads to a visual encoding where the tumor types are represented as anchor points in the scatterplot obtained by the dimensionality reduction method and the similarity to these tumor types is encoded by the distance to the anchor points. The only question remains is how to position the anchor points. A suitable choice for minimizing clutter is to place them on a unit circle. This is exactly what the RadViz approach [6] does.

RadViz projection is similar to the projection with star coordinates. In fact, RadViz is a simple non-linear transformation of the star coordinates layout. However, using anchor points as a visual depiction, it allows for an intuitive interpretation of similarity to different anchor points, which is exactly, what we want to achieve. For providing a self-contained paper, we briefly recapitulate the approach. Given m anchor points, RadViz locates the anchor points equidistantly on the unit circle. Given a data item that should be placed in the RadViz layout the m-dimensional similarity vector $\mathbf{x} \in \mathcal{R}^m$ to each of the m anchor points is computed. Then, the data item is associated with a projected point $\mathbf{p} \in \mathcal{R}^2$, whose location is computed by

$$\mathbf{p} = \frac{\sum_{i=1}^{m} \mathbf{x}_i \mathbf{u}_i}{\sum_{i=1}^{m} \mathbf{x}_i},$$

where $\mathbf{u}_i = \left(\cos(\frac{i \cdot 2\pi}{m}), \sin(\frac{i \cdot 2\pi}{m})\right)$.

The sorting of the anchor points along the unit circle has a significant impact on the readability of the RadViz visualization. Recently, Pagliosa and Telea [21] proposed an enhanced RadViz approach named RadViz++, where they sorted the anchor points based on the Pearson correlation of the multi-dimensional feature vectors that correspond to the anchor points (in our case, the metabolite concentrations of the different tumor types and normal tissue). The sorting is such that feature vectors with high correlations lead to anchor points placed next to each other on the unit circle.

Similarity Computation. What remains to be decided is how to compute the similarity of our voxels with the anchor points. This is a priori not clear and we suggest to test the two most common similarity computations for multi-dimensional points, i.e., similarity obtained from Euclidean distance and Pearson correlation.

For Euclidean similarity, we first compute the distance of each MRSI voxel to each tumor type (and normal tissue) in the multi-dimensional feature space. Then, the distance values are converted to similarity values by computing a difference between maximum distance and the distance value and dividing it by the maximum distance, leading to values in the range of [0, 1]. Since anchor points that are far away should not have an impact on the RadViz mapping of a given point, we set similarity values to zero, if the Euclidean distance is larger than three times the standard deviation of the tumor associated with the anchor point. Similarity value is set to zero if the distance is more then three time from the deviation value of that particular tumor.

For the Pearson similarity measure, we first compute the Pearson correlation between our MRSI voxel and each tumor types, which results in values in the range [−1, 1]. We neglect negative correlations by setting the similarity values to zero, while the positive correlation values are scaled by $s = \frac{r+1}{2}$, where r is the correlation value and s the similarity value.

5.2   Application Scenario, Results, and Discussion

In our application scenario we compared our findings to each of the three studies from literature. To test the effect of normalization and the choice of the similarity measure, we generate four RadViz layouts for each of the three studies, i.e., the combinations of Euclidean and Pearson similarity with and without normalization.

First Study. As mentioned above, we selected four brain regions as labels for our voxels, cf. Fig. 17. The RadViz visualizations use the same colors, see Fig. 18. Each point corresponds to one voxel. Red points correspond to tumor voxels, green points to voxel in the close surrounding of tumor, are belongs to tumor surrounding area, pink points correspond to gray-matter voxels, and purple points correspond to while-matter voxels. All not labeled voxels are shown in blue color. Those include background voxels and cerebrospinal fluid voxels. The anchor points representing the 6 tumor types and normal tissue taken form the literature (cf. Fig. 1) are labeled with black text.

Figures 18(a) and (b) show the results of Euclidean similarity measure with and without data normalization, respectively. We observe that all points clutter in the middle of the RadViz plot. One hypothesis was that the different tumor types are actually quite similar such that each voxel has a certain amount of similarity with all tumors. To test this hypothesis, we also project the mean

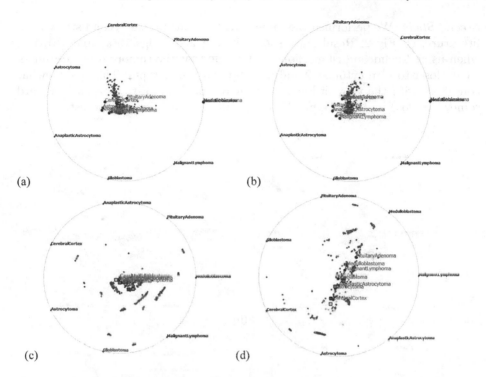

Fig. 18. RadViz visualizations for relating our MRSI data to tumor type characteristics obtained by the first study we compare against, see Fig. 1: (a) and (b) are computed using Euclidean similarity, while (c) and (d) are computed using Pearson correlation, both with (a, c) and without (b, d) normalization.

values of each tumor type in the RadViz layout. More precisely, we computed the similarity of the metabolite concentration means of a tumor type against all other tumor types (and itself). These similarity measures are then used to project the points, which are shown labeled with red text. We observe that the projected mean values of the tumor types also contribute to the clutter in the middle of the RadViz plot. Thus, we can conclude that the Euclidean similarities are indeed generally high.

We compare this against the Pearson similarity. Figures 18(c) and (d) show the results of Pearson similarity with and without data normalization, respectively. They are much less cluttered and exhibit some structures. In Fig. 18(c), tumor and its surrounding voxels are still close to the middle of the plot, while white- and gray-matter voxels show up in the middle but also in groups closer to the anchor points. Interestingly, the projected tumor types also gather in the middle, i.e., close to our tumor voxels. However, there is no observation as in what tumor type our tumor voxels are closest to. In Fig. 18(d), points are more spread, but there is still no clear trend visible.

Second Study. We performed the same investigations for the second study from literature, cf. Fig. 2. Results are shown in Fig. 2. In Figs. 19(a) and (b), i.e., when using Euclidean distance, one may argue that the tumor voxels are somewhat closer to astrocytomaG2 and meningioma tumor types. This is somewhat consistent with the view in Fig. 19(c), where anaplastic astrocytoma is a third somewhat close-by tumor type. Also, Fig. 19(d), meningioma is closest.

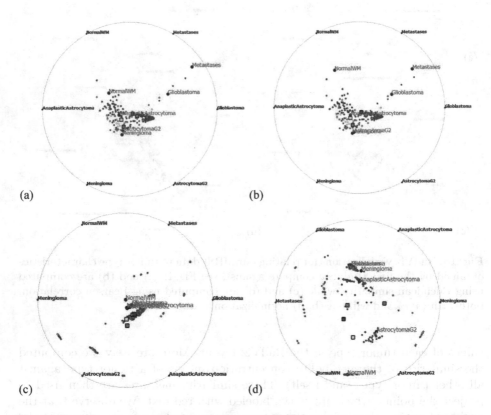

(a)   (b)

(c)   (d)

Fig. 19. RadViz visualizations for relating our MRSI data to tumor type characteristics obtained by the second study we compare against, see Fig. 2: (a) and (b) are computed using Euclidean similarity, while (c) and (d) are computed using Pearson correlation, both with (a, c) and without (b, d) normalization.

Third Study. Finally, we conducted the same investigations for the third study provided in literature, cf. Fig. 3. The same investigations were made as above. Results are presented in Fig. 20. Again, we observe that the Euclidean similarity results are rather cluttered in the middle, while the Pearson similarity exhibits more structures. For Pearson similarity, our tumor voxels are closest to meningioma and metastatic tumor types.

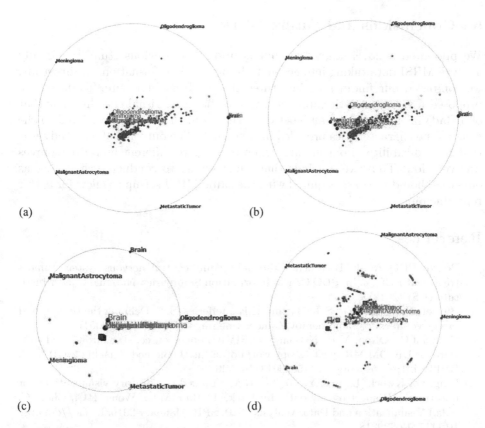

Fig. 20. RadViz visualizations for relating our MRSI data to tumor type characteristics obtained by the third study we compare against, see Fig. 3: (a) and (b) are computed using Euclidean similarity, while (c) and (d) are computed using Pearson correlation, both with (a, c) and without (b, d) normalization.

Discussion. There are multiple observations that can be made from the results presented above. First of all, we observe that the findings when comparing to different studies from literature only concur very weakly. Hence, it seems that the experimental set-up indeed matters a lot and that the reproducibility is low. Second, the data points tend to be rather close to the RadViz center, which indicates that the tumor types are somewhat close to each other. Third, Euclidean similarity led to more clutter, but the structures observed with Pearsin correlation are also hard to interpret. Fourth, the normalization should be a suitable means to make the data sets comparable, but we did not observe an obvious positive effect. Finally, our tumor voxels are not obviously closer to tumor anchor points than our normal voxels (white matter and gray matter) and the normal voxels are not obviously closer to the normal tissue anchor points.

6   Conclusions and Future Work

We presented a novel study that builds upon our previous study to visually analyze MRSI data taking into account all metabolic information. In this study, we compared our findings against three studies from literature. To do so, we proposed a visual analytics approach, where the main visual encoding was that of a RadViz layout, where all labeled voxels of our data could be related to the tumor type characteristics provided in literature. The outcome of our study was that reproducibility of our findings when relating to different literature sources was very low. Thus, we conclude that when trying to conduct such studies as ours one should use data acquired with the same MRSI set-up to allow for better reproducibility.

References

1. Board: PDQ Adult Treatment Editorial: Adult central nervous system tumors treatment PDQ®. In: PDQ Cancer Information Summaries. National Cancer Institute (US) (2018)
2. Burnet, N.G., Thomas, S.J., Burton, K.E., Jefferies, S.J.: Defining the tumour and target volumes for radiotherapy. Cancer Imaging 4(2), 153–161 (2004)
3. Crane, J.C., Olson, M.P., Nelson, S.J.: SIVIC: open-source, standards-based software for DICOM MR spectroscopy workflows. Int. J. Biomed. Imaging 2013, 1–12 (2013). https://doi.org/10.1155/2013/169526
4. Feng, D., Kwock, L., Lee, Y., Taylor, R.M.: Linked exploratory visualizations for uncertain MR spectroscopy data. In: Park, J., Hao, M.C., Wong, P.C., Chen, C. (eds.) Visualization and Data Analysis 2010. SPIE, January 2010. https://doi.org/10.1117/12.839818
5. Gujar, S.K., Maheshwari, S., Björkman-Burtscher, I., Sundgren, P.C.: Magnetic resonance spectroscopy. J. Neuro Ophthalmol. 25(3), 217–226 (2005)
6. Hoffman, P., Grinstein, G., Marx, K., Grosse, I., Stanley, E.: DNA visual and analytic data mining. In: Proceedings of IEEE Visualization 1997 (Cat. No. 97CB36155), pp. 437–441, October 1997. https://doi.org/10.1109/VISUAL.1997.663916
7. Howe, F., et al.: Metabolic profiles of human brain tumors using quantitative in vivo ^{1}H magnetic resonance spectroscopy. Magn. Reson. Med. 49(2), 223–232 (2003). https://doi.org/10.1002/mrm.10367
8. Jawad., M., Molchanov., V., Linsen., L.: Coordinated image- and feature-space visualization for interactive magnetic resonance spectroscopy imaging data analysis. In: Proceedings of the 14th International Joint Conference on Computer Vision, Imaging and Computer Graphics Theory and Applications - Volume 3: IVAPP, pp. 118–128. INSTICC, SciTePress (2019). https://doi.org/10.5220/0007571801180128
9. Johnson, H.J., McCormick, M., Ibáñez, L., Consortium, T.I.S.: The ITK Software Guide, 3rd edn. Kitware Inc. (2013). http://www.itk.org/ItkSoftwareGuide.pdf
10. Kandogan, E.: Star coordinates: a multi-dimensional visualization technique with uniform treatment of dimensions. In: Proceedings of the IEEE Information Visualization Symposium, Late Breaking Hot Topics, pp. 9–12 (2000)
11. Kinoshita, Y., Yokota, A.: Absolute concentrations of metabolites in the human brain tumors using in vitro proton magnetic resonance spectroscopy. NMR Biomed. 10(1), 2–12 (1997)

12. Li, C., Gore, J.C., Davatzikos, C.: Multiplicative Intrinsic Component Optimization (MICO) for MRI bias field estimation and tissue segmentation. Magn. Reson. Imaging **32**(7), 913–923 (2014)
13. Liu, S., Maljovec, D., Wang, B., Bremer, P., Pascucci, V.: Visualizing high-dimensional data: advances in the past decade. IEEE Trans. Vis. Comput. Graph. **23**(3), 1249–1268 (2017). https://doi.org/10.1109/TVCG.2016.2640960
14. Matkovic, K., Freiler, W., Gracanin, D., Hauser, H.: ComVis: a coordinated multiple views system for prototyping new visualization technology. In: 2008 12th International Conference Information Visualisation, pp. 215–220. IEEE (2008)
15. Maudsley, A., et al.: Comprehensive processing, display and analysis for in vivo MR spectroscopic imaging. NMR Biomed. **19**(4), 492–503 (2006)
16. McKnight, T.R., Noworolski, S.M., Vigneron, D.B., Nelson, S.J.: An automated technique for the quantitative assessment of 3D-MRSI data from patients with glioma. J. Magn. Reson. Imaging Official J. Int. Soc. Magn. Reson. Med. **13**(2), 167–177 (2001)
17. Molchanov, V., Linsen, L.: Interactive design of multidimensional data projection layout. In: Elmqvist, N., Hlawitschka, M., Kennedy, J. (ed.) EuroVis - Short Papers. The Eurographics Association (2014) https://doi.org/10.2312/eurovisshort.20141152
18. Molchanov, V., Linsen, L.: Shape-preserving star coordinates. IEEE Trans. Visual Comput. Graphics **25**(1), 449–458 (2019). https://doi.org/10.1109/TVCG.2018.2865118
19. Nunes, M., Laruelo, A., Ken, S., Laprie, A., Bühler, K.: A survey on visualizing magnetic resonance spectroscopy data. In: Proceedings of the 4th Eurographics Workshop on Visual Computing for Biology and Medicine, pp. 21–30. Eurographics Association (2014)
20. Nunes, M., et al.: An integrated visual analysis system for fusing MR spectroscopy and multi-modal radiology imaging. In: 2014 IEEE Conference on Visual Analytics Science and Technology (VAST), pp. 53–62. IEEE (2014)
21. Pagliosa, L., Telea, A.: Radviz++: improvements on radial-based visualizations. Informatics **6**(2), 16–38 (2019). https://doi.org/10.3390/informatics6020016
22. Peeling, J., Sutherland, G.: High-resolution ^{1}H NMR spectroscopy studies of extracts of human cerebral neoplasms. Magn. Reson. Med. **24**(1), 123–136 (1992). https://doi.org/10.1002/mrm.1910240113
23. Provencher, S.W.: Estimation of metabolite concentrations from localized in vivo proton NMR spectra. Magn. Reson. Med. **30**(6), 672–679 (1993)
24. Raschke, F., Jones, T., Barrick, T., Howe, F.: Delineation of gliomas using radial metabolite indexing. NMR Biomed. **27**(9), 1053–1062 (2014)
25. Reynolds, G., Wilson, M., Peet, A., Arvanitis, T.: An algorithm for the automated quantitation of metabolites in in-vitro NMR signals. Magn. Reson. Med. **56**(6), 1211–1219 (2006)
26. Rowland, B., et al.: 30th Annual Scientific Meeting Beyond the Metabolic Map: An Alternative Perspective on MRSI Data, ESMRMB 2013, p. 270 (2013)
27. Scheenen, T.W., Heerschap, A., Klomp, D.W.: Towards ^{1}H-MRSI of the human brain at 7T with slice-selective adiabatic refocusing pulses. Magn. Reson. Mater. Phys., Biol. Med. **21**(1–2), 95–101 (2008)
28. Schroeder, W., Martin, K., Lorensen, B., Avila, S.L., Avila, R., Law, C.: The Visualization Toolkit, 4th edn. Kitware Inc. (2006). https://www.vtk.org/vtk-textbook/
29. Stefan, D., et al.: Quantitation of magnetic resonance spectroscopy signals: the jMRUI software package. Meas. Sci. Technol. **20**(10), 104035 (2009)

30. Teoh, S.T., Ma, K.L.: StarClass: interactive visual classification using star coordinates. In: SDM, pp. 178–185. SIAM (2003)

31. Vanhamme, L., Sundin, T., Hecke, P.V., Huffel, S.V.: MR spectroscopy quantitation: a review of time-domain methods. NMR Biomed. **14**(4), 233–246 (2001)

32. Wilson, M., Reynolds, G., Kauppinen, R.A., Arvanitis, T.N., Peet, A.C.: A constrained least-squares approach to the automated quantitation of in vivo ^{1}H magnetic resonance spectroscopy data. Magn. Reson. Med. **65**(1), 1–12 (2011)

33. Wolf, I., et al.: The medical imaging interaction ToolKit MITK: a toolkit facilitating the creation of interactive software by extending VTK and ITK, vol. 5367, p. 5367 (2004). https://doi.org/10.1117/12.535112

Computer Vision Theory and Applications

A Self-regulating Spatio-Temporal Filter for Volumetric Video Point Clouds

Matthew Moynihan[1]([✉]) [iD], Rafael Pagés[1,2] [iD], and Aljosa Smolic[1] [iD]

[1] V-SENSE, Trinity College Dublin, Dublin, Ireland
{mamoynih,smolica}@tcd.ie
[2] Volograms, Dublin, Ireland
rafa@volograms.com
https://v-sense.scss.tcd.ie/people/
http://www.volograms.com

Abstract. The following work presents a self-regulating filter that is capable of performing accurate upsampling of dynamic point cloud data sequences captured using wide-baseline multi-view camera setups. This is achieved by using two-way temporal projection of edge-aware upsampled point clouds while imposing coherence and noise filtering via a windowed, self-regulating noise filter. We use a state of the art Spatio-Temporal Edge-Aware scene flow estimation to accurately model the motion of points across a sequence and then, leveraging the spatio-temporal inconsistency of unstructured noise, we perform a weighted Hausdorff distance-based noise filter over a given window. Our results demonstrate that this approach produces temporally coherent, upsampled point clouds while mitigating both additive and unstructured noise. In addition to filtering noise, the algorithm is able to greatly reduce intermittent loss of pertinent geometry. The system performs well in dynamic real world scenarios with both stationary and non-stationary cameras as well as synthetically rendered environments for baseline study.

Keywords: Point clouds · Upsampling · Temporal coherence · Free viewpoint video · Multiview video · Volumetric video

1 Spatio-Temporal Coherence in Volumetric Video

As the popularity of VR and AR consumer devices continues to grow, we can naturally expect an increase in the demand for engaging and aesthetic mixed reality content. The barrier to entry for creative enthusiasts and content creators has begun to decline as more digital frameworks supporting VR/AR become available, however, performance capture and reconstruction of real-world scenes still remains largely out of reach for amateur productions.

This publication has emanated from research conducted with the financial support of Science Foundation Ireland (SFI) under grant No. 15/RP/2776. We gratefully acknowledge the support of NVIDIA Corporation with the donation of the Titan Xp GPU used for this research.

© Springer Nature Switzerland AG 2020
A. P. Cláudio et al. (Eds.): VISIGRAPP 2019, CCIS 1182, pp. 391–408, 2020.
https://doi.org/10.1007/978-3-030-41590-7_16

Using Free-Viewpoint Video (FVV) or, more specifically Volumetric Video (VV), content creators have the technology to record and reconstruct performances in dynamic real-world scenarios. However, these captures are often restricted by the constraints of highly controlled studio environments, requiring dense arrays of high-resolution RGB cameras and IR depth sensors [5, 20]. The reconstruction is usually done in a frame-by-frame manner beginning with the construction of a dense point cloud via Multi-View Stereo (MVS). For such high-budget studios with very dense coverage of the subject, temporal inconsistencies in the resulting point cloud may be visually negligible after the final meshing and tracking process. Yet where low-budget content creation is concerned, such high density coverage may not be achievable and thus any spatio-temporal inconsistencies can become magnified and visually unappealing by the end of the reconstruction pipeline.

In order to address to the demand for low-cost VV and performance capture, new systems have been proposed which enable VV content creation solely on consumer-grade RGB cameras and even hand-held personal devices [30]. However, any such system which features framewise reconstruction [27] will contain spatio-temporal artifacts in the resulting volumetric sequence. This is usually a consequence of some inherent fail cases for photogrammetry-based techniques whereby the subject can contain highly reflective surfaces or a lack of textured material.

Without accounting for spatio-temporal variance, otherwise pertinent geometric features become distorted and inconsistent across VV sequences. This is especially true in the case of small or thin details such as hands or arms. An example of which can be seen in Fig. 1 where the naive frame-by-frame reconstruction fails to distinguish temporally persistent features as portions of limbs lack persistence. Furthermore large sections of geometry in relatively untextured areas may be intermittently present depending on the success of the point cloud reconstruction for that given frame. Structured noise patches can also be intermittently observed.

The proposed system is an expansion to the work presented in [24] that is able to spatio-temporally upsample a point cloud sequence captured via wide-baseline multi-view setups and further support the self-regulating noise filter metric using a new windowed approach to sampling. In summary, the following work proposes:

– A spatio-temporally coherent point cloud sequence upsampling algorithm that selectively merges point cloud projections within a variable window. The projections of which are computed iteratively using a pseudo-scene flow estimate.
– An autonomously regulated noise filter supported by a density-weighted energy term for averaging within a window of frames.

We perform a baseline comparison on the work presented in [24] as well as previous examples.

2   Previous Work

One of the fundamental processes in modern VV pipelines is spatio-temporal consistency. Ensuring this consistency across the sequence of 3D models helps reduce the impact of small geometry differences among frames and surface arti-facts, which result in temporal flickering when rendering the VV sequence. Most techniques apply a variation of on the non-rigid ICP algorithm [19,38], such as the coherent drift point method [29], performing a geometric temporal constraint to align the meshes resulting from the 3D reconstruction process on a frame-by-frame basis [14,17]. This works specially well when the 3D models acquired for every frame are detailed and accurate, as registration algorithms are not always robust to big geometry differences or loss of portions of the mesh (something that can often happen for human limbs). A good example of this is the system by Collet et al. [5]: they apply mesh tracking in the final processing stage, both to provide a smoother VV sequence and also to improve data storage efficiency as, between keyframes, only the vertex positions vary while face indices and texture coordinates remain the same. They achieve very appealing results by utilizing a sophisticated, very dense camera setup of over 100 sensors (RGB and IR), ensuring a high degree of accuracy for the reconstructed point clouds on a frame-to-frame basis. This type of temporal consistency is also key in the methods proposed by Dou et al. [7,8], where they are able to perform registration in real-time, using data coming from depth sensors. These methods ensure temporal consistency at the end of their pipeline, but differently to the method proposed, they do not address the loss of geometry in the capture stage, which can only be solved using temporal coherence at the point cloud generation.

Fig. 1. Input dense point clouds generated using an affordable volumetric video capturing platform [30]. Even after densification via multi-view stereo, the input clouds still exhibit large gaps in structure as well as patches of noise.

Mustafa et al. [25] ensure temporal consistency of their VV sequences by first, using sparse temporal dynamic feature tracking as an initial stage, followed by a shape constraint based on geodesic star convexity for the dense model. These temporal features are used to initialize a constraint which refines the alpha masks used in visual-hull carving and are not directly applied to the input point cloud. The accuracy of their results is not comparable with the methods mentioned above, but they show good performance with a reduced number of viewpoints and wide baseline. Mustafa et al. extended their work to include sequences that are not only temporally but also semantically coherent [26], and even light-field video [28].

An interesting way of pursuing spatio-temporal consistency is by using optical flow. For example, Prada et al. [31] use mesh-based optical flow for adjusting the tracking drift when generating texture atlases for the VV sequence, adding an extra layer of spatio-temporal consistency at the texturing step. It is possible to address temporal coherence by trying to use the scene flow to recover not only motion, but also depth. Examples of this are the works by Basha et al. [2] and Wedel et al. [35]. These techniques require a very dense and accurate motion estimation for every pixel to acquire accurate depth maps, together with a camera setup with a very narrow baseline. Alternatively, our system uses the temporally consistent flow proposed by Lang et al. [18] applied to multi-view sequences, allowing us to track dense point clouds across the sequence even with a wide baseline cameras configurations.

Other ways of improving incomplete 3D reconstructions, such as the ones acquired with wide baseline camera setups, include upsampling or densifying [15,36,37] them in a spatially coherent way. These systems are designed to perform upsampling for a single input point cloud, and not specifically a VV sequence, so they are unable to leverage any of the temporal information within a given sequence of point clouds. As a result, the use of such techniques alone will still suffer from temporally incoherent errors. Our system takes advantage of the geometric accuracy of the state of the art Edge-Aware Point Set Resampling technique proposed by Huang et al. [15] and supports it using the temporal information obtained from the inferred 3D scene flow along with some spatio-temporal noise filtering. The reasoning behind this approach being that increasing the density of coherent points improves the accuracy of surface reconstruction algorithms such as Poisson Surface Reconstruction [16] and thus, propagates visual improvement through th VV pipeline.

3   Proposed System

3.1   Point Cloud Generation and Upsampling

We use a low-cost VV pipeline similar to the system by [30] to generate the input clouds for the proposed algorithm. Such pipelines generally maximise the baseline between cameras in order to reduce the cost of extra hardware while still providing full coverage of the subject. The camera intrinsics are assumed

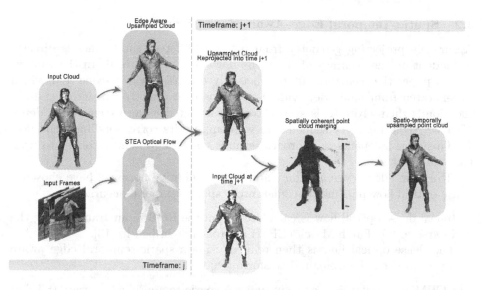

Fig. 2. Proposed pipeline: The input to the algorithm requires a sequence of temporally independent point clouds along with the corresponding RGB images and calibration data. At timeframe j, the input cloud is upsampled and projected into the subsequent frame $t+1$. This is done via an edge-aware scene flow generated from the input RGB images. Expanding on [24], this is performed iteratively across a window of frames centered about the input frame i.e. we recursively project frames within the given window toward the center frame. The output consists of a spatio-temporally coherent merge and averaging system which upsamples the input point clouds and filters against temporal noise.

to be known from prior calibration while extrinsics can be calculated automatically using sparse feature matching and incremental structure-from-motion [23]. In some cases the cameras may be handheld, whereby more advanced techniques like CoSLAM [39] can be applied to better produce dynamic poses. The input sparse clouds are further densified using multi-view stereo. The examples presented within the context of our system use the sparse point cloud estimation system by Berjón et al. [3] and are then further densified by using the unstructured MVS system of Schönberger et al. [34]. Formally, we define $S = \{s_{i=1}, ..., s_m\}$ as the set of all m video sequences, where $s_i(j)$, $j \in \{1, ..., J\}$ denotes the jth frame of a video sequence $s_i \in S$, with J frames. Then for every frame j, there will be an estimated point cloud $\mathcal{P}_j$. In a single iteration, $\mathcal{P}_j$ is taken as the input cloud which is upsampled using Edge-Aware Resampling (EAR) [15]. This initializes the geometry recovery process with a densified point cloud prior which will be temporally projected into the next time frame $j+1$ and geometrically filtered to ensure both temporal and spatial coherence. With the windowed filtering approach this iteration is performed recursively in such a way that each frame within the window is iteratively projected toward the center frame via it's respective intermediate frames (Fig. 2).

3.2   Spatio-Temporal Edge-Aware Scene Flow

Accurately projecting geometry from between different timeframes is directly dependent on the accuracy of the scene flow used to achieve it. In the context of this paper the scene flow used is actually a dense, pseudo-scene flow which is generated from multi-view videos as opposed to directly extracting it from the clouds themselves. This scene flow is calculated as an extension to dense 2D flow, thus, for every sequence s_i we compute its corresponding scene flow f_i. This view-independent approach ensures that the system is robust to wide baseline input.

To retain edge-aware accuracy and reduce additive noise we have chosen a dense optical flow pipeline that guarantees spatio-temporal accuracy:

- Initial dense optical flow is calculated from the RGB input frames using the Coarse to fine Patch Match (CPM) approach described in [13].
- The dense optical flow is then refined using a spatio-temporal edge aware filter based on the Domain Transform [18].

The CPM optical flow is used to initialize a spatio-temporal edge aware (STEA) filter which regularizes the flow across a video sequence, further improving edge-preservation and noise reduction.

While the STEA can be initialized with most dense optical flow techniques such as the popular Gunnar-Farnebäck algorithm [9], the proposed system uses the coarse-to-fine patch match algorithm by [13] as recommended in [33]. Table 1 provides a breakdown of the amount of pertinent geometry recovered via different optical flow techniques.

Table 1. Investigation by [24] on the effect of STEA filter initialization on geometry recovered expressed as % increase in points. Tested on a synthetic ground-truth sequence. Flow algorithms tested: Coarse-to-Fine Patch Match (CPM) [13], Fast Edge-Preserving Patch Match (FEPPM) [1], Pyramidial Lukas-Kanade (PyLK) [4] and Gunnar-Farnebäck (FB) [9].

STEA initialization	Area increase (%)
CPM	**37.73**
FEPPM	34.9
PyLK	34.77
FB	29.7

The STEA filter consists of the following implementation as in [18]. This implementation further builds upon the Domain Transform [11] extending into the spatial and temporal domains given the optical flow as the target application:

1. The filter is initialized as in [33], using coarse-to-fine patch match [13]. The CPM algorithm estimates optical flow as a quasi-dense nearest neighbour field (NNF) using a subsampled grid.

2. The edges of the RGB input are then calculated using the Structure Edge Detection Toolbox [6].
3. Using the calculated edges, the dense optical flow is then interpolated using Edge-Preserving Interpolation of Correspondences [32].

This dense optical flow field is then regulated by the STEA filter via multiple spatio-temporal domain iterations to reduce temporal noise. Figure 3 visualizes the intermediate stages of the flow processing pipeline.

3.3   Scene Flow Point Projection

Given known per-camera intrinsics $(C_{j_1}, ..., C_{j_m},$ at timeframe j), the set of scene flows $(f_{j_1}, ..., f_{j_m})$, and the set of point clouds $(\mathcal{P}_j, ..., \mathcal{P}_J)$, the motion of any given point across a sequence can be estimated. To achieve this, each point is back-projected $\mathbf{P}_k \in \mathcal{P}_j$ to each 2D flow f_i at that specific frame j. We check the sign of the dot product between the camera pointing vector and the normal of the point $\mathbf{P}_k$ to prune any point projections which may otherwise have been occluded for the given view. Using the flow, we can predict the position of the back-projected 2D points $\mathbf{p}_{ik}$ in sequential frames, $\mathbf{p}'_{ik}$.

The set of projected 3D points $\mathcal{P}'_j$, at frame $j + 1$, is then acquired by triangulating the flow-projected 2D points $\mathbf{p}'_{ik}$, using the camera parameters of frame $j + 1$. This is done by solving a set of overdetermined homogeneous systems in the form of $H\mathbf{P}'_k = \mathbf{0}$, where $\mathbf{P}'_k$ is the estimated 3D point and matrix H is

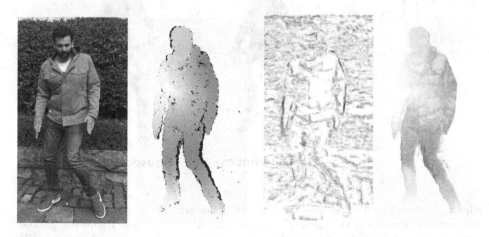

Fig. 3. From left to right, dense optical flow calculation: For a particular viewpoint, the input RGB image, (1) nearest neighbour field estimate from CPM, (2) SED detected edges, (3) interpolated dense STEA output. Conventional colour coding has been used to illustrate the orientation and intensity of the optical flow vectors. Orientation is indicated by means of hue while vector magnitude is proportional to the saturation i.e. negligible motion is represented by white, high-speed motion is shown in highly saturated color [24].

defined by the Direct Linear Transformation algorithm [12]. The reprojection error is minimized using a Gauss-Markov weighted non-linear optimisation [22].

3.4   Windowed Hausdorff Filter

The aforementioned point cloud projection framework can now be used to support the coherent merging and noise filtering process. For a given window of width w for frames $\{j_{(c-w/2)} \cdots j_c \cdots j_{(c+w/2)}\} \subset J$ where c is the center frame, we project the point cloud at each frame towards the center frame using the above method in a recursive manner. In this way structural information is retained and propagated. However, this also has the effect of accumulating any inherent noise within this window. For this reason we extend the two-way Hausdorf filter in [24] with the addition of an energy density term E_{dens}. This density term takes into account the average voxel density of the merged window of frames which is essentially the sum of the propagated clouds. Using density as a conditioning term leverages the temporal inconsistency in that statistically, occupancy due to noise is far less common than occupancy due to pertinent geometry.

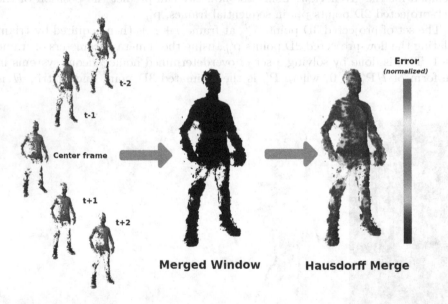

Fig. 4. The windowed merge process. Left: a 5-frame window of input clouds, Middle: the cumulative merge of the upsampled and projected input clouds. Right: the filtered merge process visualized with normalized error given by distance of each point to it's corresponding match in the input cloud. This error term is then augmented with the energy terms $E_{dynamic}$ and E_{dens}.

The coherent merged cloud $\mathcal{P}^*_{j+1}$ is given by the logical definition in Eq. 1 where $D_{\mathcal{P}'_j}$ is the summed result of projecting all point clouds within window w recursively toward the center frame j.

Given an ordered array of values $D_{\mathcal{P}'_j}$ such that $D_{\mathcal{P}'_j(k)}$ is the distance from point $\mathcal{P}_j(k)'$ to its indexed match in $\mathcal{P}_{j+1}$. We also define $D_{\mathcal{P}_{j+1}}$ as an array of distances in the direction of $\mathcal{P}_{j+1}$ to $\mathcal{P}'_j$. We then define the merged cloud to be the union of two subsets $M \subset \mathcal{P}'_j$ and $T \subset \mathcal{P}_{j+1}$ such that,

$$M \subset \mathcal{P}'_j \ \forall \ \mathcal{P}'_j(k) \ : \ D_{\mathcal{P}'_j(k)} < d_j, \ k \in \{1...j\},$$
$$T \subset \mathcal{P}_{j+1} \ \forall \ \mathcal{P}_{j+1}(k) \ : \ D_{\mathcal{P}_{j+1}}(k) < d_j, \ k \in \{1...j\}, \tag{1}$$
$$\mathcal{P}^*_{j+1} = M \cup T$$

By this definition, $\mathcal{P}^*_{j+1}$ contains only the points in $\mathcal{P}_{j+1}$ and $\mathcal{P}'_j$ whose distance to their nearest neighbour in the other point cloud is less than the computed threshold d_j. The intention of this design is effectively to remove any large outliers and incoherent points while encouraging consistent and improved point density. Figure 4 shows an example of how the coherent merge works.

3.5 Dynamic Motion Energy Term

Due to the distance-based nature of the Hausdorff-based filter, it is often observed that fast-moving objects are pruned after being projected into the next frame. This approach to filtering greatly reduces the amount of temporally inconsistent noise, but simultaneously, it over-filters dynamic objects due to the lack of spatial overlap between frames. This is especially true for sequences captured at 30 fps or less, which is often the case for affordable VV setups where bandwidth and storage are concerned. To address this issue, we supplement the distance-based threshold term with a dynamic motion energy which is designed to add bias towards fast-moving objects. This energy term is proportional to the average motion observed across the scene-flow estimates for a given timeframe. For faster-moving objects, higher confidence is assigned to clusters of fast-moving points. Given that $\mathcal{P}'_j$ is a prediction for frame $j + 1$, we validate each predicted point by back-projecting $\mathcal{P}'_j$ into the respective scene flow frames for time $j + 1$. The flow values for the pixels in each view is then averages to calculate the motion for a given pixel at that time. As in Sect. 3.3 we again filter out occluded points using the dot product of the camera pointing vector and the point normal.

3.6 Spatio-Temporal Density Term

The proposed system offers an expansion to the two-way Hausdorff-based filter presented in [24] by sampling a window of frames about the current timestamp. While the two-way filter is robust to temporal noise it isn't capable of recovering large sections of missing geometry over a spanning timeframe. As illustrated in Fig. 6, the two-way approach fails to recover much from the sequence where large patches are missing over a longer time period. To address this, the proposed system introduces a windowed approach which combines the projected information from multiple frames while retaining comparable noise filtering. In order to reduce the added noise we propose an additional energy term for the filtering threshold based on patio-temporal density within the given window. The new threshold score criteria is then given by:

$$E_{th} = d - (E_{dens} + E_{dynamic}) \qquad (2)$$

The E_{dens} term is calculated as follows:

- For a window of width w we iteratively project each frame into the current timestamp such that a single point cloud object is created consisting of the points projected from the frame range $\{t_{(c-w/2)}...t_{c}...t_{(c+w/2)}\}$

- An octree-based occupancy grid is then constructed on this object where each leaf is assigned a normalized density score. This score is the E_{dens} term for any point given its index within the occupancy grid.

Figure 4 illustrates this process for any given window. The size of this window is variable but is limited by practical limitations of computation time and the trade-off of adding multiple sources of noise. For our purposes we concluded that a window size of 5 was within practical time constraints while still providing good results. As with any filtering or averaging algorithm, there is an inherent risk of over-smoothing data and thus, such decisions may differ for various sequences depending on the degree of dynamic motion.

4   Results

In Fig. 5 we demonstrate a side-by-side comparison of the process results vs unprocessed input for two challenging yet conventional scenarios. We evaluate the system on a number of sequences captured outdoors with as little as 6 to 12 handheld devices (i.e. smartphones, tablets etc.) as well as a controlled green screen environment comprised of 12 high-end, rigidly mounted cameras (6 4K resolution, 6 Full HD). A ground-truth comparison is also presented by comparing reconstruction results against a known synthetic model within a virtual environment with rendered cameras.

4.1   Outdoor Handheld Camera Sequences

Shooting outdoors with heterogenous handheld devices can present a number of challenging factors including: non-uniform dynamic backgrounds, increased margin of error for intrinsics and extrinsics calculations, instability of automatic foreground segmentation methods and more. The cumulative effect of these factors results in temporal inconsistencies with the reconstructed point cloud sequence as well as the addition of structured noise and omission of pertinent geometry. Figure 5 (left model) shows the difference between using framewise reconstruction (a) and the proposed system (b). A significant portion of structured noise has been removed whilst also managing to fill-in gaps in the subject.

(a)   (b)   (a)   (b)

Fig. 5. An example of the proposed upsampling and filtering system. Pictured left: a sequence captured outdoors with handheld devices. Pictured right: a sequence captured in a low-cost controlled studio environment with fast-moving objects. For both sequences, (a) corresponds to the input cloud prior to filtering while (b) represents the upsampled and filtered result [24].

To further demonstrate the impact of our system targeting volumetric reconstruction, we present the effect of applying screened poisson surface reconstruction (PSR) [16] to the input point cloud. In general, the direct application of PSR creates a fully closed surface which usually creates bulging or "inflated-looking" surface meshes. Instead we use the input cloud to prune outlying faces from the PSR mesh such that the output surface mesh more accurately represents the captured data. Thus, in Fig. 6 the gaps in the input data can be visualized clearly. This figure also shows the appreciable increase in pertinent surface area after spatio-temporal upsampling.

4.2   Indoor Studio Sequences

In general, sequences shot in controlled studio environments exhibit far less temporal noise and structural inconsistencies in comparison to "in-the-wild" dynamic outdoor shots. To further test our system we introduce an extra degree of challenge in the form of multiple, fast-moving objects while still using no more than 12 cameras for full, 360-degree coverage. This introduces further difficulty due to occlusions caused when the ball passes in front of performer as well as testing the limits of the flow-based projection system. In spite of these challenges, the proposed system is still able to filter a lot of the noise generated and can recover a modest amount of missing geometry, Fig. 5, (right model).

Fig. 6. A non-sequential set of frames from an outdoor VV shoot using handheld cameras. (Top): The RGB input to the system. (Middle): The result of applying poisson reconstruction to the unprocessed, temporally incoherent point clouds. (Bottom): The same poisson reconstruction method applied to the upsampled and filtered output of the proposed system [24].

4.3   Synthetic Data Sequences

As a baseline for ground-truth quantitative benchmarking, we evaluate our system using a synthetic virtual scenario. This synthetic data consists of a short

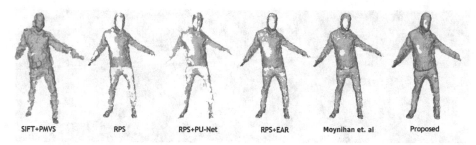

| SIFT+PMVS | RPS | RPS+PU-Net | RPS+EAR | Moynihan et. al | Proposed |

Fig. 7. A qualitative comparison of surface areas recovered from PSR meshing of point clouds from comparable systems. All meshes were created using the same octree depth for PSR and same distance threshold for outlier removal. From left to right: SIFT+PMVS [10,21], RPS [30], RPS+PU-Net [37], RPS+EAR [15], Proposed system applied in two-frame, forward direction only [24], the proposed system with windowed temporal filter centered on a window of 5 frames.

sequence featuring a human model performing a simple animated dance within a realistic environment. 12 virtual cameras were evenly spaced around a 180° arc centered about the animated character model. The images rendered from these virtual cameras provided the input to the VV systems for testing. Using this data we compare our results with those of temporally incoherent VV systems by applying PSR to the output point clouds and using the Hausdroff distance as an error metric. This is shown in Fig. 9.

We compare our results against similar framewise point cloud reconstruction systems, SIFT+PMVS [10] and RPS [30] as well as some state of the art upsampling algorithms for which we provide the method of RPS as input; PU-Net [37] and the Edge-Aware Resampling [15] method. Benchmarking against RPS+EAR also provides a form of ablation study for the effect of the proposed method as this is the approach used to initialize the system.

The proposed system demonstrates an overall improvement in quality in Table 2 yet the synthetic dataset lacks the noise which would be inherent to data captured in a real-world scenario. We would expect further improvements in such a scenario where the input error for the framewise reconstruction systems would be higher. Figure 7 qualitatively shows the effect of applying the proposed system to much noisier input data.

4.4 Flow Initialization

While practically any dense optical flow approach can be used to initialize the STEA filter in Sect. 3.2, improvements can be achieved by application-appropriate initialization. We show the results of initializing the STEA filter with CPM against other dense-flow alternatives in Table 1. The advanced edge-preservation of CPM results in it out-performing the alternatives but comparable results can achieved using GPU-based alternatives which may somewhat trade off accuracy for speed [1] (Fig. 8).

Fig. 8. An animated character model within a realistic virtual environment to generate synthetic test data [24].

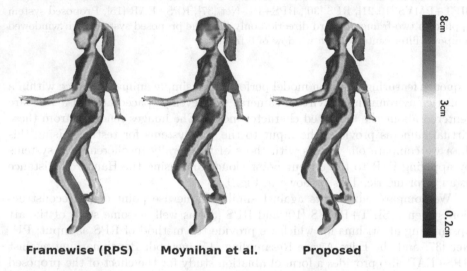

Framewise (RPS)   Moynihan et al.   Proposed

Fig. 9. Ground-truth evaluation of the proposed system against the virtual reference model using Hausdorff distance as the error metric. The left model shows a frame generated using framewise reconstruction [30], the middle model is the forward-projection, two frame filter [24], while the right shows the proposed systems [24] for a filtering window of 5 frames.

5   Limitations and Future Work

Due to the temporal nature of the algorithm, it is not possible to directly parallelize the proposed system as the most accurate scene flow is generated by providing the full length of the video sequence. Yet, if parallelism is a necessity, a compromise can be achieved in the form of a keyframe-based system whereby the input timeline is divided in reasonably-sized portions. Future work may employ some automatic keyframe detection which could maximise inter-keyframe similarity.

Table 2. Synthetic baseline comparison between the proposed method and similar state of the art approaches. Figures represent the Hausdorff distance metric with respect to the bounding box diagonal of the ground truth (%) [24].

Method	Mean error (%)	RMS error (%)
SIFT+PMVS	6.18	8.09
RPS	2.17	3.27
RPS + PU-Net	2.44	3.50
RPS + EAR	2.40	3.64
Moynihan et al.	1.78	2.72
Proposed	**1.56**	**2.30**

6 Conclusions

As the barrier to entry for VV content creation lowers, we still see a large disparity between content from affordable systems and that from high-budget studios. Sparse and dynamic, in-the-wild studio setups will always have to overcome the characteristic spatio-temporal errors of systems which continue to lower the cost to entry while maintaining creative freedom. These limitations are difficult to overcome but we have demonstrated that improvements are achievable by extending upsampling and filtering techniques into the spatio-temporal domain.

Our approach can efficiently filter temporally incoherent noise without over-correcting for otherwise pertinent geometry. We also demonstrate the ability to perform a framewise upsampling which not only creates new coherent geometry but also propagates existing, spatially-coherent geometry across a variable frame window. This expansion to [24] shows improved results over the framewise, two-way projection and filter.

The most appreciable results emerge for the most challenging sequences. Handheld, outdoor VV captures tend to be the most error prone and thus stand to benefit the most from the proposed upsampling and filtering method as can be seen in the qualitative results presented. However, despite being less susceptible to error our qualitative analysis via synthetic ground truth data shows a marked improvement over the framewise approach. We also demonstrate an improvement over the work presented in [24] with the addition of a variable temporal window that further explores the persistence of coherent geometry against noise.

Overall we present an efficient method for improving the quality of greatly constrained VV capture setups in order to meet the growing demand for affordable virtual and augmented reality content creation.

References

1. Bao, L., Yang, Q., Jin, H.: Fast edge-preserving PatchMatch for large displacement optical flow. In: Proceedings of the IEEE Conference on Computer Vision and Pattern Recognition, pp. 3534–3541 (2014)
2. Basha, T., Moses, Y., Kiryati, N.: Multi-view scene flow estimation: a view centered variational approach. Int. J. Comput. Vision 101(1), 6–21 (2013)
3. Berjón, D., Pagés, R., Morán, F.: Fast feature matching for detailed point cloud generation. In: 2016 6th International Conference on Image Processing Theory Tools and Applications (IPTA), pp. 1–6. IEEE (2016)
4. Bouguet, J.Y.: Pyramidal implementation of the affine Lucas-Kanade feature tracker. Intel Corporation (2001)
5. Collet, A., et al.: High-quality streamable free-viewpoint video. ACM Trans. Graph. (ToG) 34(4), 69 (2015)
6. Dollár, P., Zitnick, C.L.: Structured forests for fast edge detection. In: 2013 IEEE International Conference on Computer Vision (ICCV), pp. 1841–1848. IEEE (2013)
7. Dou, M., et al.: Motion2fusion: real-time volumetric performance capture. ACM Trans. Graph. (TOG) 36(6), 246 (2017)
8. Dou, M., et al.: Fusion4d: real-time performance capture of challenging scenes. ACM Trans. Graph. (TOG) 35(4), 114 (2016)
9. Farnebäck, G.: Two-frame motion estimation based on polynomial expansion. In: Bigun, J., Gustavsson, T. (eds.) SCIA 2003. LNCS, vol. 2749, pp. 363–370. Springer, Heidelberg (2003). https://doi.org/10.1007/3-540-45103-X_50
10. Furukawa, Y., Ponce, J.: Accurate, dense, and robust multiview stereopsis. IEEE Trans. Pattern Anal. Mach. Intell. 32(8), 1362–1376 (2010)
11. Gastal, E.S., Oliveira, M.M.: Domain transform for edge-aware image and video processing. ACM Trans. Graph. (ToG) 30, 69 (2011)
12. Hartley, R.I., Zisserman, A.: Multiple View Geometry in Computer Vision. Cambridge University Press, New York (2004)
13. Hu, Y., Song, R., Li, Y.: Efficient coarse-to-fine PatchMatch for large displacement optical flow. In: Proceedings of the IEEE Conference on Computer Vision and Pattern Recognition, pp. 5704–5712 (2016)
14. Huang, C.H., Boyer, E., Navab, N., Ilic, S.: Human shape and pose tracking using keyframes. In: Proceedings of the IEEE Conference on Computer Vision and Pattern Recognition, pp. 3446–3453 (2014)
15. Huang, H., Wu, S., Gong, M., Cohen-Or, D., Ascher, U., Zhang, H.: Edge-aware point set resampling. ACM Trans. Graph. 32, 9:1–9:12 (2013)
16. Kazhdan, M., Hoppe, H.: Screened poisson surface reconstruction. ACM Trans. Graph. (ToG) 32(3), 29 (2013)
17. Klaudiny, M., Budd, C., Hilton, A.: Towards optimal non-rigid surface tracking. In: Fitzgibbon, A., Lazebnik, S., Perona, P., Sato, Y., Schmid, C. (eds.) ECCV 2012. LNCS, vol. 7575, pp. 743–756. Springer, Heidelberg (2012). https://doi.org/10.1007/978-3-642-33765-9_53
18. Lang, M., Wang, O., Aydin, T.O., Smolic, A., Gross, M.H.: Practical temporal consistency for image-based graphics applications. ACM Trans. Graph. 31(4), 1–8 (2012)
19. Li, H., Adams, B., Guibas, L.J., Pauly, M.: Robust single-view geometry and motion reconstruction. ACM Trans. Graph. (ToG) 28, 175 (2009)
20. Liu, Y., Dai, Q., Xu, W.: A point-cloud-based multiview stereo algorithm for free-viewpoint video. IEEE Trans. Visual Comput. Graph. 16(3), 407–418 (2010)

21. Lowe, D.G.: Method and apparatus for identifying scale invariant features in an image and use of same for locating an object in an image, uS Patent 6,711,293, 23 March 2004
22. Luhmann, T., Robson, S., Kyle, S., Harley, I.: Close Range Photogrammetry. Wiley, New York (2007)
23. Moulon, P., Monasse, P., Marlet, R.: Adaptive structure from motion with a Contrario model estimation. In: Lee, K.M., Matsushita, Y., Rehg, J.M., Hu, Z. (eds.) ACCV 2012. LNCS, vol. 7727, pp. 257–270. Springer, Heidelberg (2013). https://doi.org/10.1007/978-3-642-37447-0_20
24. Moynihan, M., Pagéés, R., Smolic, A.: Spatio-temporal upsampling for free viewpoint video point clouds. In: Proceedings of the 14th International Joint Conference on Computer Vision, Imaging and Computer Graphics Theory and Applications - Volume 5: VISAPP, pp. 684–692. INSTICC, SciTePress (2019). https://doi.org/10.5220/0007361606840692
25. Mustafa, A., Kim, H., Guillemaut, J.Y., Hilton, A.: Temporally coherent 4D reconstruction of complex dynamic scenes. In: 2016 IEEE Conference on Computer Vision and Pattern Recognition (CVPR), pp. 4660–4669, June 2016. https://doi.org/10.1109/CVPR.2016.504
26. Mustafa, A., Hilton, A.: Semantically coherent co-segmentation and reconstruction of dynamic scenes. In: Proceedings of the IEEE Conference on Computer Vision and Pattern Recognition, pp. 422–431 (2017)
27. Mustafa, A., Kim, H., Guillemaut, J.Y., Hilton, A.: General dynamic scene reconstruction from multiple view video. In: Proceedings of the IEEE International Conference on Computer Vision, pp. 900–908 (2015)
28. Mustafa, A., Volino, M., Guillemaut, J.Y., Hilton, A.: 4D temporally coherent light-field video. In: 2017 International Conference on 3D Vision (3DV), pp. 29–37. IEEE (2017)
29. Myronenko, A., Song, X.: Point set registration: coherent point drift. IEEE Trans. Pattern Anal. Mach. Intell. 32(12), 2262–2275 (2010)
30. Pagés, R., Amplianitis, K., Monaghan, D., Ondřej, J., Smolic, A.: Affordable content creation for free-viewpoint video and VR/AR applications. J. Vis. Commun. Image Representat. 53, 192–201 (2018). https://doi.org/10.1016/j.jvcir.2018.03.012. http://www.sciencedirect.com/science/article/pii/S1047320318300683
31. Prada, F., Kazhdan, M., Chuang, M., Collet, A., Hoppe, H.: Spatiotemporal atlas parameterization for evolving meshes. ACM Trans. Graph. (TOG) 36(4), 58 (2017)
32. Revaud, J., Weinzaepfel, P., Harchaoui, Z., Schmid, C.: EpicFlow: edge-preserving interpolation of correspondences for optical flow. In: Proceedings of the IEEE Conference on Computer Vision and Pattern Recognition, pp. 1164–1172 (2015)
33. Schaffner, M., Scheidegger, F., Cavigelli, L., Kaeslin, H., Benini, L., Smolic, A.: Towards edge-aware spatio-temporal filtering in real-time. IEEE Trans. Image Process. 27(1), 265–280 (2018)
34. Schönberger, J.L., Zheng, E., Frahm, J.-M., Pollefeys, M.: Pixelwise view selection for unstructured multi-view stereo. In: Leibe, B., Matas, J., Sebe, N., Welling, M. (eds.) ECCV 2016. LNCS, vol. 9907, pp. 501–518. Springer, Cham (2016). https://doi.org/10.1007/978-3-319-46487-9_31
35. Wedel, A., Brox, T., Vaudrey, T., Rabe, C., Franke, U., Cremers, D.: Stereoscopic scene flow computation for 3D motion understanding. Int. J. Comput. Vision 95(1), 29–51 (2011)
36. Wu, S., Huang, H., Gong, M., Zwicker, M., Cohen-Or, D.: Deep points consolidation. ACM Trans. Graph. (ToG) 34(6), 176 (2015)

37. Yu, L., Li, X., Fu, C.W., Cohen-Or, D., Heng, P.A.: PU-NET: point cloud upsampling network. In: Proceedings of the IEEE Conference on Computer Vision and Pattern Recognition, pp. 2790–2799 (2018)
38. Zollhöfer, M., et al.: Real-time non-rigid reconstruction using an RGB-D camera. ACM Trans. Graph. (ToG) **33**(4), 156 (2014)
39. Zou, D., Tan, P.: CoSLAM: collaborative visual SLAM in dynamic environments. IEEE Trans. Pattern Anal. Mach. Intell. **35**(2), 354–366 (2013)

Modeling Trajectories for 3D Motion Analysis

Amani Elaoud[1(✉)], Walid Barhoumi[1,2], Hassen Drira[3],
and Ezzeddine Zagrouba[1]

[1] Institut Supérieur d'Informatique, Research Team on Intelligent Systems
in Imaging and Artificial Vision (SIIVA), LR16ES06 Laboratoire de Recherche en
Informatique, Modélisation et Traitement de l'Information et de la Connaissance
(LIMTIC), Université de Tunis El Manar, 2 Rue Bayrouni, 2080 Ariana, Tunisia
amani.elaoud@fst.utm.tn, ezzeddine.zagrouba@fsm.rnu.tn
[2] Ecole Nationale d'Ingénieurs de Carthage (ENICarthage), Université de Carthage,
45 Rue des Entrepreneurs, 2035 Tunis-Carthage, Tunisia
walid.barhoumi@enicarthage.rnu.tn
[3] IMT Lille Douai, Univ. Lille, CNRS, UMR 9189 – CRIStAL – Centre de Recherche
en Informatique Signal et Automatique de Lille, 59000 Lille, France
hassen.drira@imt-lille-douai.fr

Abstract. 3D motion analysis by projecting trajectories on manifolds in
a given video can be useful in different applications. In this work, we use
two manifolds, Grassmann and Special Orthogonal group SO(3), to analyse accurately complex motions by projecting only skeleton data while
dealing with rotation invariance. First, we project the skeleton sequence
on the Grassmann manifold to model the human motion as a trajectory.
Then, we introduce the second manifold SO(3) in order to consider the
rotation that was ignored by the Grassmann manifold on the matched
couples on this manifold. Our objective is to find the best weighted linear combination between distances in Grassmann and SO(3) manifolds
according to the nature of the input motion. To validate the proposed 3D
motion analysis method, we applied it in the framework of action recognition, re-identification and sport performance evaluation. Experiments
on three public datasets for 3D human action recognition (G3D-Gaming,
UTD-MHAD multimodal action and Florence3D-Action), on two public
datasets for re-identification (IAS-Lab RGBD-ID and BIWI-Lab RGBD-
ID) and on one recent dataset for throwing motion of handball players
(H3DD), proved the effectiveness of the proposed method.

Keywords: 3D motion analysis · Action recognition ·
Re-identification · Sport performance evaluation · Temporal modeling ·
Manifolds · Weighted distance · Human skeleton

1 Introduction

The goal of human motion analysis is to understand and study motion being
undertaken in an image [59] or a video [60]. Recent development of RGB-D

© Springer Nature Switzerland AG 2020
A. P. Cláudio et al. (Eds.): VISIGRAPP 2019, CCIS 1182, pp. 409–429, 2020.
https://doi.org/10.1007/978-3-030-41590-7_17

sensors empowers the vision solutions to move one important step towards 3D vision. In fact, 3D human motion analysis has been a popular problem statement in the vision community thanks to its large scale applications such as service robots [78], video surveillance [69], sports performance analysis [58] and clinical analysis [79]. However, RGB-D information includes many challenges such as viewpoint variations, occlusions and illumination change. In fact, 3D data can be obtained with marker motion capture systems (e.g Vicon) or with markerless motion capture systems (e.g Kinect). Vicon 3D system was utilized to compare patients and healthy controls in spatiotemporal parameters of gait with three dimensional motion analysis [64]. For instance, Kwon et al. [61] presented 3D motion analysis study to provide the changes in the lower extremity joints at different gait speeds. Nevertheless, this kind of sensor use high definition cameras while being accurate, but it is expensive and markers attached on the anatomical landmarks of participants can limit the motion. These limits justify the growing interest of recent works on low-cost markerless motion capture systems such as the Microsoft Kinect sensor. For example, Moreira et al. [62] used Kinect for upper body function assessment within breast cancer patients. Recently, a system was developed to measure active upper limb movements for patients clinical trials [63]. In the context of action recognition, authors in [75] extracted spatio-temporal depth cuboid similarity feature from depth videos.

In this work, we propose to model the 3D human motion as a weighted linear combination on the Grassmann and SO(3) manifolds. The proposed method, which was introduced in [80], is tested herein for action recognition, re-identification and athlete performance evaluation within the context of inter-preting similarity of throwing motions. Comparatively to our previous work [80], the proposed method was validated with a new context, which is performance evaluation with an innovative 3D dataset for handball player. Furthermore, the re-identification was validated in our previous work [52] in order to perform the modeling and the analysis of human motion using skeleton data. In this work, we introduce 3D motion analysis using skeleton information focusing on power of RGB-D sensor to measure the similarity between two motion sequences. For each sequence, we have one trajectory on the Grassmann manifold, which is invari-ant to the rotation, and a second trajectory on the special orthogonal SO(3). Indeed, we firstly project on the Grassmann manifold and we calculate the dis-tance on this manifold. Then, we introduce special orthogonal group SO(3) for the couples of points that were matched in the Grassmann manifold in order to incorporate the rotation ignored by the Grassmann manifold. We combine these two distances while testing different weights to find the best weighted lin-ear combination according to the nature of the input motion, such as an action contains or not a rotation. For re-identification, we compute distances between adjacent joints which correspond to parts lengths. Then, we combine distance on Grassmann manifold and distance between joints. It is worth noting that the proposed method adapts the Dynamic Time Warping (DTW) [12] in order to match temporal sequences with different sizes. In fact, DTW confirmed it ability of accurate modeling of variations within model sequences for motion analysis. For instance, [13] applied an evolution algorithm to select an optimal subset of joints for action representation and classification, while using DTW for sequence matching. More recently, a fast dynamic time warping was investigated in [1] for human action recognition by defining adequate procedures for applying the Globally Optimal Reparameterization Algorithm (GORA) to characterize and compare signals in the form of real trajectories and video sequences.

The rest of this paper is organized as follows. In Sect. 2, we briefly present the related work on 3D motion analysis from RGB-D data captured by Kinect sensor. In Sect. 3, we suggest to model trajectories for 3D motion analysis. Then, in Sect. 4, we show experimental results to demonstrate the effectiveness of the proposed method. Finally, we present conclusions and future work in Sect. 5.

2   Related Work

In this section, we summarize some works that dealt with motion analysis applied in many topics such as human action recognition, re-identification and sport evaluation. Different methods are relying on motion analysis from RGB data, such as texture-based features [2] and silhouette shapes [4]. For example, texture features to characterize the observed movements were used for recognition of human actions [70]. Similarly, Ahad et al. [71] focused on human motions as a texture pattern. Otherwise, Blank et al. [72] represented actions as space-time shapes. Recently, shape-based features were proposed for action recognition using spectral regression discriminant analysis [73]. For re-identification, Wu et al. [81] extracted color histogram features and texture features for person re-identification to get a deep feature representation. A shape descriptor of the head-torso region of persons' silhouettes was used for the analysis of pedestrians and its applications to person re-identification [82]. Within sport evaluation, team sport analysis integrated analytical visualizations of soccer player movement data to identify weaknesses of opposing teams [60]. An analysis technique was also introduced with a floating colored diamond-shape over heads for a player [83]. However, factors such as biometric changes, camera movement and cluttered background limit the performance of human motion representation with RGB data. Thus, 3D human motion analysis methods are more and more adapted, notably while using 3D data captured by the Kinect sensing device [14].

3D motion analysis methods using Kinect depth sensor are mostly based on skeleton joints, depth maps, or both. As an example of depth-based methods, Yang et al. [74] recognized actions from sequences of depth maps based on histograms of oriented gradients. Authors in [15] used Depth Motion Maps (DMMs) to capture the motion cues of actions and Local Binary Patterns (LBPs) to represent the features. Similarly, [16] presented a real-time method for human action recognition based on depth motion maps. Moreover, descriptors for depth maps were proposed using a histogram capturing the distribution of the surface normal orientation in the 4D volume of time [17]. Authors in [18] clustered hypersurface normals in a depth sequence to form the polynormal for extracting the local motion and shape information. Space-Time Occupancy Patterns (STOP) were also proposed as descriptors for classifying human action from depth sequences [19]. Furthermore, depth re-identification was introduced with recurrent network for learning shape information from low-resolution depth images [86]. Similarly, Karianakis et al. [87] proposed a recurrent deep neural network with temporal attention for person re-identification with commodity depth sensors. Within sports fields' performance assessment based on 3D motion analysis can be used to record and evaluate the performance of athletes, which is beneficial for their further improvement in order to gain a competitive advantage. The objective is to find the gaps and propose corrections. Most of existing works that were interested in athletes' videos were limited to athlete segmentation [5] or detection [6]. Only few works evaluate sport performance with 3D information because of the lack of RGB-D benchmark datasets [7]. For example, Ting et al. [88] presented badminton performance analysis using Kinect sensor to improve the athletes' motion.

Differently to depth data, skeleton data contains human representation with the locations of human key joints in the 3D space leading to the emergence of many recent works on skeleton-based 3D action recognition [20]. For example, in order to recognize human actions, [21] applied histograms of 3D joint locations in skeleton estimation from Kinect sensors. Similarly, [22] extracted the 3D joint positions for converting skeletons into histograms. Recently, [23] proposed Hierarchical Recurrent Neural Network (HRNN) for skeleton-based action recognition. Furthermore, learning discriminative trajectorylet detector sets were adapted for capturing dynamic and static information of the skeletal joints [24]. More recently, [25] used Convolutional Neural Networks (ConvNets) to learn the spatio-temporal information of a skeleton sequence. In [26], action recognition from skeleton data was performed via analogical generalization over qualitative representations. In [27], authors utilized the temporal position of skeletal joints and the Fisherpose method in order to create a feature vector for recognizing the pose of the body in each frame. Moreover, an image classification approach was presented in [28] to transform the 3D skeleton videos to color images. For re-identification, skeleton-based features were used as on anthropometric measurement of distances between joints and geodesic distances on body surface [89]. Preis et al. [85] proposed distances between joints and dynamic features (stride length and speed) for people identification. Recently, authors in [52] assumed 3D re-identification by modeling skeleton information on Grassmann manifold. For performance evaluation of athletes' motions, Tsou and Wu [90] estimated the energy expenditure during aerobics exercises using Kinect-based skeleton. Recent work [58] introduced a RGB-D dataset in order to compare and evaluate handball players' performances using main angles extracted from skeleton data.

Otherwise, with hybrid methods for 3D human motion analysis, the idea is to benefit from both types of data: depth and skeleton [30]. For example, authors in [76] incorporated the skeleton information into depth maps to extract geometric information of human body. Two descriptors were proposed in [33], one of joint angle similarities and another using a modified HOG algorithm, such that features were derived from skeleton tracking of the human body and depth maps. For action recognition, authors [34] used an heterogeneous set of depth- and skeleton- based features for multipart learning in depth videos. In a recent work, a learning model combining features from depth and skeleton data was proposed for view-invariant appearance representation of human body-parts [36]. The work of [77] presented two kinds of features such as skeleton cues and depth information: the first one to measure the local motions and 3D joint positions and the second one to estimate the intensity difference among shape structure variation. For re-identification, depth shape descriptor and skeleton-based feature were combined in order to deal with person re-identification in the cases of clothing change and extreme illumination [84]. However, most of hybrid methods are suffering with the long computational time that is why there are few works with hybrid methods notably for re-identification and sport evaluation.

Thus, the 3D skeleton data represented human body as an articulated system of rigid segments connected by joints and encoded motion information in the scene into a smaller set of features-point trajectories. The success of manifold-based skeleton representations was performed using a finite number of salient points. In fact, using the geometry of non-Euclidean spaces gives rise to the notion of manifolds performing dimensionality reduction [38] in order to easily manipulate this considerable number of landmarks. As a consequence, various computer vision applications are dealing with motion interpretation based on manifold analysis [39], such as tracking [41], face recognition [40] and action recognition [42]. In particular, the issue of 3D human action recognition has been recently considered while using skeleton manifolds. For instance, the skeletons were represented in the Kendall's shape space [44] for 3D action recognition [43]. In [45], 3D skeletal sequences were encoded within dictionary learning on the Kendall's shape space. In the Grassmann manifold, human motion was modeled by 3D body skeletons over non-Euclidean spaces [46]. Similarly, [35] performed a kernelized Grassmann manifold learning using multigraph embedding method. In [47], 3D geometric relationships were extracted from skeleton joints with feature representation on the Lie group. For skeleton-based action recognition, authors in [49] presented the Lie group structure into a deep network architecture to learn more appropriate Lie group features. Differently, authors of [48] focused on skeleton joints within a Riemannian manifold while considering the temporal evolution. The works of [50,51] performed a Riemannian representation of the distance trajectories for a real-time human action recognition.

3 Proposed Method

A human motion video can be considered as a sequence represented by 3D joint positions. In fact, we adopt a sparse representation of human motion videos while coding skeletal trajectories. Firstly, we propose to model the variability of skeleton sequences as group action on underlying space representing the data. Then, we compare the quotient spaces resulting of action of this group. In fact, we present the skeleton sequences as trajectories in the Stiefel manifold (the set of k-dimensional orthonormal bases in $\mathbb{R}^n$ where $(k < n)$). The action of the

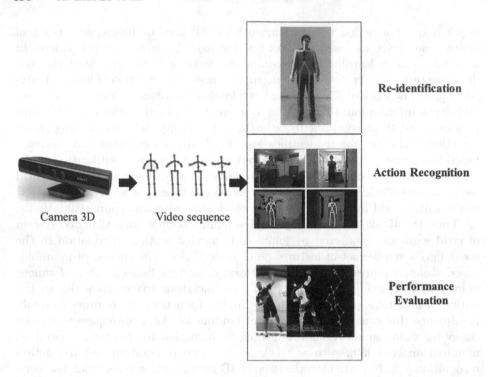

Fig. 1. 3D human motion analysis using skeleton data for re-identification, action recognition and performance evaluation.

rotation group SO(3) is considered as a quotient space of Stiefel manifold in order to remove the rotation. Thus, we obtain the quotient space that represents the skeleton independently of the rotation in $\mathbb{R}^3$ and makes the comparison of skeletons invariant to the rotation. Moreover, the rotations between corresponding skeletons in the $SO(3)$ represents another metric that consider the ignored rotation. Define the underlying manifolds, it will be the first step in our discussion. In fact, the special orthogonal group SO(n) is a matrix Lie group formed by the set of all $n \times n$ rotation matrices. It is obtained (1) by considering the subset of orthogonal matrices with a determinant equals to $+1$.

$$SO(n) = \left\{ Y \in \mathbb{R}^{n \times n} \ / \ Y^t Y = I \ \ and \ \ det(Y) = 1 \right\} \tag{1}$$

Stiefel manifold is the set of k-dimensional orthonormal bases in $\mathbb{R}^n$ where $(k < n)$. It is a quotient space of the special orthogonal group SO(n), given by $V(\mathbb{R}^n) = SO(n)/SO(n-k)$. Grassmann manifold $G(\mathbb{R}^n)$ is defined as the set of k-dimensional linear subspaces of $\mathbb{R}^n$. It is a quotient space of Stiefel manifold $V(\mathbb{R}^n)$, represented by $V(\mathbb{R}^n)/SO(k)$. Given P_1 and P_2 ($P_1, P_2 \in G(20,3)$) two matrices of size $n \times k$ with orthonormal columns, the distance on the Grassmann manifold is the geodesic distance, which is the length of shortest curve in the manifold.

3.1   Projecting Motions on Manifolds

The proposed method is validated within three applications: action recognition, re-identification and sport evaluation (Fig. 1). The proposed method has as input the skeleton sequence of the studied test person and the skeleton sequences of all persons in the training set for action recognition and re-identification. For sport evaluation, the proposed method has as input the skeleton sequence of a beginner handball player and the skeleton sequences of all expert handball players within the dataset, in order to identify the most similar expert handball players to the studied beginner. Given the fact that the used sequences are of different sizes, for both action recognition and re-identification, the used sequence is composed of two sets: the first one Set_1 for input test sequence T_i and the second set Set_2 for training sequences $T_1...T_n$. Otherwise, for sport evaluation, the used sequence contains one set Set_1 for beginner handball player sequence T_i and a second set Set_2 for expert handball sequences $T_1...T_n$. Our objective is to compare the input sequence T_i in Set_1 with all composed sequences $T_1...T_n$ in set Set_2 in order to identify the most similar one to T_i among $T_1...T_n$.

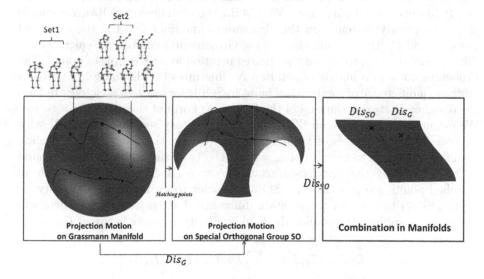

Fig. 2. Outline of the proposed method for 3D motion analysis.

Two trajectories represent the sequence motion as illustrated in Fig. 2. We start our proposed method by the comparison of the trajectories on the Grassmann manifold. In fact, each frame in the skeleton sequence is presented by a point in the trajectory such that the projection of skeleton data on the Grassmann manifold is mainly performed using Singular Value Decomposition (SVD)-based orthogonalization [52]. Indeed, all data points on the Grassmann manifold are projected on $\mathbb{R}^{Skel}$ (Skel, which denotes the number of landmarks that are given with each skeleton, is equal to 20 for Kinect V1 and to 25 for Kinect V2).

Each frame in a skeleton sequence motion is presented by a matrix M_l of size $Skel \times 3$, where $l \in \{1, \ldots, n\}$ such that n denotes the number of frames in the studied sequence. To obtain the closest distance, we calculate the distance on the Grassmann manifold $Dist_G$ for one trajectory sequence of test T_i with all trajectories sequences of training T_t (2). We chose to use the DTW algorithm that resolves the problem of temporal alignment and measures the similarity between sequences with different sizes. It is worth noting that the Grassmann manifold is characterized by ignoring rotation, what motivated to resort in addition to special orthogonal group.

$$Dist_G(T_t, T_i) = \sum_{l=1}^{n} Dist_{Grass}(T_t(l), T_i(l)) = \sum_{l=1}^{n} \theta_l^2 \qquad (2)$$

where T_t is training trajectory and T_i is test trajectory.

The sequence motion is represented by the difference of rotation between two skeletons with 3D joints. Use the special orthogonal group SO(3) is the second representation in this work as illustrated in Fig. 2. Indeed, we use a trajectory composed of difference of rotations such that a 3×3 rotation matrix defined on SO(3) (3-dimensional subspace). We obtain the distance on SO(3), for couple of points already matched on the Grassmann manifold (3). For the couple of point P_1 and $P_{2'}$ that are matched on the Grassmann manifold, we calculate the difference of rotation to obtain the ignored rotation for this couple of points. The Frobenius norm of a matrix of rotation N illustrates the distance on SO(3). In our case, a difference of rotation is a point in SO(3) for any two skeletons. In fact, the distance $Dist_{SO}$ is the sum of the Frobenius norm of the matrix of rotation. We calculate the distance on SO(3) for one trajectory sequence of test T_i with all trajectories sequences of training $T_1 \ldots T_n$ (3). By this way, the Grassmann manifold was used to match points within the two studied trajectories T_i and T_t by considering the rotation-invariant distance $Dist_G$. Then, the couple of matched points are processed on SO(3) in order to evaluate the similarity by using $Dist_{SO}$ to consider the rotation difference. Lastly, the two distances will be linearly combined to produce the final similarity measure between T_i and T_t.

$$Dist_{SO}(T_t, T_i) = \sum_{l=1}^{n} Dist_{SO}(T_t(l), T_i(l))$$
$$= \sum_{l=1}^{n} \|N(l)\|_F \qquad (3)$$

3.2   Combination in Manifolds

The proposed method is based on combining the distance $Dist_G$ on the Grassmann manifold with the distance $Dist_{SO}$ on SO(3) in order to find a trade-off between these two manifolds considering the assessment of similarity between two sequences according to the nature of their actions (including rotation or not). We proceed firstly by normalizing the distances D ($Dist_G$ and $Dist_{SO}$)

within trajectories in set Set_1 and trajectories in set Set_2 in order to bring all new values D' into the range $[0, 1]$ while restricting the range of values. Then, we test different values of each distance weight, α and $1 - \alpha$, on the training set in order to find the best weighted linear combination for action recognition. In fact, for evaluation performance, we test different values of each distance weight, α and $1 - \alpha$, for one half of beginner handball players. The selection of the tested values of α is strongly depending on the nature of the studied actions (4). For example, if the action does not contain a rotation, α has a value that is generally close to 1. For the first case, the defined weighted linear combination is applied on the test set in order to recognize the input action. Indeed, to recognize an unknown test trajectory T_i, a classification by the nearest neighbour distance is used to decide the nature of the action within the sequence T_t (4). This step is driven by a decision rule for the dataset while finding the weight that maximizes the recognition action accuracy. For the other case, the defined weighted linear combination is applied on the second half for beginner handball players in order to identify the most similar expert handball player.

$$\arg\min_{1 \leq l \leq n} \alpha.Dist_G(T_t, T_i) + (1 - \alpha).Dist_{SO}(T_t, T_i), \tag{4}$$

In this work, 3D motion analysis is validated within the context of three applications: action recognition, re-identification and performance evaluation. We calculate the distance between two trajectories T_t and T_i in order to involve the identification of different actions from 3D videos, to calculate the dissimilarity between persons and to improve sport performance. More precisely, the 3D motion was performed while $\beta_3 = 0$ in the case of action recognition, $\beta_2 = 0$ in the case of re-identification and $\beta_3 = 0$ for the case of performance evaluation (5). It is worth noting that we already presented in a previous work [52] motion analysis of skeleton for person re-identification. The 3D motion is modeled by projecting skeleton information on Grassmann manifold. In fact, we compute distances between adjacent joints which correspond to parts lengths. Then, we combine distance on Grassmann manifold and distance between joints to re-identify a given person.

$$d(T_t, T_i) = \beta_1 Dist_G(T_t, T_i) + \beta_2 Dist_{SO}(T_t, T_i) + \beta_3 Dist_L(T_t, T_i) \tag{5}$$

where $Dist_L$ denotes the distance between adjacent skeleton joints.

4   Experimental Validation

To evaluate the proposed 3D motion analysis within the context of action recognition, we used three standard 3D human action datasets: G3D-Gaming, UTD-MHAD multimodal action and Florence3D-Action. G3D-Gaming dataset [53] contains 663 gaming motions sequences captured using Kinect V1 camera and is composed of 10 subjects performing 20 different gaming motions (*e.g.* golf swing, tennis swing forehand, tennis swing backhand, tennis serve, throw bowling ball...). Each one repeats every action more than two times. For comparison purpose, likewise [54] and [49] the cross-subject protocol was adopted for this dataset. Indeed, we used the cross-subject test setting, in which five subjects were used for training and five other subjects were used for testing. All the reported results for G3D-Gaming dataset were averaged over ten different random combinations of training and test subjects. Besides, the recognition accuracy is defined by the average of the 20 actions within the testing set. UTD-MHAD dataset [57] consists of four temporally synchronized data modalities: RGB videos, depth videos, skeleton positions, and inertial signals from a Kinect V1 camera and a wearable inertial sensor for a comprehensive set of 27 human actions. The dataset is composed of 861 data sequences from 27 human actions performed by 8 subjects with 4 repetitions. In fact, the dataset contains different actions that cover sport actions (*e.g.* "bowling", "tennis serve" and "baseball swing" ...), hand gestures (*e.g.* "draw X", "draw triangle", and "draw circle"...), daily activities (*e.g.* "knock on door", "sit to stand" and "stand to sit"...) and training exercises (*e.g.* "arm curl", "lung" and "squat"...). In order to follow the same evaluation setting of [55] and [57], the cross-subject protocol was adopted. The data from the subject numbers 1, 3, 5, 7 were used for the training while the subject numbers 2, 4, 6, 8 were used for the testing. Florence3D-Action [56] is composed of 9 actions (wave, drink from a bottle, answer phone, clap, lace, sit down, stand up, read watch, bow) performed by 10 subjects, each subject repeats every action two or three times. It includes 215 action sequences captured using a Kinect V1 camera. In fact, we followed the cross-subject test setting in which half of the subjects was used for the training and the remaining half was used for the testing.

The results achieved were numerically compared with methods that used skeletal representations while validating on the aforementioned datasets. For the G3D-Gaming dataset, we compared with three methods that are based on skeleton manifolds. Indeed, the two first compared methods [47,54] adopted learning methods using pairwise transformations of skeletal joints on Lie group and the last method [49] used deep learning, also on Lie Groups, for skeleton-based action recognition. Moreover, for the second used dataset UTD-MHAD multimodal, the proposed method was compared with two works of [55,57]. The first method [55] used convolutional neural networks based on joint trajectory maps for action recognition and the second method [57] integrated depth and inertial sensor data in addition to the skeleton data. For Florence3D-action dataset, the proposed method was compared with two methods that are based on Riemannian representations. In fact, authors in [54] presented trajectories on Lie groups and authors in [45] proposed the motion of skeletal shapes as trajectories on the Kendall's shape.

It is worthy noting that we tested different values of α with the training set to extract the best weighted linear combination. While validating using the G3D-Gaming dataset, we tried different values of α with the training set and we observed a growth in performance with value of α equals to 1. We applied this combination on the set of test and 13 actions ('PunchRight', 'PunchLeft', 'KickRight', 'KickLeft', 'Defend', 'Tennis serve', 'Walk', 'Run', 'Climb', 'Crouch',

Table 1. Recognition rate of the proposed method for the G3D-Gaming dataset.

PunchRight	100%
PunchLeft	100%
KickRight	100%
KickLeft	100%
Defend	100%
GolfSwing	50%
TennisSwingForehand	90%
TennisSwingBackhand	75%
TennisServe	100%
ThrowBowlingBall	90%
AimAndFireGun	65%
Walk	100%
Run	100%
Jump	95%
Climb	100%
Crouch	100%
Steer	100%
Wave	95%
Flap	100%
Clap	100%

'Steer', 'Flap' and 'Clap') were well distinguished with values of the recognition accuracy equal to 100%. However, the worst result was 50% for the 'GolfSwing' action (Table 1). The recorded results proved that our method outperforms the state of the art methods with value of accuracy 93% against 87.23% with [47], 87.95% with [54] and 89.10% with [49].

For the UTD-MHAD dataset, the best value of α is equal to 0.9. Then, we applied this weighted linear combination on the testing set to perform action recognition. In fact, this combination provides 100% accuracy for 18 actions ('Clap', 'Throw', 'Arm cross', 'Basketball shoot', 'Draw X', 'Draw circle', 'Bowling', 'Boxing', 'Arm curl', 'Push', 'Catch', 'Pickup and throw', 'Jog', 'Walk', 'Sit to stand', 'Stand to sit', 'Lunge', 'Squat') thanks to the accurate modeling information exploitation in the proposed method. However, the worst value was 84.65% for the 'Draw circle counter clockwise' action (Table 2). The main confusions concern very similar actions of 'Draw circle counter clockwise' and 'Draw circle (clockwise)'. We show that the proposed method outperforms the state of the art methods with value of 95.37% against 79.10% for [55] and 85.10% for [57].

With the Florence3D-Action dataset, we tried different values of α with the training set and we observed a growth in performance with value of α equals

Table 2. Recognition rate of the proposed method for the UTD-MHAD dataset.

Swipe left	93%
Swipe right	89%
Wave	88%
Clap	100%
Throw	100%
Arm cross	100%
Basketball shoot	100%
Draw X	100%
Draw circle (clockwise)	100%
Draw circle (counter clockwise)	84.65%
Draw triangle	87%
Bowling	100%
Boxing	100%
Baseball swing	87%
Tennis swing	90 %
Arm curl	100%
Tennis serve	90%
Push	100%
Knock	88%
Catch	100%
Pickup and throw	100%
Jog	100%
Walk	100%
Sit to stand	100%
Stand to sit	100%
Lunge	100%
Squat	100%

to 1. We applied this combination on the set of test and four actions ('clap', 'light place', 'sit down', 'stand up') were well distinguished with values of the recognition accuracy equal to 100%. However, the worst result was 78% for the 'answer phone' action (Table 3). Results demonstrate that the proposed method outperforms many state of the art methods with value of 92.59% against 91.4% for [54]. But, we performed less well against [45] with value of 93.04%. This confirms once again the relevance of adopting the geometry of manifolds for dealing with 3D human action recognition. Nevertheless, for making a statistical decision and comparing the studied methods while performing solid argument that is supported by statistical analysis, we used the p-value (or probability value). In fact, a level of 0.05 indicates that a 5% risk is used as the cutoff for

significance. If the p-value is lower than 0.05, we reject the null hypothesis that there is no difference between the two compared methods and we conclude that a significant difference exists (*i.e.* below 0.05 it is significant; over 0.05 it is not significant). In our case, we recorded a significant difference with p-value below 0.05 for the two datasets G3D-Gaming and UTD-MHAD multimodal. However, there is no significant difference for Florence3D-Action dataset, with a p-value over 0.05 against [55] as well as against [45].

Table 3. Recognition rate of the proposed method for the Florence3D-Action dataset.

Wave	88.88%
Drink from a bottle	88.88%
Answer phone	77.77%
Clap	100%
Light place	100%
Sit down	100%
Stand up	100%
Read watch	88.88%
Bow	88.88%

Table 4. Comparison of the rank-1 of the proposed method against [91] using testing sets of the IAS-Lab: RGBD-ID dataset "TestingA" and "TestingB".

	TestingA	TestingB
[91]	25.25%	62%
Proposed method	45.45%	72.72%

Furthermore, we validated the proposed 3D motion analysis method in the framework of re-identification while using two challenging RGB-D datasets: IAS-Lab RGBD-ID [91] and BIWI-Lab RGBD-ID [91]. The first dataset is composed of 33 sequences of 11 different people, where 11 sequences compose the training set and the remaining 22 sequences form the testing set. It contains different sequences of people with two sets of testing (TestingA, TestingB). The second dataset contains 50 training sequences and 56 testing sequences of 50 different people, performing a certain routine of motions in front of a Kinect sensor. Two testing sets have been collected: Still sequence and Walking sequence. We used the Cumulative Match Curve (CMC) which is widely adopted for the evaluation of re-identification system performance. In fact, we compared our results with the results reported in [91] on the IAS-Lab RGBD-ID dataset using its two test sets: TestingA and "TestingB". We show in Table 4 the recorded rank-1 is better for the proposed method than the one of [91], 45.45% for "TestingA" of IAS-Lab against 25.5%, and 72.72% for "TestingB" of IAS-Lab against 62%. For BIWI-Lab RGBD-ID dataset, the rank-1 of the proposed method outperforms the ones of [91] and [92]. Indeed, using the test set Still of BIWI-Lab RGBD-ID, we recorded a value of 28.57% for the proposed method against 26.6% for [91] and 11.6% for [92]. However, the rank-1 of the proposed method records similar results compared to [91,92] (Table 5).

Table 5. Comparison of the rank-1 of the proposed method against [91] and [92] using Walking and Still testing sets of the BIWI-Lab RGBD-ID dataset.

	Still	Walking
[91]	28.57%	12%
[92]	11..6%	21%
Proposed method	25.57%	8%

Table 6. Results of the recognition accuracy (Acc) of the proposed method with H3DD for different values of α within the Set_1.

α	0	0.1	0.2	0.3	0.4	0.5	0.6	0.7	0.8	0.9	1
Acc (%)	92.14	76.75	78.92	79.75	80.26	80.16	79.44	79.33	79.33	79.33	79.13

Then, we validated sport evaluation with H3DD dataset [58]. This dataset is composed of 3D video sequences of throwing motion for handball players. Each video has been captured by Kinect V2 containing a PNG frame of size 1920×1980 showing the RGB information (Fig. 3), a PNG frame of size 514×424 presenting the depth information and a 'txt' file containing the skeleton information illustrating the 3D spatial coordinates (x, y, z) for 25 joints. The number of subjects is 44 beginners and 18 experts. We compare our results with results of [58]. The recent work of [58] proposed to compare beginners and experts handball players performing the same throwing. Using skeleton data given 3D joints, authors in [58] used the main angles responsible for throwing performance to compare the throwing motion between two athletes with different levels. While validating using the H3DD dataset, we tried different values of α with the one half of beginner players set and we noticed a growth in performance with value of α equals to 0 because athletes have very similar throwing motion and only rotation in this action is slightly different (Table 6). In fact, we tested this combination on the second half of beginner players set and we reach values of results similar to the ones recorded by [58]. Thus, we find 17 true results and 5 false results (Table 7). For example, we found the same result to the one given in [58], the beginner number 23 has expert number 17 as the most similar expert, however the beginner number 27 has expert number 1 as the most similar expert

Table 7. Verification results of the proposed method with H3DD.

Beginner	23	24	25	26	27	28	29	30	31	32	33	34	35	36	37	38	39	40	41	42	43	44
Expert [58]	17	13	12	2	12	9	5	6	4	16	14	16	18	12	10	6	2	16	14	5	5	10
Expert Proposed Method	17	13	12	2	1	9	17	8	4	16	14	15	18	12	14	6	2	16	14	5	5	10
Verification	✓	✓	✓	✓	×	✓	×	×	✓	✓	✓	×	✓	✓	×	✓	✓	✓	✓	✓	✓	✓

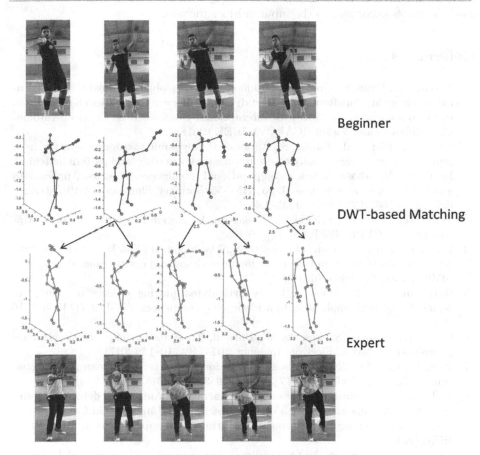

Beginner

DWT-based Matching

Expert

Fig. 3. Comparing beginner *vs* expert.

but expert number must be 12. The objective is to improve athlete performance and gain a competitive advantage. For example, in Fig. 3, a comparison can be suggested between a beginner and an expert, such that DTW matches the similar frames to propose corrections and suggestions to reach the expert level.

5   Conclusion

We presented 3D motion analysis with weighted linear combination validated in human action recognition, re-identification and sport evaluation. The rapid development of depth sensors (e.g., Microsoft Kinect) has attracted more attention to revisit computer vision problems using RGB-D data. In fact, 3D motion analysis was proposed using different distances within manifolds. The classification is based on closest neighbor method. Machine learning techniques will be investigated in the future to improve the classification performance. Moreover, we are studying the possibility of choosing dynamically the ponderation of the used distances according to the input subject motion.

References

1. Mitchel, T., Ruan, S., Gao, Y., Chirikjian, G.: The globally optimal reparameterization algorithm: an alternative to fast dynamic time warping for action recognition in video sequences. In: 2018 15th International Conference on Control, Automation, Robotics and Vision (ICARCV). IEEE (2018)
2. Susan, S., Mittal, M., Bansal, S., Agrawal, P.: Dynamic texture recognition from multi-offset temporal intensity co-occurrence matrices with local pattern matching. In: Verma, N., Ghosh, A. (eds.) Computational Intelligence: Theories, Applications and Future Directions-Volume II, pp. 545–555. Springer, Singapore (2019). https://doi.org/10.1007/978-981-13-1135-2_41
3. Wang, H., Kläser, A., Schmid, C., Cheng-lin, L.: Action recognition by dense trajectories. In: CVPR. IEEE (2011)
4. Islam, S., Qasim, T., Yasir, M., Bhatti, N., Mahmood, H., Zia, M.: Single-and two-person action recognition based on silhouette shape and optical point descriptors. SIViP **12**(5), 853–860 (2018)
5. Barhoumi, W.: Detection of highly articulated moving objects by using co-segmentation with application to athletic video sequences. SIViP **9**(7), 1705–1715 (2015)
6. Carey, P., Bennett, S., Lasenby, J., Purnell, T.: Aerodynamic analysis via foreground segmentation. Electron. Imaging **2017**(16), 10–14 (2017)
7. Kim, Y., Kim, D.: Real-time dance evaluation by markerless human pose estimation. Multimedia Tools Appl. **77**(23), 31199–31220 (2018)
8. Ladjailia, A., Bouchrika, I., Merouani, H., Harrati, N.: Automated detection of similar human actions using motion descriptors. In: 16th International Conference on Sciences and Techniques of Automatic Control and Computer Engineering (STA). IEEE (2015)
9. Carreira, J., Zisserman, A.: Quo vadis, action recognition? A new model and the kinetics dataset. In: Proceedings of the IEEE Conference on Computer Vision and Pattern Recognition (2017)
10. Barmpoutis, P., Stathaki, T., Camarinopoulos, S.: Skeleton-based human action recognition through third-order tensor representation and spatio-temporal analysis. Inventions. **4**, 9 (2019)
11. Pers, J., Bon, M., Vuckovic, G.: CVBASE 06 dataset. http://vision.fe.uni-lj.si/cvbase06/dataset.html
12. Sakoe, H., Chiba, S.: Dynamic programming algorithm optimization for spoken word recognition. IEEE Trans. Acoust. Speech Signal Process. **26**, 43–49 (1978)
13. Chaaraoui, A., Padilla-lópez, J., Climent-pérez, P., Flórez-revuelta, F.: Evolutionary joint selection to improve human action recognition with RGB-D devices. Expert Syst. Appl. **41**, 786–794 (2014)

14. Han, J., Shao, L., Xu, D., Shotton, J.: Enhanced computer vision with microsoft kinect sensor: a review. IEEE Trans. Cybern. **43**, 1318–1334 (2013)
15. Chen, C., Jafari, R., Kehtarnavaz, N.: Action recognition from depth sequences using depth motion maps-based local binary patterns. In: Proceedings of the IEEE Winter Conference on Applications of Computer Vision, Waikoloa Beach, HI, pp. 1092–1099, January 2015
16. Chen, C., Liu, K., Kehtarnavaz, N.: Real-time human action recognition based on depth motion maps. J. Real-Time Image Proc. **12**(1), 155–163 (2016)
17. Oreifej, O., Liu, Z.: HoN4D: histogram of oriented 4D normals for activity recognition from depth sequences. J. Real-time Image Process. **12**, 155–163 (2016)
18. Yang, X., Tian, Y.: Super normal vector for activity recognition using depth sequences. In: Proceedings of the IEEE Conference on Computer Vision and Pattern Recognition (2014)
19. Vieira, A.W., Nascimento, E.R., Oliveira, G.L., Liu, Z., Campos, M.F.M.: STOP: space-time occupancy patterns for 3D action recognition from depth map sequences. In: Alvarez, L., Mejail, M., Gomez, L., Jacobo, J. (eds.) CIARP 2012. LNCS, vol. 7441, pp. 252–259. Springer, Heidelberg (2012). https://doi.org/10.1007/978-3-642-33275-3_31
20. Li, B., He, M., Cheng, X., Chen, Y., Dai, Y.: Skeleton based action recognition using translation-scale invariant image mapping and multi-scale deep CNN. In: IEEE International Conference on Multimedia & Expo Workshops (ICMEW). IEEE (2017)
21. Xia, L., Chen, C., Aggarwal, J.: View invariant human action recognition using histograms of 3D joints. In: IEEE Computer Society Conference on Computer Vision and Pattern Recognition Workshops. IEEE (2012)
22. Thanh, T., Chen, F., Kotani, K., Le, H.: Extraction of discriminative patterns from skeleton sequences for human action recognition. In: IEEE RIVF International Conference on Computing & Communication Technologies, Research, Innovation, and Vision for the Future. IEEE (2012)
23. Du, Y., Wang, W., Wang, L.: Hierarchical recurrent neural network for skeleton based action recognition. In: Proceedings of the IEEE Conference on Computer Vision and Pattern Recognition (2015)
24. Qiao, R., Liu, L., Shen, C., Vandenhengel, A.: Learning discriminative trajectorylet detector sets for accurate skeleton-based action recognition. Pattern Recogn. **66**, 202–212 (2017)
25. Hou, Y., Li, Z., Wang, P., Li, W.: Skeleton optical spectra-based action recognition using convolutional neural networks. IEEE Trans. Circ. Syst. Video Technol. **28**, 807–811 (2018)
26. Chen, K., Forbus, K.: Action recognition from skeleton data via analogical generalization. In: 30th International Workshop on Qualitative Reasoning (2017)
27. Ghojogh, B., Mohammadzade, H., Mokari, M.: Fisherposes for human action recognition using Kinect sensor data. IEEE Sens. J. **18**, 1612–1627 (2018)
28. Li, B., He, M., Dai, Y., Cheng, X., Chen, Y.: 3D skeleton based action recognition by video-domain translation-scale invariant mapping and multi-scale dilated CNN. Multimedia Tools Appl., 1–21 (2018)
29. Shahroudy, A., Wang, G., Ng, T.: Multi-modal feature fusion for action recognition in RGB-D sequences. In: 6th International Symposium on Communications, Control and Signal Processing (ISCCSP). IEEE (2014)
30. Elmadany, N., He, Y., Guan, L.: Information fusion for human action recognition via Biset/Multiset globality locality preserving canonical correlation analysis. IEEE Trans. Image Process. **27**, 5275–5287 (2018)

31. Ofli, F., Chaudhry, R., Kurillo, G., Vidal, R., Bajcsy, R.: Berkeley MHAD: a comprehensive multimodal human action database. In: IEEE Workshop on Applications of Computer Vision (WACV). IEEE (2013)
32. Zhu, Y., Chen, W., Guo, G.: Fusing spatiotemporal features and joints for 3D action recognition. In: Proceedings of the IEEE Conference on Computer Vision and Pattern Recognition Workshops (2013)
33. Ohn-bar, E., Trivedi, M.: Joint angles similarities and HOG2 for action recognition. In: Proceedings of the IEEE Conference on Computer Vision and Pattern Recognition Workshops (2013)
34. Shahroudy, A., Ng, T., Yang, Q., Wang, G.: Multimodal multipart learning for action recognition in depth videos. IEEE Trans. Pattern Anal. Mach. Intell. **38**, 2123–2129 (2016)
35. Rahimi, S., Aghagolzadeh, A., Ezoji, M.: Human action recognition based on the Grassmann multi-graph embedding. Signal Image Video Process. **13**, 1–9 (2018)
36. Rahmani, H., Bennamoun, M.: Learning action recognition model from depth and skeleton videos. In: Proceedings of the IEEE International Conference on Computer Vision (2017)
37. Bakr, N., Crowley, J.: Histogram of oriented depth gradients for action recognition. In: The Computing Research Repository (CoRR), pp. 1801–09477 (2018)
38. Cherian, A., Sra, S.: Riemannian dictionary learning and sparse coding for positive definite matrices. IEEE Trans. Neural Netw. Learn. Syst. **28**, 2859–2871 (2017)
39. Efros, A., Torralba, A.: Guest editorial: big data. Int. J. Comput. Vision **119**, 1–2 (2016)
40. Harandi, M., Shirazi, S., Sanderson, C., Lovell, B.: Graph embedding discriminant analysis on Grassmannian manifolds for improved image set matching. In: CVPR, Colorado Springs, CO, USA, pp. 2705–2712, June 2011
41. Hu, H., Ma, B., Shen, J., Shao, L.: Manifold regularized correlation object tracking. IEEE Trans. Neural Netw. Learn. Syst. **29**, 1786–1795 (2018)
42. Chen, X., Weng, J., Lu, W., Xu, J., Weng, J.: Deep manifold learning combined with convolutional neural networks for action recognition. IEEE Trans. Neural Netw. Learn. Syst. **29**, 3938–3952 (2018)
43. Amor, B., Su, J., Srivastava, A.: Action recognition using rate-invariant analysis of skeletal shape trajectories. IEEE Trans. Pattern Anal. Mach. Intell. **38**, 1–13 (2016)
44. Kendall, D.: Shape manifolds, procrustean metrics, and complex projective spaces. Bull. London Math. Soc. **16**, 81–121 (1984)
45. Tanfous, A., Drira, H., Amor, B.: Coding Kendall's shape trajectories for 3D action recognition. In: Proceedings of the IEEE Conference on Computer Vision and Pattern Recognition (2018)
46. Slama, R., Wannous, H., Daoudi, M., Srivastava, A.: Accurate 3D action recognition using learning on the Grassmann manifold. Pattern Recogn. **48**, 556–567 (2015)
47. Vemulapalli, R., Arrate, F., Chellappa, R.: Human action recognition by representing 3D skeletons as points in a lie group. In: Proceedings of the IEEE Conference on Computer Vision and Pattern Recognition (2014)
48. Devanne, M., Wannous, H., Berretti, S., Pala, P., Daoudi, M., Delbimbo, A.: 3-D human action recognition by shapshape analysis of motion trajectories on riemannian manifold. IEEE Trans. Cybern. **45**, 1340–1352 (2015)
49. Huang, Z., Wan, C., Probst, T., Vangool, L.: Deep learning on lie groups for skeleton-based action recognition. In: Proceedings of the IEEE Conference on Computer Vision and Pattern Recognition (2017)

50. Meng, M., Drira, H., Daoudi, M., Boonaert, J.: Human-object interaction recognition by learning the distances between the object and the skeleton joints. In: 11th IEEE International Conference and Workshops on Automatic Face and Gesture Recognition (FG), vol. 7. IEEE (2015)
51. Meng, M., Drira, H., Boonaert, J.: Distances evolution analysis for online and offline human object interaction recognition. Image Vis. Comput. **70**, 32–45 (2018)
52. Elaoud, A., Barhoumi, W., Drira, H., Zagrouba, E.: Analysis of skeletal shape trajectories for person re-identification. In: Blanc-Talon, J., Penne, R., Philips, W., Popescu, D., Scheunders, P. (eds.) ACIVS 2017. LNCS, vol. 10617, pp. 138–149. Springer, Cham (2017). https://doi.org/10.1007/978-3-319-70353-4_12
53. Bloom, V., Makris, D., Argyriou, V.: G3D: a gaming action dataset and real time action recognition evaluation framework. In: IEEE Computer Society Conference on Computer Vision and Pattern Recognition Workshops. IEEE (2012)
54. Vemulapalli, R., Chellapa, R.: Rolling rotations for recognizing human actions from 3D skeletal data. In: Proceedings of the IEEE Conference on Computer Vision and Pattern Recognition (2016)
55. Wang, P., Li, Z., Hou, Y., Li, W.: Action recognition based on joint trajectory maps using convolutional neural networks. In: Proceedings of the 24th ACM International Conference on Multimedia. ACM (2016)
56. Seidenari, L., Varano, V., Berretti, S., Bimbo, A., Pala, P.: Recognizing actions from depth cameras as weakly aligned multi part bag-of-poses. In: Proceedings of the IEEE Conference on Computer Vision and Pattern Recognition Workshops (2013)
57. Chen, C., Jafari, R., Kehtarnavaz, N.: UTD MHAD: a multimodal dataset for human action recognition utilizing a depth camera and a wearable inertial sensor. In: IEEE International Conference on Image Processing (ICIP). IEEE (2015)
58. Elaoud, A., Barhoumi, W., Zagrouba, E., Agrebi, B.: Skeleton-based comparison of throwing motion for handball players. J. Ambient Intell. Hum. Comput., 1–13 (2019)
59. Lowney, C., Hsung, T., Morris, D., Khambay, B.: Quantitative dynamic analysis of the nasolabial complex using 3D motion capture: a normative data set. J. Plast. Reconstr. Aesthetic Surg. **71**, 1332–1345 (2018)
60. Stein, M., et al.: Bring it to the pitch: combining video and movement data to enhance team sport analysis. IEEE Trans. Vis. Comput. Graph. **24**, 13–22 (2017)
61. Kwon, J., Son, S., Lee, N.: Changes of kinematic parameters of lower extremities with gait speed: a 3D motion analysis study. J. Phys. Ther. Sci. **27**, 477–479 (2015)
62. Moreira, R., Magalhães, A., Oliveira, H.: A Kinect-based system for upper-body function assessment in breast cancer patients. J. Imaging **1**, 134–155 (2015)
63. Chen, X., et al.: Feasibility of using Microsoft Kinect to assess upper limb movement in type III spinal muscular atrophy patients. PLoS ONE **12**, e0170472 (2017)
64. Mirek, E., Rudzińska, M., Szczudlik, A.: The assessment of gait disorders in patients with Parkinson's disease using the three-dimensional motion analysis system Vicon. Neurol. Neurochir. Pol. **41**, 128–133 (2007)
65. Elaiwat, S., Bennamoun, M., Boussaïd, F.: A spatio-temporal RBM-based model for facial expression recognition. Pattern Recogn. **49**, 152–161 (2016)
66. Li, B., Mian, A., Liu, W., Krishna, A.: Using Kinect for face recognition under varying poses, expressions, illumination and disguise. In: IEEE Workshop on Applications of Computer Vision (WACV). IEEE (2013)
67. Saleh, Y., Edirisinghe, E.: Novel approach to enhance face recognition using depth maps. In: International Conference on Systems, Signals and Image Processing (IWSSIP). IEEE (2016)

68. Nambiar, A., Bernardino, A., Nascimento, J., Fred, A.: Towards view-point invariant person re-identification via fusion of anthropometric and gait features from kinect measurements. In: VISIGRAPP (5: VISAPP) (2017)

69. Patruno, C., Marani, R., Cicirelli, G., Stella, E., D'orazio, T.: People re-identification using skeleton standard posture and color descriptors from RGB-D data. Pattern Recogn. **89**, 77–90 (2019)

70. Kellokumpu, V., Zhao, G., Pietikäinen, M.: Recognition of human actions using texture descriptors. Mach. Vis. Appl. **22**, 767–780 (2011)

71. Ahad, M., Islam, M., Jahan, I.: Action recognition based on binary patterns of action-history and histogram of oriented gradient. J. Multimodal User Interfaces **10**, 335–344 (2016)

72. Blank, M., Gorelick, L., Shechtman, E., Irani, M., Basri, R.: Actions as space-time shapes. In: Tenth IEEE International Conference on Computer Vision (ICCV 2005) Volume 1, vol. 2. IEEE (2005)

73. Selvam, G., Gnanadurai, D.: Shape-based features for reliable action recognition using spectral regression discriminant analysis. Int. J. Sig. Imaging Syst. Eng. **9**, 379–387 (2016)

74. Yang, X., Zhang, C., Tian, Y.: Recognizing actions using depth motion maps-based histograms of oriented gradients. In: Proceedings of the 20th ACM international conference on Multimedia. ACM (2012)

75. Xia, L., Aggarwal, J.: Spatio-temporal depth cuboid similarity feature for activity recognition using depth camera. In: Proceedings of the IEEE Conference on Computer Vision and Pattern Recognition (2013)

76. Ji, X., Cheng, J., Feng, W., Tao, D.: Skeleton embedded motion body partition for human action recognition using depth sequences. Sig. Process. **143**, 56–68 (2018)

77. Jalal, A., Kim, Y., Kim, Y., Kamal, S., Kim, D.: Robust human activity recognition from depth video using spatiotemporal multi-fused features. Pattern Recogn. **61**, 295–308 (2017)

78. Wang, K., Tobajas, P., Liu, J., Geng, T., Qian, Z., Ren, L.: Towards a 3D passive dynamic walker to study ankle and toe functions during walking motion. Rob. Auton. Syst. **115**, 49–60 (2019)

79. Nazarahari, M., Noamani, A., Ahmadian, N., Rouhani, H.: Sensor-to-body calibration procedure for clinical motion analysis of lower limb using magnetic and inertial measurement units. J. Biomech. **85**, 224–229 (2019)

80. Elaoud, A., Barhoumi, W., Drira, H., Zagrouba, E.: Weighted linear combination of distances within two manifolds for 3D human action recognition. In: VISIGRAPP (VISAPP) (2019)

81. Wu, S., Chen, Y., Li, X., Wu, A., You, J., Zheng, W.: An enhanced deep feature representation for person re-identification. In: IEEE Winter Conference on Applications of Computer Vision (WACV). IEEE (2016)

82. Nambiar, A., Bernardino, A., Nascimento, J.: Shape context for soft biometrics in person re-identification and database retrieval. Pattern Recogn. Lett. **68**, 297–305 (2015)

83. Stein, M., et al.: Director's cut: analysis and annotation of soccer matches. IEEE Comput. Graph. Appl. **36**, 50–60 (2016)

84. Wu, A., Zheng, W., Lai, J.: Robust depth-based person re-identification. IEEE Trans. Image Process. **26**, 2588–2603 (2017)

85. Preis, J., Kessel, M., Werner, M., Linnhoff-popien, C.: Gait recognition with Kinect. In: 1st International Workshop on Kinect in Pervasive Computing, New Castle, UK (2012)

86. Nikolaos, K., Zicheng, L., Yinpeng, C.: Person depth ReID: robust person re-identification with commodity depth sensors. Corr. abs/1705.0988 (2017)
87. Karianakis, N., Liu, Z., Chen, Y., Soatto, S.: Reinforced temporal attention and split-rate transfer for depth-based person re-identification. In: Ferrari, V., Hebert, M., Sminchisescu, C., Weiss, Y. (eds.) ECCV 2018. LNCS, vol. 11209, pp. 737–756. Springer, Cham (2018). https://doi.org/10.1007/978-3-030-01228-1_44
88. Ting, H., Tan, Y., Lau, B.: Potential and limitations of Kinect for badminton performance analysis and profiling. Indian J. Sci. Technol. **9**, 1–5 (2016)
89. Barbosa, I.B., Cristani, M., Del Bue, A., Bazzani, L., Murino, V.: Re-identification with RGB-D sensors. In: Fusiello, A., Murino, V., Cucchiara, R. (eds.) ECCV 2012. LNCS, vol. 7583, pp. 433–442. Springer, Heidelberg (2012). https://doi.org/10.1007/978-3-642-33863-2_43
90. Tsou, P., Wu, C.: Estimation of calories consumption for aerobics using Kinect based skeleton tracking. In: IEEE International Conference on Systems, Man, and Cybernetics. IEEE (2015)
91. Munaro, M., Basso, A., Fossati, A., Vangool, L., Menegatti, E.: 3D reconstruction of freely moving persons for re-identification with a depth sensor. In: IEEE International Conference on Robotics and Automation (ICRA). IEEE (2014)
92. Munaro, M., Fossati, A., Basso, A., Menegatti, E., Van Gool, L.: One-shot person re-identification with a consumer depth camera. In: Gong, S., Cristani, M., Yan, S., Loy, C.C. (eds.) Person Re-Identification. ACVPR, pp. 161–181. Springer, London (2014). https://doi.org/10.1007/978-1-4471-6296-4_8
93. Wang, J., Liu, Z., Wu, Y., Yuan, J.: Learning actionlet ensemble for 3D human action recognition. IEEE Trans. Pattern Anal. Mach. Intell. **36**, 914–927 (2013)
94. Mian, A., Bennamoun, M., Owens, R.: On the repeatability and quality of key-points for local feature-based 3d object retrieval from cluttered scenes. Int. J. Comput. Vision **89**, 348–361 (2010)
95. Rahmani, H., Mahmood, A., Huynh, D., Mian, A.: Histogram of oriented principal components for cross-view action recognition. IEEE Trans. Pattern Anal. Mach. Intell. **38**, 2430–2443 (2016)

Quantitative Comparison of Affine Feature Detectors Based on Quadcopter Images

Zoltán Pusztai(✉)📍, Gergő Gál📍, and Levente Hajder📍

Department of Algorithms and Applications, Eötvös Loránd University,
Pázmány Péter stny. 1/C, Budapest 1117, Hungary
{puzsaai,qa11v5,hajder}@inf.elte.hu

Abstract. Affine correspondences are in the focus of work in many research groups nowadays. Components of projective geometry, e.g. homography or fundamental matrix, can be recovered more accurately when not only point but affine correspondences are exploited. This paper quantitatively compares state-of-the-art affine covariant feature detectors based-on outdoor images taken by a quadcopter-mounted camera. Accurate Ground Truth (GT) data can be calculated from the restricted flight path of the quadcopter. The GT data consist of not only affine transformation but feature locations as well. Quantitative comparison and in-depth analysis of the affine covariant feature detectors are also presented.

Keywords: Affine covariant detector · Quantitative evaluation · Affine covariant detectors · Quadcopter · Ground truth generation

1 Introduction

Feature detectors have been studied since the born of computer vision. First, point-related local detectors are developed to explore properties of camera motion and epipolar geometry. Local features describe only a small area of the image, thus they can be effectively used to find point correspondences, even in the presence of high illumination, viewpoint change or occlusion. Affine feature detectors can detect the local affine warp of the detected point regions. The affine transformation can be used to solve the basic task of epipolar geometry (e.g. to detect camera motion, object detection or 3D reconstruction) using less correspondences than general point features. This paper deals with the quantitative comparison of affine feature detectors using video sequences, taken by a quadcopter, captured in a real-world environment.

EFOP-3.6.3-VEKOP-16-2017-00001: Talent Management in Autonomous Vehicle Control Technologies – The Project was supported by the Hungarian Government and co-financed by the European Social Fund. This research was also supported by Thematic Excellence Programme, Industry and Digitization Subprogramme, NRDI Office, 2019.

A. P. Cláudio et al. (Eds.): VISIGRAPP 2019, CCIS 1182, pp. 430–455, 2020.
https://doi.org/10.1007/978-3-030-41590-7_18

Interest feature detectors have been studied in a long period of computer vision. The well-known Harris [9] corner detector or Shi-Tomasi detector [19] have been published more than two decades ago. Since then, new point feature detectors have been implemented, e.g. SIFT [12], SURF [6], KAZE [2], BRISK [11] and so on. Correspondences are made using a feature location and a feature descriptor. The latter describes the local small area of the feature with a vector in a compact and distinguish way. These descriptor vectors can be used for feature matching over the successive image. Features, whose descriptor vectors are close to each other, potentially yield a match.

While point-based features estimate only point correspondences, affine feature detectors can extract the affine transformations around the feature centers as well. An affine transformation contains the linear approximation of the warp around the point correspondences. In other words, this is a 2 by 2 linear transformation matrix which transforms the local region of the feature to that of the corresponding feature. Each affine transformation contains enough information for estimating the normal vector of the tangent plane at the corresponding 3D location of the feature. These additional constraints can be used to estimate fundamental matrix, camera movement or other epipolar properties using less number of features, than using point correspondences.

The aim of this paper is to quantitatively compare the feature detectors and the estimated affine transformations, using real-world video sequences captured by a quadcopter. The literature of previous work is not rich. Maybe the most significant work was published by Mikolajczyk et al. [13]. They compare several affine feature detectors using real-world images. However, in their comparison, either the camera has fixed location or the scenes are planar, thus, the images are related by homographies. The error for the affine transformations is computed by the overlapping error of the related affine ellipses or by the repeatability score. Even trough the authors compared the detectors using several different noise types (blur, JPEG compression, light change) we think that the constraints of non-moving camera or planar scene yield very limited test cases. A comprehensive study can be found in [21], however the paper does not contain any real-world test. Recently Pusztai et al. proposed a technique to quantitatively compare feature detectors and descriptors using a structured-light 3D scanner [15]. However, the testing data consist of small objects and rotation movement only. Tareen et al. [20] also published a comparison of the most popular feature detector and descriptor algorithms. Their ground-truth (GT) data generation is twofold: (i) They use the Oxfordian dataset, that was also applied in the work of [13], (ii) they generated test images, including GT transformations, carried out by different kinds of affine transformations for the original images: translation, rotation and scale, and the transformed images are synthesized by common bilinear interpolation. In the latter case, the processed images are not taken by a camera, thus the input data for the comparison is not really realistic.

The literature of affine feature comparisons is not extensive, despite the fact that it is an important subject. Most detectors are compared to others, using only a small set of images, and parameters tuned to achieve the best results. In real world applications, the best parameter set may differ from the laboratory experience. Thus, more comparisons have to be made using real-world video sequences and various camera movements.

It is shown here that the affine invariant feature detectors can be evaluated quantitatively on real-world test sequences if images are captured by a quadcopter-mounted camera. The main contributions of the paper are twofold: (i) First, the GT affine transformation generation is shown in case of several special movements of a quadcopter where affine transformations and corresponding point locations can be very accurately determined. To the best of our knowledge, this is the first study in which the ground truth data is generated using real images of a moving copter. (ii) Then several affine covariant feature detectors are quantitatively compared using the generated GT data. Both point locations and the related affine transformations are examined in the comparison. This paper is based on our conference publication [16], however, more rival algorithms are compared, and robustification of planar motion estimation is also discussed here.

The structure of this paper is as follows. First, the rival methods are theoretically described in Sect. 2. Then the ground truth data generation methods are overviewed for different drone movements and camera orientations. Section 4 deals with outlier filtering for image pairs in which a plane is observed that is parallel to the camera images. Section 5 contains the methodology of the evaluation. The test results are discussed in Sect. 6, and Sect. 7 concludes the research.

2 Overview of Affine-Covariant Detectors

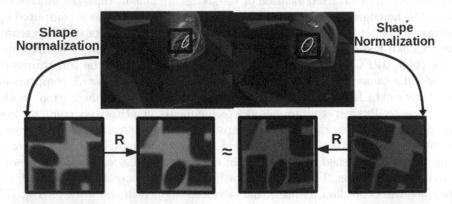

Fig. 1. The affine transformation of corresponding features. The ellipses in the top images show the estimated affine shapes around the features. The shape normalization and orientation assignment are done independently for the affine shapes. The affine transformation is calculated from the normalization and orientation parameters. It approximately transforms the area of the first ellipse into the second one.

In this section, the affine transformations and the affine covariant detectors are briefly introduced. The detectors aim to separately find discriminate features in the images. If a feature is found, then the affine shape can be determined which is usually visualized as an ellipse. Figure 1 shows the local affine regions of a corresponding feature pair in successive images. The methods to detect discriminate features and their affine shapes vary from detector to detector. They are briefly introduced as follows:

Harris-Laplace, Harris-Affine. The methods introduced by [14] are based on the well-known Harris detector [9]. Harris uses the so-called second moment matrix to extract features in the images. The matrix is as follows:

$$M(\mathbf{x}) = \sigma_D^2 G(\sigma_I) * \begin{bmatrix} f_x^2(\mathbf{x}, \sigma_D) & f_x(\mathbf{x}, \sigma_D) f_y(\mathbf{x}, \sigma_D) \\ f_x(\mathbf{x}, \sigma_D) f_y(\mathbf{x}, \sigma_D) & f_y^2(\mathbf{x}, \sigma_D) \end{bmatrix},$$

where

$$G(\sigma) = \frac{1}{2\pi\sigma^2} exp\left(-\frac{|\mathbf{x}|^2}{2\sigma^2}\right), \quad \text{and} \quad f_x(\mathbf{x}, \sigma_D) = \frac{\partial}{\partial \mathbf{x}} G(\sigma_D) * f(\mathbf{x}).$$

The matrix (M) contains the gradient distribution around the feature, σ_D, σ_I are called the differentiation scale and integration scale, respectively. A local feature point is found, if the term $det(M) - \lambda trace(M)$ is higher than a pre-selected threshold. This means that both of the eigenvalues of M are large, which indicates a corner in the image.

After the location of the feature is found, a characteristic scale selections needs to be carried out. The circular Laplace operator is used for this purpose. The characteristic scale is found if the similarity of the operator and the underlying image structure is the highest. The final step of this affine invariant detector is to determine the second scale of the feature points using the following iterative estimation:

1. Detect the initial point and corresponding scale.
2. Estimate the affine shape using M.
3. Normalize the affine shape into a circle using $M^{1/2}$.
4. Detect new position and scale in the normalized region.
5. Goto (2), if the eigenvalues of M are not equal.

The iteration always converges, the obtained affine shape is described as an ellipse.

The scale and shape selections described above can be applied to any point feature. Mikolajczyk also proposed the **Hessian-Laplace** and **Hessian-Affine** detectors, which use the Hesse matrix for extracting features, instead of the Harris. The matrix is as follows:

$$H(\mathbf{x}) = \begin{bmatrix} f_{xx}(\mathbf{x}, \sigma_D) & f_{xy}(\mathbf{x}, \sigma_D) \\ f_{xy}(\mathbf{x}, \sigma_D) & f_{yy}(\mathbf{x}, \sigma_D) \end{bmatrix}, \tag{1}$$

where $f_{xx}(\mathbf{x}, \sigma_D)$ is the second order Gaussian smoothed image derivatives. The Hesse matrix can be used to detect blob-like features.

Edge-Based Regions (EBR). [22] introduced a method to detect affine covariant regions around the corners. The Harris corner detector is used along with standard Canny edge detector. Affine regions are found where two edges meet at a corner. The corner point (**p**) and two points moving along the two edges (**p**$_1$ and **p**$_2$) define a parallelogram. The final shape is found, where the region yields extremum in the following function:

$$f(\Omega) = abs\left(\frac{|(\mathbf{p} - \mathbf{p}_g)(\mathbf{q} - \mathbf{p}_g)|}{|(\mathbf{p} - \mathbf{p}_1)(\mathbf{p} - \mathbf{p}_2)|}\right) \times \frac{M_{00}^1}{\sqrt{M_{00}^2 M_{00}^0 - (M_{00}^1)^2}}, \qquad (2)$$

where

$$M_{pq}^n = \int_\Omega I^n(x,y)x^p y^q \, dx \, dy, \quad \mathbf{p}_g = \left(\frac{M_{00}^1}{M_{00}^1}, \frac{M_{01}^1}{M_{00}^1}\right),$$

and **p**$_g$ is the center of gravity. The parallelogram regions are then converted to ellipses.

Intensity-Extrema-Based Regions (IBR). While EBR find features at the corners and edges, IBR extracts affine regions based on intensity properties. First, the image is smoothed, then the local extrema is selected using non-maximum suppression. These points cannot be detected precisely, however they are robust to monotonic intensity transformations. Rays are cast from the local extremum to every direction, and the following function is evaluated on each ray:

$$f_I(t) = \frac{abs(I(t) - I_0)}{max\left(\frac{\int_0^t abs(I(t) - I_0) \, dt}{t}, d\right)} \qquad (3)$$

where, t, $I(t)$, I_0 and d are the arc length along the ray, the intensity at position, the intensity extremum and a small number to prevent dividing by 0, respectively. The function yields extremum where the intensity suddenly changes. The points along the cast rays define an usually irregularly-shaped region, which is replaced by an ellipse having the same moments up to the second order.

TBMR. Tree-Based Morse Regions is introduced in [23]. This detector is motivated by Morse theory, selecting critical regions as features, using the Min and Max-tree. TBMR can be seen as a variant of MSER, however, TBMR is invariant to illumination change and needs less number of parameters.

SURF. SURF is introduced in [6]. It is the fast approximation of SIFT [12]. The Hessian matrix is roughly approximated using box filters, instead of Gaussian filters. This makes the detector relatively fast comparing to the others. Despite the approximations, SURF can find reliable features.

Affine Image Synthesis. Affine SIFT [24] is the extension of the well-known SIFT [12] detector. SIFT can detect only scale invariant features. In contract, ASIFT is detects fully affine covariant feature by generating synthetic images. The image is transformed by an affine transformation into a synthetic one by an affine mapping. Then, SIFT is used to extract the scale-invariant features.

Finally, the affine shape of the feature is given by combining the scale of the detected feature and the affine map of the synthetic image. This procedure of feature detection can be combined with other feature detectors as well. This combination with the AKAZE [1], BRISK [11] and ORB [17] feature detectors yields AAKAZE, ABRISK and AORB, respectively.

3   GT Data Generation

In order to quantitatively compare the performance of the detectors and the reliability of affine transformations, ground truth (GT) data is needed. Many comparison databases [7,13,25] are based on homography, which is extracted from the observed planar object. Images are taken from different camera positions, and the GT affine transformations are calculated from the homography. Instead of this, our comparison is based on images captured by a quadcopter in a real world environment and the GT affine transformation is extracted directly from constrained movement of the quadcopter. This work is motivated by the fact, that the affine parameters can be calculated more precisely from the constrained movements, than from the homography. Moreover, if the parameters of the motion are estimated, the GT location of the features can be determined as well. Thus, it is possible to compare not only the affine transformations, but the locations of the features, additionally. In this section, these restricted movements, and the computations of the corresponding affine transformations are introduced.

3.1   Rotation

(a) Example images from the sequence.

(b) Colored boxes indicate corresponding areas computed by the rotation parameters. Images are the same as in the first row.

Fig. 2. Images are taken while the quadcopter is rotating around its vertical axis, while its position is fixed. Figure from Pusztai and Hajder [16, Figure 2].

In case of the rotation movement, the quadcopter stays at the same position, and rotates around its vertical axis. Thus, the translation vector of the movement is equivalent to **0** all the time. The rotation matrices between the images can be computed if the degrees and center of the rotation are known. Example images are given in Fig. 2. The first row shows the images taken by the quadcopter, and the second row shows colored boxes, which are related by affine transformations.

Let α_i be the degree of rotation in radians, then the rotation matrix is defined as follows:

$$\mathbf{R}_i = \begin{bmatrix} \cos \alpha_i & -\sin \alpha_i \\ \sin \alpha_i & \cos \alpha_i \end{bmatrix}. \tag{4}$$

This matrix describes the transformation of corresponding affine shapes, in case of the rotation movement. Let us assume that corresponding feature points in the images are given, then the relation of the corresponding features can be expressed as follows:

$$\mathbf{p}_i^f = \mathbf{v} + \mathbf{R}_f \left(\mathbf{p}_i^1 - \mathbf{v} \right), \tag{5}$$

where $\mathbf{p}_p^f$, α_i, $\mathbf{v}$ are the p-th feature location in the f-th image, the degree of rotation between the first to the f-th image and the center of rotation, respectively. The latter one is considered constant during the rotation movement.

To estimate the degree and center of the rotation, the Euclidean distances of the selected and estimated features have to be minimized. The cost function describing the error of estimation is as follows:

$$\sum_{f=2}^{F} \sum_{i=1}^{P} \left\| \mathbf{p}_i^f - \mathbf{R}_f \left(\mathbf{p}_i^1 - \mathbf{v} \right) - \mathbf{v} \right\|_2^2, \tag{6}$$

where F and P are the number of frames and selected features, respectively. The minimization can be solved by an alternation algorithm. The alternation itself consist of two steps: (i) estimation of the center of rotation $\mathbf{v}$ and (ii) estimation of rotation angles α_i.

Estimation of Rotation Center. The problem of rotation center estimation can be formalized as $\mathbf{A}\mathbf{v} = \mathbf{b}$, where

$$\mathbf{A} = \begin{bmatrix} \mathbf{R}_2 - \mathbf{I} \\ \vdots \\ \mathbf{R}_F - \mathbf{I} \end{bmatrix}, \quad \text{and} \quad \mathbf{b} = \begin{bmatrix} \mathbf{R}_2 \mathbf{p}_1^1 - \mathbf{p}_1^2 \\ \vdots \\ \mathbf{R}_F \mathbf{p}_P^1 - \mathbf{p}_p^F \end{bmatrix}. \tag{7}$$

The optimal solution in the lest-squares sense is given by the pseudo-inverse of A:

$$\mathbf{v} = \left(\mathbf{A}^\mathbf{T} \mathbf{A} \right)^{-1} \mathbf{A}^T \mathbf{b} \tag{8}$$

Estimation of Rotating Angles. The rotation angles are separately estimated for each image. The estimation can be written in a linear form $\mathbf{C}\mathbf{x} = \mathbf{d}$ subject to $\mathbf{x}^T \mathbf{x} = 1$, where

$$\mathbf{C} \begin{bmatrix} \cos \alpha_i \\ \sin \alpha_i \end{bmatrix} = \mathbf{d}. \tag{9}$$

The coefficient matrix $\mathbf{C}$ and vector $\mathbf{d}$ are as follows:

$$\mathbf{C} = \begin{bmatrix} x_1^1 - u & v - y_1^1 \\ y_1^1 - v & x_1^1 - u \\ \vdots & \vdots \\ x_P^1 - u & v - y_P^1 \\ y_P^1 - v & x_P^1 - u \end{bmatrix} \quad \mathbf{d} = \begin{bmatrix} x_1^f - u \\ y_1^f - v \\ \vdots \\ x_P^f - u \\ y_P^f - v \end{bmatrix}, \tag{10}$$

where $\left[x_i^f, y_i^f\right]^T = \mathbf{p}_i^f$ and $[u, v]^T = \mathbf{v}$.

The optimal solution for this problem is given by one of the roots of a four degree polynomial as it is written in the appendix.

Convergence. The steps described above are repeated one after the other, iteratively. The speed of convergence does not matter for our application. However, we empirically found that it convergences after a few iterations, when the center of the image is used as the initial value for the center of the rotation.

Ground Truth Affine Transformations. The affine transformations are easy to determine: they are equal to the rotation: $\mathbf{A} = \mathbf{R}$.

3.2 Uniform Motion: Front View

For this test case, the quadcopter moves along a straight line and does not rotate. Thus, the rotation matrix relative to the camera movement is equal to the identity. The camera faces to the front, thus the Focus of Expansion (FOE) can be computed. The FOE is the projection of the spatial line of movement at the infinity to the camera image. It is also an epipole in the image, meaning that all epipolar lines intersect at the FOE. The epipolar lines can be determined from the projections of corresponding spatial points. Figure 3 shows an example image sequence for this motion. In the second row, the red dot indicates the calculated FOE, and the blue lines mark the epipolar lines.

(a) A few images of the sequence.

(b) Red dot marks the FOE, blue lines are the epipolar ones. Images are the same as in the first row.

Fig. 3. The images are taken while the quadcopter is moving parallel to the ground. Figure from Pusztai and Hajder [16, Figure 3]. (Color figure online)

The maximum likelihood estimation of the FOE can be solved by a numerical minimization of the sum of squared orthogonal distances from the projected points and the measured epipolar lines [10]. Thus, the cost function to be minimized contains the sum of all feature distance to the related epipolar line. It can be formalized as follows:

$$\sum_{f=1}^{F}\sum_{i=1}^{P}\left(\left(\mathbf{p}_i^f - \mathbf{m}\right)^T \begin{bmatrix} -\sin\beta_i \\ \cos\beta_i \end{bmatrix}\right)^2, \tag{11}$$

where $\mathbf{m}$ and β_i are the FOE and the angle between the epipolar line and the horizontal axis. Note, that the expression $[-\sin\beta_i, \cos\beta_i]^T$ is the normal vector of the i-th epipolar line. This cost function can be minimized with an alternation, iteratively refining the FOE and angles of epipolar lines. The center of the image is used as the initial value for the FOE, then the epipolar lines can be calculated as it is described in the following subsection.

Estimation of Epipolar Lines. The epipolar lines intersect at the FOE, because a pure translation motion is considered. Moreover, the epipolar lines connecting corresponding features with the FOE are the same along the images. Each angle between the epipolar lines and the horizontal (image) axis can be computed as a homogeneous system of equations $\mathbf{Ax} = 0$ with constraint $\mathbf{x}^T\mathbf{x} = 1$ as follows:

$$\mathbf{A} = \begin{bmatrix} y_i^1 - m_y & m_x - x_i^1 \\ \vdots & \vdots \\ y_i^F - m_y & m_x - x_i^F \end{bmatrix}, \quad \mathbf{x} = \begin{bmatrix} \cos\beta_i \\ \sin\beta_i \end{bmatrix}, \tag{12}$$

where $[m_x, m_y] = \mathbf{m}$ and $\left[x_i^f, y_i^f\right] = \mathbf{p}_i^f$ are the coordinates of the FOE and that of the selected features, respectively.

The solution which minimizes the cost function is obtained as the eigenvector ($\mathbf{v}$) of the smallest eigenvalue of matrix $\mathbf{A}^T\mathbf{A}$. Then, $\beta_i = atan2(v_y, v_x)$.

Estimation of the Focus of Expansion. The FOE is located where the epipolar lines of the features intersect, thus the estimation of the FOE can be formalized as a linear system of equations $\mathbf{Am} = \mathbf{b}$, where

$$\mathbf{A} = \begin{bmatrix} -\sin\beta_1 & \cos\beta_1 \\ -\sin\beta_1 & \cos\beta_1 \\ \vdots & \vdots \\ -\sin\beta_1 & \cos\beta_1 \\ -\sin\beta_2 & \cos\beta_2 \\ \vdots & \vdots \\ \vdots & \vdots \\ -\sin\beta_N & \cos\beta_N \end{bmatrix}, \quad \text{and} \quad \mathbf{b} = \begin{bmatrix} y_1^1\cos\beta_1 - x_1^1\sin\beta_1 \\ y_1^2\cos\beta_1 - x_1^2\sin\beta_1 \\ \vdots \\ y_1^F\cos\beta_1 - x_1^F\sin\beta_1 \\ y_2^1\cos\beta_2 - x_2^1\sin\beta_2 \\ \vdots \\ \vdots \\ y_N^F\cos\beta_N - x_N^F\sin\beta_N \end{bmatrix}. \tag{13}$$

This system of equations can be solved with the Pseudo inverse of $\mathbf{A}$. Thus $\mathbf{v} = \left(\mathbf{A}^T\mathbf{A}\right)^{-1}\mathbf{A}\mathbf{b}$.

The Fundamental Matrix. The steps explained above are iteratively repeated until convergence. The GT affine transformation is impossible to compute in an unknown environment, because it depends on the surface normal [5], and the normals are varying from feature to feature. However, some constraints can be achieved if the fundamental matrix is known. The fundamental matrix describes the transformation of epipolar lines between stereo images in a static environment. It is the composition of the camera matrices and the parameters of camera motion as follows:

$$\mathbf{F} = \mathbf{K}^{-T}\mathbf{R}[\mathbf{t}]_x\mathbf{K}^{-1}, \tag{14}$$

where $\mathbf{K}$, $\mathbf{R}$ and $[\mathbf{t}]_x$ are the camera matrix, rotation matrix and the matrix representation of the cross product, respectively. The fundamental matrix can be computed as $\mathbf{F} = [\mathbf{v}]_x$ from the FOE if the relative motion of cameras contains only translation [10]. If the fundamental matrix for two images is known, the closest valid affine transformation can be determined [4]. These closest ones are labelled as ground-truth transformations in our experiments. The details of the comparison using the fundamental matrix can be found in Sect. 5.2.

3.3   Uniform Motion: Bottom View

This motion is the same as described in the previous section. However, the camera observes the ground, instead of facing forward. Since the movement is parallel to the image plane, it can be considered as a degenerate case of the previous one, because the FOE is located at the infinity. In this scenario the features of the ground are related by a pure translation between the images as the ground is planar. The projections of the same corresponding features form a line in the camera images, but these epipolar lines are parallel to each other, and also parallel to the motion of the quadcopter. Figure 4 shows example images of this motion and colored boxes related by the affine transformations.

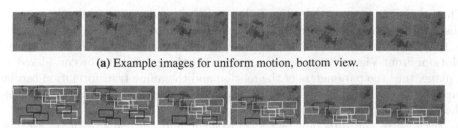

(a) Example images for uniform motion, bottom view.

(b) The colored boxes indicates the same areas computed by the motion parameters.

Fig. 4. The images are taken while the quadcopter is moving forward. Figure from Pusztai and Hajder [16, Figure 4].

Let us denote the angle between the epipolar lines and the horizontal axis by γ, and l^i denotes a point which lies on the i-th epipolar line. The cost function to be minimized contains the squared distances of the measured points to the related epipolar lines:

$$\sum_{f=1}^{F}\sum_{i=1}^{P}\left(\left(\mathbf{p}_i^f - \mathbf{l}^i\right)^T \begin{bmatrix} -\sin\gamma \\ \cos\gamma \end{bmatrix}\right)^2. \tag{15}$$

The alternation minimizes the error with refining the angle of the epipolar lines (γ) first, then their points ($\mathbf{l}^i$). The first part of the alternation can be written as a homogeneous system of equations $\mathbf{A}\mathbf{x} = 0$ with respect to $\mathbf{x}^T\mathbf{x} = 1$, similarly to Eq. 12, but it contains all points for all images:

$$\begin{bmatrix} x_1^f - l_x^1 \ y_1^f - l_y^1 \\ \vdots \quad \vdots \\ x_2^f - l_x^2 \ y_2^f - l_y^2 \end{bmatrix} \begin{bmatrix} -\sin\gamma \\ \cos\gamma \end{bmatrix} = 0, \tag{16}$$

the solution is obtained as the eigenvector of the smallest eigenvalue of matrix $\mathbf{A}^T\mathbf{A}$, then $\gamma = atan2(v_x, v_y)$.

The second step of the alternation refines the points located on the epipolar lines. The equation can be formed as $Ax = b$, where

$$\begin{bmatrix} \cos\gamma \ \sin\gamma \\ \vdots \quad \vdots \\ \cos\gamma \ \sin\gamma \end{bmatrix} \begin{bmatrix} l_x^i \\ l_y^i \end{bmatrix} = \begin{bmatrix} x_i\cos\gamma + y_i\sin\gamma \\ \vdots \\ x_i\cos\gamma + y_i\sin\gamma \end{bmatrix}, \tag{17}$$

the point on the line is given by the pseudoinverse of $\mathbf{A}$, $\mathbf{l} = \left(\mathbf{A}^T\mathbf{A}\right)^{-1}\mathbf{A}\mathbf{b}$.

Ground Truth Affine Transformations. The GT affine transformations for forward motion with a bottom-view camera is a simple identity:

$$\mathbf{A} = \mathbf{I} \tag{18}$$

3.4 Scaling

This motion is generated by the sinking of the quadcopter, while the camera observes the ground. The direction of the motion is approximately perpendicular to the camera plane, thus this can be considered as a special case of the Uniform Motion: Front View. The only difference is that the ground can be considered as a plane, thus the parameters of the motion and the affine transformation can be precisely calculated. Figure 5 first row shows an example images captured during the motion.

Because of the direction of the movement and the camera plane are not perpendicular, the FOE and the epipolar lines for the corresponding selected features can be computed. It is also true, that during the sinking of the quadcopter, the features move along their corresponding epipolar line. See Sect. 3.2 for the computation of the FOE and epipolar lines.

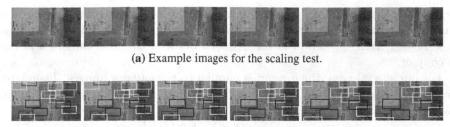

(a) Example images for the scaling test.

(b) The colored boxes indicate the same areas computed by the scaling parameters.

Fig. 5. The images are taken while the quadcopter is sinking. Figure from Pusztai and Hajder [16, Figure 5].

The corresponding features are related in the images by the parameter of the scaling. Let s be the scaling parameter, then the distances between the features and the FOE are related by s. This can be formalized as follows:

$$\left(\mathbf{p}_i^2 - \mathbf{v}\right) = s_2 \left(\mathbf{p}_i^1 - \mathbf{v}\right) \quad i \in [1, P]. \tag{19}$$

Thus, the parameter of the scale is given, by the average of the fraction of the distances:

$$s_2 = \frac{1}{P} \sum_{i=1}^{P} \frac{\left\|\mathbf{p}_i^2 - \mathbf{v}\right\|_2}{\left\|\mathbf{p}_i^1 - \mathbf{v}\right\|_2}. \tag{20}$$

Ground Truth Affine Transformations. The GT affine transformations for the scaling is trivially a simple scaled identity:

$$\mathbf{A}_j = s\mathbf{I} \tag{21}$$

4 Outlier Filtering

Outlier filtering is one of the main challenges for motion estimation, for which the input data are the detected affine correspondences. In this section, it is shown how outliers can be filtered out for quadcopters images if the ground is observed by the applied camera.

4.1 Homography Estimation for Planes Parallel to the Camera Image

For quadcopter images, when the vehicle is parallel to the ground, the camera pose between two images can be computed by the algorithm as follows.

For point correspondences, the problem can be solved by the 2-point method of Saurer et al. [18]. In this section, we show how this method can be extended by the exploitation of affine transformations.

For planes, the corresponding point locations are transformed by a homography. If the normal vector of a plane is denoted by $\mathbf{n}$, the relative pose between a calibrated camera image pair is represented by the rotation matrix $\mathbf{R}$ and translation vector $\mathbf{t} = [t_x \quad t_y \quad t_z]^T$, then the homography $\mathbf{H}$ can be written as $\mathbf{H} = \mathbf{R} - (1/d)\mathbf{t}\mathbf{n}^T$, where d is the distance between the first camera and the image pair.

For our special case, the ground is parallel to the camera images. Therefore, plane normal is $\mathbf{n} = [0 \quad 0 \quad 1]$, and rotation has represented by one DoF, an angle Θ, as

$$\mathbf{R} = \mathbf{R}_z = \begin{bmatrix} \cos(\Theta) & -\sin(\Theta) & 0 \\ \sin(\Theta) & \cos(\Theta) & 0 \\ 0 & 0 & 1 \end{bmatrix} \tag{22}$$

Considering the equation $\mathbf{H} = \mathbf{R_z} - [t_x, t_y, t_z]^T [0, 0, 1]$, the homography can be reformulated as

$$\mathbf{H} = \begin{bmatrix} \cos(\Theta) & -\sin(\Theta) & -t_x \\ \sin(\Theta) & \cos(\Theta) & -t_y \\ 0 & 0 & 1-t_z \end{bmatrix} = \begin{bmatrix} h_1 & -h_2 & h_3 \\ h_2 & h_1 & h_4 \\ 0 & 0 & h_5 \end{bmatrix}. \tag{23}$$

A homography for this special motion has only 4 DoFs as there are five unknown elements, but the values in h_1 and h_2 are not independent.

A homography represents the transformation of the ground plane between two images. If the corresponding plane coordinates are denoted by $\mathbf{x} = [x, y, 1]^T$ and $\mathbf{x}' = [x', y', 1]^T$, the cross product is carried out in order to cope with the scale ambiguity of the transformation by the homography as

$$\mathbf{x}' \times \mathbf{H}\mathbf{x} = \mathbf{0}. \tag{24}$$

Thus,

$$\begin{bmatrix} x' \\ y' \\ 1 \end{bmatrix} \times \begin{bmatrix} h_1x - h_2y + h_3 \\ h_2x + h_1y + h_4 \\ h_5 \end{bmatrix} = \mathbf{0}. \tag{25}$$

After taking the cross product with $\mathbf{x}'$:

$$\begin{bmatrix} y'h_5 - h_2x - h_1y - h_4 \\ h_1x - h_2y + h_3 - x'h_5 \\ x'h_2x + x'h_1y + x'h_4 - y'h_1x + y'h_2y - y'h_3 \end{bmatrix} = \mathbf{0}. \tag{26}$$

Only the first two rows have to be considered as the third one is the linear combination of those. Therefore, the following linear system of equations is obtained:

$$\begin{bmatrix} -y & -x & 0 & -1 & y' \\ x & -y & 1 & 0 & -x' \end{bmatrix} \begin{bmatrix} h_1 \\ h_2 \\ h_3 \\ h_4 \\ h_5 \end{bmatrix} = \mathbf{0}. \tag{27}$$

Thus, each point correspondence gives two equations for the homography estimation. As the scale of the solution is arbitrary, in other words, a homography can be determined up to an unknown scale, at least two point pairs are required in order to solve the system. This is why the authors of the method [18] call it as the 2-pt algorithm. The scale of the solution can be computed from the constraint $h_1^2 + h_2^2 = 1$.

The main idea here is that an affine transformation gives another four constraints, therefore, homography can be estimated from a single correspondence, moreover, the problem is over-determined if the point correspondences are applied as well. Barath et al. [3] showed that an affine transformation between two corresponding patches gives four equations for homography estimation. Their formulas can be rewritten for the discussed problem as

$$
\begin{bmatrix}
1 & 0 & 0 & 0 & -a_{11} \\
0 & -1 & 0 & 0 & -a_{12} \\
0 & 1 & 0 & 0 & -a_{21} \\
1 & 0 & 0 & 0 & -a_{22}
\end{bmatrix}
\begin{bmatrix}
h_1 \\ h_2 \\ h_3 \\ h_4 \\ h_5
\end{bmatrix} = \mathbf{0},
\tag{28}
$$

where affine transformation $\mathbf{A}$ is written as

$$
\mathbf{A} = \begin{bmatrix} a_{11} & a_{12} \\ a_{21} & a_{22} \end{bmatrix}.
$$

In Eqs. 27 and 28, six homogeneous linear equations are given for five unknowns with only 4 DoFs, therefore the problem is over-determined. Optimal solution for this algebraic problem can be computed by finding the eigenvalue of the product of the coefficient matrix by its transpose from the left [10]. The scale of the solution is given by the constraint $h_1^2 + h_2^2 = 1$.

4.2   Outlier Filtering

In computer vision, usually RANSAC [8]-based robust algorithms are applied to filter the contaminated data. A minimal model estimator is required for RANSAC first, input data are randomly selected. This procedure is repeated many times, then the best model is selected and inliers are separated. Inliers are the data points that supports the obtained model.

In our case, RANSAC is not required as only one local affinity determines a model. The model can be estimated from the linear equations defined in Eqs. 27 and 28. Model estimation can be calculated for each sample, then the angle of rotation Θ can be computed from the elements h_1 and h_2 of the homography matrix $\mathbf{H}$. If this estimation is repeated, the correct angle can be estimated by simple techniques, e.g. histogram voting, as the estimated correct angles are close to the correct value, while incorrect values are uniformly distributed.

5   Evaluation Method

The evaluation of the feature detectors is twofold. In the first comparison, the detection of features location and affine parameters are evaluated. To compare the feature locations and affine parameters, the camera motion needs to be known. These can be calculated for each motion described in the previous section, however, for the Uniform Motion: Front View, it is not possible. Thus, a second comparison is carried out which uses only the fundamental matrix instead of the motion parameters.

5.1   Affine Evaluation

Location Error. The location of the GT feature can be determined by the location of the same feature in the previous image and the parameters of the motion. The calculation of the motion parameter differs from motion to motion, the details can be found in the related sections. The error of the feature detection is the Euclidean distance of the GT and the estimated features:

$$Err_{det}(\mathbf{P}_{estimated}, \mathbf{P}_{GT}) = \|\mathbf{P}_{estimated} - \mathbf{P}_{GT}\|_2^2, \tag{29}$$

where P_{GT}, $P_{estimated}$ are the GT and estimated feature points, respectively.

Affine Error. While the error of the feature detection is based on the Euclidean distance, the error of affine transformation is calculated using the Frobenius norm of the difference matrix of the estimated and GT affine transformation. It can be formalized as follows:

$$Err_{aff}(\mathbf{A}_{estimated}, \mathbf{A}_{GT}) = \|\mathbf{A}_{estimated} - \mathbf{A}_{GT}\|_F, \tag{30}$$

where A_{GT}, $A_{estimated}$ are the GT and estimated affine transformations, respectively. The Frobenius norm is chosen, because it has a geometrical meaning of the related affine transformations, see [4] for details.

5.2   Fundamental Matrix Evaluation

For the motion named as Uniform Motion: Front View, the affine transformations and parameters of the motion can not be estimated. The objects, which are closer to the camera, modify more pixels between the successive images than objects located at the distance. Only the direction of the moving features can be calculated. That is the epipolar line that goes through the feature and the FOE. However, the fundamental matrix can be precisely calculated from FOE, $\mathbf{F} = [\mathbf{v}]_x$ and it can be used to refine the affine transformations.

The refined affine transformations are considered as GT, and used for the calculation of affine errors, as it is described in the previous section. Barath et al. [4] introduced an algorithm to find the closest affine transformations corresponding to the fundamental matrix. The paper states that it can be determined by solving a six-dimensional linear problem if the Lagrange-multiplier technique is applied for the constraints for the fundamental matrix.

For quantifying the quality of an affine transformation, the closest valid affine transformation [4] is computed first. Then the difference matrix between the original and closest valid affine transformations is computed. The quality is given by the Frobenius norm of this difference matrix.

6 Comparison

Sixteen affine feature detectors have been quantitatively compared in our tests[1]. The implementations are downloaded from the website of Visual Geometry Group, University of Oxford[2], except TBMR, which is available on the website of one of the authors[3], and ASIFT is available online[4]. Most of the methods are introduced in Sect. 2. Additionally to those, HARHES and SEDGELAP are added to the comparison. HARHES is the composition of HARAFF and HESAFF, while SEDGELAP finds shapes along the edges, using the Laplace operator.

After the features are extracted from the images, feature matching is run considering SIFT descriptors. The ratio test published by Lowe [12] was used for outlier filtering. Finally, the affine transformations are calculated for the filtered matches. The detectors determine only the elliptical area of the affine shapes without orientation. The orientation is assigned to the areas using the SIFT descriptor in a separate step. Finally, the affine transformation of the matched feature is given by:

$$A = A_2 R_2 (A_1 R_1)^{-1}, \tag{31}$$

where A_i, R_i $i \in [1, 2]$ are the local affine areas defined by ellipses, and the related rotation matrices assigned by the SIFT descriptor, respectively.

Table 1 summarizes the number of features, number of matched features and running time of the detectors on the tests. SEDGELAP and HARHES find the most features, however, the high number of features makes the matching more complicated and time consuming. In general, feature detectors using the synthetic images (ASIFT, ABRIK, AORB, etc.) find 10 to 40 thousand features, EBR, IBR and MSER find hundreds, the Harris based methods (HARAFF, HARLAP) find approximately ten to twenty thousand, and the Hessian based methods (HESAFF, HESLAP) find a few thousands of features. The ratio test [12], used for outlier filtering, excludes some feature matches. The second row of each test sequence in Table 1 shows the number of features after matching and outlier filtering. Note that approximately 50% of features are lost due to the ratio test. The running time of the methods are shown in the third rows of each test sequence. These times highly depend on the image resolution, which is 5MP in our tests. Each implementation was run on the CPU, using one core of the machine. Obviously, MSER is the fastest method, followed by SURF and TBMR. The Hessian based methods need approximately 1 s to process an image, while Harris based methods need 1.5 or 2 more times. The slowest are IBR, EBR, and the methods that use synthetic images. Note that, by parallelism and/or GPU implementations, the running times may show different results.

---

[1] In the previous version of this work [16], only eleven rival methods were compared.
[2] http://www.robots.ox.ac.uk/vgg/research/affine/index.html.
[3] http://laurentnajman.org/index.php?page=tbmr.
[4] http://www.ipol.im/pub/art/2011/my-asift/.

Table 1. Number of matched features and required running time for test sequences. The columns are the methods and the rows (triplets) are the test sequences. **First row** of each sequence shows the number of detected features. The number of matched features are shown in the **second row**. The **third row** of each test sequence contains the required running times.

		aakaze	abrisk	akaze	hessian	aorb	asift	EBR	HARAFF	HARHES	HARLAP	HESAFF	HESLAP	IBR	MSER	SEDGELAP	SURF	TBMR
Scaling	# All	13121	17991	**45702**	20511	44366		217	18720	21198	19033	3866	3994	1277	336	20132	783	4122
	# Matched	10206	9859	**38318**	9753	33307		91	8369	9544	8534	1702	1757	678	235	10071	636	2259
	Running time (s)	8.27	4.1	9.15	1.38	6.93	20.43	3.27	2.32	2.06	1.19	0.97	7.59	**0.41**	3.29		0.65	0.68
Rotation	# All	8405	11194	**30213**	18837	28245		114	14266	12686	12795	2655	2751	892	223	13173	508	3141
	# Matched	4661	4534	**21889**	4283	18202		33	5235	5943	5294	1070	1111	395	133	5872	390	1447
	Running time (s)	8.24	4.09	6.8	1.36	6.68	18.01	2.21	1.69	1.56	0.91	0.77	6.27	**0.35**	2.41		0.54	0.54
Bottomview1	# All	3669	5941	13552	**16219**	13712		59	7514	8122	7629	1211	1216	441	92	7222	204	2688
	# Matched	2848	3254	**11639**	6792	10120		27	3450	3730	3503	522	524	226	64	3743	169	1049
	Running time (s)	8.22	3.97	4.40	1.31	6.43	18.09	1.69	1.37	1.37	1.30	0.80	5.52	**0.38**	1.85		0.51	0.67
Bottomview2	# All	2840	4264	12776	**14695**	11710		26	12484	13020	12615	1119	1135	588	117	9461	139	3051
	# Matched	1985	2012	**9771**	5261	7523		8	4451	4681	4506	418	426	241	68	4094	107	1037
	Running time (s)	8.19	4.06	3.87	1.32	6.46	20.49	2.22	1.65	1.65	1.64	0.78	5.25	**0.36**	2.11		0.57	0.95
FrontView1	# All	28503	**46388**	24383	20910	40908		78	10128	13639	10415	4708	5032	695	233	14869	803	2083
	# Matched	21929	**26333**	20448	11040	30059		36	4623	6476	4758	2444	2583	416	164	7650	653	1058
	Running time (s)	8.42	4.38	6.61	1.43	6.81	17.97	3.16	2.12	2.12	1.66	1.19	8.47	**0.29**	2.67		0.81	0.76
FrontView2	# All	19504	16227	16825	**19696**	19527		59	9210	11193	9369	2108	2812	592	50	12345	511	1856
	# Matched	**16582**	10672	15157	11579	15618		30	4615	5764	4703	1192	1617	398	38	7275	447	1034
	Running time (s)	8.31	4.11	5.44	1.37	6.52	15.56	2.07	1.65	1.48	1.01	0.87	6.58	**0.29**	2.31		0.65	0.55

6.1   Affine Evaluation

The first test considers the estimated camera motions and affine transformations introduced in Sect. 3. The quantitative evaluation is twofold, since the localization of features and accuracy of affine transformation can both be evaluated.

Four test sequences are captured for this kind of comparison. One for the scaling, one for the rotation and two for the Uniform Motion: Bottom View motions. See Fig. 2 for the rotation, Fig. 5 for the scale, first row of Figs. 4 and 8 for the Uniform Motion: Bottom View test images.

Figure 6 summarizes the errors for the detectors. The average and median error of feature detection can be seen on the bar-charts, where the left vertical axis mark the Euclidean distance between the estimated and GT features, measured in pixels. The average and median error of the affine transformations are visualized as green and black lines, respectively. The measure is given by the Frobenius norm of the difference of the GT and estimated affine transformations. The quantity of the error can be seen at the right vertical axis of the charts.

Since the images in BottomView2, Scaling and Rotation datasets consist of a dominant plane and some other regions like bushes and trees, the application of homography based outlier filtering presented in Sect. 4 is possible. For all image pairs, the homography matrix for each affine correspondence is calculated. For each correspondence of an image pair, a plane can be derived. For all other correspondences in the image pair, points are transformed with the estimated homography from the first image to the second one. The Euclidean distances of the transformed and original points on the second image are calculated. Points whose Euclidean distance is smaller than 20 pixels are considered to be coplanar with the selected point, they are the inliers. Other points are considered to be outliers. The plane with the largest number of inlier points is considered to be the most dominant plane detected on the image pair.

With the application of the plane based outlier filtering, both the mean and median errors are decreased for all detectors except EBR. In case of EBR, there are too few detected points for the outlier filtering to work. In some image pairs, there is no dominant plane with more than one inlier point, in this case, all points are considered as outliers.

The charts of Fig. 7 indicate similar results for the test sequences compared to Fig. 6. The detection error is always the lowest for the ASIFT, AHessian and the Hessian- and Harris-based methods in average, indicating that these methods can find features more accurately than others. This measure is the highest for the AORB, EBR, IBR, MSER and TBMR. The averages can even be higher than 20 pixels for these methods. Remark that the median values are always lower than the averages, because despite of the outlier filtering, the false matches can yield large detection error which distorts the averages. Surprisingly, the affine error of AHESSIAN, HESLAP and SURF is the lowest, while that of the EBR, IBR and TBMR is the highest in all test cases.

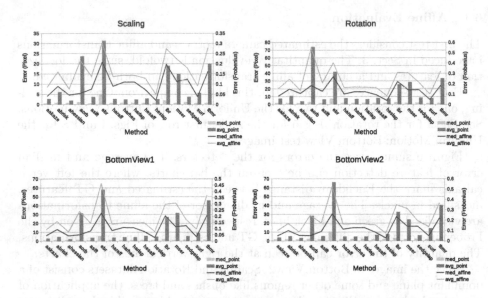

Fig. 6. Errors for quantitative evaluation without outlier filtering. The error of feature detection is measured in pixels, and can be seen on the left axis. The error of affine transformations is given by the Frobenius norm, it is plotted on the right axis. The average and median values for both the affine and detection errors are shown.

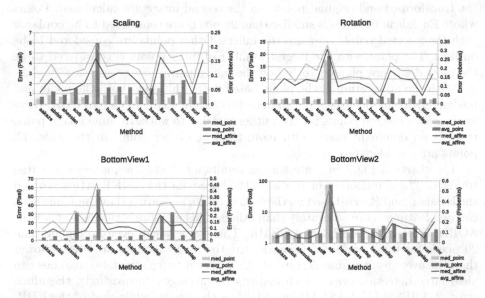

Fig. 7. Errors for quantitative evaluation with outlier filtering. The error of feature detection is measured in pixels, and can be seen on the left axis. The error of affine transformations is given by the Frobenius norm, it is plotted on the right axis. The average and median values for both the affine and detection errors are shown.

6.2   Fundamental Matrix Evaluation

In case of the fundamental matrix evaluation, the quadcopter moves forward, perpendicular to the image plane. Objects are located at different distances from the camera, thus the GT affine transformation and GT position of features can not be recovered. In this comparison, the fundamental matrix is used to refine the estimated affine transformations. Then, the error is measured by Frobenius norm of the difference matrix of the estimated and refined affine transformation.

Two test scenarios are considered. The example images of the first one can be seen in Fig. 3, where the height of the quadcopter was approximately 2 m. The images of the second scenario are shown in the second row of Fig. 8, these images are taken at around the top of the trees.

Figure 9 shows the error of the affine transformations. Two test sequences are captured, each shows similar result. The AHESSIAN, HARAFF, HARLAP and SURF affine transformation yield the least affine errors. The characteristic of the errors is similar to that in the previous comparison.

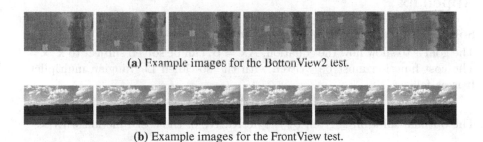

(a) Example images for the BottonView2 test.

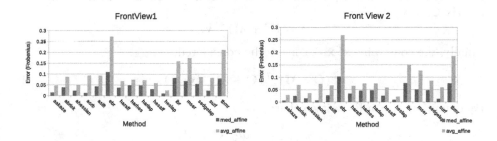

(b) Example images for the FrontView test.

Fig. 8. The images are taken while the quadcopter is moving parallel to the ground. **First row:** The camera observes the ground. **Second row:** It faces to the front. Figure from Pusztai and Hajder [16, Figure 7].

Fig. 9. The error of fundamental matrix evaluation.

7 Conclusions

The main goal of this paper is to quantitatively compare the state-of-the-art affine matchers on realistic images taken by a quadcopter-mounted camera. We have proposed novel test sequences for this purpose. The generated ground truth data is very accurate due to two reasons: (i) The motion of the quadcopter is constrained, camera motion has only a few degrees of freedom, therefore motion parameters can be accurately estimated, and (ii) the epipolar geometry between two images determines two parameters [4] for each affine transformation, therefore only two degrees of freedom remain.

The conclusion of our tests states that the ranking of the compared algorithms does not significantly depend on the sequences. As a side effect, point matching is also compared in the paper. Based on the results, the most accurate affine detectors/matchers are Harris-based, Hessian-based, SURF and ASIFT algorithms.

Appendix

Solving Constrained Linear Equations

The goal is to show how the equation, $\mathbf{A}\mathbf{x} = \mathbf{b}$, can be solved subject to $\mathbf{x}^T\mathbf{x} = 1$. The cost function must be written with the so-called Lagrangian multiplier λ. It is as follows:

$$J = (\mathbf{A}\mathbf{x} - \mathbf{b})^T(\mathbf{A}\mathbf{x} - \mathbf{b}) + \lambda\mathbf{x}^T\mathbf{x}.$$

The optimal solution is given by the derivative of the cost function w.r.t $\mathbf{x}$.

$$\frac{\partial J}{\partial \mathbf{x}} = 2\mathbf{A}^T(\mathbf{A}\mathbf{x} - \mathbf{b}) + 2\lambda\mathbf{x} = 0.$$

Therefore the optimal solution is as follows:

$$\mathbf{x} = (\mathbf{A}^T\mathbf{A} + \lambda\mathbf{I})^{-1}\mathbf{A}^T\mathbf{b}.$$

For the sake of simplicity, we introduce the vector $\mathbf{v} = \mathbf{A}^T\mathbf{b}$ and the symmetric matrix $\mathbf{C} = \mathbf{A}^T\mathbf{A}$, then:

$$\mathbf{x} = (\mathbf{C} + \lambda\mathbf{I})^{-1}\mathbf{v}.$$

Finally, the constraint $\mathbf{x}^T\mathbf{x} = 1$ has to be considered:

$$\mathbf{v}^T(\mathbf{C} + \lambda\mathbf{I})^{-T}(\mathbf{C} + \lambda\mathbf{I})^{-1}\mathbf{v} = 1.$$

By definition, it can be written that:

$$(\mathbf{C} + \lambda\mathbf{I})^{-1} = \frac{adj(\mathbf{C} + \lambda\mathbf{I})}{det(\mathbf{C} + \lambda\mathbf{I})}.$$

If

$$\mathbf{C} = \begin{bmatrix} c_1 & c_2 \\ c_3 & c_4 \end{bmatrix},$$

then

$$\mathbf{C} + \lambda\mathbf{I} = \begin{bmatrix} c_1 + \lambda & c_2 \\ c_3 & c_4 + \lambda \end{bmatrix}$$

The determinant and adjoint matrix of $C + \lambda I$ can be written as:

$$det(\mathbf{C} + \lambda\mathbf{I}) = (c_1 + \lambda)(c_4 + \lambda) - c_2 c_3$$

and

$$adj(\mathbf{C} + \lambda\mathbf{I}) = \begin{bmatrix} c_4 + \lambda & -c_2 \\ -c_3 & c_1 + \lambda \end{bmatrix}$$

$$\begin{bmatrix} c_4 + \lambda & -c_2 \\ -c_3 & c_1 + \lambda \end{bmatrix} \begin{bmatrix} v_1 \\ v_2 \end{bmatrix} = \begin{bmatrix} v_1\lambda + c_4 v_1 - c_2 v_2 \\ v_2\lambda + c_1 v_2 - c_3 v_1 \end{bmatrix}.$$

Furthermore, the expression $\mathbf{v}^T(\mathbf{C} + \lambda\mathbf{I})^{-T}(\mathbf{C} + \lambda\mathbf{I})^{-1}\mathbf{v} = 1$ can be rewritten as

$$\mathbf{v}^T \frac{adj^T(\mathbf{C} + \lambda\mathbf{I})adj(\mathbf{C} + \lambda\mathbf{I})}{det(\mathbf{C} + \lambda\mathbf{I})det(\mathbf{C} + \lambda\mathbf{I})}\mathbf{v} = 1,$$

$$\mathbf{v}^T adj^T(\mathbf{C} + \lambda\mathbf{I})adj(\mathbf{C} + \lambda\mathbf{I})\mathbf{v} = det^2(\mathbf{C} + \lambda\mathbf{I}).$$

Both sides of the equation contain polynomials. The degrees of the left and right sides are $2n - 2$ and $2n$, respectively. If the expression in the sides are subtracted by each other, a polynomial of degree $2n$ is obtained. Note that, $n = 2$ in the discussed case, i.e planar motion. The optimal solution is obtained as the real roots of this polynomial. The vector corresponding to the estimated λ_i, $i \in 1, 2$, is calculated as $\mathbf{g}_i = (\mathbf{L} + \lambda_i\mathbf{I})^{-1}r$. Then the vector with minimal norm $\|\mathbf{F}\mathbf{g}_i - \mathbf{h}\|$ is selected as the optimal solution of the problem.

Quadcopter Movement

This section proves, that in case of a pure translation motion of the quadcopter, the motion of features in the images depends on global parameters, which can be accurately approximated independently of the location of the features. The pure translation motion denotes that the camera itself does not rotate, the intrinsic parameters does no change and the extrinsics are related by a translation. Let the first camera position considered as the origin and let $\tilde{\mathbf{X}} = [X, Y, Z, 1]^T$ and $\mathbf{K}$ be the homogeneous coordinate of arbitrary point in the ground and the camera matrix, respectively. Furthermore, let $\mathbf{t}$ be the translation vector of the movement. Projecting $\tilde{\mathbf{X}}$ to the two camera images gives:

$$u^1 = \mathbf{K}[\mathbf{I}|0]\,\tilde{\mathbf{X}}, \tag{32}$$

$$u^2 = \mathbf{K}[\mathbf{I}|\mathbf{t}]\,\tilde{\mathbf{X}} = [\mathbf{K}|\mathbf{Kt}]\,\tilde{\mathbf{X}}. \tag{33}$$

The dehomogenized image coordinates are as follows:

$$\mathbf{u}^1 = \begin{bmatrix} u_x^1/u_z^1 \\ u_y^1/u_z^1 \end{bmatrix} = \begin{bmatrix} (Xf_x + Zc_x)/Z \\ (Yc_y + Zc_y)/Z \end{bmatrix},$$

$$\mathbf{u}^2 = \begin{bmatrix} u_x^2/u_z^2 \\ u_x^2/u_z^2 \end{bmatrix} = \begin{bmatrix} (Xf_x + Zc_x + f_x t_x + c_x t_z)/(Z + t_z) \\ (Yf_y + Zc_y + f_y t_y + c_y t_z)/(Z + t_z) \end{bmatrix} \tag{34}$$

$$= \begin{bmatrix} (Xf_x + Zc_x)/(Z + t_z) \\ (Yc_y + Zc_y)/(Z + t_z) \end{bmatrix} + \begin{bmatrix} (f_x t_x + c_x t_z)/(Z + t_z) \\ (f_y t_y + c_y t_z)/(Z + t_z) \end{bmatrix}$$

$$= \frac{Z}{Z + t_z} u^1 + \frac{1}{Z + t_z} \begin{bmatrix} f_x t_x + c_x t_z \\ f_y t_y + c_y t_z \end{bmatrix}. \tag{35}$$

Equation 35 indicates that the movement of the image pixel $\mathbf{u}^1$ indicated by $\mathbf{t}$ pure translation motion depends on $\mathbf{t}$, the intrinsic parameters and the distance of the related 3D location along the Z axis. Note that $Z + t_z \neq 0$, because that would mean that the spatial features are located in the camera plane in the second image. Let $\mathbf{m}$ denote the projection of $\mathbf{t}$ to the image plane as it follows:

$$\tilde{\mathbf{m}} = \mathbf{Kt},$$

$$\mathbf{m} = \frac{1}{t_z} \begin{bmatrix} f_x t_x + c_x t_z \\ f_y t_y + c_y t_z \end{bmatrix}. \tag{36}$$

$\mathbf{m}$ is called the Focus of Expansion (FoE). Subtracting the FoE from both sides of the Eq. 35 gives:

$$\mathbf{u}^2 - \mathbf{m} = \frac{Z}{Z + t_z} u^1 + \frac{1}{Z + t_z} \begin{bmatrix} f_x t_x + c_x t_z \\ f_y t_y + c_y t_z \end{bmatrix} - \mathbf{m}$$

$$= \frac{Z}{Z + t_z} u^1 + \frac{1}{Z + t_z} \begin{bmatrix} f_x t_x + c_x t_z \\ f_y t_y + c_y t_z \end{bmatrix} - \frac{1}{t_z} \begin{bmatrix} f_x t_x + c_x t_z \\ f_y t_y + c_y t_z \end{bmatrix}$$

$$= \frac{Z}{Z + t_z} u^1 + \frac{t_z}{(Z + t_z)t_z} \begin{bmatrix} f_x t_x + c_x t_z \\ f_y t_y + c_y t_z \end{bmatrix} - \frac{Z + t_z}{(Z + t_z)t_z} \begin{bmatrix} f_x t_x + c_x t_z \\ f_y t_y + c_y t_z \end{bmatrix}$$

$$= \frac{Z}{Z + t_z} u^1 + \frac{-Z}{(Z + t_z)t_z} \begin{bmatrix} f_x t_x + c_x t_z \\ f_y t_y + c_y t_z \end{bmatrix}$$

$$= \frac{Z}{Z + t_z} \left(u^1 - \frac{1}{t_z} \begin{bmatrix} f_x t_x + c_x t_z \\ f_y t_y + c_y t_z \end{bmatrix} \right)$$

$$= \frac{Z}{Z + t_z} (u^1 - \mathbf{m}). \tag{37}$$

Which indicates that all features move along a ray casted from the FoE to the feature. The speed of the movement depends on the location of the feature, the distance of the ground and the length of the translation vector. In the quadcopter test, the ground is considered to be planar, thus all of the features have the same depth. The FoE depends on only the motion of the quadcopter. These indicates that the parameters of the movement of the features in the images can be measured globally for each image pair. An exploitation case is given, if the translation **t** is parallel to the image plane. The case is described in the next section.

Degenerate Case: Parallel Movement. In case of the parallel movement, the translation vector is parallel to the image plane, which indicates that $\mathbf{t} = [t_x, t_y, 0]$. The FoE is located in the infinity and cannot be computed. However, in this case, Eq. 35 changes to:

$$\mathbf{u}^2 = \mathbf{u}^1 + \frac{1}{Z}\begin{bmatrix} f_x t_x \\ f_y t_y \end{bmatrix} \tag{38}$$

The equation above indicates that all of the features move in the same direction. Similarly to the previous case, if the quadcopter observes the ground then all features have the same depth. Thus, Z is constant and the movement of the features depends only on $\mathbf{t}$ and the focal lengths. Therefore, the translation is global in the sense that it is independent of the feature locations, thus it can be estimated for an image pair.

Fundamental Matrix from the FoE

This section proves that the fundamental matrix $\mathbf{F}$ can be computed from the FoE, if the camera motion is a pure translation.

$$\mathbf{F} = \mathbf{K}_1^{-T}\mathbf{E}\mathbf{K}_1^{-1} = \mathbf{K}_1^{-T}[\mathbf{t}]_x\mathbf{R}\mathbf{K}_1^{-1}. \tag{39}$$

Since the motion contains only translation, the $\mathbf{R}$ rotation matrix is the identity. The definition of the matrix representation of the cross product and the definition of the FoE from Eq. 36 gives:

$$
\begin{aligned}
\mathbf{F} = \mathbf{K}_1^{-T}[\mathbf{t}]_x\mathbf{I}\mathbf{K}_1^{-1} &= \begin{bmatrix} 1/f_x & 0 & 0 \\ 0 & 1/f_y & 0 \\ -c_x/f_x & -c_y/f_y & 1 \end{bmatrix}\begin{bmatrix} 0 & -t_z & t_y \\ t_z & 0 & -t_x \\ -t_y & t_x & 0 \end{bmatrix}\begin{bmatrix} 1/f_x & 0 & -c_x/f_x \\ 0 & 1/f_y & -c_y/f_y \\ 0 & 0 & 1 \end{bmatrix} \\
&= \begin{bmatrix} 0 & -t_z/(f_x f_y) & (c_y t_z + f_y t_y)/(f_x f_y) \\ t_z/(f_x f_y) & 0 & -(c_x t_z + f_x t_x)/(f_x f_y) \\ -(t_y + (c_y t_z)/f_y)/f_x & (t_x + (c_x t_z)/f_x)/f_y & 0 \end{bmatrix} \\
&= \frac{1}{f_x f_y}\begin{bmatrix} 0 & -t_z & f_y t_y + c_y t_z \\ t_z & 0 & -(f_x t_x + c_x t_z) \\ -(f_y t_y + c_y t_z) & (f_x t_x + c_x t_z) & 0 \end{bmatrix} \sim [\tilde{\mathbf{m}}]_x. \tag{40}
\end{aligned}
$$

References

1. Alcantarilla, P.F., Nuevo, J., Bartoli, A.: Fast explicit diffusion for accelerated features in nonlinear scale spaces. In: British Machine Vision Conference (BMVC) (2013)

2. Alcantarilla, P.F., Bartoli, A., Davison, A.J.: Kaze features. In: Fitzgibbon, A., Lazebnik, S., Perona, P., Sato, Y., Schmid, C. (eds.) Computer Vision - ECCV 2012 (2012)
3. Barath, D., Hajder, L.: A theory of point-wise homography estimation. Pattern Recogn. Lett. **94**, 7–14 (2017)
4. Barath, D., Matas, J., Hajder, L.: Accurate closed-form estimation of local affine transformations consistent with the epipolar geometry. In: Proceedings of the British Machine Vision Conference 2016, BMVC 2016, York, UK, 19–22 September 2016 (2016)
5. Barath, D., Molnár, J., Hajder, L.: Optimal surface normal from affine transformation. In: VISAPP 2015 - Proceedings of the 10th International Conference on Computer Vision Theory and Applications, Berlin, Germany, 11–14 March 2015, vol. 3, pp. 305–316 (2015)
6. Bay, H., Ess, A., Tuytelaars, T., Van Gool, L.: Speeded-up robust features (SURF). Comput. Vis. Image Underst. **110**(3), 346–359 (2008)
7. Cordes, K., Rosenhahn, B., Ostermann, J.: High-resolution feature evaluation benchmark. In: Wilson, R., Hancock, E., Bors, A., Smith, W. (eds.) CAIP 2013. LNCS, vol. 8047, pp. 327–334. Springer, Heidelberg (2013). https://doi.org/10.1007/978-3-642-40261-6_39
8. Fischler, M., Bolles, R.: Random sampling consensus: a paradigm for model fitting with application to image analysis and automated cartography. Commun. Assoc. Comp. Mach. **24**, 358–367 (1981)
9. Harris, C., Stephens, M.: A combined corner and edge detector. In: Proceedings of Fourth Alvey Vision Conference, pp. 147–151 (1988)
10. Hartley, R., Zisserman, A.: Multiple View Geometry in Computer Vision, 2nd edn. Cambridge University Press, New York (2003)
11. Leutenegger, S., Chli, M., Siegwart, R.Y.: BRISK: binary robust invariant scalable keypoints. In: Proceedings of the 2011 International Conference on Computer Vision, ICCV 2011, pp. 2548–2555 (2011)
12. Lowe, D.G.: Distinctive image features from scale-invariant keypoints. Int. J. Comput. Vision **60**(2), 91–110 (2004)
13. Mikolajczyk, K., et al.: A comparison of affine region detectors. Int. J. Comput. Vision **65**(1), 43–72 (2005)
14. Mikolajczyk, K., Schmid, C.: An affine invariant interest point detector. In: Heyden, A., Sparr, G., Nielsen, M., Johansen, P. (eds.) ECCV 2002. LNCS, vol. 2350, pp. 128–142. Springer, Heidelberg (2002). https://doi.org/10.1007/3-540-47969-4_9
15. Pusztai, Z., Hajder, L.: Quantitative comparison of affine invariant feature matching. In: Proceedings of the 12th International Joint Conference on Computer Vision, Imaging and Computer Graphics Theory and Applications (VISIGRAPP 2017), VISAPP, Porto, Portugal, 27 February–1 March 2017, vol. 6, pp. 515–522 (2017)
16. Pusztai, Z., Hajder, L.: Quantitative affine feature detector comparison based on real-world images taken by a quadcopter. In: Proceedings of the 14th International Joint Conference on Computer Vision, Imaging and Computer Graphics Theory and Applications, VISIGRAPP, pp. 704–715 (2019)
17. Rublee, E., Rabaud, V., Konolige, K., Bradski, G.: ORB: an efficient alternative to SIFT or SURF. In: 2011 International Conference on Computer Vision, pp. 2564–2571 (2011)
18. Saurer, O., Vasseur, P., Boutteau, R., Demonceaux, C., Pollefeys, M., Fraundorfer, F.: Homography based egomotion estimation with a common direction. IEEE Trans. Pattern Anal. Mach. Intell. **39**(2), 327–341 (2017)

19. Shi, J., Tomasi, C.: Good features to track. In: Conference on Computer Vision and Pattern Recognition, CVPR 1994, Seattle, WA, USA, 21–23 June 1994, pp. 593–600 (1994)
20. Tareen, S.A.K., Saleem, Z.: A comparative analysis of SIFT, SURF, KAZE, AKAZE, ORB, and BRISK. In: International Conference on Computing, Mathematics and Engineering Technologies (iCoMET) (2018)
21. Tuytelaars, T., Mikolajczyk, K.: Local invariant feature detectors: a survey. Found. Trends. Comput. Graph. Vis. **3**(3), 177–280 (2008)
22. Tuytelaars, T., Van Gool, L.: Matching widely separated views based on affine invariant regions. Int. J. Comput. Vision **59**(1), 61–85 (2004)
23. Xu, Y., Monasse, P., Géraud, T., Najman, L.: Tree-based morse regions: a topological approach to local feature detection. IEEE Trans. Image Process. **23**(12), 5612–5625 (2014)
24. Yu, G., Morel, J.M.: ASIFT: an algorithm for fully affine invariant comparison. Image Process. Line **1**, 11–38 (2011)
25. Zitnick, C.L., Ramnath, K.: Edge foci interest points. In: IEEE International Conference on Computer Vision, ICCV 2011, Barcelona, Spain, 6–13 November 2011, pp. 359–366 (2011)

An MRF Optimisation Framework for Full 3D Reconstruction of Scenes with Complex Reflectance

Gianmarco Addari[✉] and Jean-Yves Guillemaut

Centre for Vision, Speech and Signal Processing (CVSSP), University of Surrey,
Guildford GU2 7XH, United Kingdom
{g.addari,j.guillemaut}@surrey.ac.uk

Abstract. The ability to digitise real objects is fundamental in applications such as film post-production, cultural heritage preservation and video game development. While many existing modelling techniques achieve impressive results, they are often reliant on assumptions such as prior knowledge of the scene's surface reflectance. This considerably restricts the range of scenes that can be reconstructed, as these assumptions are often violated in practice. One technique that allows surface reconstruction regardless of the scene's reflectance model is Helmholtz Stereopsis (HS). However, to date, research on HS has mostly been limited to 2.5D scene reconstruction. In this paper, a framework is introduced to perform full 3D HS using Markov Random Field (MRF) optimisation for the first time. The paper introduces two complementary techniques. The first approach computes multiple 2.5D reconstructions from a small number of viewpoints and fuses these together to obtain a complete model, while the second approach directly reasons in the 3D domain by performing a volumetric MRF optimisation. Both approaches are based on optimising an energy function combining an HS confidence measure and normal consistency across the reconstructed surface. The two approaches are evaluated on both synthetic and real scenes, measuring the accuracy and completeness obtained. Further, the effect of noise on modelling accuracy is experimentally evaluated on the synthetic dataset. Both techniques achieve sub-millimetre accuracy and exhibit robustness to noise. In particular, the method based on full 3D optimisation is shown to significantly outperform the other approach.

Keywords: 3D reconstruction · Helmholtz stereopsis · Markov random fields

1 Introduction

Reconstructing real world objects provides an efficient mechanism for producing digital assets. This finds application in many industries such as film post-production, cultural heritage and gaming. Recent advances in 3D modelling have been substantial, but many open problems still remain. This paper focuses on the specific problem of modelling scenes exhibiting complex surface reflectance (e.g. glossy surfaces). Existing techniques often operate under the assumption that the reflectance of the scene (its

© Springer Nature Switzerland AG 2020
A. P. Cláudio et al. (Eds.): VISIGRAPP 2019, CCIS 1182, pp. 456–476, 2020.
https://doi.org/10.1007/978-3-030-41590-7_19

Bidirectional Reflectance Distribution Function (BRDF)) is known a priori, and furthermore it is often assumed to be Lambertian, which simplifies the reconstruction process. These assumptions restrict the range of scenes that can be reconstructed or, when not satisfied, produce inaccurate results.

In recent years, a new technique called Helmholtz Stereopsis (HS) was proposed to attempt to tackle this problem. The method enforces a principle from physics known as Helmholtz reciprocity, which allows to perform reconstruction regardless of the scene's BRDF. Work performed on this technique has, however, been mostly limited to depth estimation [23,35], which results in only a partial (2.5D) scene reconstruction. In the case where full 3D reconstruction was explored [5], no global optimality was guaranteed due to the reliance on gradient descent to perform the optimisation. A robust approach that does not require a close initialisation to the target surface and that offers strong guarantees on global optimality has yet to be proposed.

In this paper, we propose two new approaches for full 3D reconstruction of scenes with arbitrary unknown surface reflectance, thus contributing two valuable additions to the state of the art in the field. The first contribution is the introduction of an approach for 3D HS reconstruction via fusion of multiple 2.5D reconstructions obtained from a small number of viewpoints (six considered in this paper) and inferred via MRF optimisation defined over a viewpoint-dependent 2D grid. The second contribution is a volumetric approach which directly performs the optimisation in the 3D domain and where each voxel in the grid corresponds to a node in the MRF graph. In this approach the optimal labelling identifies not only inside and outside voxels, but also the surface position inside the boundary voxels. In contrast with previous work which normally ignores visibility, both approaches introduced incorporate a novel visibility aware formulation, which is essential to enable generalisation to full 3D. Both approaches are shown to allow for accurate reconstruction and to be robust to noise.

The two approaches share some commonalities being both based on MRF optimisation and achieving full 3D modelling. They can either operate independently (providing two different strategies to generalise HS to the 3D domain) or be combined into a single modelling pipeline where they complement each other (e.g. with the volumetric approach serving as a refinement process to the output of the fused view-dependent approach). To the authors' knowledge, these are the first MRF formulations of HS for full 3D reconstruction. This paper extends our previous work initially reported in [1]. In particular, we provide substantially improved evaluation to better assess the performance and robustness of the two techniques.

The paper is structured as follows. Section 2 provides an overview of related work, with an emphasis on how the different approaches are able to handle surface reflectance. The following three sections introduce the different approaches. A similar formalism is adopted when presenting the different approach in order to highlight their commonalities and differences. Then the different approaches are experimentally evaluated on both synthetic and real datasets with different metrics considered to analysis performance. Finally, conclusions and avenues for future work are discussed.

2   Related Work

In this section, we start by providing an overview of the most common 3D reconstruction approaches. Then we provide a more detailed description of Helmholtz stereopsis which is used in this paper.

2.1   Overview of Main 3D Reconstruction Approaches

Shape from Silhouettes (SfS) was proposed in [15] and consists in using 2D silhouette data to reconstruct a Visual Hull (VH) of the object. Despite having recently been improved upon in recent works [17,20], SfS suffers from being unable to reconstruct concavities.

Classic binocular and multi-view stereo approaches [25,26] allow for more complex geometric reconstructions than SfS, but are limited by the scene BRDF, which needs to be Lambertian. As this is often not the case, assuming the wrong BRDF can lead to incorrect reconstructions. Some more recent multi-view techniques [18,21,22] attempt to jointly estimate the geometry and reflectance models of the scene by calculating iteratively the scene shape from the current estimated reflectance and vice versa, which ultimately constrains each calculation to the accuracy with which the other parameter was estimated.

Finally, Photometric Stereo (PS) [32] consists in computing the normals of a scene given a set of inputs with stationary point of view but varying lighting. Despite the possibility of reconstructing non-Lambertian scenes, in PS the BRDF needs to be known a priori. State-of-the-art work includes [29], where a shiny, textureless object is reconstructed using shadows and varying illumination, under the simplifying assumption of a Lambertian reflectance model. In [4] image derivatives are used to reconstruct surfaces with unknown BRDF, albeit being limited to isotropic BRDFs. In [6] a generic form of BRDF is used to compute shapes, each point on the surface is considered to be a mixture of previously calculated BRDFs, however this is limited to a maximum of two materials per point. In [9], PS is computed for isotropic and anisotropic reflectance models by considering the following characteristics of the BRDF: the diffuse component, the concentration of specularities and the resulting shadows. Despite achieving great results, the set-up in this paper is extremely complex and a large number of different light directions are needed to perform the surface reconstruction.

2.2   Helmholtz Stereopsis

The use of Helmholtz reciprocity [30] for 3D reconstruction was first proposed by Magda et al. in [19] where the principle is used to recover the geometry of scenes with arbitrary and anisotropic BRDF in a simplified scenario. The proposed technique was then further developed into HS in [35] by utilising it to perform normal estimation as well. In the classical HS formulation maximum likelihood is used to determine the depth of each point in the scene, while the normals are estimated using Singular Value Decomposition (SVD). No integration is performed over the surface, resulting in discontinuities and a noisy reconstruction. To enforce Helmholtz reciprocity, a set-up similar to the one shown in Fig. 1 needs to be used. Let us consider a camera and an

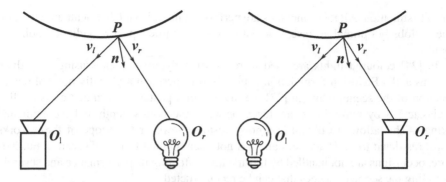

Fig. 1. Camera/light pair positioning in HS.

isotropic, unit-strength point light source, respectively positioned at O_l and O_r. For a given surface point P, the irradiance measured at its projection in the left image (i_l) can be expressed as:

$$i_l = f_r(v_r, v_l) \cdot \frac{n \cdot v_r}{\|O_r - P\|^2} \tag{1}$$

where f_r indicates the BRDF at point P with incident light direction v_r and viewing direction v_l. n is the surface normal at P and $\frac{1}{\|O_r - P\|^2}$ accounts for the light falloff. If the light and camera are interchanged, a corresponding equation for the irradiance of the projection of P in the right image (i_r) is obtained where $f_r(v_l, v_r) = f_l(v_r, v_l)$, due to the Helmholtz reciprocity constraint. By performing a substitution, the reflectance term can be removed and the following equation, independent from the surface BRDF, is obtained:

$$\left(i_l \frac{v_l}{\|O_l - P\|^2} - i_r \frac{v_r}{\|O_r - P\|^2} \right) \cdot n = w \cdot n = 0 \tag{2}$$

Utilising multiple camera light pairs (at least three) allows to obtain a matrix W where each row is a w vector. By minimising the product $W \cdot n$ it is possible to obtain an estimate of the normal at point P. To do so the matrix is decomposed using SVD:

$$SVD(W) = U \Sigma V^T \tag{3}$$

where Σ is a diagonal matrix and U and V are orthogonal matrices. The last column of V gives an estimate of the normal, while the non zero terms in the diagonal matrix ($\sigma_1, \sigma_2, \sigma_3$) can be used to compute a quality measure of the normal.

In [33] and [27] it is demonstrated that HS can be performed with as low as a single pair of reciprocal images, however the assumption of a C_1 continuous surface is made, making it impossible to reconstruct surfaces that present discontinuities. Further work include [8] where HS is applied to rough and textured surfaces by integrating the image intensities over small areas, [11] where it is shown how performing radiometric scene calibration allows for a vast improvement in the normal accuracy calculated using HS, [34] where geometric and radiometric calibration are performed by exploiting correspondences between the specular highlights in the images and finally [7] where an alternative radiometric distance is proposed to perform a maximum likelihood surface normal estimation. All these methods are performed in 2.5D and do not attempt to compute a globally optimal surface, calculating instead an occupancy likelihood point by point.

In [24] coloured lights are used to reconstruct dynamic scenes using only three cameras and a Bayesian formulation is proposed to perform a globally optimal reconstruction of the scene, while in [23] the maximum a posteriori formulation is applied to classic HS by enforcing consistency between the points depth and the estimated normals. This allows to obtain less noisy results, however the scope of these works is still restricted to 2.5D surfaces and has not been applied to full 3D scenes. Furthermore, occlusions are not handled by this method, affecting its performance and severely restricting the scope of scenes that can be reconstructed.

HS was first applied in the 3D domain in [31], where it is used to complement structured light consistency, which is unable to obtain high frequency details on the reconstructed surface and [5], where a variational formulation is presented to reconstruct a 3D mesh using gradient descent optimisation. In Weinmann's work HS is only used as a refinement step on areas of the surface where fine details are present and a very complex set-up consisting of a light dome is used, which makes this method extremely difficult to deploy and constrained to a specific set of scenes. In Delaunoy's work, instead, the set-up is simply a turntable, a pair of fixed lights and a fixed camera, which makes it easier to reproduce. However, because the optimisation employed is based on gradient descent, global optimality is not guaranteed and the method could get trapped in local minima if a proper initialisation is not provided.

In contrast, the proposed method performs multiple 2.5D reconstructions from different view-points, which are then fused together to obtain a full 3D model of the scene. This is followed by an MRF optimisation on the 3D reconstruction, which, compared to gradient descent, is less reliant on having a good initialisation and benefits from some optimality guarantees depending on the choice of algorithm used to optimise the energy function. During both steps self-occlusions are taken into consideration by performing an approximate visibility check.

3   Framework Overview

Here we provide an overview of the general framework introduced to achieve full 3D reconstruction of surfaces with complex surface reflectance. In particular, this section identifies the fundamental steps and how the two proposed approaches are integrated into a coherent modelling framework.

Figure 2 provides an overview of the general pipeline introduced. The inputs to this method are calibrated Helmholtz reciprocal pairs of images of the object and the silhouettes for each view. The reciprocal camera pairs can be arbitrary positioned in the scene and are assumed to be fully calibrated (i.e. both intrinsic and extrinsic are known). The first step consists in defining a voxel grid that contains the whole object, and reconstructing its VH applying SfS to the silhouettes of the entire set of images captured.

The next step reconstructs a set of separate views of the object using a visibility aware Bayesian formulation of HS initialised using the VH. The approach extends the

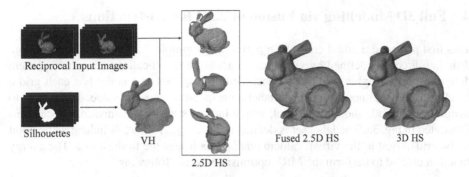

Fig. 2. Pipeline overview [1].

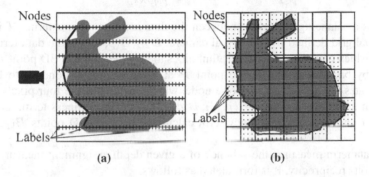

Fig. 3. Simplified representation of labelling in the 2.5D (a) and 3D (b) methods, with an example of labelling solution indicated in red in each case [1]. (Color figure online)

method from [23] with the use of additional information on visibility provided by the VH to select a subset of cameras for each view and point independently. Selecting camera visibility correctly is a critical step since the cameras are not all placed on a plane as in other 2.5D Helmholtz methods, but can instead be placed anywhere in the 3D space surrounding the object, depending on how the dataset was collected. In the proposed approach, reconstruction is performed from the six viewing directions defined by the cardinal axes of the reference frame. These define six orthographic virtual cameras and provide sufficient coverage to reconstruct the complete scene.

To eliminate redundancies and possible inconsistencies, the partial surfaces are fused together using Poisson surface reconstruction. The resulting surface is then used to initialise the final step of the reconstruction pipeline. In this final step, the problem is defined as a volumetric MRF, where each voxel corresponds to a node and the labelling defines whether the node is outside, inside, or on the surface of the reconstructed scene. A refined 3D model is obtained by MRF optimisation using a tailored Iterative Conditional Modes (ICM) algorithm.

The following two sections provide a detailed description of the novel approaches introduced which are core contributions of this paper.

4   Full 3D Modelling via Fusion of 2.5D Reconstructions

This first proposed method consists in performing multiple 2.5D reconstructions from different directions defined by virtual cameras. More specifically, each virtual camera defines an orthographic grid, where each cell corresponds to a node. For each grid a label set is then defined, where each label corresponds to a depth value. The goal is to assign to each node the correct label, which indicates where the surface is located as illustrated in Fig. 3a. The label set is defined as $l_0, ..., l_{d-1}$, where l_0 indicates the point in the grid closest to the virtual camera and l_{d-1} indicates the farthest one. The energy function utilised to perform the MRF optimisation is the following:

$$E(l) = (1 - \alpha) \sum_{p \in \mathcal{I}} D_{2D}(B(p, l_p)) + \alpha \sum_{p,q \in \mathcal{N}_{2D}} S_{2D}(B(p, l_p), B(q, l_q)) \quad (4)$$

where α is a balancing parameter between the data and smoothness terms, $\mathcal{I}$ is the 2-dimensional grid defined by the virtual camera, $D_{2D}(B(p, l_p))$ is the data term of the function, which corresponds to the Helmholtz saliency measured at 3D point $B(p, l_p)$, obtained by back-projecting image point P at the depth corresponding to label l_p. $\mathcal{N}_{2D}$ indicates the neighbourhood of a node, which consists of the four pixels directly adjacent in the image and $S_{2D}(B(p, l_p), B(q, l_q))$ is the smoothness term, and it corresponds to the normal consistency term calculated between 3D points $B(p, l_p)$ and $B(q, l_q)$.

The data term measures the saliency of a given depth assignment measured based on Helmholtz reciprocity. It is formulated as follows:

$$D_{2D}(\boldsymbol{P}) = \begin{cases} 1, & \text{if } |vis(\boldsymbol{P})| < min_{vis} \\ e^{-\mu \times \frac{\sigma_2(P)}{\sigma_3(P)}}, & \text{otherwise} \end{cases} \quad (5)$$

where $vis(\boldsymbol{P})$ indicates the set of reciprocal pairs of cameras from which point $\boldsymbol{P}$ is visible, min_{vis} is a variable set to enforce the minimum number of reciprocal pairs of cameras that make a normal estimate reliable, μ is assigned the value $0.2 \ln(2)$ to replicate the same weight used in [23] and σ_2 and σ_3 are two values from the diagonal matrix obtained performing SVD on matrix W as described previously (Eq. 3).

Unlike previous 2.5D HS approaches, our formulation incorporates a visibility term. This is an important contribution to generalise the approach and make it robust to occlusions when dealing with complex full 3D scenes. The first criterion to determine visibility is to only consider the cameras whose axis stands at an angle smaller than $80°$ with respect to the virtual camera axis. Then occlusions are computed by approximating each point's visibility based on the visibility of its closest point on the surface of the VH. If an intersection is found between the VH and the segment connecting the camera center to the approximated point, the camera and its reciprocal are considered to be occluded and are therefore discarded. Figure 4 illustrates the approach to visibility computation.

A smoothness term is used to regularise the labelling problem. Leveraging the unique ability of HS to infer both depth and surface normal, we define a smoothness term that is tailored to HS. More specifically, we use the distance based DNprior [23],

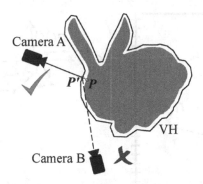

Fig. 4. Points on the surface (P) are approximated to the closest point on the VH (P') before occlusions are taken into consideration for visibility [1].

which enforces a smooth surface that is consistent with the normals obtained through HS and was to be related to integrability. This term is defined as follows.

$$S_{2D}(P, Q) = \begin{cases} \frac{1}{2}(\delta_{P,Q}^2 + \delta_{Q,P}^2), & \text{if } \delta_{P,Q} \text{ and } \delta_{Q,P} < t \\ t^2, & \text{otherwise} \end{cases} \tag{6}$$

where t is the maximum threshold for $\delta_{P,Q}$. $\delta_{P,Q}$ is the distance between point P and the projection of Q, perpendicular to its estimated normal, on the grid pixel where P lies as illustrated in Fig. 5a, and it is calculated as follows:

$$\delta_{P,Q} = \frac{PQ \cdot n(Q)}{n(Q) \cdot C} \tag{7}$$

where PQ is the vector connecting P and Q, $n(Q)$ indicates the estimated normal at point Q and C is the virtual camera axis. Whenever $\delta_{P,Q}$ or $\delta_{Q,P}$ are greater than a threshold t dependent on the reconstruction resolution, this term is truncated to t^2 in order to avoid heavy penalties where a strong discontinuity is present on the surface.

Tree Reweighted Message Passing (TRW) [13] is used to minimise the energy function thus recovering a depth map from each viewing direction. The different depth maps are then fused together using Poisson surface reconstruction [12]. This produces a complete 3D representation of the scene and improves reconstruction accuracy by reducing the inconsistencies between depth maps and filtering out artefacts present in individual depth maps. The obtained surface can be used as initialisation to the approach presented in the following section which directly performs full 3D optimisation.

5   Full 3D Modelling via Volumetric Optimisation

In this second approach, the optimisation is performed directly on a 3D volumetric representation. After obtaining the initialisation surface (e.g. the VH, or the results from the previous approach), an orthographic grid is defined, fully encompassing the object, and

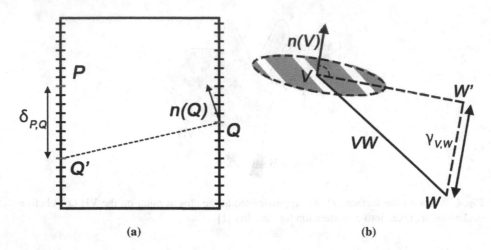

Fig. 5. Illustration of how S_{2D} (a) and S_{3D} (b) are computed [1].

the MRF graph is defined with each node corresponding to a voxel. During the optimisation, each voxel will be assigned a label from the following set: $\{I, O, L_0, ..., L_{R-1}\}$, where I and O respectively correspond to inside and outside voxels, while L_0 to L_{R-1} are used for surface voxels, and indicate specifically where the surface intersects the voxel. In this paper, surface labels are defined by regularly sampling the interior of a voxel as shown in Fig. 3b, although other sampling strategies could also be considered. The following energy function is used to perform the optimisation on the surface:

$$E(L) = (1 - \beta) \sum_{v \in \mathcal{V}} D_{3D}(v, L_v) + \beta \sum_{v,w \in \mathcal{N}_{3D}} S_{3D}(v, L_v, w, L_w) \qquad (8)$$

where β is a weight to balance the effects of the data and smoothness terms, $\mathcal{V}$ is the 3D grid and $\mathcal{N}_{3D}$ is the neighbourhood composed by the six voxels directly adjacent to the current one. $M(v, L_v)$ indicates the position of the surface point at node v when assigned label L_v.

As previously, the data term measures the saliency of a given label assignment. The formulation distinguishes different cases depending on whether the point is inside, outside or on the surface. The saliency of a surface is measured based on the consistency of the constraints derived from Helmholtz reciprocity, while inside and outside points incur no cost. The term is defined as follows:

$$D_{3D}(v, L_v) = \begin{cases} 0, & L_v \in \{I, O\} \\ e^{-\mu \times \frac{\sigma_2(M(v, L_v))}{\sigma_3(M(v, L_v))}}, & \text{otherwise} \end{cases} \qquad (9)$$

As in the previous section, a smoothness term is introduced to regularise the depth and surface normal variations across the surface. We adopt a pairwise smoothness term formulation which measures smoothness between pairs of neighbouring voxels. Similarly to the data term, the formulation of this term requires us to distinguish different cases depending on whether the point is inside, outside or on the surface. The term is defined as follows:

$$S_{3D}(v, L_v, w, L_w) = \begin{cases} \Gamma(M(v, L_v), M(w, L_w)), & L_v, L_w \in \{L_0, \ldots, L_{R-1}\} \\ \infty, & L_v, L_w \in \{I, O\}, \ L_v \neq L_w \\ 0, & \text{otherwise} \end{cases}$$

(10)

where

$$\Gamma(V, W) = \frac{1}{2}(\gamma_{V,W}^2 + \gamma_{W,V}^2)$$

(11)

indicates the normal consistency in the 3D optimisation between points V and W. $\gamma_{V,W}$ is calculated as follows:

$$\gamma_{V,W} = |VW \cdot n(W)|.$$

(12)

VW is the vector connecting points V and W, while $n(W)$ indicates the unit normal estimated via HS at point W. This term consists of the distance between W and the plane perpendicular to $n(V)$ intersecting point V. In Fig. 5b an illustration of how this term is calculated is shown. The ∞ term is here used to constrain inside and outside voxels to be separated by surface voxels, thus avoiding an empty solution where all nodes are either labelled to be inside or outside. The normal consistency term does not require truncation in the 3D domain.

Having defined the energy, optimisation is performed using a tailored version of ICM [3]. ICM is an exhaustive search algorithm that iterates through an MRF graph and changes one variable at a time by trying to optimise its local neighbourhood cost. In its classic formulation, ICM would not work in this scenario because of the constraint on the surface. Namely, changing the label of a surface node to be either outside or inside would result in a hole on the surface, which is currently prevented by having an infinite weight when outside and inside voxels are found to be neighbours. However, by changing two neighbouring variables at a time, and considering all neighbouring nodes to at least one of the two variables, the surface can be shifted close to its optimal solution through multiple iterations. Only tuples where one node is on the current surface of the reconstruction are considered, and for each tuple and their neighbours, all possible configurations are considered, selecting the solution with the lowest energy. Since the problem is initialised close to the actual surface, this step typically converges after a small number of iterations. Finally, the nodes labelled to be on the surface are extracted together with their Helmholtz estimated normals and are integrated using Poisson surface reconstruction to obtain a mesh representation.

6   Experimental Evaluation

In this section, we provide a brief description of the datasets utilised to performed the evaluation, followed by a discussion of the obtained results. For brevity, the methods evaluated will be labelled as follows: 'VH' for the visual hull obtained through SfS, '2.5D HS' for the single partial reconstructions obtained using the approach described in Sect. 4, 'Fused 2.5D HS' for the complete reconstruction obtained fusing the '2.5D

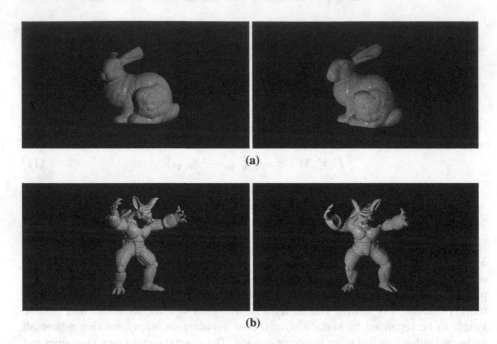

(a)

(b)

Fig. 6. Two reciprocal pairs from the synthetic dataset: 'Bunny' (a) and 'Armadillo' (b).

HS' results, and finally '3D HS' for the method described in Sect. 5. The ground truth is instead indicated as 'GT'.

6.1 Datasets

To test the methodology, both synthetic and real scenes were used. It is important to note that, when generating synthetic scenes, commonly used rendering pipelines often break Helmholtz reciprocity, providing images that are not physically plausible. To address this, synthetic images were rendered using the modified Phong reflectance model [16], which combines a diffuse and a specular part. In Fig. 6 two reciprocal pairs of images are shown for the Stanford Bunny [28] and Armadillo scenes [14]. These scenes were chosen because they both present elongated thin structures, namely the ears of the bunny and the limbs and claws of the armadillo; strong specularities with no textures; and numerous self occlusions. The synthetic scenes were also distorted to measure the robustness of the methodology against noisy input data. In Fig. 7, a close up of the distorted images used for the experiments are shown.

Each scene is composed of 40 reciprocal pairs of images captured from a set of viewpoints obtained sampling a sphere around the object. The images are rendered at a resolution of 1920×1080. Using synthetic scenes allows for a quantitative evaluation of the methodology, by comparing the results obtained against the ground truth data.

The real dataset from [5] is composed of two scenes called Dragon and Fish. Two reciprocal pairs of images are shown in Fig. 8. These two datasets are challenging due to the strong specularities present on the surface of 'Fish' and the numerous self occlusions

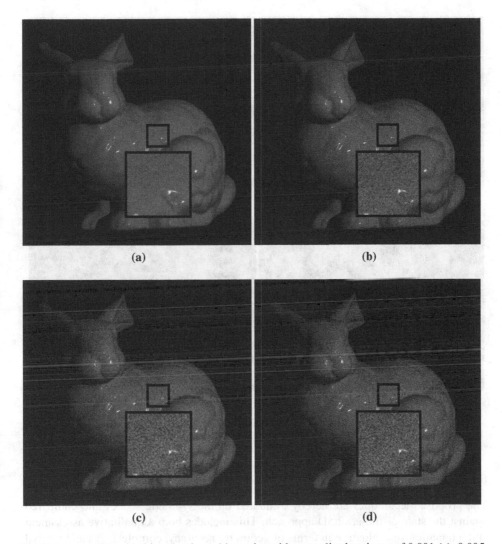

Fig. 7. Images with added input noise (white noise with normalised variance of 0.001 (a), 0.005 (b), 0.010 (c), 0.020 (d) of the maximum image range). Noise effect on the images is highlighted in the rectangles.

in 'Dragon'. It must be noted that no ground truth data or laser scan is available for these datasets, and thus only a qualitative evaluation could be performed. The resolution of the images from these scenes is 1104×828 and they are all from a single ring of cameras positioned on top of the objects, which means that one side of the object is always completely occluded. This is relevant to show how the reconstruction method used is able to deal with a lack of data.

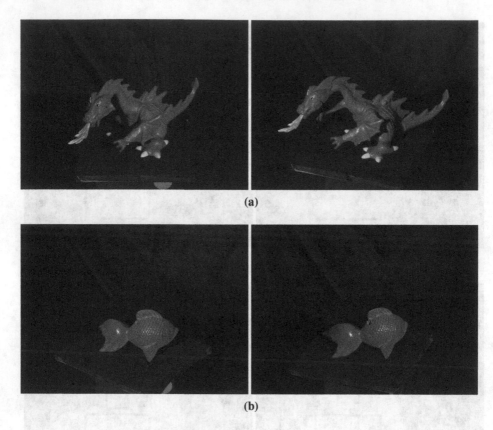

Fig. 8. Two reciprocal pairs from the real datasets: 'Dragon' (a) and 'Fish' (b) presented in [5].

6.2   Results with Synthetic Scenes

The proposed techniques are hereby evaluated on the synthetic scenes and compared against the state-of-the-art 2.5D approach. This includes both a qualitative assessment and a quantitative evaluation in terms of geometry accuracy, completeness and normal accuracy. The '2.5D HS' was not included in the quantitative evaluation because it only produces partial reconstructions and therefore any comparison against the other methods would be unfair. In the completeness analysis different thresholds are used depending on the accuracy achieved on the scene, to measure whether the reconstruction presents holes or areas where a strong accuracy drop is present.

Results are shown in Fig. 9, including multiple views and the ground truth for the different scenes. In the first row, the results obtained using 'VH' are shown, followed by some of the partial reconstructions obtained using '2.5D HS'. It can be noted that self occlusions in this technique cannot be properly handled as only the surface visible from the virtual camera is reconstructed in each view. Moreover, concave surfaces perpendicular to the virtual camera axis cannot be properly reconstructed as well, as shown in the upper left result of the 'Bunny'. Both of these issues result in holes in the partial reconstructions, which will later be filled by integrating multiple views. After

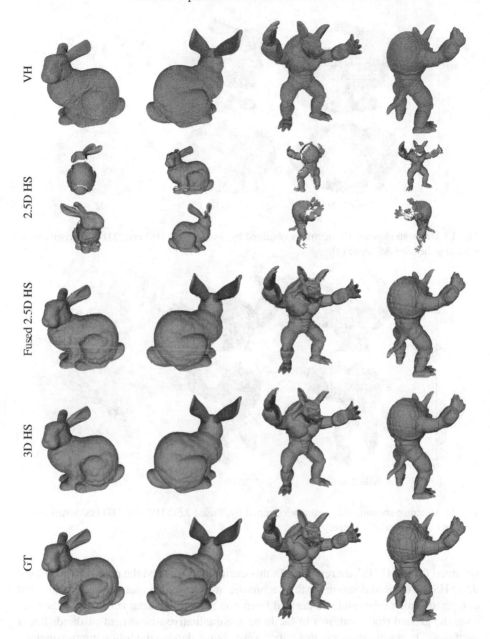

Fig. 9. Results from the 'Armadillo' and 'Bunny' scenes.

performing the integration, the results obtained from 'Fused 2.5D HS' are shown on the next row. While this method allows for a full-3D reconstruction, it can be prone to artefacts whenever non matching surfaces are fused together. This is partially mitigated by the Poisson surface reconstruction implementation. In the following line the results

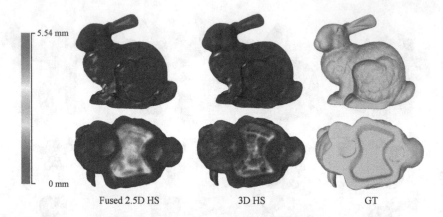

Fig. 10. Heatmap showing the accuracy obtained by 'Fused 2.5D HS' and '3D HS' when reconstructing the 'Bunny' scene [1].

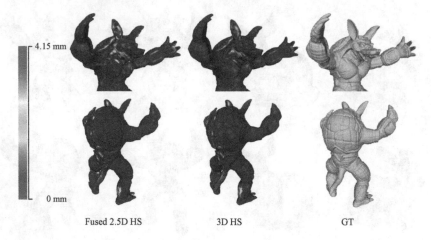

Fig. 11. Heatmap showing the accuracy obtained by 'Fused 2.5D HS' and '3D HS' when reconstructing the 'Armadillo' scene.

obtained from '3D HS' are reported. In this evaluation, we used the results from 'Fused 2.5D HS' as the initialisation for this technique, which yielded more precise results and corrected some of the artefacts derived from non matching partial surfaces. In the final row, the ground truth is shown to facilitate the qualitative assessment of the different methods. The results shown in this paper were obtained using the following parameters: $\{\alpha = 0.3, \beta = 0.4, t = 3 \times r\}$, where r is the edge length of a pixel in the reference frame.

In Figs. 10 and 11, heatmaps are used to highlight some details of the results where the reconstruction accuracy is improved by '3D HS' with respect to 'Fused 2.5D HS'. In particular, concavities with a strong error are improved upon by using '3D HS', some

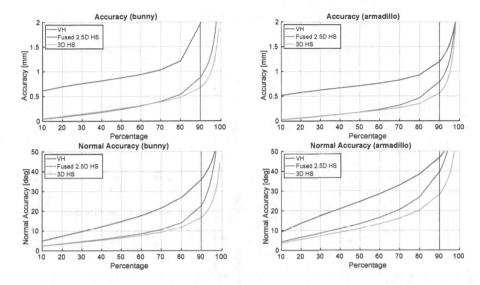

Fig. 12. Graphs representing the results on the 'Bunny' and 'Armadillo' scene, including accuracy and normal accuracy with respect to the percentage of the surface taken into consideration.

notable parts where this can be observed are the ears of the animals in both scenes and the concavity at the base of the 'Bunny' scene.

In Fig. 12, the performance of these methods is objectively measured in terms of geometric accuracy and normal accuracy, showing how these vary for a growing percentage of the surface points. In both scenes the proposed techniques significantly outperform 'VH', even in the 'Armadillo' scene where little to no concavities are present. Both proposed techniques achieve sub-millimetre geometric accuracy at 90% on both scenes, and in particular '3D-HS' obtains the most accurate results in terms of geometric and normal accuracy. To assess the robustness of the approaches, Fig. 13 contains the graphs for geometric accuracy at 90%, normal accuracy at 90% and completeness at different levels of input noise. In the 'Armadillo' scene, which is characterised by a high number of self-occlusions, '3D HS' obtains exceptional results across all inputs when compared with 'Fused 2.5D HS'. In terms of completeness, both techniques are able to reconstruct the scene properly, without any holes or parts with significant loss of accuracy. In terms of normal accuracy, '3D HS' outperforms 'Fused 2.5D HS', and in particular in the 'Armadillo' scene, which presents high frequency details where it is hard to obtain very precise normals.

The geometric and normal accuracy performance degrades almost linearly with the introduction of noise, still maintaining good results when the input images are distorted with strong Gaussian noise with a normalised variance of 0.02. This indicates that the approaches are both robust to noise. In particular, the normal accuracy does not vary significantly, showing how HS normal estimation is robust to noise.

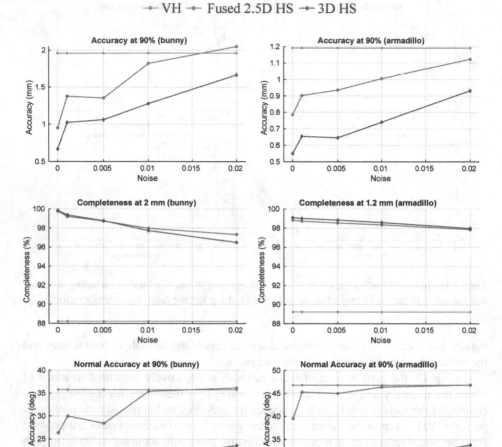

Fig. 13. Graphs representing the results on the 'Bunny' and 'Armadillo' scenes when degraded by noise [1].

6.3 Results with Real Scenes

Finally, the techniques were tested on the real scenes. The results are shown in Fig. 14. Since no precise silhouette information could be retrieved, the 'VH' reconstruction is imprecise and results in a weak initialisation for the proposed techniques. The next results are obtained fusing the '2.5D HS' partial reconstructions, and as it can be observed they present some artefacts. These are then corrected and improved upon in the '3D HS' results, obtained performing full-3D optimisation on top of the 'Fused 2.5D HS' reconstruction. The results are followed by the input image corresponding to the view used for the results.

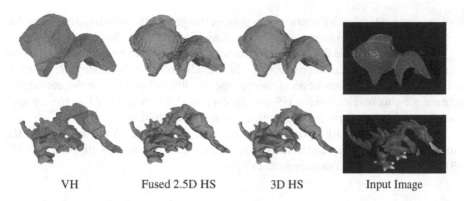

| VH | Fused 2.5D HS | 3D HS | Input Image |

Fig. 14. Results from the 'Fish' and 'Dragon' scenes [1].

While it is not possible to perform an objective evaluation of these results due to lack of ground truth data, the results can be compared qualitatively. Both methods are able to reconstruct the two objects despite the many self occlusions present in the 'Dragon' scene, and the strong specular finish that characterises the 'Fish' object. In particular, the results from '3D HS' are much cleaner than the ones obtained by 'Fused 2.5D HS'. Some artefacts are corrected while details such are thin structures are correctly preserved.

7   Conclusions

In this paper we presented the first MRF framework for full-3D reconstruction that relies on HS to handle scenes with arbitrary surface reflectance. Two methodologies were proposed which can be used either independently or sequentially in a pipeline where the volumetric approach is used to refine the fused view-dependent output.

The first approach consists in performing partial reconstructions from multiple viewpoints and fusing them together. A coarse VH is used to initialise the technique. Each partial surface is then obtained through an MRF optimisation enforcing the coherence of the normals estimated through HS. In contrast, the second method is volumetric and based on enforcing Helmholtz reciprocity to locate the surface and the normal consistency between neighbouring surface voxels.

Both methods were able to achieve sub-millimetre accuracy on the synthetic scenes analysed in this paper, while showing robustness to strong noise. Results on real data demonstrate that the proposed approaches are able to reconstruct scenes which exhibit complex reflectance properties and intricate geometric detail.

8   Future Work

In future work, we will investigate ways to address some of the limitations of the current formation with a view to further improving the accuracy and efficiency of the modelling pipeline. In the viewpoint-dependent approach, the reconstruction could be performed from the viewpoint of each camera using a perspective grid instead of a fixed orthographic one. This would allow to increase scene sampling, while better defining visibility.

The way in which the final surface is put together could also be improved by avoiding use of Poisson surface reconstruction which tends oversmooth the surface. An alternative could be to implement an algorithm specifically tailored to use the confidence measure weights obtained using HS and the estimated normals. In [23] a similar approach was used to perform surface integration in a 2.5D scenario.

Finally, different MRF optimisation strategies could be explored to increase the implementation efficiency and accuracy. Some examples are the lazy flipper [2], TRW [13] and higher order cliques approaches [10].

References

1. Addari, G., Guillemaut, J.Y.: An MRF optimisation framework for full 3D Helmholtz stereopsis. In: 14th International Joint Conference on Computer Vision, Imaging and Computer Graphics Theory and Applications, pp. 725–736, January 2019. https://doi.org/10.5220/0007407307250736
2. Andres, B., Kappes, J.H., Beier, T., Köthe, U., Hamprecht, F.A.: The lazy flipper: efficient depth-limited exhaustive search in discrete graphical models. In: Fitzgibbon, A., Lazebnik, S., Perona, P., Sato, Y., Schmid, C. (eds.) ECCV 2012. LNCS, Part VII, vol. 7578, pp. 154–166. Springer, Heidelberg (2012). https://doi.org/10.1007/978-3-642-33786-4_12
3. Besag, J.: On the statistical analysis of dirty pictures. J. Roy. Stat. Soc. B **48**(3), 48–259 (1986)
4. Chandraker, M., Bai, J., Ramamoorthi, R.: On differential photometric reconstruction for unknown, isotropic brdfs. IEEE Trans. Pattern Anal. Mach. Intell. **35**(12), 2941–2955 (2013). https://doi.org/10.1109/TPAMI.2012.217
5. Delaunoy, A., Prados, E., Belhumeur, P.N.: Towards full 3D Helmholtz stereovision algorithms. In: Kimmel, R., Klette, R., Sugimoto, A. (eds.) ACCV 2010. LNCS, Part I, vol. 6492, pp. 39–52. Springer, Heidelberg (2011). https://doi.org/10.1007/978-3-642-19315-6_4
6. Goldman, D.B., Curless, B., Hertzmann, A., Seitz, S.M.: Shape and spatially-varying BRDFs from photometric stereo. IEEE Trans. Pattern Anal. Mach. Intell. **32**(6), 1060–1071 (2010). https://doi.org/10.1109/TPAMI.2009.102. http://dx.doi.org/10.1109/TPAMI.2009.102
7. Guillemaut, J., Drbohlav, O., Illingworth, J., Sára, R.: A maximum likelihood surface normal estimation algorithm for Helmholtz stereopsis. In: VISAPP 2008: Proceedings of the Third International Conference on Computer Vision Theory and Applications 2008, vol. 2, pp. 352–359 (2008)
8. Guillemaut, J.Y., Drbohlav, O., Sára, R., Illingworth, J.: Helmholtz stereopsis on rough and strongly textured surfaces. In: 3DPVT, pp. 10–17. IEEE Computer Society (2004)
9. Han, T., Shen, H.: Photometric stereo for general BRDFs via reflection sparsity modeling. IEEE Trans. Image Process. **24**(12), 4888–4903 (2015). https://doi.org/10.1109/TIP.2015.2471081
10. Ishikawa, H.: Higher-order clique reduction without auxiliary variables. In: 2014 IEEE Conference on Computer Vision and Pattern Recognition, pp. 1362–1369, June 2014. https://doi.org/10.1109/CVPR.2014.177
11. Janko, Z., Drbohlav, O., Sara, R.: Radiometric calibration of a Helmholtz stereo rig. In: Proceedings of the IEEE Computer Society Conference on Computer Vision and Pattern Recognition, vol. 1, pp. I-166, January 2004

12. Kazhdan, M., Bolitho, M., Hoppe, H.: Poisson surface reconstruction. In: Proceedings of the Fourth Eurographics Symposium on Geometry Processing, SGP 2006, pp. 61–70. Eurographics Association (2006). http://dl.acm.org/citation.cfm?id=1281957.1281965

13. Kolmogorov, V.: Convergent tree-reweighted message passing for energy minimization. IEEE Trans. Pattern Anal. Mach. Intell. **28**(10), 1568–1583 (2006). https://doi.org/10.1109/TPAMI.2006.200. http://dx.doi.org/10.1109/TPAMI.2006.200

14. Krishnamurthy, V., Levoy, M.: Fitting smooth surfaces to dense polygon meshes. In: Proceedings of the 23rd Annual Conference on Computer Graphics and Interactive Techniques, SIGGRAPH 1996, pp. 313–324 (1996). https://doi.org/10.1145/237170.237270, http://doi.acm.org/10.1145/237170.237270

15. Laurentini, A.: The visual hull concept for silhouette-based image understanding. IEEE Trans. Pattern Anal. Mach. Intell. **16**(2), 150–162 (1994). https://doi.org/10.1109/34.273735. http://dx.doi.org/10.1109/34.273735

16. Lewis, R.R.: Making shaders more physically plausible. In: Fourth Eurographics Workshop on Rendering, pp. 47–62 (1994)

17. Liang, C., Wong, K.Y.K.: 3D reconstruction using silhouettes from unordered viewpoints. Image Vision Comput. **28**(4), 579–589 (2010). https://doi.org/10.1016/j.imavis.2009.09.012. http://dx.doi.org/10.1016/j.imavis.2009.09.012

18. Lombardi, S., Nishino, K.: Radiometric scene decomposition: scene reflectance, illumination, and geometry from RGB-D images. CoRR abs/1604.01354 (2016). http://arxiv.org/abs/1604.01354

19. Magda, S., Kriegman, D.J., Zickler, T.E., Belhumeur, P.N.: Beyond lambert: reconstructing surfaces with arbitrary brdfs. In: ICCV (2001)

20. Nasrin, R., Jabbar, S.: Efficient 3D visual hull reconstruction based on marching cube algorithm. In: 2015 International Conference on Innovations in Information, Embedded and Communication Systems (ICIIECS), pp. 1–6, March 2015. https://doi.org/10.1109/ICIIECS.2015.7193189

21. Nishino, K.: Directional statistics BRDF model. In: 2009 IEEE 12th International Conference on Computer Vision, pp. 476–483, September 2009. https://doi.org/10.1109/ICCV.2009.5459255

22. Oxholm, G., Nishino, K.: Shape and reflectance estimation in the wild. IEEE Trans. Pattern Anal. Mach. Intell. **38**(2), 376–389 (2016). https://doi.org/10.1109/TPAMI.2015.2450734

23. Roubtsova, N., Guillemaut, J.: Bayesian Helmholtz stereopsis with integrability prior. IEEE Trans. Pattern Anal. Mach. Intell. **40**(9), 2265–2272 (2018). https://doi.org/10.1109/TPAMI.2017.2749373

24. Roubtsova, N., Guillemaut, J.Y.: Colour Helmholtz stereopsis for reconstruction of dynamic scenes with arbitrary unknown reflectance. Int. J. Comput. Vision **124**(1), 18–48 (2017). https://doi.org/10.1007/s11263-016-0951-0

25. Seitz, S.M., Curless, B., Diebel, J., Scharstein, D., Szeliski, R.: A comparison and evaluation of multi-view stereo reconstruction algorithms. In: Proceedings of the 2006 IEEE Computer Society Conference on Computer Vision and Pattern Recognition, CVPR 2006, vol. 1, pp. 519–528. IEEE Computer Society, Washington, DC, USA (2006). https://doi.org/10.1109/CVPR.2006.19, http://dx.doi.org/10.1109/CVPR.2006.19

26. Szeliski, R., et al.: A comparative study of energy minimization methods for markov random fields with smoothness-based priors. IEEE Trans. Pattern Anal. Mach. Intell. **30**(6), 1068–1080 (2008)

27. Tu, P., Mendonca, P.R.S.: Surface reconstruction via Helmholtz reciprocity with a single image pair. In: 2003 IEEE Computer Society Conference on Computer Vision and Pattern Recognition, 2003. Proceedings, vol. 1, pp. I-541–I-547, June 2003. https://doi.org/10.1109/CVPR.2003.1211401

28. Turk, G., Levoy, M.: Zippered polygon meshes from range images. In: Proceedings of the 21st Annual Conference on Computer Graphics and Interactive Techniques, SIGGRAPH 1994, pp. 311–318. ACM, New York (1994). https://doi.org/10.1145/192161.192241, http://doi.acm.org/10.1145/192161.192241

29. Vogiatzis, G., Hernandez, C., Cipolla, R.: Reconstruction in the round using photometric normals and silhouettes. In: Proceedings of the 2006 IEEE Computer Society Conference on Computer Vision and Pattern Recognition, CVPR 2006, vol. 2, pp. 1847–1854. IEEE Computer Society, Washington, DC, USA (2006). https://doi.org/10.1109/CVPR.2006.245, http://dx.doi.org/10.1109/CVPR.2006.245

30. Von Helmholtz, H., Southall, J.P.: Helmholtz's treatise on physiological optics, vol. 1. Optical Society of America, New York (1924). Translation

31. Weinmann, M., Ruiters, R., Osep, A., Schwartz, C., Klein, R.: Fusing structured light consistency and Helmholtz normals for 3D reconstruction. In: British Machine Vision Conference, September 2012. accepted for publication

32. Woodham, R.J.: Photometric method for determining surface orientation from multiple images. Opt. Eng. **19**(1), 191139 (1980). https://doi.org/10.1117/12.7972479. http://dx.doi.org/10.1117/12.7972479

33. Zickler, T.E., Ho, J., Kriegman, D.J., Ponce, J., Belhumeur, P.N.: Binocular Helmholtz stereopsis. In: Proceedings Ninth IEEE International Conference on Computer Vision, vol. 2, pp. 1411–1417, October 2003. https://doi.org/10.1109/ICCV.2003.1238655

34. Zickler, T.: Reciprocal image features for uncalibrated Helmholtz stereopsis. In: IEEE Computer Vision and Pattern Recognitiion, pp. II: 1801–1808 (2006). http://www.cs.virginia.edu/~mjh7v/bib/Zickler06.pdf

35. Zickler, T.E., Belhumeur, P.N., Kriegman, D.J.: Helmholtz stereopsis: exploiting reciprocity for surface reconstruction. Int. J. Comput. Vision **49**(2), 215–227 (2002). https://doi.org/10.1023/A:1020149707513. http://dx.doi.org/10.1023/A:1020149707513

Robustifying Direct VO to Large Baseline Motions

Georges Younes[1,2], Daniel Asmar[2(✉)], and John Zelek[1]

[1] University of Waterloo, Waterloo, ON, Canada
{gyounes,jzelek}@uwaterloo.ca
[2] American University of Beirut, Beirut, Lebanon
da20@aub.edu.lb

Abstract. While Direct Visual Odometry (VO) methods have been shown to outperform feature-based ones in terms of accuracy and processing time, their optimization is sensitive to the initialization pose typically seeded from heuristic motion models. In real-life applications, the motion of a hand-held or head-mounted camera is predominantly erratic, thereby violating the motion models used, causing *large baselines* between the initializing pose and the actual pose, which in turn negatively impacts the VO performance.

As the camera transitions from a leisure device to a viable sensor, robustifying Direct VO to real-life scenarios becomes of utmost importance. In that pursuit, we propose FDMO, a hybrid VO that makes use of Indirect residuals to seed the Direct pose estimation process. Two variations of FDMO are presented: one that only intervenes when failure in the Direct optimization is detected, and another that performs both Indirect and Direct optimizations on every frame. Various efficiencies are introduced to both the feature detector and the Indirect mapping process, resulting in a computationally efficient approach. Finally, An experimental procedure designed to test the resilience of VO to large baseline motions is used to validate the success of the proposed approach.

Keywords: Monocular odometry · Hybrid · Direct · Indirect

1 Introduction

The Visual Odometry (VO) problem formulates camera pose estimation as an iterative optimization of an objective function. Central to each optimization step is data association, where cues (features) from a new image are corresponded to those found in previous measurements. The type of cues used split VO systems along three different paradigms: Direct, Indirect, or a hybrid of both, with each using its own objective function and exhibiting dissimilar but often complementary traits [22].

An underlying assumption to all paradigms is the convexity of their objective functions, allowing for iterative Newton-like optimization methods to converge. However, none of the objective functions are convex; and to relax this limitation,

© Springer Nature Switzerland AG 2020
A. P. Cláudio et al. (Eds.): VISIGRAPP 2019, CCIS 1182, pp. 477–496, 2020.
https://doi.org/10.1007/978-3-030-41590-7_20

VO systems assume local convexity and employ motion models to perform data association, as well as to seed the optimization. Typical motion models include a constant velocity model (CVMM) [16] and [20], or a zero motion model [11], or, in the case the CVMM fails, a combination of random motions [6].

In real-life applications (*e.g.* Augmented Reality), the motion of a hand-held or head-mounted camera is predominantly erratic, easily violating the assumed motion models, and effectively reducing the VO performance from what is typically reported in most benchmark experiments. In fact, erratic motions can be viewed as a special case of large motions that induces discrepancies between the assumed motion model and the actual camera motion. The error induced is also further amplified when a camera is arbitrarily accelerating or decelerating with low frame-rates, causing errors in the VO data association and corresponding optimization. Similar effects are observed when the computational resources are slow, and VO cannot add keyframes in time to accommodate fast motions; VO is then forced to perform pose estimation across keyframes separated by relatively large distances.

The impact large motions can have depends on various components of a VO; namely, on the resilience of the data association step, on the radius of convergence of the objective function, and on the ability of the VO system to detect and account for motion model violations.

In an effort to handle large baseline motions, we proposed a Feature-Assisted Direct Monocular Odometry (FDMO) [23] that performs photometric image alignment for pose estimation at frame-rate, and invokes an Indirect pose estimation only when tracking failure is detected. The reason for not using Indirect measurements on every frame in FDMO is to avoid the large computational costs associated with extracting features; as a consequence, FDMO maintains the computational efficiency of Direct methods but requires a heuristic approach to detect local minima in the Direct objective function optimization.

This paper builds on the use of both Direct and Indirect measurements presented in FDMO [23] to *alleviate the need for a heuristic failure detection approach*. More specifically, the expensive computational cost associated with feature extraction is overcome by an efficiently parallelizable feature detector, allowing the use of both types of measurements sequentially on every frame, and causing important modifications to the overall architecture of the VO system.

The contributions of the Feature assisted Direct Odometry proposed in this paper includes:

- The ability to use both Direct and Indirect residuals efficiently on every frame via a computationally cheap feature detector.
- Resilience to large baseline motions.
- Achieves the sub-pixel accuracy of Direct methods.
- Maintains the robustness of Direct methods to feature-deprived environments.
- A computationally efficient Indirect mapping approach.
- An experimental procedure designed to evaluate the robustness of VO to large baseline motions.

2   State of the Art VO Paradigms

Direct Systems: use pixels with intensity gradients as image cues. While data association is not an explicit step in Direct methods, it is implicitly encoded within the photometric objective function, which is best described as a minimization of the intensity difference between a reference image taken at time $t-1$, and the that extracted from a new image taken at time t, over the relative pose $T_{I_t, I_{t-1}}$ between the two images:

$$\underset{T_{I_t, I_{t-1}}}{argmin} \quad I_t(\pi(T_{I_t, I_{t-1}}(X))) - I_{t-1}(x), \tag{1}$$

where x is the 2D location of a pixel in the image and X is its associated 3D coordinates.

However, due to the nature of intensity values—can arbitrarily match to random other pixels with a similar local intensity gradient—the resulting photometric objective function is highly non-convex, and relies on a 'good' starting point (provided by the motion model) to correctly compute the intensity difference [2], and thereby converge to its global minimum. To reduce the susceptibility of Direct methods to violations of the motion model, Direct systems either require a high frame-rate camera [11], or employ a multi-scale pyramidal implementation [6,11] and [7]. On the positive side, Direct methods are less susceptible to failure in feature-deprived environments, and do not require a time-consuming feature extraction and matching step. More importantly, since the alignment takes place at the pixel intensity level, the photometric residuals can be interpolated over the image domain, resulting in an image alignment with sub-pixel accuracy, and relatively less drift than feature-based odometry methods [15].

DSO [6] is currently considered the state of the art in Direct methods, and as such will be the topic of further discussion in the performed experiments.

Indirect Systems: extract salient image locations with high dimensional descriptors as cues. Since the descriptors used are somewhat invariant to rotation, translation, scale, and illumination changes, the explicit data association step in indirect methods is robust to violations of the motion model used. However, to speed up the feature matching process, real-time indirect methods restrict feature matching to take place on a small search window surrounding the projected feature locations estimated from the motion model. While this step introduces dependency on the motion model, it can also be used to measure its reliability: a low number of features matched typically means the motion model does not correspond to the actual camera motion, and accordingly should trigger recovery methods that are independent of the motion model. On the downside, the robustness of the descriptor matching process relies on the distinctiveness of each feature from the other; a condition that becomes more difficult to satisfy the higher the number of features extracted in each scene, and thereby favouring sparse scene representations. Furthermore, as a result of their discretized image representation space, indirect solutions offer inferior accuracy when compared to direct methods, as the image domain cannot be interpolated for sub-pixel accuracy.

ORB SLAM 2 [20], considered the state of the art in Indirect systems, employs the aforementioned strategy to detect violations in the motion model, and as such will be used as a representative of Indirect methods in the experiments performed below.

Hybrid Systems: Realizing the complementary properties of the different type of cues (Table 1), recent trends in VO have witnessed the extension of traditional Direct and Indirect systems to include both types, either in a joint formulation (*i.e.* within the same objective function) [24], or in loosely coupled formulations that uses different cues for tracking, mapping, or failure recovery [11,17] and [23]. Most Hybrid systems attempt to improve VO by either modifying the objective function used, extracting more information from the images or by explicitly accounting for the limitations of specific cues, yielding various properties and performance differences under varying conditions. In particular, the main goal of the Hybrid system proposed in this work is to make Direct VO more robust against large baseline motions while maintaining computational efficiency.

Table 1. Comparison between Direct and Indirect methods. The more of the symbol +, the higher the attribute.

Trait	Feature-based	Direct
Large baseline	+++	+
Computational efficiency	++	+++
Robustness to feature deprivation	+	+++
Recovered scene point density	+	+++
Accuracy	+	+++
Optimization non-convexity	+	++
Global localization	+++	None
Map reuse	+++	None

3   Related Work

While various hybrid (Direct and Indirect) systems were previously proposed in the literature, few manage to successfully integrate the advantages of both paradigms. More specifically, the weaknesses of [11,17] and [1] are discussed in FDMO [23] and the interested reader is referred to that discussion. More recent hybrid systems were also proposed in [13,18] and [24]. However, Gao *et al.* [13] uses the Indirect features for the sole purpose of loop closure detection and as such does not improve on the frame-to-frame pose estimation process. Lee *et al.* [18] runs both DSO and ORB

SLAM 2 on top of each other, with the frame-to-frame pose estimation first performed by DSO and then refined by ORB SLAM. Unfortunately, the end result is a degraded performance over using DSO alone due to the various limitations of Indirect methods. Younes *et al.* [24] achieves state-of-the-art performance by using both Direct and Indirect residuals simultaneously in a tightly coupled optimization; however, to maintain computational efficiency, [24] uses the photometric residuals to triangulate the Indirect features, which limits the systems' ability to triangulate features from multiple pyramid levels; thereby reducing the robustness of Indirect features to large baseline motions.

Finally, FDMO [23] is of particular importance to this paper, and as such it is further detailed in the proposed section.

4   Proposed System

To capitalize on the advantages of Direct methods while inheriting the resilience of Indirect methods to large baseline motions, we propose a feature-assisted Direct approach, where the output of the Indirect formulation is used to seed the Direct optimization. Since Indirect methods require a computationally expensive feature extraction process, and to reduce its impact on frame rate, we proposed in FDMO [23] to perform frame-to-frame pose estimation using the Direct residuals only. The Indirect features are only extracted when a new keyframe is added or when failure is detected. However, detecting failure is equivalent to identifying a local minimum within the photometric optimization, for which we employ a heuristic test that is not guaranteed to always detect such minima.

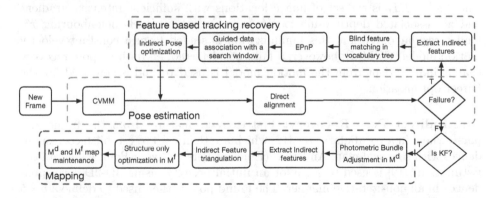

Fig. 1. FDMO tracking, mapping and feature-based recovery flow charts.

In this work, we address FDMO's dependence on a heuristic failure detection test by using both Direct and Indirect residuals on every frame. To overcome the computational expenses of extracting features, we propose an efficient and parallelizable alternative to the feature detector employed in typical Indirect methods. Finally, an Indirect map quality feedback from the frame-to-frame feature matches is used to introduce various efficiencies in the mapping process, resulting in a 50% faster Indirect mapping process while maintaining the same performance.

4.1   Notation

To maintain consistency with [23], the superscript d is be assigned to all Direct-based measurements and f for all feature-based (Indirect) measurement. I_{f_i} refers to the image of frame i and $T_{f_i,KF}$ is the SE(3) transformation relating frame i to the latest keyframe KF. We also make the distinction between z referring to depth measurements associated with a 2D point x and Z referring to the Z-coordinate of a 3D point. Finally, the symbol π is used to denote the pinhole projection function mapping a point from the camera coordinate frame to the image coordinate frame. Finally we assign M^d to the set of local Direct keyframes and M^f to the set of local Indirect keyframes.

4.2   Feature Assisted Direct Monocular Odometry (FDMO)

FDMO is made of three main processes, tracking, mapping and an auxiliary feature-based recovery process as shown in Fig. 1.

Pose Estimation: newly acquired frames are tracked by minimizing a photometric objective function:

$$\underset{T_{f_i,KF}}{\operatorname{argmin}} \sum_{x^d} \sum_{x \in N(x^d)} Obj(I_{f_i}(\pi(x, z, T_{f_i,KF}) - I_{KF}(x))), \tag{2}$$

where $x^d \in \Omega I_f$ is the set of image locations with sufficient intensity gradient and an associated depth value z. $N(x^d)$ is the set of pixels neighbouring x^d. Under regular operation, this minimization is seeded from a constant velocity motion model (CVMM); however, if failure is detected, an Indirect pose recovery that is independent from the motion model is used to find a new seed for the Direct optimization.

The Indirect Pose Recovery: first extracts features from the frame and parses them in a vocabulary tree [12] which are then used to establish correspondences with the local map. An EPnP (Efficient Perspective-n-Point Camera Pose Estimation) [19] is used to solve for an initial pose T_{f_i} using 3D-2D correspondences in an non-iterative manner. The found pose is then used to define a 5×5 search window in f_i surrounding the projected locations of all 3D Indirect map points X^f which are finally used to refine the pose estimation by minimizing a geometric re-projection error:

$$\underset{T_{f_i}}{\operatorname{argmin}} \sum_{X^f \in M^f} Obj(\pi(X^f, T_{f_i}) - obs), \tag{3}$$

where obs $\in \mathbb{R}^2$ is the features' matched location in f_i, found through descriptor matching.

Local Mapping: is composed of two main components, namely Direct and feature-based that are executed sequentially as shown in Fig. 1. When a new keyframe is added, a photometric-based bundle adjustment is used, as described in [6], to ensure the local optimality of the added keyframe. ORB Features are then extracted using ORB SLAM's feature detector [20] which are then used with the keyframe pose from the photometric bundle adjustment to match other features from M^f. The features that match to previously added Indirect map points are used to add more observations which are then refined in a structure-only optimization, while new Indirect map points are triangulated from 2D-2D matches established via Epipolar search lines.

Finally, a structure only optimization is performed on every Indirect map point x_j^f:

$$\underset{X_j^f}{\operatorname{argmin}} \sum_i Obj(x_{i,j}^f - \pi(X_j^f, T_{KF_i^f})), \tag{4}$$

where j spans all 3D map points observed in the local Indirect keyframes set and i spans all the matches of a particular map point X_j^f We use ten iterations of Gauss-Newton to minimize the normal equations associated with (4) which yield the following update rule per 3D point X_j^f per iteration:

$$X_j^{t+1} = X_j^t - (J^T W J)^{-1} J^T W e, \tag{5}$$

where e is the stacked reprojection residuals e_i associated with a point X_j and its found match x_i in keyframe i. J is the stacked Jacobians of the reprojection error which is found by stacking:

$$J_i = \begin{bmatrix} \frac{f_x}{Z} & 0 & -\frac{f_x X}{Z^2} \\ 0 & \frac{f_y}{Z} & -\frac{f_y Y}{Z^2} \end{bmatrix} R_{KF_i} \tag{6}$$

and R_{KF_i} is the 3×3 orientation matrix of the keyframe observing the 3D point X_j. Similar to ORB-SLAM, W is a block diagonal weight matrix that down-weighs the effect of residuals computed from feature matches found at high pyramidal levels and is computed as

$$W_{ii} = \begin{bmatrix} Sf^{2n} & 0 \\ 0 & Sf^{2n} \end{bmatrix} \tag{7}$$

where Sf is the scale factor used to generate the pyramidal representation of the keyframe (we use $Sf = 1.2$) and n is the pyramid level from which the feature was extracted ($0 < n < 8$). The Huber norm is also used to detect and remove outliers.

5   FDMO on Every Frame

While FDMO operates as a Direct odometry for frame-to-frame pose estimation, FDMO-f extracts Indirect features and perform both Direct and Indirect optimizations on every frame. As a result, a modified architecture that exploits the availability of Indirect data at every frame is required.

5.1   Feature Extraction

Several design considerations are taken into account when designing a feature detector for a SLAM algorithm. In particular, the detected keypoints should be repeatable, discriminable, and homogeneously distributed across the image. ORB SLAM, state of the art in Indirect Odometry, takes into account these considerations by extracting features using an octomap, which adapts the FAST [21] corner detector thresholds to different image regions. However, this process is computationally involved; for example, it takes 12 ms on our current hardware to extract 1500 features along with their ORB descriptors from 8 pyramid levels. Unfortunately, this means that the feature extraction alone requires more time than the entire Direct pose estimation process. Several attempts at parallelizing ORB SLAM's feature extraction process were made in the literature; Zhang *et al.* [25] attempted parallelizing the extraction process on a CPU resulting in no speedups, and ended up adopting a CPU - GPU acceleration to reduce the computational cost by 40 %. Similar results were also reported in [4].

In contrast to the aforementioned methods, we forego the adaptive octomap approach in favor of an efficiently parallelizable feature detector implementation on a CPU. Our proposed feature detector first computes the image pyramid levels, which are then distributed across parallel CPU threads. Each thread operates on its own pyramid level independent of the others. The following describes the operations performed by each thread: FAST corners [21] are first extracted with a low cutoff threshold, resulting in a large number of noisy corners with an associated *corner-ness* score (the higher the score the more discriminant). The corners are then sorted in descending order of their scores and accordingly added as features, with each added corner preventing other features from being added in a region of 11 × 11 pixels around its location. This ensures that the most likely repeatable corners are selected, while promoting a homogeneous distribution across the image. The area 11 × 11 is chosen slightly smaller than the ORB descriptor patch size of 15 × 15 pixels to ensure small overlap between the feature descriptors, thereby improving their discriminability. The features orientation angles are then computed and a Gaussian kernel is applied before extracting their ORB descriptors.

When compared to the 21 ms required by ORB SLAM's detector, our proposed feature detector extracts the same number of features in 4.4 ms using the same CPU, making feature extraction on every frame feasible. A side by side comparison between the extracted features from ORB SLAM and our proposed detector are shown in Fig. 2.

5.2   Pose Estimation

Unlike FDMO, FDMO-f extracts and uses Indirect features on every frame. The CVMM from frame-to-frame pose is usually accurate enough to establish feature correspondences with the local map using a search window. However, if few matches are found, the motion-model-independent pose recovery described in Sect. 4.2 is used to obtain a more accurate pose for feature matching to take

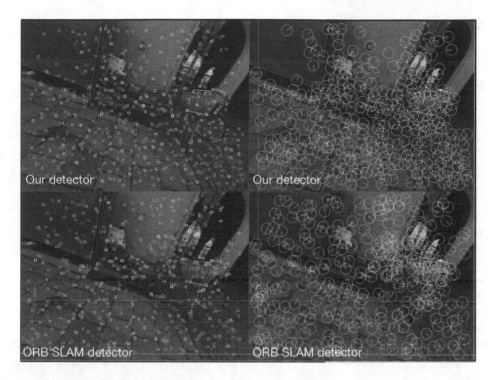

Fig. 2. The top left image shows the features detected by our approach whereas the bottom left shows the features detected by ORB SLAM. The different colors correspond to the different pyramid levels. The right images show the image areas that were used to compute the descriptors of the features from the lowest pyramid level. It can be seen how our feature detector results in a better feature distribution; in contrast, ORB SLAM's detector extracts a large number of features from the same area (right part of the image) resulting in a large overlap between the feature descriptors.

place. The frame pose is then optimized using the Indirect features as described in Eq. 3 before being used to seed the direct image alignment process which ensures a sub-pixel level accuracy of the pose estimation process. FDMO-f pose estimation process is summarized in Fig. 3.

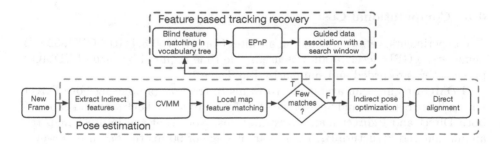

Fig. 3. FDMO-f pose estimation flow chart.

5.3   Local Mapping

Similar to FDMO, FDMO-f uses hybrid keyframes that contains both Direct and Indirect measurements. However, unlike FDMO, keyframe insertion is triggered from two sources, either from (1) the Direct optimization using the criteria described in [6], or from (2) the Indirect optimization by monitoring the ratio of the inlier feature matches in the latest frame to that of the latest keyframe $r = \frac{inliers\ in\ inf_i}{inliers\ in\ KF}$; if r drops below a threshold (0.8), a keyframe is added, thus ensuring an ample amount of reliable Indirect features present in the local Indirect map M^f.

While all added keyframes are used to expand the set of direct map points x^d, they contribute differently to the Indirect mapping process depending on which criteria was used to create the keyframe. In particular, only keyframes that are triggered from the Indirect inlier ratio are used to triangulate new Indirect map points X^f. Keyframes that were not selected for Indirect triangulation are used to provide constraints on the previously added Indirect map points in the structure-only optimization. As a result, the modified mapping process is significantly more efficient than that of FDMO which did not have frame-to-frame feedback on the quality of the Indirect map, forcing it to triangulate new Indirect map points on every added keyframe. The final mapping process of FDMO-f is shown in Fig. 4.

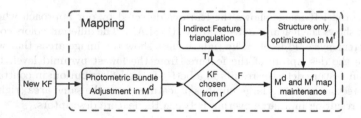

Fig. 4. FDMO-f mapping flow chart.

6   Quantitative Evaluation

6.1   Computational Cost

The experiments were conducted on an Intel Core i7-8700 3.4 GHZ CPU, 32 GB memory; no GPU acceleration was used. The time required by each of FDMO-f processes was recorded and summarized in Table 2.

FDMO-f requires an average of 16.13 ms to track a frame using both Direct and Indirect residuals and an average of 73.58 ms to add a keyframe and expands both Direct and Indirect map representations. Note that the Indirect map triangulation and Structure BA require an average of 90 ms to compute, however

Table 2. Computational time (ms) breakdown for every processes in FDMO-f.

	Process	Mean Time
Tracking	Direct data preparation	0.8
	Features and descriptors extraction	4.24
	Feature matching and indirect pose optimization	7.11
	Direct pose refinement	3.98
Mapping	Photometric bundle adjustment	30.76
	Indirect map maintenance	10.11
	Indirect map triangulation and structure BA	12.45
	Direct map maintenance and marginalization	20.26

it is only called once every twenty keyframes on average, thereby significantly reducing its average computational cost.

The mean computational costs for DSO, FDMO-f and ORB SLAM 2 are shown in Table 3.

Table 3. Computational time (ms) for processes in DSO, FDMO-f and ORB-SLAM2.

Process	DSO	FDMO-f	ORB-SLAM2
Tracking (averaged over number of frames)	9.05	16.13	20.05
Mapping (averaged over number of keyframes)	50.64	73.58	198.52

6.2 Datasets

TUM MONO Dataset. [8] Contains 50 sequences of a camera moving along a path that begins and ends at the same location. The dataset is photometrically calibrated: camera response function, exposure times and vignetting are all available; however, ground truth pose information is only available for two small segments at the beginning and end of each sequence; fortunately, such information is enough to compute translation, rotation, and scale drifts accumulated over the path, as described in [8].

EuRoC MAV Dataset. [5] Contains 11 sequences of stereo images recorded by a drone mounted camera. Ground truth pose for each frame is available from a Vicon motion capture system.

6.3 Frame Drop Experiment

The frame drop experiment, as shown in Fig. 5, consists of inducing artificial motions by dropping a number of frames, and measuring the accuracy of the relative pose found across the gap. Given that in monocular systems the measured

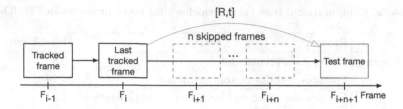

Fig. 5. Depiction of the forward frame drop experiment where frames are skipped towards the future and the measured [R,T] between the last tracked frame and a test frame are compared against the ground truth.

poses are up to a random scale, to compute the error between the measured values and the ground truth, we compute the scale as $S = \frac{\|T_i^{gt}\|_2}{\|T_i^{vo}\|_2}$, where $\|T_i^{gt}\|_2$ is the L2-norm of the ground-truth translation vector of the ith frame (last tracked frame) and $\|T_i^{vo}\|_2$ is its VO measured counterpart. The percentage translation error can then be computed as:

$$E_t = 100 \times \frac{\mid (S\|T_i^{vo} - T_{i+n+1}^{vo}\|_2 - \|T_i^{gt} - T_{i+n+1}^{gt}\|_2 \mid}{\|T_i^{gt} - T_{i+n+1}^{gt}\|_2}, \tag{8}$$

where T_{i+n+1} is the translation vector of the $(i+n+1)^{th}$ frame corresponding to the first frame after skipping n frames. As for rotations, we compute the angle between the ith and $(i+n+1)^{th}$ frames as the geodesic angle separating them on a unit sphere, which is computed as:

$$\theta = arccos(2\langle q_i, q_{i+n+1}\rangle^2 - 1), \tag{9}$$

where q is the normalized quaternion representation of the frame rotation and $\langle \cdot, \cdot \rangle$ is the inner product. Equation 9 is derived from [14, eq. (23)] using simple trigonometric identities. The rotation error percentage is then computed as $E_r = 100 \times \frac{|\theta^{vo} - \theta^{gt}|}{\theta^{gt}}$.

The experiment is repeated until VO fails, each time starting from the same Frame i but increasing the number of dropped frames n (*e.g.* $n = 0, 5, 10$, etc.). We record and plot for every n dropped frames E_t against the ground truth translation $\|T_i^{gt} - T_{i+n+1}^{gt}\|_2$ and E_r against the ground truth rotation angle θ^{gt}. Note that it is essential to measure the relative pose rather than the absolute pose for both E_r and E_t so that any previously accumulated drift does not have an effect on the obtained results. The same can be said about estimating the scale using the latest tracked frame so that accumulated scale drift does not distort the reported measurements.

Since the relative poses are used, aligning the two paths, which might yield over-optimistic results, is not required.

Frame Drop Experiment Variation. While the suggested experiment skips frames towards previously unseen parts of the scene (future frames), one could argue that the camera can randomly move towards previously observed parts as well. Therefore, to account for such motions, we propose a variation of the first experiment shown in Fig. 6, where the dropped frames are taken in the opposite direction, after reaching F_{i+n+1} without any interruption. To distinguish between both experiments, we refer to the first experiment as the forward experiment (as in skipping frames towards forward in time), and the second as the backward experiment. Note that the same equations as before hold for both types of experiments.

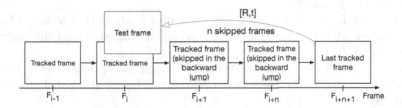

Fig. 6. The backward frame drop experiment where frames are skipped towards the past and the measured $[R,T]$ between the last tracked frame and a test frame are compared against the ground truth.

Frame Drop Experiment Analysis. While the effect of violating the motion model could be simulated by manually changing the motion model values instead of dropping frames, we chose to drop frames for two reasons. First, varying the motion model is equivalent to measuring the convergence basin of the objective function; however, it is not indicative of the system's ability to handle changes in the observed scene, which is usually the case in large motions. Second, the act of dropping frames models another real-life problem; namely, when an image feed is broadcast from a device to be processed on a ground station, connectivity issues may arise, causing frame drops.

Other aspects of the experiment must also be considered. First, the observed scene plays an important role; as one can reason, a far-away scene requires larger motions to yield small changes in the image plane, whereas a close scene may easily go out of the field of view causing instant failure. Therefore, it is meaningless to report, for example, a camera motion of one meter causes a 30% error, without taking into account the scene depth. However, for the sake of comparing various VO systems against each other, it is sufficient to compare their results when observing the same scene and undergoing the same frame drops, as long as it is established that these results will be different and cannot be compared to other scenes.

Second, the camera motion also plays an important role; in particular, a camera moving along the same direction as its optical axis may exhibit different performance than a camera undergoing motion perpendicular to its optical axis, or than a camera that is purely-rotating. While our experiment can be applied to random motions, it is more insightful to separate the three cases and run the experiment on each, as such we have identified three locations in different sequences from the EuroC dataset [5], where the camera motion is one of the three cases, and accordingly report on the found results.

Frame Drop Experiment Results. As mentioned earlier, the experiment was repeated three times for different camera motions as follows:

- Experiment A: Camera motion parallel to its optical axis, performed in sequence MH01 with the forward experiment starting at frame 200 and the backward at frame 250.
- Experiment B: Camera motion perpendicular to its optical axis, performed in sequence MH03 with the forward starting at frame 950 and the backward at frame 1135.
- Experiment C: Pure rotation motion (not along the optical axis), performed in sequence MH02 with the forward starting at frame 510 and the backward at frame 560.

The experiment was repeated in the three locations for ORB-SLAM 2 as a representative of Indirect systems, DSO [6] for Direct systems and FDMO [23] for Hybrid systems. The reported results are shown in Fig. 7.

Frame Drop Results Discussion. DSO performed consistently the worst across the three experiments (Fig. 7-A,B,C). It is interesting to note that DSO's behavior was very similar in both forward and backward variations, *i.e.* it could not handle any violations to its motion model, irrespective of the observed scene. The reported result is expected due to the highly non-convex objective function in DSO. Meanwhile, ORB SLAM 2 and FDMO performed consistently better in all experiments on the backward variation than on the forward. This is due to the resilience of their data association step to large motions *i.e.*, where they were capable of matching image features to old map points. On the other hand, their reduced performance in the forward variation is mainly attributed to feature deprivation, as the new part of the scene had not been mapped yet.

On another note, a notable performance decrease in ORB SLAM 2 is observed before complete failure (oscillatory behavior shown in Fig. 7-B), which is indicative of its failure recovery procedure localizing the frames at erroneous poses.

Finally, FDMO was able to track in both variations of the experiment an angle of 30° before failure as shown in Fig. 7; whereas ORB SLAM 2 was able to track 58° before failure in the backward variation of the experiment. The reduced performance in FDMO is due to its need to track the latest frame with respect to the latest keyframe in its map, as opposed to ORB SLAM 2 which tracks the latest frame with respect to the local map without the need for tracking with respect to a particular keyframe.

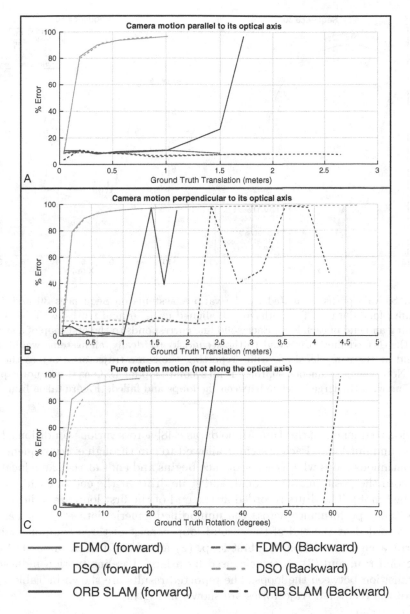

Fig. 7. The results reported by the experiment in three different locations of the EuroC dataset [5] with each location corresponding to a different camera motion. Figure adapted from [23].

6.4   Two Loop Experiment

In this experiment, we investigate the quality of the estimated trajectory by comparing ORB-SLAM2, DSO, and FDMO. We allow all three systems to run

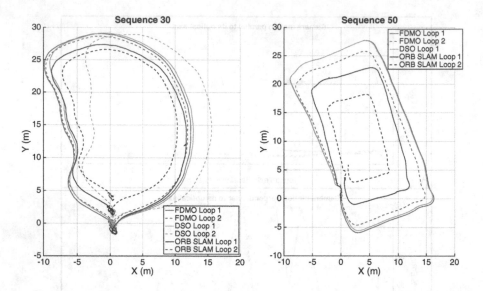

Fig. 8. Sample paths estimated by the various systems on Sequences 30 and 50 of the Tum_Mono dataset. The paths are all aligned using ground truths available at the beginning and end of each loop. Each solid line corresponds to the first loop of a system while the dashed line correspond to the second loop. Ideally, all systems would start and end at the same location, while reporting the same trajectories across the two loops. Note that in Sequence 50, there is no second loop for DSO as it was not capable of dealing with the large baseline between the loops and failed. Figure taken from [23].

on various sequences of the Tum_Mono dataset [8] across various conditions, both indoors and outdoors. Each system is allowed to run through every sequence for two continuous loops where each sequence begins and ends at the same location. We record the positional, rotational, and scale drifts at the end of each loop, as described in [8]. The drifts recorded at the end of the first loop are indicative of the system's performance across the unmodified generic datasets, whereas the drifts recorded at the end of the second loop consist of three components: (1) the drift accumulated from the first loop, (2) an added drift accumulated over the second run, and (3) an error caused by a large baseline motion induced at the transition between the loops. The reported results are shown in Table 4 and some of the recovered trajectories are shown in Fig. 8.

6.5 Qualitative Assesment

Figure 9 compares the resilience of FDMO and ORB-SLAM2 to feature-deprived environments. Since FDMO exploits the sub-pixel accurate localized direct keyframes to propagate its feature-based map, it is capable of generating accurate and robust 3D landmarks that have a higher matching rate even in low textured environments. In contrast, ORB-SLAM2 fails to propagate its map causing tracking failure.

Table 4. Measured drifts after finishing one and two loops over various sequences from the Tum Mono dataset. The alignment drift (meters), rotation drift (degrees) and scale($\frac{m}{m}$) drifts are computed similar to [8]. Table first appeared in [23].

		Sequence 20		Sequence 25		Sequence 30		Sequence 35		Sequence 40		Sequence 45		Sequence 50	
		Loop 1	Loop 2	Loop 1	Loop 2	Loop 1	Loop 2	Loop 1	Loop 2	Loop 1	Loop 2	Loop 1	Loop 2	Loop 1	Loop 2
Alignment	FDMO	**0.752**	**1.434**	**0.863**	**1.762**	**0.489**	**1.045**	**0.932**	2.854	**2.216**	**4.018**	**1.344**	**2.973**	**1.504**	**2.936**
	DSO	0.847	–	0.89	3.269	0.728	5.344	0.945	–	2.266	4.251	1.402	8.702	1.813	–
	ORB SLAM	4.096	8.025	3.722	8.042	2.688	4.86	1.431	**2.846**	–	–	8.026	12.69	6.72	13.56
Rotation	FDMO	1.4	**1.192**	**1.154**	**2.074**	0.306	**0.317**	**1.425**	**6.246**	**3.877**	**6.524**	0.522	**5.595**	**0.448**	**1.062**
	DSO	1.607	–	1.278	7.699	**0.283**	18.9	2.22	–	4.953	19.89	**0.462**	23.17	0.594	–
	ORB SLAM	26.92	53.28	2.373	4.647	2.982	4.549	3.676	6.498	–	–	3.707	7.375	3.243	6.668
Scale	FDMO	1.079	1.161	**1.113**	**1.238**	**1.033**	**1.071**	1.072	1.211	**1.109**	**1.219**	**1.082**	1.106	**1.107**	**1.224**
	DSO	1.089	–	1.116	1.424	1.045	1.109	**1.067**	–	1.118	1.226	1.084	**1.023**	1.133	–
	ORB SLAM	**1.009**	**1.019**	1.564	2.403	1.199	1.373	1.094	**1.206**	–	–	1.867	2.574	1.7	2.675

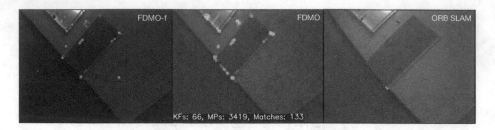

Fig. 9. Features matched of FDMO (left) and ORB-SLAM2 (right) in a feature deprived environment (sequence 40 of the Tum_mono dataset).

7   Discussion and Conclusion

We have presented a Feature-assisted Direct odometry that combines the advantages of both Direct and Indirect measurements at every frame. By seeding the Direct optimization with an Indirect pose estimate, we reduce the negative effects of initializations far from the optimum pose, while maintaining sub-pixel accuracy and robustness to feature-deprived environments. The computational cost typically associated with feature extraction is addressed via an efficiently parallelizable feature detector. The success of the approach is demonstrated in the proposed experiments which reveal various insights into the behavior of different VO systems that are typically not discussed in traditional experiments.

Finally, the potential of using both Direct and Indirect residuals is not limited to the achieved results. In particular, keeping track of a global Direct map is memory inefficient and computationally exhaustive; typical Direct methods are either limited to odometry [6] which keeps track of a small subset of the map only, or rely on Indirect feature encoding to query the global map [9]. The inherent availability of the Indirect features in FDMO provides an efficient global scene representation that can be used to ensure global consistency of the reconstructed map via loop closure.

The brightness constancy assumption is another limitation of Direct methods. In particular, during regular operation, a camera is allowed to adjust its exposure to accommodate varying lighting conditions. However, for the sake of brightness constancy in Direct measurements, auto exposure is typically turned off so that the camera does not lose track of the map. In an effort to increase the robustness of Direct Odometry to varying lighting conditions, [10] proposed an offline photometric calibration process to estimate the camera response function and vignetting map. If a photometric calibration for a camera is then available along with the exposure times per frame, they can be used to map the image intensities to the scene irradiance, which is independent of the camera response, thereby exploiting auto-exposure to maintain the brightness constancy assumption. However, obtaining a photometric calibration and exposure times is a daunting task. Bergmann et al. [3] showed that if features can be reliably matched between photometrically distorted images in a video sequence, they can be used to recover both the photometric calibration and the exposure time per frame.

The required feature matches are inherently available in FDMO through its Indirect features that use ORB descriptors to establish correspondences across frames.

While we don't make use of a global map representation nor the online photometric calibration capabilities of the Indirect measurements, they remain part of our future work and are expected to improve results further.

References

1. Ait-Jellal, R., Zell, A.: Outdoor obstacle avoidance based on hybrid stereo visual SLAM for an autonomous quadrotor MAV. In: IEEE 8th European Conference on Mobile Robots (ECMR) (2017)
2. Baker, S., Matthews, I.: Lucas-Kanade 20 years on: a unifying framework. Int. J. Comput. Vision **56**(3), 221–255 (2004)
3. Bergmann, P., Wang, R., Cremers, D.: Online photometric calibration of auto exposure video for realtime visual odometry and slam. IEEE Robot. Autom. Lett. (RA-L) **3**, 627–634 (2018)
4. Bourque, D.: CUDA-Accelerated ORB-SLAM for UAVs. Worcester Polytechnic Institute (2017)
5. Burri, M., et al.: The euroc micro aerial vehicle datasets. Int. J. Robot. Res. (2016). https://doi.org/10.1177/0278364915620033
6. Engel, J., Koltun, V., Cremers, D.: Direct sparse odometry. IEEE Trans. Pattern Anal. Mach. Intell. **PP**(99), 1 (2017). https://doi.org/10.1109/TPAMI.2017.2658577
7. Engel, J., Stuckler, J., Cremers, D.: Large-scale direct slam with stereo cameras. In: 2015 IEEE/RSJ International Conference on Intelligent Robots and Systems (IROS), pp. 1935–1942, September 2015. https://doi.org/10.1109/IROS.2015.7353631
8. Engel, J., Usenko, V., Cremers, D.: A photometrically calibrated benchmark for monocular visual odometry. arXiv:1607.02555, July 2016
9. Engel, J., Schöps, T., Cremers, D.: LSD-SLAM: large-scale direct monocular SLAM. In: Fleet, D., Pajdla, T., Schiele, B., Tuytelaars, T. (eds.) ECCV 2014. LNCS, vol. 8690, pp. 834–849. Springer, Cham (2014). https://doi.org/10.1007/978-3-319-10605-2_54
10. Engel, J., Usenko, V.C., Cremers, D.: A photometrically calibrated benchmark for monocular visual odometry. CoRR abs/1607.02555 (2016). http://arxiv.org/abs/1607.02555
11. Forster, C., Pizzoli, M., Scaramuzza, D.: SVO: fast semi-direct monocular visual odometry. In: IEEE International Conference on Robotics and Automation (ICRA) (2014)
12. Galvez-López, D., Tardos, J.D.: Bags of binary words for fast place recognition in image sequences. IEEE Trans. Robot. **28**(5), 1188–1197 (2012). https://doi.org/10.1109/TRO.2012.2197158
13. Gao, X., Wang, R., Demmel, N., Cremers, D.: LDSO: direct sparse odometry with loop closure. In: 2018 IEEE/RSJ International Conference on Intelligent Robots and Systems (IROS), pp. 2198–2204 (2018). https://doi.org/10.1109/IROS.2018.8593376
14. Huynh, D.Q.: Metrics for 3D rotations: comparison and analysis. J. Math. Imaging Vis. **35**(2), 155–164 (2009). https://doi.org/10.1007/s10851-009-0161-2

15. Irani, M., Anandan, P.: About direct methods. In: Triggs, B., Zisserman, A., Szeliski, R. (eds.) IWVA 1999. LNCS, vol. 1883, pp. 267–277. Springer, Heidelberg (2000). https://doi.org/10.1007/3-540-44480-7_18
16. Klein, G., Murray, D.: Parallel tracking and mapping for small AR workspaces. In: 6th IEEE and ACM International Symposium on Mixed and Augmented Reality, pp. 1–10, November 2007
17. Krombach, N., Droeschel, D., Behnke, S.: Combining feature-based and direct methods for semi-dense real-time stereo visual odometry. In: Chen, W., Hosoda, K., Menegatti, E., Shimizu, M., Wang, H. (eds.) IAS 2016. AISC, vol. 531, pp. 855–868. Springer, Cham (2017). https://doi.org/10.1007/978-3-319-48036-7_62
18. Lee, S.H., Civera, J.: Loosely-coupled semi-direct monocular slam. IEEE Robot. Automat. Lett. 4(2), 399–406 (2019). https://doi.org/10.1109/LRA.2018.2889156
19. Lepetit, V., Moreno-Noguer, F., Fua, P.: EPnP: an accurate o(n) solution to the PnP problem. Int. J. Comput. Vision 81(2), 155–166 (2009)
20. Mur-Artal, R., Montiel, J.M.M., Tardos, J.D.: ORB-SLAM: a versatile and accurate monocular SLAM system. IEEE Trans. Robot. PP(99), 1–17 (2015)
21. Rosten, E., Drummond, T.: Machine learning for high-speed corner detection. In: Leonardis, A., Bischof, H., Pinz, A. (eds.) ECCV 2006. LNCS, vol. 3951, pp. 430–443. Springer, Heidelberg (2006). https://doi.org/10.1007/11744023_34
22. Younes, G., Asmar, D., Shammas, E., Zelek, J.: Keyframe-based monocular SLAM: design, survey, and future directions. Robot. Auton. Syst. 98, 67–88 (2017). https://doi.org/10.1016/j.robot.2017.09.010
23. Younes, G., Asmar, D.C., Zelek, J.: FDMO: feature assisted direct monocular odometry. In: Proceedings of the 14th International Joint Conference on Computer Vision, Imaging and Computer Graphics Theory and Applications. INSTICC (2019)
24. Younes, G., Asmar, D.C., Zelek, J.S.: A unified formulation for visual odometry. CoRR abs/1903.04253 (2019). http://arxiv.org/abs/1903.04253
25. Zhang, C., Liu, Y., Wang, F., Xia, Y., Zhang, W.: VINS-MKF: a tightly-coupled multi-keyframe visual-inertial odometry for accurate and robust state estimation. Sensors 18(11) (2018). https://doi.org/10.3390/s18114036

Localization and Grading of Building Roof Damages in High-Resolution Aerial Images

Melanie Böge, Dimitri Bulatov, and Lukas Lucks[✉]

Fraunhofer IOSB Institute of Optronics, System Technologies and Image Exploitation, Gutleuthausstr. 1, 76275 Ettlingen, Germany
{melanie.boege,dimitri.bulatov,lukas.lucks}@iosb.fraunhofer.de

Abstract. According to the United States National Centers for Environmental Information (NCEI), 2017 was one of the most expensive year of losses due to numerous weather and climate disaster events. To reduce the expenditures handling insurance claims and interactive adjustment of losses, automatic methods analyzing post-disaster images of large areas are increasingly being employed. In our work, roof damage analysis was carried out from high-resolution aerial images captured after a devastating hurricane. We compared the performance of a conventional (Random Forest) classifier, which operates on superpixels and relies on sophisticated, hand-crafted features, with two Convolutional Neural Networks (CNN) for semantic image segmentation, namely, SegNet and DeepLabV3+. The results vary greatly, depending on the complexity of the roof shapes. In case of homogeneous shapes, the results of all three methods are comparable and promising. For complex roof structures the results show that the CNN based approaches perform slightly better than the conventional classifier; the performance of the latter one is, however, most predictable depending on the amount of training data and most successful in the case this amount is low. On the building level, all three classifiers perform comparable well. However, an important prerequisite for accurate damage grading of each roof is its correct delineation. To achieve it, a procedure on multi-modal registration has been developed and summarized in this work. It allows adjusting freely available GIS data with actual image data and it showed a robust performance even in case of severely destroyed buildings.

Keywords: Damage detection · Superpixels · Feature extraction · Random forest CNN · DeepLabv3+ · SegNet · Building footprint alignment

1 Motivation

Regarding urban infrastructure, buildings and roads are essential parts for many applications as city planning, civil security and disaster management. However,

A. P. Cláudio et al. (Eds.): VISIGRAPP 2019, CCIS 1182, pp. 497–519, 2020.
https://doi.org/10.1007/978-3-030-41590-7_21

buildings are not only relevant components of urban infrastructure. They gain even more importance if people call these buildings *home*. Numerous weather and climate disasters destroy thousands of buildings per year. While homeless people face desperation in these times, helpers in need must find their way through the chaos and – often far from it all – insurance companies need to start their work. Especially if people are still able to fix damages on their homes, a fast processing of the latter instance is necessary. Affected people need the money of their insurance companies to hold their homes and to avoid greater damage to them. The insurance companies are also interested in a fast and as easy as possible processing to reduce their costs in damage assessment. This damage assessment typically takes place by loss-adjusters that have to visit every insured building in order to record the incurred loss. Without support, this is a very time-consuming and potentially dangerous task.

In this paper, we will focus on roof damages after storm events, like Hurricane Irma 2017 in the south-east of the USA. The presented work, is a continuation of the results published in [25]. We will reprocess the introduced methods and compare the achieved results to deep-learning approaches. Furthermore, we will analyze in this paper the necessary number of training data for all approaches. We will extend our tests for the transferability of training data in order to analyze some damage details during evaluation.

Altogether, we will introduce an approach to

1. identify buildings of the portfolio in airborne images of a city as high-throughput manner,
2. localize the damage on every roof,
3. determine the degree of roof damage for each building to enable a damage ranking.

This automatic execution of damage assessment allows the evaluation of whole cities at once, prevents time-consuming and dangerous roof climbing for loss adjusters and enable a very much faster and prioritized processing of insured objects.

Image-based detection of roof damages or other anomalies is a pattern recognition problem. Here, we leverage state-of-the-art tools for the analysis of patterns and textures. We will solve the task to estimate the roof damage by using trained classifies: Once a conventional Random Forest Classifier that is trained by specific image features of choice and second two CNN based approaches – the SegNet and DeepLabv3+. We chose Random Forest as representative of conventional classifiers since it is quite robust against redundant features [16] and against over-fitting. The CNNs are chosen in order to test two operationally different neural network. For one thing, a comparably simple but well established neural network depending on only a few parameters and for the other thing a much more complex network. Especially the latter one provided reliable results on *PASCAL VOC 2012* benchmark [13], concerning tasks of semantic image segmentation. Hence, we want to convey its performance in the context of damage detection. In contrast to other studies on the field of CNNs, our data and training set only consist of a comparably small number of buildings with even fewer references for damaged and undamaged regions.

The developed method is applied to two different portfolios. Only post-loss (i.e. post-event) data is available. Since damages of the roof area shall be analyzed, the outline of this roof area is required. This information is typically not freely available since it depends on image properties as height (scaling) and angle (oblique view) of admission. To evaluate the performance thoroughly, we will work with images presenting the buildings in nearly nadir view and work with precisely fitting roof outlines annotated by experts.

2 Related Work

The approach of [43] propose to determine the amount of damage by a change detection regarding pre- and post-loss data by considering differences in average intensities and variances of image color values. Others, like [34] use texture analysis to differentiate between damaged and undamaged parts of the roof. The challenge is to figure out, which radiometric properties in the images stem from temporal differences and which are consequences of damage caused e.g. by natural disasters. Various studies have investigated morphological profiles, structural and radiometric features [29] or correlation coefficients from a co-occurrence matrix as in [31]. Another study [18] also generated data with annotated geo-referenced changes on the ground that is often used as training data for semi-supervised techniques. However, all images in this dataset must be re-sampled to a uniform scale, resulting in rather low resolution. Using groups of neighbored pixels enables the introduction of features analyzing larger-area features as linear segments [21] or measure certain metrics such as normalized differential vegetation index (NDVI) [15], segment properties [22], etc.

The approach of [35] used pre-event satellite images only to localize the buildings and to project the building footprints onto a high-resolution post-event image. To correlate buildings in multi-sensor data, the authors applied Support Vector Machines (SVMs) over the composed Hue, Saturation, Value (HSV) indexes of the pixels and the 128 entries of the dense Scale-Invariant Feature Transform (SIFT) descriptor [24]. This approach bridges the gap to techniques that do not need pre-event images (that are rarely available in desired resolution) as the study of [32]. These techniques rely on high-dimensional sets of features often without explicit semantic meaning. The classification tools currently considered as state of the art are, therefore, Convolutional Neural Networks (CNNs). They are particularly suitable for a wide class of classification tasks, because they perform the two steps feature selection and decision-making simultaneously, relying merely on a large pool of training examples. Since this training data is often available online, CNNs as black-boxes do not require too much additional expertise and are therefore increasingly becoming popular. In the context of damage assessment, CNNs were applied by [14] to a dataset containing buildings destroyed by flooding. Either pairs of pre- and post-event color images (if the former ones were available) or only post-event images were analyzed with quite a high accuracy, without assessing intermediate damage grades.

Many studies have shown a well-designed approach with a standard classifier can produce results similar to those obtained using CNNs [10, 14] and emphasized the necessity of three-dimensional (3D) data to improve the results [39].

To reduce the amount of training data that is often necessary, model assumptions were introduced. The study of [32] for example assumes that typical shadows around buildings are missing if they are destroyed. This work therewith relies on nearly completely destroyed buildings since smaller damages to the roofs as blown-away shingles will not change the shadow of a building. In a subsequent study [26], spectral, spatial, and morphologic features were combined to achieve building-wise detection rates around 90%.

The previously mentioned study of [39] introduced an algorithm also to localize damage. In this approach, the image is subdivided into superpixels, which are approximately equally-sized image segments that ideally coincide with the image edges. A patch is formed around each superpixel to be the input of a CNN based evaluation. Contemporaneously, 3D features resulting from the eigenvectors of the structure tensors of different radii (as in [41]) are calculated in a point cloud and projected onto the superpixels. Moreover, to integrate features from different modalities, a multiple-kernel-learning framework [39] was investigated. As a result of this work, the authors reported the difficulty of differentiation between complex structures and damage without considering 3D information.

All approaches lead us to the conclusion that a precise localization of roof damages from high-resolution aerial images is possible and can support the work of loss adjusters in multiple ways. Pre-event images are not really necessary to solve that task although they are helpful since the appearance of different natural disasters will contain different types of destruction [12]. Still, we will want to get along without pre-event images. To cope with this, our algorithm is designed to compensate for a small number of training examples by using a hand-tailored, purpose-based, and quickly extractable feature set.

3 Preliminaries

Our procedure for damage detection depends on some preliminary work. For the classification based on Random Forest, this area will be decomposed into smaller sub-areas of similar properties (superpixels). Furthermore, the selection of training data is an essential step to differentiate between damaged and undamaged roof parts.

3.1 Superpixel Decomposition

The decision whether a roof is damaged or not will be derived from characteristics, to be referred to as features. A simple yes-no decision is not sufficient for a comprehensive loss assessment for which it is also necessary to localize the damaged roof parts and therefrom to derive the damage degree of the roof. This localization of roof damage is possible if for every locality (pixels or superpixels) on the roof the decision *damaged or undamaged* is made. On the finest scale, this means to assign each pixel to one of these classes; however, in this paper we will work partly with superpixels. This decomposition is carried out for three reasons: On the one hand to reduce computation time for the Random Forest classifier; on the other hand for stabilizing reasons at feature extraction (see Sect. 4.1). Further, also features can be considered that are not only derived from a pixel information itself but also include neighboring properties like the presence of straight lines traversing the roof. For superpixel decomposition, we used the implementation of compact superpixels described in [38]. A data term prohibits a superpixel to leave the area that it is supposed to cover and a smoothness term ensures compact borders. The chosen approach might not be the fastest one among those introduced in literature but the parameters are reliably controllable. In the obtained superpixel decomposition, we further merge neighboring superpixels that are too small and remove those that lie outside of or on borders of the roof outline.

(a) Examples for damaged roofs

(b) Examples for undamaged roofs

Fig. 1. High diversity of roof structures for damaged as well as undamaged roofs. Illustration as in [25] of data set D1 (see Sect. 5).

3.2   Training Data Acquisition

The classification result and its accuracy depend on the selection of training data. It is important to use appropriate representatives that include all types of damage to be detected: We unify blown away shingles, broken roofs and covering tarpaulins as well as tree branches as an indicator for roof damage. All of these appearances shall be detected. At the same time, undamaged roofs can also consist of structures of high diversity: Some roofs are just covered by a shingle texture. Many roofs are covered by other superstructures, such as solar panels, air conditioners and chimneys. Additionally, different kinds of roof material should not produce false alarms. Therefore, it is essential to include as many types of damaged and undamaged roof areas as possible into the training data. Some examples are shown in Fig. 1.

We interactively selected buildings evenly distributed across the data set. Within these buildings, experts selected and labeled representative areas as reference data for the two classes undamaged and damaged. The amount of necessary training data depends on the variety of roofs characteristics. The more roof materials are present and the more superstructures a roof includes, the more training data is required; the same applies to the diversity of damages.

However, the training examples do not have to be recreated for every data set. Often, large-area evaluations can be fastened by splitting up into subareas or datasets of different timestamps have to be evaluated. If these datasets consist of similar architecture and damage characteristics, the training data may be reused. We will test this way of proceeding in Sect. 5.2.

4   Methods

One important goal of this work is to compare the performance of a conventional classifier with one (or more precisely two) based on CNNs. We chose Random Forest as a conventional classification tool since it is a relatively fast classifier and provides good results with only a few training data. As CNN-based classifiers DeepLabv3+ and SegNet are applied. All classifiers rely on feature sets, which comprise the properties of roof area. While CNN based classifiers choose their own features and only need a set of training examples as input, Random Forest additionally depends on predefined features.

4.1   Classification with Random Forest

We use a Random Forest classifier [3] to combine powerful features in order to differentiate between damaged and undamaged superpixels. The information content of the features is learned using previously extracted training data (Sect. 3.2). The classification takes place by evaluation of a set of decision trees whose number was determined by adding trees successively and analyzing the out-of-bag error. We found out that 40 trees were sufficient. For each superpixel, the probability of being damaged corresponds to the percentage of decision trees voting for this class. If this probability exceeds 0.5, the superpixel is classified as damaged.

As features, 52 were chosen for Random Forest classifier, to decide between damaged and undamaged roofs. The images directly derived some of these features: unfiltered color channels (RGB) and combinations thereof like saturation, (pseudo-)NDVI, and Opponent Gaussian Color Space (OGCS) [17]. Further, differential morphological profiles were applied. They became quite popular in remote sensing because of their invariance to changes in contrast and shape of a characteristic region. For detection of edges and blobs, the rotationally invariant MR8 filter bank [37] is used. Intact regions are often represented by homogeneous textures. As indicator for homogeneity, the entropy is used.

Additionally, we used higher-level features in cooperation of model assumptions on a more global scale like lines or repetitive structures. The evaluation of patterns is a strong indicator of damage or intactness. Long lines are often found on undamaged roofs while short lines either indicate inventive architecture or much more probable: damage. The number of line segments traversing one superpixel is denoted as the lineness measure [7]. Similarly, a regular texture indicates undamaged roof area even if the average gradient norm is high. In order to characterize texture properties, a modified version of the histogram of oriented gradient (HOG) features was used [11]. The gradient orientations modulo π are collected and weighted by their occurrence. Therefore, a discretization with a step length of 0.1π is used. The histogram is normalized and smoothed [30] to enable the estimation and evaluation of a cumulative distribution. We use only the first three entries of the ascending sorted histogram to highlight superpixels with only a few dominant gradient orientations. Some examples of features per superpixel are shown in Fig. 2. The features are obviously different for damaged and undamaged roof regions.

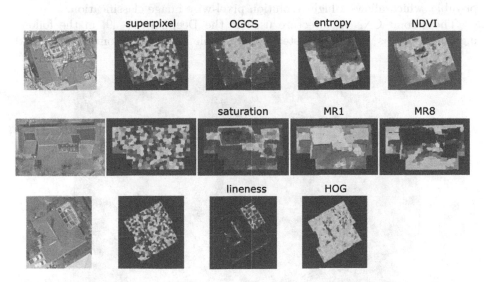

Fig. 2. Alarm rates of selected features shown for three exemplary buildings. The first six feature images clearly delineate the damaged parts of the roof from undamaged areas. The lineness and HOG reliably indicate the positions of roof edges and texture salience. Image source [25].

Features are usually computed pixelwise and then assigned to a superpixel using a statistical measure, like mean or variance over all pixels contained in one superpixel, see [6] for a deeper insight into the theoretical framework. Higher-level features are assigned directly to the superpixels.

There are studies discussing redundant and irrelevant features (e.g. [16,40]), but in our case a feature evaluation had no advantage. The reason for this is that the algorithm always selects the most meaningful feature at every split node of the decision tree.

4.2 Classification with CNNs

Besides the classification of the superpixels by Random Forest, the damage of the roof areas will be investigated by deep learning methods that work on pixel level. Due to the numerous researches in this field [44], a huge number of different methods are available. For this study, we focus on two methods of the field of semantic image segmentation. In contrast to object detection and image classification, these methods are not assigning an entire input image to one target class, but in a pixel-wise classification and localization of the objects in the image.

First, a relatively small network with a simple encoder-decoder structure called SegNet [1] is used. The encoder module primarily consists of the VGG16 network with the difference that the indices of the max-pooling layer are transferred to the un-pooling layer in the decoder. Thus, a nonlinear up-sampling is possible, which allows a high-resolution pixel-wise image classification.

The second CNN architecture used is the DeepLabv3+ [9] in the following shortened as DeepLab. Instead of a simple cascade of convolutional and

Fig. 3. Visualization of damaged buildings on a map. Damage grades are color coded to enable further evaluation. Green: intact buildings, yellow: light damage, orange: medium damage, red: heavy damage. (Color figure online)

pooling layers, the encoder of the network contains parallel atrous convolution layers with different rates (so-called Atrous Spatial Pyramid Pooling) to generate some meaningful multi-scale features. The atrous layers allow features with high context information without reducing the spatial resolution too much. The decoder of the network contains some simple upsampling and convolutional layers, which refine the output of the encoder and some low-level features to a fine and an an accurate segmentation.

How mentioned in Sect. 3.2 our training data consist of labeled areas in single buildings. To avoid regions without labels to influence the loss function during the training of the network, we mask these samples. For both networks, we use binary cross-entropy as a loss function.

4.3   Damage Grading

Since the damage is localized by assigning each pixel or superpixel either to class damaged or to class undamaged, it is now possible to compute the ratio of damaged and undamaged roof area. This ratio reflects the damage of a building and supports a prioritization in claims settlement. Damage categories can be derived from the percentage of damaged superpixels. These categories are: intact, light damage, medium damage, and heavy damage as visualized in Fig. 3. The map visualization turned out to be a helpful instrument for insurance companies. First of all, these insurance companies have a visualization of all insured buildings in their portfolio. A further color coding for the degree of damage provides a visual impression of the location of the insured buildings, enables estimations of compensation payments, and makes the routing of loss-adjusters more efficient.

5   Results and Discussion

We demonstrate the usability of our algorithms by analyzing two different data sets of cities impacted by Hurricane Irma 2017:

D1: 421 randomly chosen buildings of Rockport, Texas, with homogeneous roof structures suffering heavy damage; ground resolution: 7.5 cm.
D2: 416 randomly chosen buildings of Marco Island, Florida, with inhomogeneous roof structures and less damage; ground resolution: 15 cm.

Because of data privacy, we work with fictive portfolios of randomly chosen buildings. The relatively small number of evaluated buildings enabled a qualitative comparison with a manual reference set that was provided by experts.

D1 consists mostly of buildings of similar roof architecture. Hurricane Irma caused heavy damage to many of these buildings. For that reason many roofs are already covered by tarpaulins. The high diversity of roof structures in D2 and the lesser amount of damage makes this data set the more challenging one. Furthermore, this data set includes many pool houses next to dwelling houses. These pool houses typically have glass roofs that allow the view through the roof to the floor below. This is the hardest part for our classifiers since they have to differentiate floor from roof structure.

For evaluation with Random Forest classifier, the extent of each superpixel was determined to cover approximately (0.75×0.75) m^2 of roof area. Due to ambiguous boundary values, superpixels on the border of the roof outline were excluded from the evaluation. In the following images, these superpixels are colored gray.

For D1 and D2, precisely fitting roof outlines were available.

5.1   Damage Detection

In order to assess the accuracy of the procedure, we applied a three-fold cross-validation on widely labeled areas of each building. We will analyze the *overall accuracy* as well as *precision* and *recall*. *Precision* indicates the ratio of correctly predicted (damaged or undamaged) samples; *recall* the ratio of correctly retrieved damaged or undamaged samples. The partitioning in training and test set takes place randomly on a per building level to avoid an over-fitting of the classifiers by learning samples of each present damage.

The accuracy of all classifiers is provided in Table 1. In general, all methods achieve reliable results. All methods have a high overall accuracy in the range of approximately 89% to 94% for D1 and 94% to 98% for D2. In detail, DeepLab achieves the most precise results. The results of Random Forest are around 5% less accurate. At first sight, this amount seems to be a significant difference, but in fact, we can partially trace this amount to grouping pixels to superpixels and using winner-takes-all strategy. Surveys have shown that the amount of mis-assignment of pixels due to superpixel grouping can be up to 3%.

However, the significance of the overall accuracy is limited by the fact that within the data sets, the classes are strongly unbalanced. A more detailed view can be obtained by looking at the precision and recall, especially for the damage class. For this class, all three methods achieve a high precision compared to a little lower recall. Although the methods tend to miss some damaged areas, the samples classified as damaged are very reliable.

To compare the classifiers among themselves rather than to evaluate each classifier, we use Kappa-statistic [8] to compare the overall accuracy to the expected accuracy. As one can see for D1, all classifiers provide similar κ values. Our conventional classifier sticks with SegNet. For D2, the conventional classifier struggles with the damage detection of pools. The only way to keep the performance of the conventional classifier up with the CNN approaches is to ignore the pools. In this evaluation, the accuracy of Random Forest is similar to that of SegNet.

To provide also a visual impression, we will show selected examples that include typical properties of the data sets. An example of lighter damage caused by blown away shingles and an example of heavy damage in form of a broken roof is shown in Fig. 4. The red-green color plots show the damage map of the roof. Green and red regions denote undamaged respectively damaged roof areas. The color map plots show the probability of damage. The darker red the more likely

Table 1. Confusion matrices for results of damage detection in D1 and D2 using Random Forest (RF), DeepLab (DL) and SegNet (SN) classifier. In case of D2 RF (-) indicates the evaluation without taking account of pools.

D1/RF

reference	predicted undam.	dam.	rec. (%)
undam.	30672	2109	92.1
dam.	4069	17667	81.4
prec. (%)	88.2	87.3	88.7

$\kappa = 0.8$

D1/DL

reference	predicted undam.	dam.	rec. (%)
undam.	2647295	43872	98.4
dam.	231170	1361561	85.5
prec. (%)	92.0	96.9	93.6

$\kappa = 0.9$

D1/SN

reference	predicted undam.	dam.	rec. (%)
undam.	2548150	143017	94.7
dam.	304196	1288535	81.0
prec. (%)	89.3	90.0	89.6

$\kappa = 0.8$

D2/RF

reference	predicted undam.	dam.	rec. (%)
undam.	120202	1829	98.5
dam.	6771	4798	41.5
prec. (%)	94.7	72.4	93.6

$\kappa = 0.5$

D2/RF (-)

reference	predicted undam.	dam.	rec. (%)
undam.	105238	1800	98.3
dam.	3332	6305	65.4
acc. (%)	96.9	77.8	95.6

$\kappa = 0.7$

D2/DL

reference	predicted undam.	dam.	rec. (%)
undam.	32808952	108530	99.7
dam.	774190	2010329	72.2
prec. (%)	97.7	94.9	97.5

$\kappa = 0.8$

D2/SN

reference	predicted undam.	dam.	rec. (%)
undam.	32575261	342221	99.0
dam.	980817	1803702	64.8
prec. (%)	97.1	84.1	96.3

$\kappa = 0.7$

an area is damaged. The darker blue the more likely an area is undamaged. Both types of damage can be reliably detected. The color coding of the probabilities indicate that a differentiation of blown away shingles and roof destruction might be possible in future work (Sect. 7).

The examples show that DeepLab provides the most accurate results. The evaluation of Random Forest is obviously influenced by the size of the superpixels. Therefore, the result strongly depends on the default size of the superpixels and the quality of segmentation that should not extend the average damage size in order not to over-smooth damages. However, also the results of SegNet appear a little fringy. Further, this classifier often predicts damage at building borders. A possible explanation for this is that the field of view of the features used for borders also includes ground areas. These could influence the feature computation and the result.

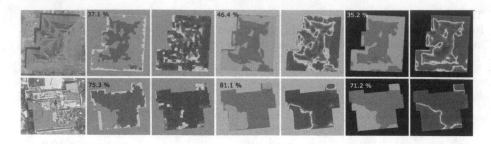

Fig. 4. D1: Detection of lighter damage (blown away shingles) and heavy damage (broken roof); orthophoto, damage map and probability map of certainties. The colored background codes for the applied methods: blue for Random Forest, ocher SegNet, and black DeepLab. (Color figure online)

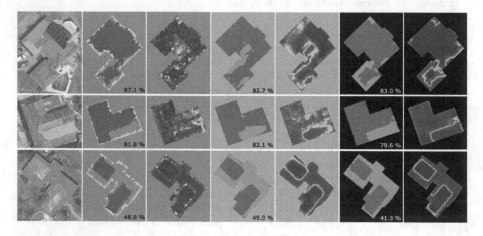

Fig. 5. D1: Recognition of tarpaulins as area of damage using Random Forest (blue background), SegNet (ocher background) and DeepLab (black background) classification. The percentage values indicate the amount of damage. (Color figure online)

Most notably for D1 are tarpaulins covering many of the roofs. These tarpaulins usually mean damage and can be recognized very well with all classifiers (see Fig. 5).

Data set D2 was much more challenging due to the small amount of damage and the high diversity of roof structures and superstructures. Also, the pool houses belonging to each building demand a sophisticated evaluation since the glassy roofs allow a view to the floor. Here the biggest differences between the classifiers happen and all classifiers have difficulties with the damage detection of these areas. Examples for damage detection in D2 are given in Fig. 6. Again, DeepLab classification achieves the best results; but in the damage maps of all classifiers, the roof damage is detected correctly. The classification with Random Forest tends to miss some damaged area and therewith delivers a smaller damage

Fig. 6. D2: Damage detection maps with Random Forest (blue background), SegNet (ocher background) and DeepLab (black background) with corresponding damage rates for buildings with and without pool houses. (Color figure online)

ratio compared to DeepLab. In the case of SegNet classification this is very case-sensitive: Sometimes, less damage is detected, while in other cases the degree of damage is extracted significantly more reliable. In the last example of Fig. 6, the areas with roof damage are detected much more extensively.

To compare the performance of the conventional classifier to the CNN-based approaches on a pixel level, we also tested the performance of Random Forest without superpixels. The evaluation of 100 buildings in D1 required 12570000 pixels to be analyzed compared to 62600 superpixels what was less than 1% and greatly affects computation time. The computation time of these 100 buildings

superpixelwise   pixelwise

Fig. 7. Comparison between superpixelwise and pixel-by-pixel evaluation often leads to a less precise result since information including neighborhoods and surroundings are omitted. Image source [25].

took only 6% by using superpixels compared to a pixel-by-pixel approach. Further, going on without superpixels causes the loss of neighborhood relations for feature computation and results in an even worse overall accuracy by 10%. An example is shown in Fig. 7.

5.2   Amount of Necessary Training Data

We further evaluate the required amount of training data for each classifier. Therefore, we consider a steady test set containing one-third of available reference data for D1. The remaining two-thirds of reference data are used as training data and partitioned in subsets to analyze the necessary amount of training data to satisfactorily evaluate the test set. Again, we will analyze the overall accuracy as well as precision and recall by successively increasing the number of training data by 10%. Figure 8 shows these values for damage detection by Random Forest, DeepLab and SegNet. Please note that only the discrete points at multiples of 5% are measured. The dotted light colored lines are depicted only to provide a better visual impression of trends. A continuous decreasing or increasing correlation between multiples of 5% is not necessarily given. The evaluation shows that Random Forest already converges at 70% (of one third) of training data. To use more training data does not improve the performance significantly. It is difficult to draw a conclusion about accuracy convergence from the figures for CNNs, especially regarding precision and recall. This confirmed the initial reservations against using CNNs [25] since they typically rely on huge training data sets. However, they must be credited with the fact that they provided good results in damage detection, even with a small amount of training data.

Recurring events or similar tasks often require repeating evaluations on different data sets. As one can see from Sect. 3.2, labeling training data examples is the most time-consuming step. This is why we decided to investigate further research in the transferability of our models from one dataset to the other. Due to the lack of more reference data, we used the training data or models of D1 to evaluate D2 and vice versa. Even though the data sets are very different in their appearance concerning roof structures and damage, we got promising results. For similar patterns of damage and roof structure, the transfer worked quite

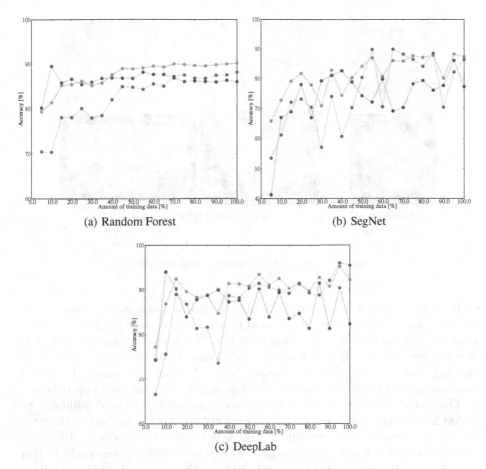

(a) Random Forest

(b) SegNet

(c) DeepLab

Fig. 8. Overall accuracy (green), precision (red) and recall (blue) for Random Forest, SegNet and DeepLab. The x-axis shows the amount of training data of one third of reference data that is used for training. (Color figure online)

well, so e.g. blown away shingles were detected. Examples are shown in Fig. 9. Of course the procedure failed to evaluate pools in D2 correctly since in training data of D1 no pools were present. Vice versa, the procedure had problems to recognize the blue or very homogeneous tarpaulins in D1 while using training data of D2 without tarpaulins.

6 Ongoing Work: Footprint Alignment

For an automatic detection of roof damages, it is often necessary to have an exact alignment of the observed area. For Random Forest classification, this delineation is necessary to prevent false alarms (e.g. if the surrounding ground is included into the area of interest) or to miss damages (if parts of the roof are not

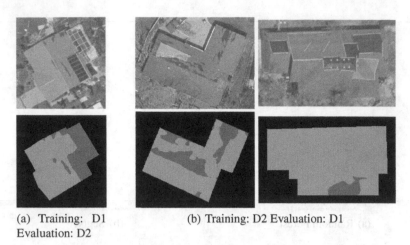

(a) Training: D1
Evaluation: D2

(b) Training: D2 Evaluation: D1

Fig. 9. Examples for transfer of training data.

evaluated). Although CNNs would be able to detect damages on a whole image at once without delineation of an area of interest, the correct pixel-based results would be obviously worsened by incorrect building-wise assessments (Sect. 4.3). Besides, the correct delineation allows to concentrate on the regions around the building and eases the job on training data selection enormously. Further, a common area of interest simplifies to compare both classification approaches.

One could argue that the outlines of vast majority of insured buildings are stored in cadastral maps and can be made available for the evaluator. However, due to many reasons, these outlines are often not sufficiently accurate [19]. These reasons are: Non-identity of ground plans with roof outlines, especially in non-nadir images, the fact the measurements had mostly been carried out by humans, and the eventual small registration errors committed by preparation of the post-event images.

In order to develop a fully automatic damage detection tool, automatic alignment of footprints is an indispensable component of the workflow. Developments in this field have already been successful [5].

Among the approaches undertaken in the past for alignment of GIS-based outlines with the actual image data, we can first mention the interactive procedures, such as [4], which allow a good quality control, but are cumbersome when it comes to processing large scenes. Active contour models [27,28], such as snakes, are helpful for the evolution of already available approximate values, but are often an overkill if alignment transformations can be described by a few parameters. For rigid alignment transformation, such as pure translation, matching key-points, for instance, [24] followed by RANSAC is probably the fastest strategy, however, not always feasible if the modalities of two images (multi-spectral vs. binary mask) are too different. Convolutional neural networks (CNNs) are increasingly being applied for outlining [27,33] and aligning [36,42] buildings. These methods rely on large amounts of training data and, since it was not trivial to retrieve such ground truth data for many destroyed buildings, we decided to leave CNN-based alignment methods for future work and focus on variational approaches. In [20], intensity values are interpreted as samples of two random processes and are linked by a probability density function (such as Mutual Information). The transformation is computed pixelwise and a regularization term is added to penalize transformations of neighboring pixels. After [2] considered a convex combination of a color-based and a gradient-based error term for pose estimation of a single frame, [5] recently minimized a similar function in order to obtain the unknown translation parameters for every building. Because of the robustness of this approach and its overall good performance, we will stick to the method and give the necessary details on its implementation and evaluation in the following.

We use free available building footprints from Open Street Map. Their outlines are often similar to the roof outlines but due to non-nadir views do not satisfy the position of the roof. For that reason, the building outline will be aligned to the roof outline by considering dominant color information and image edges.

This procedure was already mentioned in [25] and presented more detailed in [5]. We assume that there exists a transformation φ for every building footprint $\mathcal{P}$ to align it to its corresponding roof. We search for it in an image fragment around the footprint that is denoted by $\mathcal{J}$. The size of $\mathcal{J}$ is the size of $\mathcal{P}$ in pixels enlarged by a typical search radius, to be computed beforehand. Edges of $\mathcal{J}$ shall coincide with $\mathcal{P}(\varphi)$. Up to eight-dimensional representations of φ are required to map every affine respectively projective transformation. In [5], φ is only a two-dimensional vector.

The alignment process works on image patches. Every patch contains one building with some surrounding area of dimension $(j \times k)$. The image pixels are denoted by q_{xy}, $1 \leq x \leq j$, $1 \leq y \leq k$. The available building footprint $\mathcal{P}$ is rasterized to the image patch. For every patch a mask

$$\mathcal{M} = \mathcal{M}(\mathcal{P}) = \begin{pmatrix} m_{11} & m_{12} & \\ & m_{22} & \ddots \\ & & & m_{jk} \end{pmatrix} \tag{1}$$

is created with $m_{xy} = 1$ for all image pixels that lie inside $\mathcal{P}$, 0 for all outside pixels and 2 for border pixels. Since freely available building footprints already overlap with the region of interest in most cases, we consider a known search range for φ. For the same reason, we can assume good starting values for the alignment function that is represented by a minimization problem of an energy function. We apply a modified Mutual Information approach as target function that rewards homogeneity of the dominant color $\mathbf{f} \in \mathbb{R}^3$ sampled from a 3D histogram of 16 bins (for 8-bit images) overall color values of image pixels satisfying $\mathcal{M} = 1$. In some cases, the dominant color $\mathbf{f}$ cannot be determined unambiguously; that is, if roofs are destroyed or contain many superstructures (e.g. chimneys and dormers) as well as their shadows. Using the L_1-norm in the target function makes the minimization in these cases more robust. According to how likely a pixel is supposed to belong to a building, we further introduce a weighting term w_f to the energy function. This term ideally reflects the probability of separation between buildings and their surroundings, especially vegetation

$$w_f(q_{xy}) = \begin{cases} 1 & \text{if } m_{xy} > 1, \\ 0 & \text{if } m_{xy} = 0. \end{cases} \tag{2}$$

Apart from color information, image edges are strong indicators for coinciding outlines. We introduce a term that is significantly higher at border pixels than inside $\mathcal{P}$. A corresponding penalty term w_∇ takes its minimum at a border, a small positive value inside $\mathcal{P}$ and a smoothly decreasing function outside $\mathcal{P}$

$$w_\nabla(q_{xy}) = \begin{cases} -1 & \text{if } m_{xy} = 2, \\ 0.01 & \text{if } m_{xy} = 1, \\ -e^{\frac{-d(q_{xy})}{\sigma}} \text{ OR } 0 & \text{if } m_{xy} = 0. \end{cases} \tag{3}$$

The overall accuracy is higher by choosing the first term for $m_{xy} = 0$ but for destroyed buildings the registration is worse. But this includes a paradox dilemma. Especially for destroyed buildings a precisely fitting roof outline is necessary to solve the task of roof damage evaluation. Therefore, we are not directly interested in an *overall* accuracy that would be particularly high by choosing the first option if the data set contains only a few destroyed buildings. Hence the first option should be used for registration problems in general if no destroyed buildings are expected; in our case the choice of the second option (0) is much more useful.

The function d can be computed by morphological operations to determine the distance to the next pixel labeled as *inside*, σ is a constant around 0.5. Before applied, w_∇ is smoothed with a Gaussian function in order to take rasterized lines into account. The smoothed function will be denoted by $\tilde{w}$.

The energy function can be summarized as follows

$$E(\varphi) = \sum_{xy} \alpha w_f(q_{xy}) \|\mathcal{J}_f(q_{xy})\| + (1-\alpha)\, \tilde{w}_\nabla(q_{xy})\, \|\nabla \mathcal{J}(q_{xy})\|, \tag{4}$$

where α is a balance parameter to be explored in the experiment section. While the weighting terms are supposed to fit the outline $\mathcal{P}$ to the building, both normalizing terms pull the outline to the relevant building.

For minimizing the L_1-norm of cost function (4), a gradient-free Nelder-Mead method is used as introduced in [23]. In order not to stagnate in a local minimum, we use a multi-level approach as in [42] to obtain a pyramid decomposition of the image.

In [5] the author also points out that in a few cases, registration nevertheless stagnates in a local minimum or does not converge and the outline lies far away from its expected position. This can happen if many parallel lines are present in the image and the footprint stagnates at the wrong image edge. Because of this, the author proposes to replace each determined transformation by the median of the determined transformations of its neighbors using a kNN search.

In our case, the introduced data sets (see Sect. 5) unfortunately contain sometimes image mosaicing artifacts. On edges of mis-stitched areas with dissected buildings are often present and will produce further problems with kNN post-processing. For that reason, we will not follow the instructions of [5] and skip this post-processing step.

Results. In order to demonstrate the functionality of the alignment we just evaluate 259 buildings of D2. We reduced the number because for these 259 buildings freely available outlines from OSM fitted the shape of roof outlines well. Therewith, we can relate the freely available building outlines to the outlines of roofs that were manually annotated by experts. This enabled us to test the algorithm without the necessity to insert further nodes and edges to the polygonal chains in order to change the shape course of the building outline at all. The extension of the method to change the course of the polygon will be part of future work. Figure 10 shows an exemplary concept.

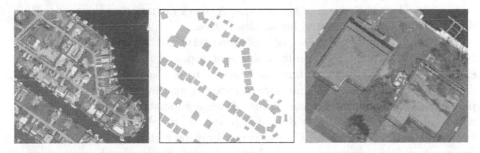

Fig. 10. Alignment of building outlines: Often, for a scene (left), building outlines are available (middle). For many reasons they may not fit the roof outlines (right, red line). By using registration techniques, the building footprint can be aligned to the roof (right, green line). (Color figure online)

The initial deviations of building outlines from roof outlines were quite high – up to 0.3 m. After alignment, this deviation can be significantly be reduced to 0.3 m. Using root-mean-squared error, the deviation after alignment is 7.5 cm that corresponds almost to pixel accuracy. Regarding the distribution of errors, these scatter for destroyed roofs more than for intact ones. This is not surprising since damaged roofs more often contain occasional gradient discontinuities that lead to confusion with usual texture elements. For more details concerning the choice of parameters, see [5].

7   Conclusion and Outlook

We presented three procedures for semi-automatically detecting and localizing roof damage in post-event aerial images. For data sets with similar roof shapes, all three methods (Random Forest, SegNet and DeepLab) provide good results. In contrast, there are large differences for buildings with complex roof structures – especially with glass roofs. The performance of DeepLab was better than that of an approach based on conventional classifier and superpixels. Nevertheless, Random Forest is able to work with less training data and performs more reliable, depending on the available amount of references. The resolution of the images, the precise outlines of the roof area, and the diversity of roofs in the data directly affects the performance of the classifiers immensely. Although our procedure was proposed for damage detection, the method was already successful and could be employed for other related applications, such as roof analysis, installation of solar panels or roof window detection.

Furthermore, more work should be invested to test and analyze the transferability of training data with a larger reference set. Reference data of higher diversity and a larger amount of training data should provide better results.

Future work has to be done concerning the topic of footprint registration to roof outlines. The procedure should be extended by more general rigid transformations (euclidean, affine, projective) as well as the possibility to insert new vertices to adapt the course of the outline. A direct deviation of the roof outline directly from the image would be an alternative.

Differentiating between types of roof damage is a challenging task that should be investigated further. We unified different types of damage as blown away shingles, broken roofs, tarpaulins and trees into a single class; also the high diversity of roof structures and superstructures as ventilation facilities, solar panels and many more. Because some of these superstructures cast shadows, which also can be mistaken as damaged roof area, we plan to analyze in future work whether a separation of these types into finer classes is useful. Here, a greater benefit is expected from including 3D information and multi-spectral imagery. In future work, it would be useful to differentiate between these types of damaged and possibly of undamaged roof areas by including 3D information and near-infrared measurements. Nevertheless, the results obtained in our work are extremely important for insurance companies to make a first quick estimation of the incurred loss. This estimation allows to accelerate the processes in which the insurance companies can provide the compensation to those people whose homes were damaged and who, in general, cannot wait to begin the clearing-up operations and rebuilding work.

References

1. Badrinarayanan, V., Kendall, A., Cipolla, R.: SegNet: a deep convolutional encoder-decoder architecture for image segmentation. IEEE Trans. Pattern Anal. Mach. Intell. **39**(12), 2481–2495 (2017)
2. Bodensteiner, C., Hebel, M., Arens, M.: Accurate single image multi-modal camera pose estimation. In: Kutulakos, K.N. (ed.) ECCV 2010. LNCS, vol. 6554, pp. 296–309. Springer, Heidelberg (2012). https://doi.org/10.1007/978-3-642-35740-4_23
3. Breiman, L.: Random forests. Mach. Learn. **45**(1), 5–32 (2001)

4. Brooks, R., Nelson, T., Amolins, K., Hall, G.B.: Semi-automated building footprint extraction from orthophotos. Geomatica **69**(2), 231–244 (2015)
5. Bulatov, D.: Alignment of building footprints using quasi-nadir aerial photography. In: Felsberg, M., Forssén, P.-E., Sintorn, I.-M., Unger, J. (eds.) SCIA 2019. LNCS, vol. 11482, pp. 361–373. Springer, Cham (2019). https://doi.org/10.1007/978-3-030-20205-7_30
6. Bulatov, D., Häufel, G., Lucks, L., Pohl, M.: Land cover classification in combined elevation and optical images supported by OSM data, mixed-level features, and non-local optimization algorithms. Photogram. Eng. Remote Sens. **85**(3), 179–195 (2019)
7. Bulatov, D., Solbrig, P., Gross, H., Wernerus, P., Repasi, E., Heipke, C.: Context-based urban terrain reconstruction from UAV-videos for geoinformation applications. ISPRS Int. Arch. Photogram. Remote Sens. Spat. Inf. Sci. **3822**, 75–80 (2011)
8. Carletta, J.: Assessing agreement on classification tasks: the kappa statistic. Comput. Linguist. **22**(2), 249–254 (1996)
9. Chen, L.C., Zhu, Y., Papandreou, G., Schroff, F., Adam, H.: Encoder-decoder with atrous separable convolution for semantic image segmentation. In: Proceedings of the European Conference on Computer Vision (ECCV), pp. 801–818 (2018)
10. Cooner, A.J., Shao, Y., Campbell, J.B.: Detection of urban damage using remote sensing and machine learning algorithms: revisiting the 2010 Haiti earthquake. Remote Sens. **08–00868**(10), 1–17 (2016)
11. Dalal, N., Triggs, B.: Histograms of oriented gradients for human detection. In: Proceedings of the IEEE Conference on Computer Vision and Pattern Recognition, vol. 1, pp. 886–893. IEEE (2005)
12. Dell'Acqua, F., Gamba, P.: Remote sensing and earthquake damage assessment: experiences, limits, and perspectives. Proc. IEEE **100**(10), 2876–2890 (2012)
13. Everingham, M., Van Gool, L., Williams, C.K.I., Winn, J., Zisserman, A.: The PASCAL Visual Object Classes Challenge 2012 (VOC2012) Results. http://www.pascal-network.org/challenges/VOC/voc2012/workshop/index.html
14. Fujita, A., Sakurada, K., Imaizumi, T., Ito, R., Hikosaka, S., Nakamura, R.: Damage detection from aerial images via convolutional neural networks. In: 2017 Fifteenth IAPR International Conference on Machine Vision Applications (MVA), pp. 5–8. IEEE (2017)
15. Gamba, P., Dell'Acqua, F., Odasso, L.: Object-oriented building damage analysis in VHR optical satellite images of the 2004 tsunami over Kalutara, Sri Lanka. In: Urban Remote Sensing Joint Event, 2007, pp. 1–5. IEEE (2007)
16. Genuer, R., Poggi, J.M., Tuleau-Malot, C.: Variable selection using random forests. Pattern Recogn. Lett. **31**(14), 2225–2236 (2010)
17. Geusebroek, J.M., Van den Boomgaard, R., Smeulders, A.W.M., Geerts, H.: Color invariance. IEEE Trans. Pattern Anal. Mach. Intell. **23**(12), 1338–1350 (2001)
18. Gueguen, L., Hamid, R.: Large-scale damage detection using satellite imagery. In: Proceedings of the IEEE Conference on Computer Vision and Pattern Recognition (CVPR), pp. 1321–1328 (2015)
19. Haklay, M.: How good is volunteered geographical information? A comparative study of OpenStreetMap and ordnance survey datasets. Environ. Plann. B Plann. Des. **37**(4), 682–703 (2010)
20. Hermosillo, G., Chefd'Hotel, C., Faugeras, O.: Variational methods for multimodal image matching. Int. J. Comput. Vision **50**(3), 329–343 (2002)

21. Huyck, C.K., Adams, B.J., Cho, S., Chung, H.C., Eguchi, R.T.: Towards rapid citywide damage mapping using neighborhood edge dissimilarities in very high-resolution optical satellite imagery-application to the 2003 Bam, Iran, earthquake. Earthquake Spectra **21**(S1), 255–266 (2005)
22. Im, J., Jensen, J., Tullis, J.: Object-based change detection using correlation image analysis and image segmentation. Int. J. Remote Sens. **29**(2), 399–423 (2008)
23. Lagarias, J.C., Reeds, J.A., Wright, M.H., Wright, P.E.: Convergence properties of the Nelder-Mead Simplex method in low dimensions. SIAM J. Optim. **9**(1), 112–147 (1998)
24. Lowe, D.G.: Distinctive image features from scale-invariant keypoints. Int. J. Comput. Vision **60**(2), 91–110 (2004)
25. Lucks, L., Pohl, M., Bulatov, D., Thönessen, U.: Superpixel-wise assessment of building damage from aerial images. In: International Conference on Computer Vision Theory and Applications (VISAPP), pp. 211–220 (2019)
26. Ma, J., Qin, S.: Automatic depicting algorithm of earthquake collapsed buildings with airborne high resolution image. In: 2012 IEEE International Geoscience and Remote Sensing Symposium (IGARSS), pp. 939–942. IEEE (2012)
27. Marcos, D., et al.: Learning deep structured active contours end-to-end. In: Conference on Computer Vision and Pattern Recognition (CVPR), pp. 8877–8885. IEEE (2018)
28. Peng, J., Zhang, D., Liu, Y.: An improved snake model for building detection from urban aerial images. Pattern Recogn. Lett. **26**(5), 587–595 (2005)
29. Pesaresi, M., Gerhardinger, A., Haag, F.: Rapid damage assessment of built-up structures using VHR satellite data in tsunami-affected areas. Int. J. Remote Sens. **28**(13–14), 3013–3036 (2007)
30. Pohl, M., Meidow, J., Bulatov, D.: Simplification of polygonal chains by enforcing few distinctive edge directions. In: Sharma, P., Bianchi, F.M. (eds.) SCIA 2017. LNCS, vol. 10270, pp. 3–14. Springer, Cham (2017). https://doi.org/10.1007/978-3-319-59129-2_1
31. Rathje, E.M., Woo, K.S., Crawford, M., Neuenschwander, A.: Earthquake damage identification using multi-temporal high-resolution optical satellite imagery. In: Proceedings of the IEEE on Geoscience and Remote Sensing Symposium, vol. 7, pp. 5045–5048. IEEE (2005)
32. Sirmacek, B., Unsalan, C.: Damaged building detection in aerial images using shadow information. In: 4th International Conference on Recent Advances in Space Technologies, pp. 249–252. IEEE (2009)
33. Tasar, O., Maggiori, E., Alliez, P., Tarabalka, Y.: Polygonization of binary classification maps using mesh approximation with right angle regularity. In: IGARSS 2018–2018 IEEE International Geoscience and Remote Sensing Symposium, pp. 6404–6407. IEEE (2018)
34. Tomowski, D., Klonus, S., Ehlers, M., Michel, U., Reinartz, P.: Change visualization through a texture-based analysis approach for disaster applications. In: ISPRS Proceedings on Advanced Remote Sensing Science, vol. XXXVIII, pp. 263–269 (2010)
35. Tu, J., Li, D., Feng, W., Han, Q., Sui, H.: Detecting damaged building regions based on semantic scene change from multi-temporal high-resolution remote sensing images. ISPRS Int. J. Geo-Inf. **6**(5), 131 (2017)
36. Vargas-Muñoz, J., Marcos, D., Lobry, S., Dos Santos, J.A., Falcão, A.X., Tuia, D.: Correcting misaligned rural building annotations in open street map using convolutional neural networks evidence. In: International Geoscience and Remote Sensing Symposium (IGARSS), pp. 1284–1287. IEEE (2018)

37. Varma, M., Zisserman, A.: A statistical approach to texture classification from single images. Int. J. Comput. Vision **62**(1–2), 61–81 (2005)
38. Veksler, O., Boykov, Y., Mehrani, P.: Superpixels and supervoxels in an energy optimization framework. In: Proceeding on European Conference on Computer Vision, pp. 211–224 (2010)
39. Vetrivel, A., Gerke, M., Kerle, N., Nex, F., Vosselman, G.: Disaster damage detection through synergistic use of deep learning and 3D point cloud features derived from very high resolution oblique aerial images, and multiple-kernel-learning. ISPRS J. Photogram. Remote Sens. **140**, 45–59 (2017)
40. Warnke, S., Bulatov, D.: Variable selection for road segmentation in aerial images. ISPRS Int. Arch. Photogram. Remote Sens. Spat. Inf. Sci. **42**, 297–304 (2017)
41. Weinmann, M., Jutzi, B., Hinz, S., Mallet, C.: Semantic point cloud interpretation based on optimal neighborhoods, relevant features and efficient classifiers. ISPRS J. Photogram. Remote Sens. **105**, 286–304 (2015)
42. Zampieri, A., Charpiat, G., Girard, N., Tarabalka, Y.: Multimodal image alignment through a multiscale chain of neural networks with application to remote sensing. In: European Conference on Computer Vision (ECCV) (2018)
43. Zhang, J.F., Xie, L.L., Tao, X.X.: Change detection of remote sensing image for earthquake-damaged buildings and its application in seismic disaster assessment. J. Nat. Disasters **11**(2), 59–64 (2002)
44. Zhu, X.X., et al.: Deep learning in remote sensing: a comprehensive review and list of resources. IEEE Geosci. Remote Sens. Mag. **5**(4), 8–36 (2017). https://doi.org/10.1109/MGRS.2017.2762307

Semantic Image Completion Through an Adversarial Strategy

Patricia Vitoria[✉][ID], Joan Sintes, and Coloma Ballester[ID]

Department of Information and Communication Technologies,
University Pompeu Fabra, Barcelona, Spain
{patricia.vitoria,coloma.ballester}@upf.edu, joansintesmarcos@gmail.com

Abstract. Image completion or image inpainting is the task of filling in missing regions of an image. When those areas are large and the missing information is unique such that the information and redundancy available in the image is not useful to guide the completion, the task becomes even more challenging. This paper proposes an automatic semantic inpainting method able to reconstruct corrupted information of an image by semantically interpreting the image itself. It is based on an adversarial strategy followed by an energy-based completion algorithm. First, the data latent space is learned by training a modified Wasserstein generative adversarial network. Second, the learned semantic information is combined with a novel optimization loss able to recover missing regions conditioned by the available information. Moreover, we present an application in the context of face inpainting, where our method is used to generate a new face by integrating desired facial attributes or expressions from a reference face. This is achieved by slightly modifying the objective energy. Quantitative and qualitative top-tier results show the power and realism of the presented method.

Keywords: Generative models · Wasserstein GAN · Image inpainting · Semantic understanding

1 Introduction

When looking at a censored photograph or at a portrait whose eyes, nose and/or mouth are occluded, our brain has no difficulty in hallucinating a plausible completion of the face. Moreover, if the visible parts of the face are familiar to us because, e.g., they remind us a friend or a celebrity, we mentally conceive a face having the whole set of attributes: the visible ones and the ones we infer. However, as an automatic computer vision task, it remains a challenging task.

The mentioned task falls into the so-called image inpainting where the goal is to recover missing information of an image in a realistic manner. Its applications are numerous and range from automatizing cinema post-production tasks enabling, e.g., the deletion of annoying objects, to new view synthesis generation for, e.g., broadcasting of sport events. Classical methods use the available

© Springer Nature Switzerland AG 2020
A. P. Cláudio et al. (Eds.): VISIGRAPP 2019, CCIS 1182, pp. 520–542, 2020.
https://doi.org/10.1007/978-3-030-41590-7_22

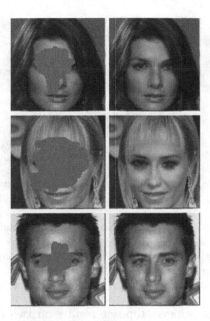

Fig. 1. Several inpainted images resulting from the proposed algorithm using different masks.

redundancy of the incomplete input image: smoothness priors in the case of geometry-oriented approaches and self-similarity principles in the non-local or exemplar-based ones. Nevertheless, they fail in recovering large regions when the available image redundancy is not useful, such as the cases of Fig. 1. In this paper we capitalize on the understanding of more abstract and high level information that learning strategies may provide. We propose an extension of the semantic image inpainting model proposed in [34] to automatically complete any region of an image including those challenging cases. It consists of a deep learning based strategy which uses generative adversarial networks (GANs) to learn the image space and an appropriate loss for an inpainting optimization algorithm which outputs a semantically plausible completion where the missing content is conditioned by the available data. Our method can be applied to recover any incomplete image no matter the shape and the size of the holes of information.

In this work, following our previous paper [34], we will train an improved version of the Wasserstein GAN (WGAN) to implicitly learn a data latent space and subsequently to generate new samples from it. We incorporate in this paper the PatchGAN network structure [20] in the discriminator as well as we update the generator to handle images with higher resolution. The new discriminator is able to take decisions over patches of the image and it is combined with the original global discriminator that takes a decision for the whole image itself. With this purpose, we define a new energy function able to generate the missing content conditioned to a reference image. We deploy it on hallucinating faces conditioned by reference facial attributes or expressions. We quantitatively and qualitatively

(a)   (b)

(c)   (d)

Fig. 2. Image inpainting results using three different approaches. (a) Input images, each with a big hole or mask. (b) Results obtained with the non-local method [14]. (c) Results with the local method [15]. (d) Our semantic inpainting method. Figure retrieved from our previous work [34].

show that our proposal achieves top-tier results on two datasets: CelebA and Street View House Numbers. Additionally, we show qualitative results in our application for face hallucination. The code has been made publicly available.

The remainder of the paper is organized as follows. In Sect. 2, we review some preliminaries and state-of-the-art related work on that topic focusing first on generative adversarial networks and then on inpainting methods and face complation. Section 3 details both methods for image inpainting and face hallucination. In Sect. 4, we present quantitative and qualitative assessments of all parts of the proposed method. Section 5 concludes the paper.

2   Preliminaries and State-of-the-Art Related Work

2.1   Generative Adversarial Networks

GAN [16] learning strategy is based on a game theory scenario between two networks, the generator and the discriminator, competing against each other. The goal of the generator is to generate samples of data from an implicitly learned probability distribution that is aimed to be as closer as possible as the probability distribution the real data. On the other hand, the discriminator tries to distinguish real from fake data. To do so, the discriminator, denote here by D, is trained to maximize the probability of correctly distinguish between real examples and samples created by the generator, denoted by G, while G is trained to fool the discriminator and to minimize $\log(1-D(G(z)))$ by generating realistic examples. In other words, D and G play the following min-max game with value function $V(G, D)$ defined as follows:

$$\min_G \max_D V(D, G) = \min_G \max_D \mathbb{E}_{x \sim P_{real}(x)}[\log D(x)] + \mathbb{E}_{z \sim p_z(z)}[\log(1 - D(G(z)))],$$

$$(1)$$

where P_{real} denotes the probability distribution the real data, and p_z represents the distribution of the latent variables z. The authors of [31] introduced convolutional layers to the GANs architecture, and proposed the so-called Deep Convolutional Generative Adversarial Network (DCGAN) able to learn more complex data.

GANs have been applied with success to many computer vision related tasks such as image colorization [8], text to image synthesis [32], super-resolution [21], image inpainting [6,11,39], and image generation [17,25,28,31], to mention just a few. They have also been applied to other modalities such as speech, audio and text. However, three difficulties still persist as challenges. One of them is the quality of the generated images and the remaining two are related to the well-known instability problem in the training procedure. For instance, two problems can appear: vanishing gradients and mode collapse.

Aiming a stable training of GANs, several authors have promoted the use of the Wasserstein GAN (WGAN). WGAN minimizes an approximation of the Earth-Mover (EM) distance or Wasserstein-1 metric between two probability distributions. The authors of [2] analyzed the properties of this distance. They showed that one of the main benefits of the Wasserstein distance is that it is continuous. This property allows to robustly learn a probability distribution by smoothly modifying the parameters through gradient descend. Moreover, the Wasserstein or EM distance is known to be a powerful tool to compare probability distributions with non-overlapping supports, in contrast to other distances such as the Kullback-Leibler divergence and the Jensen-Shannon divergence (used in the DCGAN and other GAN approaches) which produce the vanishing gradients problem, as mentioned above. Using the Kantorovich-Rubinstein duality, the Wasserstein distance between two distributions, say a *real* distribution P_{real} and an estimated distribution P_g, can be computed as

$$W(P_{real}, P_g) = \sup \mathbb{E}_{x \sim P_{real}} [f(x)] - \mathbb{E}_{x \sim P_g} [f(x)] \tag{2}$$

where the supremum is taken over all the 1-Lipschitz functions f (notice that, if f is differentiable, it implies that $\|\nabla f\| \leq 1$). Let us notice that f in Eq. (2) can be thought to take the role of the discriminator D in the GAN terminology. In [2], the WGAN is defined as the network whose parameters are learned through optimization of

$$\min_{G} \max_{D \in \mathcal{D}} \mathbb{E}_{x \sim P_{real}} [D(x)] - \mathbb{E}_{x \sim P_G} [D(x)] \tag{3}$$

where $\mathcal{D}$ denotes the set of 1-Lipschitz functions. Under an optimal discriminator (called a *critic* in [2]), minimizing the value function with respect to the generator parameters minimizes $W(P_{real}, P_g)$. To enforce the Lipschitz constraint, the authors proposed to use an appropriate weight clipping. The resulting WGAN solves the vanishing problem, but several authors [1,17] have noticed that weight clipping is not the best solution to enforce the Lipschitz constraint and it causes optimization difficulties. For instance, the WGAN discriminator ends up learning an extremely simple function and not the real distribution. Also, the clipping threshold must be properly adjusted. Since a differentiable function is 1-Lipschitz if it has gradient with norm at most 1 everywhere, [17] proposed an alternative to weight clipping by adding a gradient penalty term constraining the L^2 norm of the gradient while optimizing the original WGAN during training called WGAN-GP. The WGAN-GP minimization loss is defined as

$$\min_{G} \max_{D \in \mathcal{D}} \mathbb{E}_{\tilde{x} \sim P_{real}} [D(\tilde{x})] - \mathbb{E}_{x \sim P_G} [D(x)] - \lambda \mathbb{E}_{\tilde{x} \sim P_{\tilde{x}}} \left[(\| \nabla_{\tilde{x}} D(\tilde{x}) \|_2 - 1)^2 \right] \quad (4)$$

As in [17], $P_{\tilde{x}}$ is implicitly defined sampling uniformly along straight lines between pairs of point sampled from the data distribution P_{real} and the generator distribution P_G. Let us notice that the minus before the gradient penalty term in (4) corresponds to the fact that, in practice, when optimizing with respect to the discriminator parameters, one minimizes the negative of the loss instead of maximizing it.

In this work, we leverage the mentioned WGAN-GP improved with a new design of the generator and discriminator architectures.

2.2 Image Inpainting

In general, image inpainting methods found in the literature can be classified into two groups: model-based approaches and deep learning approaches. In the former, two main groups can be distinguished: local and non-local methods. In local methods, images are modeled as functions with some degree of smoothness (see [5,7,9,26] to mention but a few of the related literature). These methods show good performance in propagating smooth level lines or gradients but fail in the presence of texture or large missing regions. Non-local methods exploit the self-similarity prior by directly sampling the desired texture to perform the synthesis (e.g., [3,10,13]. They provide impressive results while inpainting textures and repetitive structures even in the case of large regions to recover. However, as both type of methods use the redundancy present in known parts of the incomplete input image, through smoothness priors in the case of geometry-based and through self-similarity principles in the non-local or patch-based ones and thus fail in completing singular information.

In the last decade, most of the state-of-the-art methods are based on deep learning approaches [11,29,38–40]. The authors of [29] (see also [33,35]) modified the original GAN architecture by inputting the image context instead of random noise to predict the missing patch. [39] proposes an algorithm which generates the missing content by conditioning on the available data given a trained generative model, while [38] adapts multi-scale techniques to generate high-frequency details on top of the reconstructed object to achieve high resolution results. The work [19] adds a local discriminator network that considers only the filled region to emphasize the adversarial loss on top of the global discriminator. This additional network, called the local discriminator (L-GAN), facilitates exposing the local structural details. Also, the authors of [11] design a discriminator that aggregates the local and global information by combining a global GAN and a Patch-GAN.

2.3   Face Completion

Several works focus on the task of face completion. For example, the classical work of [18] completes a face by computing the least squares solution giving the appropriate optimal coefficients combining a set of prototypes of shape and texture. [12] uses a spectral-graph-based filling-in algorithm to fill-in the occluded regions of a face. The occluded region is automatically detected and reconstructed in [23] by using GraphCut-based detection and confidence-oriented sampling, respectively. For a detailed account of the face completion and hallucination literature we refer to [37]. Deep learning strategies are used in [22]. Their method is based on a GAN, two adversarial loss functions (local and global) and a semantic parsing loss.

3   Proposed Approach

We construct from our previous work on semantic image inpainting [34] which is built on two main blocks. First, given a dataset of (non-corrupted) images, it trains the proposed version of the WGAN [34] to implicitly learn a data latent space to subsequently generate new random samples from the dataset. Once the model is trained, given an incomplete image and the trained generative model, an iterative minimization procedure is performed to infer the missing content of the incomplete image by conditioning the resulting image on the known regions. This procedure consists on searching the closest encoding of the corrupted data in the latent manifold by minimizing the proposed loss which is made of a combination of contextual, through image values and image gradients, and prior losses.

In this paper we introduce two updates in the WGAN architecture. The first one allows us to train with images of higher resolution. The second improvement is aimed at the discriminator, where instead of having a single decision for each input image, the discriminator additionally takes decisions over patches of the image.

Additionally, we propose a method for conditional face completion. As in the previous method, given a dataset of images of faces, we train our network in order to learn the data distribution. Then, given an image of a face, say y_1 (that can be either complete or incomplete), and the previously trained generative model, we perform an energy-based optimization procedure to generate a new image which is similar to y_1 but has some meaningful visage portions (such as the eyes, mouth or nose) similar to a reference image y_2 by conditioning the generated image through the objective loss function.

3.1   Adversarial Based Learning of the Data Distribution

The adversarial architecture presented in [34] was built on the top of the WGAN-GP [17]. Several improvements were proposed in [34] to increase the stability of the network:

- Deep networks can learn more complex, non-linear functions, but are more difficult to train. One of the improvements were the introduction of residual learning in both the generator and discriminator which eases the training of these networks, and enables them to be substantially deeper and stable. Instead of hoping each sequence of layers to directly fit a desired mapping, we explicitly let these layers fit a residual mapping. Therefore, the input x of the residual block is recast into $F(x) + x$ at the output. Figures 3 and 4 show the layers that make up a residual block in our model. Figures 5 and 6 display a visualization of the architecture of the generator (Fig. 6) and of the discriminators (Fig. 5).
- The omition of batch normalization in the discriminator. To not introduce correlation between samples, it uses layer normalization [4] as a drop-in replacement for batch normalization in the discriminator.
- Finally, the ReLU activation is used in the generator with the exception of the output layer which uses the Tanh function. Within the discriminator ReLU activation are also used. This is in contrast to the DCGAN, which makes use of the LeakyReLu.

Residual Block (up)

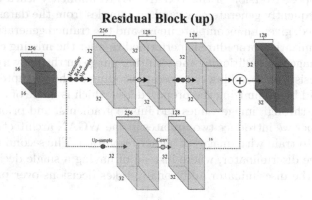

Fig. 3. An example of the residual block used in the generator. Figure retrieved from our previous work [34].

Residual Block (down)

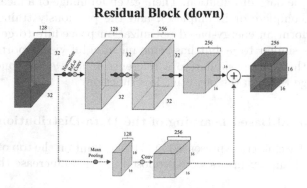

Fig. 4. An example of the residual block used in the discriminator. Figure retrieved from [34].

Additionally to the previous mentioned changes applied in the work [34], further modifications have been applied: an extra discriminator architecture based on PatchGAN and a modification on the model architecture to deal with higher resolution images.

PatchGAN Discriminator. Inspired by the Markovian architecture (PatchGAN [20]) we have added a new discriminator in our adversarial architecture. The PatchGAN discriminator keeps track of the high-frequency structures of the generated image by focusing on local patches. Thus, instead of penalizing at the full image scale, it tries to classify each patch as real or fake. Hence, rather than giving a single output for each input image, it generates a decission value for each patch.

3.2 Semantic Image Inpainting

Once the model is trained until the mapping from the data latent space to uncorrupted data has been properly estimated, semantic image completion can

Original Discriminator

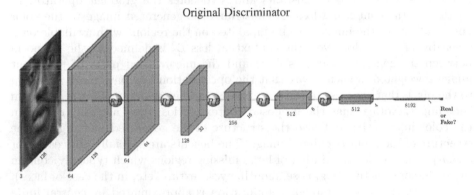

PatchGAN - Discriminator

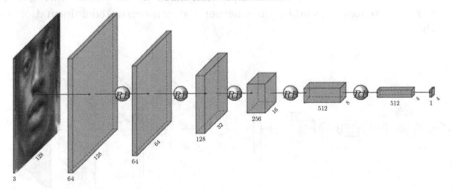

Fig. 5. Overview of the original discriminator architecture (above) and PatchGAN Discriminator (below). In our model we use both types of discriminators. RB stands for Residual Block.

be performed. More precisely, after traning, the generator is able to take a random vector z drawn from p_z and generate an image mimicking samples from P_{real}. In order to perform inpainting or completion of an incomplete image, the aim is to recover the encoding $\hat{z}$ that is closest to the corrupted image while being constrained to the learned encoding manifold of z. Then, when $\hat{z}$ is found, the damaged areas can be restored by using the trained generator G on $\hat{z}$.

We formulate the process of finding $\hat{z}$ as an optimization problem. Let y be a damaged image and M a binary mask of the same spatial size as the input image y, where the white pixels (that is, the pixels i such that $M(i) = 1$) determine the uncorrupted areas of y. The closest encoding $\hat{z}$ can be defined as the optimum of the following optimization problem with the loss defined as [34]:

$$\hat{z} = arg\,min_{z}\{\mathcal{L}_c(z|y, M) + \eta\mathcal{L}_p(z)\} \tag{5}$$

where $\mathcal{L}_p$ stays for prior loss and $\mathcal{L}_c$ for contextual loss defined as

$$\mathcal{L}_c(z|y, M) = \alpha W\|M(G(z) - y)\| + \beta W\|M(\nabla G(z) - \nabla y)\| \tag{6}$$

where α, β, η are positive constants and ∇ denotes the gradient operator. In particular, the contextual loss $\mathcal{L}_c$ constrains the generated image to the color and gradients of the image y to be inpainted on the regions with available data given by $M \equiv 1$. Moreover, the contextual loss $\mathcal{L}_c$ is defined as the L^1 norm between the generated samples $G(z)$ and the uncorrupted parts of the input image y weighted in such a way that the optimization loss pays more attention to the pixels that are close to the corrupted area when searching for the optimum encoding $\hat{z}$. Notice, that the proposed contextual loss does not only constrain the color information but also the structure of the generated image given the structure of the input corrupted image. The benefits are specially noticeable for a sharp and detailed inpainting of large missing regions which typically contain some kind of structure (e.g. nose, mouth, eyes, texture, etc, in the case of faces). In practice, the image gradient computation is approximated by central finite differences. In the boundary of the inpainting hole, we use either forward or backward differences depending on whether the non-corrupted information is available.

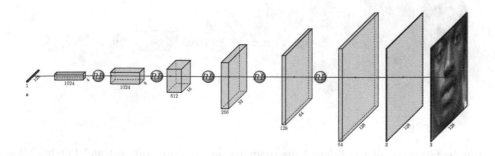

Fig. 6. Overview of the generator architecture.

The weight matrix W is defined for each uncorrupted pixel i as

$$W(i) = \begin{cases} \sum_{j \in N_i} \dfrac{(1 - M(j))}{|N_i|} & if M(i) \neq 0 \\ 0 & if M(i) = 0 \end{cases} \qquad (7)$$

where N_i denotes a local neighborhood or window centered at i, and $|N_i|$ denotes its cardinality, i.e., the area (or number of pixels) of N_i. This weighting term has also been used by [39]. In order to compare our results with them, we have fixed the window size to the value used by them (7×7) for all the experiments.

Finally, the prior loss $\mathcal{L}_p$ is defined such as it favours realistic images, similar to the samples that are used to train our generative model, that is,

$$\mathcal{L}_p(z) = -D_{w_1}(G_\theta(z)) - D_{w_2}^{patch}(G_\theta(z)) \qquad (8)$$

where D_{w_1} and $D_{w_2}^{patch}$ are the output of the discriminator D and D^{patch} with parameters w_1 and w_2 given the image $G_\theta(z)$ generated by the generator with parameters θ and input vector z. In other words, the prior loss is defined as the second WGAN loss term in (3) penalizing unrealistic images. Without $\mathcal{L}_p$ the mapping from y to z may converge to a perceptually implausible result. Therefore z is updated to fool the discriminator and make the corresponding generated image more realistic.

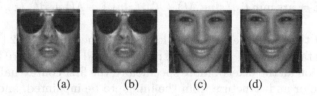

(a)   (b)   (c)   (d)

Fig. 7. Images (b) and (d) show the results obtained after applying Poisson editing (Eq. (9) in the text) to the inpainting results shown in (a) and (c), respectively. Figure retrieved from [34].

The parameters α, β and η in Eq. (6) allow to balance among the three losses. With the defined contextual, gradient and prior losses, the corrupted image can be mapped to the closest z in the latent representation space, denoted by $\hat{z}$. z is randomly initialized with Gaussian noise of zero mean and unit standard deviation and updated using back-propagation on the total loss given in the Eq. (6). Once $G(\hat{z})$ is generated, the inpainting result can be obtained by overlaying the uncorrupted pixels of the original damaged image to the generated image. Even so, the reconstructed pixels may not exactly preserve the same intensities of the surrounding pixels although the content and structure is correctly well aligned. To solve this problem, a Poisson editing step [30] is added at the end of the pipeline in order to reserve the gradients of $G(\hat{z})$ without mismatching intensities of the input image y. Thus, the final reconstructed image $\hat{x}$ is equal to:

$$\hat{x} = \arg\min_{x} \|\nabla x - \nabla G(\hat{z})\|_2^2$$
$$\text{such that } x(i) = y(i) \text{ if } M(i) = 1 \tag{9}$$

In Fig. 7 two examples can be seen where visible seams are appreciated in (a) and (c), but less in (b) and (d) after applying Poisson editing (9).

3.3 Conditional Face Completion

Let G be the previously defined generator mapping the noise vector z belonging to the latent space to an image $G(z)$ as obtained in Sect. 3.1. In order to generate a new face by integrating the desired facial attributes or expressions from a reference face, similar to the previous outline, we formulate the process of finding the closed encoding of the corrupted data in the latent manifold as an optimization problem. Let y_1 be an image of a face, y_2 a reference image, M a binary mask of the same spatial size as the image where the white pixels ($M(i) = 1$) determine the area to preserve of y_1 (last row, second column, of Fig. 12 displays an example of M). We define the closest encoding $\hat{z}$ to y_1 conditioned by y_2 as the optimum of following optimization problem with the new proposed loss:

$$\hat{z} = \arg\min_{z} \mathcal{L}_{c_1}(z|y_1, M) + \mathcal{L}_{c_2}(z|y_2, \mathbb{I} - M) + \beta\tilde{\mathcal{L}}_p(z) \tag{10}$$

where the first two terms are contextual losses as defined in Sect. 3.2 that penalize on complementary regions (given by the masks M and $\mathbb{I} - M$, where $\mathbb{I}$ is constant and equal to 1 on all the pixels). More specifically, the first contextual loss favours to maintain color and structure from the image to be inpainted, and the second contextual loss favours to maintain structure from the reference image. The third term is the prior loss as defined in Sect. 3.2. Let us write them again now in this conditional face completion context:

$$\mathcal{L}_{c_1}(z|y_1, M) = \alpha_1 W_1 \|M(G(z) - y_1)\| + \alpha_2 W_1 \|M(\nabla G(z) - \nabla y_1)\|, \tag{11}$$

$$\mathcal{L}_{c_2}(z|y_2, \mathbb{I} - M) = \alpha_3 W_2 \|(\mathbb{I} - M)(\nabla G(z) - \nabla y_2)\|, \tag{12}$$

$$\tilde{\mathcal{L}}_p(z) = -D_w(G_\theta(z)) \tag{13}$$

where W_1 and W_2 denote the weights defined for each pixel i and its neighborhood N_i as

$$W_1(i) = M(i) \sum_{j \in N_i} \frac{(1 - M(j))}{|N_i|} \tag{14}$$

$$W_2(i) = (\mathbb{I} - M)(i) \left(1 - \sum_{j \in N_i} \frac{M(j)}{|N_i|}\right) \tag{15}$$

4   Experimental Results

In this section we present qualitative and quantitative results of the proposed methods. We will show qualitative and quantitative results of our inpainting method proposed in [34]. The results will be compared with the ones obtained by [39] as both algorithms use first a GAN procedure to learn semantic information from a dataset and, second, combine it with an optimization loss for inpainting in order to infer the missing content. Additionally, further visual results in higher resolution images will be shown. To conclude, results on conditional face completion will be presented.

For all the experiments we use a fixed number of epochs equal to 10, batch size equal to 64, learning rate equal to 0.0001 and exponential decay rate for the first and second moment estimates in the Adam update technique, $\beta_1 = 0, 0$ and $\beta_2 = 0, 9$, respectively. Training the generative model required three days using an NVIDIA QUADRO P6000.

During the inpainting stage, the window size used to compute $W(i)$ in (7) is fixed to 7×7 pixels. In our algorithm, we use back-propagation to compute $\hat{z}$ from the latent space. We make use of an Adam optimizer and restrict z to fall into $[-1, 1]$ in each iteration, which we found it produces more stable results. In that stage we used the Adam hyperparameters learning rate, α, equal to 0.03 and the exponential decay rate for the first and second moment estimates, $\beta_1 = 0, 9$ and $\beta_2 = 0, 999$, respectively. After initializing with a random 128 dimensional vector z drawn from a normal distribution, we perform 1000 iterations.

The assessment is given on two different datasets in order to check the robustness of our method: the CelebFaces Attributes Dateses [24] and the Street View House Numbers (SVHN) [27]. CelebA dataset contains a total of 202.599 celebrity images covering large pose variations and background clutter. We split them into two groups: 201,599 for training and 1,000 for testing. In contrast, SVHN contains only 73,257 training images and 26,032 testing images. SVHN images are not aligned and have different shapes, sizes and backgrounds. The images of both datasets have been cropped with the provided bounding boxes and resized to only 64×64 pixel size.

Remark that we have trained the proposed improved WGAN by using directly the images from the datasets without any mask application. Afterwards, our semanting inpainting method is evaluated on both datasets using the inpainting masks. Notice that our algorithm can be used with any type of inpainting mask.

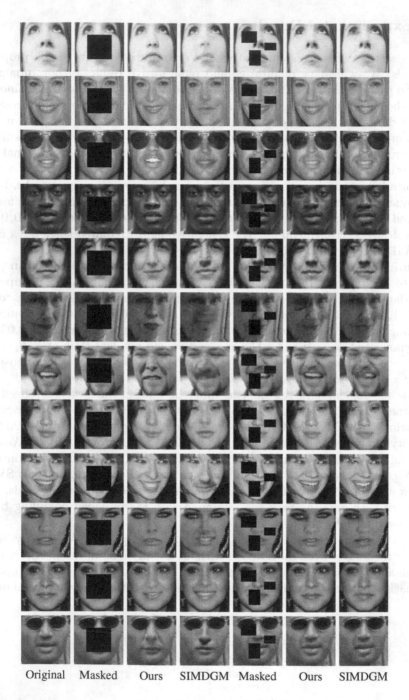

Original   Masked   Ours   SIMDGM   Masked   Ours   SIMDGM

Fig. 8. Inpainting results on the CelebA dataset: Qualitative comparison with the method [39] (fourth and seventh columns, referenced as SIMDGM), using the two masks shown in the second and fifth columns, is also displayed.

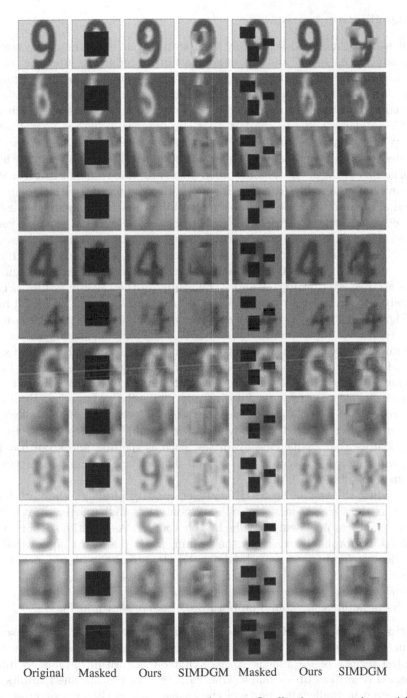

Original   Masked   Ours   SIMDGM   Masked   Ours   SIMDGM

Fig. 9. Inpainting results on the SVHN dataset: Qualitative comparison with the method [39] (fourth and seventh columns, referenced as SIMDGM), using the two masks shown in the second and fifth columns, is also displayed.

4.1   Qualitative Assessment

In [34] we have analyzed separately each step of our algorithm: The training of the adversarial model and the minimization procedure to infer the missing content. Since the inpainting result of the latter strongly depends on what the generative model is able to produce, a good estimation of the data latent space is crucial for our task. Notice that the CelebA dataset will be better estimated than SVHN dataset due to the fact that the number of images as well as the diversity of the dataset directly affects the prediction of the latent space and the estimated underlying probability density function (pdf). In contrast, as bigger the variability of the dataset, more spread is the pdf which difficult its estimation.

To evaluate the proposed inpainting method, a comparison with the semantic inpainting method by [39] was performed. While training our model, we use the proposed architecture (see Sect. 3.1) where the model takes a random vector, of dimension 128, drawn from a normal distribution. In contrast, [39] uses the DCGAN architecture where the generative model takes a random 100 dimensional vector following a uniform distribution between $[-1, 1]$. Some qualitative results are displayed in Figs. 8 and 9. Focusing on the CelebA results (Fig. 8), obviously the algorithm by [39] performs better than local and non-local methods (Fig. 2) since it also makes use of adversarial models. However, although it is able to recover the semantic information of the image and infer the content of the missing areas, in some cases it keeps producing results with lack of structure and detail which can be caused either by the generative model or by the procedure to search the closest encoding in the latent space. It will be further analyzed with a quantitative ablation study. Since the proposed method takes into account not only the pixel values but also the structure of the image, this kind of problems are solved. In many cases, our results are as realistic as the real images. Notice that challenging examples, such as the first and sixth row from Fig. 8, which image structures are not well defined, are not properly recovered with our method nor with [39].

Regarding the results on SVHN dataset (Fig. 9), although they are not as realistic as the CelebA ones, the missing content is well recovered even when different numbers may semantically fit the context. As mentioned before, the lack of detail is probably caused by the training stage, due to the large variability of the dataset (and the size of the dataset). Despite of this, let us notice that our results outperform qualitatively the ones obtained by [39]. This may indicate that our algorithm is more robust when using smaller datasets than [39]. Some examples of failure cases found on both datasets are shown in Fig. 11.

Additional Results in Higher Resolution Images. Figure 10 shows several resulting higher resolution images after applying the proposed algorithm in the corrupted regions of the image. Notice, that our algorithm is able to inpaint any region regardless of its shape. One can see that the obtained results look realistic even in challenging parts of the image such as the eyes or nose. Also, it obtains good results when the observer does not see all the face, such in the middle example in the second row.

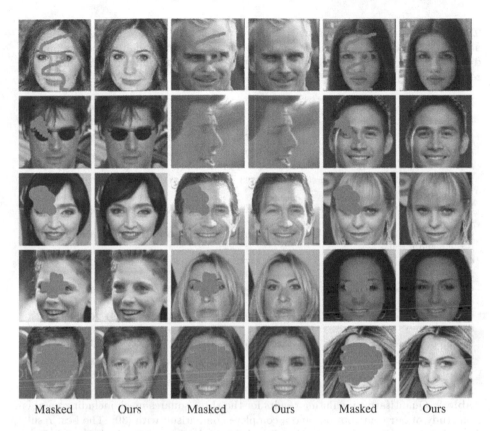

Masked   Ours   Masked   Ours   Masked   Ours

Fig. 10. Inpainting results on the CelebA dataset using the proposed architecture able to create images with higher resolution.

4.2 Quantitative Analysis and Evaluation Metrics

The goal of semantic inpainting is to fill-in the missing information with realistic content. However, with this purpose, there are many correct possibilities to semantically fill the missing information apart from the ground truth solution. Thus, in order to provide a thorough analysis and quantify the quality of our method in comparison with other methods, an ablation study was presented in [34]. We include it here for the sake of completeness. Different evaluation metrics were used: First, metrics based on a distance with respect to the ground truth and, second, a perceptual quality measure that is acknowledged to agree with similarity perception of the human visual system.

In the first case, considering the real images from the database as the ground truth reference, the most used evaluation metrics are the Peak Signal-to-Noise Ratio (PSNR) and the Mean Square Error (MSE). Notice, that both MSE and PSNR, will choose as best results the ones with pixel values closer to the ground truth.

Table 1. Quantitative inpainting results for the central square mask, including an ablation study of our contributions in comparison with [39]. The best results for each dataset are marked in bold and the best results for each method are underlined. Table retrieved from [34].

Loss formulation	CelebA dataset			SVHN dataset		
	MSE	PSNR	SSIM	MSE	PSNR	SSIM
[39]	872.8672	18.7213	0.9071	1535.8693	16.2673	0.4925
[39] adding gradient loss with $\alpha = 0.1$, $\beta = 0.9$ and $\eta = 1.0$	832.9295	18.9247	0.9087	1566.8592	16.1805	0.4775
[39] adding gradient loss with $\alpha = 0.5$, $\beta = 0.5$ and $\eta = 1.0$	862.9393	18.7710	0.9117	1635.2378	15.9950	0.4931
[39] adding gradient loss with $\alpha = 0.1$, $\beta = 0.9$ and $\eta = 0.5$	<u>794.3374</u>	<u>19.1308</u>	<u>0.9130</u>	<u>1472.6770</u>	<u>16.4438</u>	<u>0.5041</u>
[39] adding gradient loss with $\alpha = 0.5$, $\beta = 0.5$ and $\eta = 0.5$	876.9104	18.7013	0.9063	1587.2998	16.1242	0.4818
Our proposed loss with $\alpha = 0.1$, $\beta = 0.9$ and $\eta = 1.0$	855.3476	18.8094	0.9158	631.0078	20.1305	**0.8169**
Our proposed loss with $\alpha = 0.5$, $\beta = 0.5$ and $\eta = 1.0$	**785.2562**	**19.1807**	**0.9196**	743.8718	19.4158	0.8030
Our proposed loss with $\alpha = 0.1$, $\beta = 0.9$ and $\eta = 0.5$	862.4890	18.7733	0.9135	**622.9391**	**20.1863**	0.8005
Our proposed loss with $\alpha = 0.5$, $\beta = 0.5$ and $\eta = 0.5$	833.9951	18.9192	0.9146	703.8026	19.6563	0.8000

Table 2. Quantitative inpainting results for the three squares mask including an ablation study of our contributions and a complete comparison with [39]. The best results for each dataset are marked in bold and the best results for each method are underlined. Table retrieved from [34].

Method	CelebA dataset			SVHN dataset		
	MSE	PSNR	SSIM	MSE	PSNR	SSIM
[39]	622.1092	20.1921	0.9087	1531.4601	16.2797	0.4791
[39] adding gradient loss with $\alpha = 0.1$, $\beta = 0.9$ and $\eta = 1.0$	584.3051	20.4644	0.9067	1413.7107	16.6272	0.4875
[39] adding gradient loss with $\alpha = 0.5$, $\beta = 0.5$ and $\eta = 1.0$	600.9579	20.3424	0.9080	1427.5251	16.5850	0.4889
[39] adding gradient loss with $\alpha = 0.1$, $\beta = 0.9$ and $\eta = 0.5$	580.8126	20.4904	0.9115	1446.3560	16.5281	<u>0.5120</u>
[39] adding gradient loss with $\alpha = 0.5$, $\beta = 0.5$ and $\eta = 0.5$	<u>563.4620</u>	<u>20.6222</u>	0.9103	<u>1329.8546</u>	<u>16.8928</u>	0.4974
Our proposed loss with $\alpha = 0.1$, $\beta = 0.9$ and $\eta = 1.0$	424.7942	21.8490	0.9281	168.9121	25.8542	0.8960
Our proposed loss with $\alpha = 0.5$, $\beta = 0.5$ and $\eta = 1.0$	380.4035	22.3284	0.9314	221.7906	24.6714	**0.9018**
Our proposed loss with $\alpha = 0.1$, $\beta = 0.9$ and $\eta = 0.5$	**321.3023**	**23.0617**	**0.9341**	**154.5582**	**26.2399**	0.8969
Our proposed loss with $\alpha = 0.5$, $\beta = 0.5$ and $\eta = 0.5$	411.8664	21.9832	0.9292	171.7974	25.7806	0.8939

In the second case, in order to evaluate perceived quality, the Structural Similarity index (SSIM) [36] is used to measure the similarity between two images. It is considered to be correlated with the quality perception of the human visual system.

Given these metrics the obtained results are compared with the one proposed by [39] as it is the method more similar to ours. Tables 1 and 2 show the numerical performance of our method and [39]. To perform an ablation study of all our contributions and a complete comparison with [39], Tables 1 and 2 not only show the results obtained by their original algorithm and our proposed algorithm, but also the results obtained by adding our new gradient-based term $\mathcal{L}_g(z|y, M)$ to their original inpainting loss. We present the results varying the trade-off effect between the different loss terms (weights α, β, η).

By looking at the numerical results it can be seen that the proposed algorithm always performs better than the semantic inpainting method by [39]. For the case of the CelebA dataset, the average MSE obtained by [39] is equal to 872.8672 and 622.1092, respectively, compared to our results that are equal to 785.2562 and 321.3023, respectively. It is highly reflected in the results obtained using the SVHN dataset, where the original version of [39] obtains an MSE equal to 1535.8693 and 1531.4601, using the central and three squares mask respectively, and our method 622.9391 and 154.5582. On the one side, the proposed WGAN structure is able to create a more realistic latent space and, on the other side, the proposed loss takes into account essential information in order to recover the missing areas.

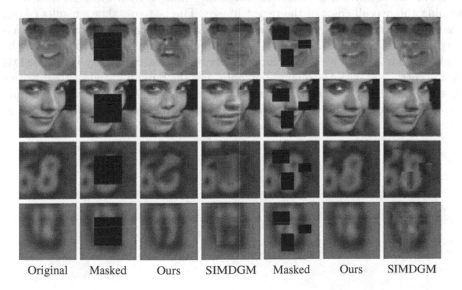

| Original | Masked | Ours | SIMDGM | Masked | Ours | SIMDGM |

Fig. 11. Some failure cases in CelebA and SVHN dataset.

Regarding the accuracy results obtained with the SSIM measure, can be seen that the results obtained by the proposed method always have a better perceived quality than the ones obtained by [39]. In some cases, the values are close to the double, for example, in the case where the training dataset is small, namely, SVHN.

To conclude, the proposed method is more stable in smaller datasets such in the case of SVHN. Also, by decreasing the number of samples in the dataset does not mean to reduce the quality of the inpainted images in the proposed method. Contrary with what is happening in the case of [39]. Finally, in the cases where we add the proposed loss to the algorithm [39], in most of the cases the MSE, PSNR and SSIM improves. This fact clarifies the big importance of the gradient loss in order to perform semantic inpainting.

4.3   Conditional Face Completion

In this section, we evaluate our algorithm in the CelebA dataset [24] that consists on 202,599 face images, which are aligned and cropped to have pixel size equal to 64×64.

Some qualitative and quantitative results are shown in Fig. 12. Our algorithm outputs a face hallucination of one of the images y_1 displayed in the first row, having as a reference the portion displayed in the second column of the image y_2 of the first column. More often than not, the results look natural and the combination of the target image together with its reference is plausible. As can be seen, our algorithm is robust in combining images with different skin tone, keeping the overall color of the target image. The last row shows results of our baseline semantic completion method showing that it can perceptually halluci-nate a plausible completion without any reference image. In order to quantify the quality of our results, we have computed the Structural Similarity Index (SSIM) that is correlated with the quality perception of the human visual system. Notice that the SSIM is computed with respect to the target image y_1 although ours results are a combination of two images. Even so, the resulting SSIM is high in all the cases (above 0.85) which translates to a high perceived quality.

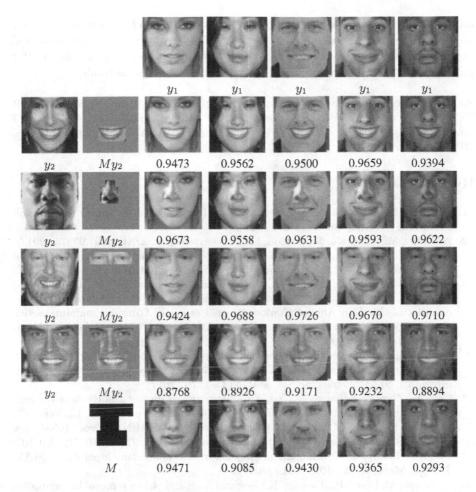

Fig. 12. Face halucination results obtained either using a reference image (from second to fourth rows) or no reference (last row). First row: image y_1 to change or complete, respectively. First column: reference image. Second column: inpainting mask together with the reference region (last row does not have any reference image). The corresponding SSIM is showed under each image.

5   Conclusions

This paper proposes a semantic inpainting method based on an adversarial strategy. The method performs in two phases. First, the data latent manifold is learned by training a proposed improved version of the WGAN. Then, we propose a conditional objective loss. This loss is able to properly infer the missing content having into account the structure and pixel value of the data present on the image. Moreover, it takes also into account the perceptual realism of the reconstructed image. Additionally, a new loss is presented able to perform personalized face completion based on semantic image inpainting. By iteratively minimizing this new loss we are able to generate an image similar to a target image together with meaningful characteristics of a reference one. The presented experiments show the capabilities of the proposed method that is able to infer more realistic content for incomplete images than classical methods.

Acknowledgements. The authors acknowledge partial support by MICINN/FEDER UE project, reference PGC2018-098625-B-I00 and by H2020-MSCA-RISE-2017 project, reference 777826 NoMADS. We also thank NVIDIA for the Quadro P6000 GPU donation.*

References

1. Adler, J., Lunz, S.: Banach wasserstein gan. In: Advances in Neural Information Processing Systems, pp. 6754–6763 (2018)
2. Arjovsky, M., Chintala, S., Bottou, L.: Wasserstein gan. arXiv:1701.07875 (2017)
3. Aujol, J.F., Ladjal, S., Masnou, S.: Exemplar-based inpainting from a variational point of view. SIAM J. Math. Anal. **42**(3), 1246–1285 (2010)
4. Ba, J.L., Kiros, J.R., Hinton, G.E.: Layer normalization. arXiv:1607.06450 (2016)
5. Bertalmio, M., Sapiro, G., Caselles, V., Ballester, C.: Image inpainting. In: Proceedings of the 27th Annual Conference on Computer Graphics and Interactive Techniques, SIGGRAPH 2000, pp. 417–424. ACM Press/Addison-Wesley Publishing Co., New York (2000). https://doi.org/10.1145/344779.344972
6. Burlin, C., Le Calonnec, Y., Duperier, L.: Deep image inpainting (2017)
7. Cao, F., Gousseau, Y., Masnou, S., Pérez, P.: Geometrically guided exemplar-based inpainting. SIAM J. Imaging Sci. **4**(4), 1143–1179 (2011)
8. Cao, Y., Zhou, Z., Zhang, W., Yu, Y.: Unsupervised diverse colorization via generative adversarial networks. In: Ceci, M., Hollmén, J., Todorovski, L., Vens, C., Džeroski, S. (eds.) ECML PKDD 2017, Part I. LNCS (LNAI), vol. 10534, pp. 151–166. Springer, Cham (2017). https://doi.org/10.1007/978-3-319-71249-9_10
9. Chan, T., Shen, J.H.: Mathematical models for local nontexture inpaintings. SIAM J. Appl. Math. **62**(3), 1019–1043 (2001)
10. Criminisi, A., Pérez, P., Toyama, K.: Region filling and object removal by exemplar-based inpainting. IEEE Trans. IP **13**(9), 1200–1212 (2004)
11. Demir, U., Unal, G.: Patch-based image inpainting with generative adversarial networks. arXiv preprint arXiv:1803.07422 (2018)
12. Deng, Y., et al.: Graph laplace for occluded face completion and recognition. IEEE Trans. IP **20**(8), 2329–2338 (2011)
13. Fedorov, V., Arias, P., Facciolo, G., Ballester, C.: Affine invariant self-similarity for exemplar-based inpainting. In: Proceedings of the 11th Joint Conference on Computer Vision, Imaging and Computer Graphics Theory and Applications, pp. 48–58 (2016)
14. Fedorov, V., Facciolo, G., Arias, P.: Variational framework for non-local inpainting. Image Process. Line **5**, 362–386 (2015). https://doi.org/10.5201/ipol.2015.136
15. Getreuer, P.: Total variation inpainting using split bregman. Image Process. Line **2**, 147–157 (2012). https://doi.org/10.5201/ipol.2012.g-tvi
16. Goodfellow, I., et al.: Generative adversarial nets. In: Advances in Neural Information Processing Systems, pp. 2672–2680 (2014)
17. Gulrajani, I., Ahmed, F., Arjovsky, M., Dumoulin, V., Courville, A.C.: Improved training of wasserstein gans. In: Advances in Neural Information Processing Systems, pp. 5769–5779 (2017)

18. Hwang, B.W., Lee, S.W.: Reconstruction of partially damaged face images based on a morphable face model. IEEE TPAMI **25**(3), 365–372 (2003)
19. Iizuka, S., Simo-Serra, E., Ishikawa, H.: Globally and locally consistent image completion. ACM Trans. Graph. **36**(4), 107:1–107:14 (2017). https://doi.org/10.1145/3072959.3073659. http://doi.acm.org/10.1145/3072959.3073659
20. Isola, P., Zhu, J.Y., Zhou, T., Efros, A.A.: Image-to-image translation with conditional adversarial networks. In: Proceedings of the IEEE Conference on Computer Vision and Pattern Recognition, pp. 1125–1134 (2017)
21. Ledig, C., et al.: Photo-realistic single image super-resolution using a generative adversarial network. In: CVPR, vol. 2, p. 4 (2017)
22. Li, Y., Liu, S., Yang, J., Yang, M.H.: Generative face completion. In: CVPR, vol. 1, p. 3 (2017)
23. Lin, D., Tang, X.: Quality-driven face occlusion detection and recovery. In: IEEE CVPR
24. Liu, Z., Luo, P., Wang, X., Tang, X.: Deep learning face attributes in the wild. In: ICCV (2015)
25. Mao, X., Li, Q., Xie, H., Lau, R.Y., Wang, Z., Smolley, S.P.: Least squares generative adversarial networks. In: ICCV, pp. 2813–2821 (2017)
26. Masnou, S., Morel, J.M.: Level lines based disocclusion. In: Proceedings of IEEE ICIP (1998)
27. Netzer, Y., Wang, T., Coates, A., Bissacco, A., Wu, B., Ng, A.Y.: Reading digits in natural images with unsupervised feature learning. In: NIPS Workshop on Deep Learning and Unsupervised Feature Learning
28. Nguyen, A., Yosinski, J., Bengio, Y., Dosovitskiy, A., Clune, J.: Plug & play generative networks: conditional iterative generation of images in latent space. arXiv:1612.00005 (2016)
29. Pathak, D., Krahenbuhl, P., Donahue, J., Darrell, T., Efros, A.A.: Context encoders: feature learning by inpainting. In: CVPR, June 2016
30. Pérez, P., Gangnet, M., Blake, A.: Poisson image editing. In: ACM SIGGRAPH 2003 (2003)
31. Radford, A., Metz, L., Chintala, S.: Unsupervised representation learning with deep convolutional generative adversarial networks. arXiv:1511.06434 (2015)
32. Reed, S., Akata, Z., Yan, X., Logeswaran, L., Schiele, B., Lee, H.: Generative adversarial text to image synthesis. In: Proceedings of The 33rd International Conference on Machine Learning, pp. 1060–1069 (2016)
33. Song, Y., et al.: Contextual-based image inpainting: infer, match, and translate. In: Ferrari, V., Hebert, M., Sminchisescu, C., Weiss, Y. (eds.) ECCV 2018, Part II. LNCS, vol. 11206, pp. 3–18. Springer, Cham (2018). https://doi.org/10.1007/978-3-030-01216-8_1
34. Vitoria., P., Sintes., J., Ballester., C.: Semantic image inpainting through improved wasserstein generative adversarial networks. In: Proceedings of the 14th International Joint Conference on Computer Vision, Imaging and Computer Graphics Theory and Applications - Volume 4: VISAPP, pp. 249–260. INSTICC, SciTePress (2019). https://doi.org/10.5220/0007367902490260
35. Vo, H.V., et al.: Structural inpainting. In: 2018 ACM Multimedia Conference (2018)
36. Wang, Z., Bovik, A.C., Sheikh, H.R., Simoncelli, E.P.: Image quality assessment: from error visibility to structural similarity. IEEE Trans. IP **13**(4), 600–612 (2004). https://doi.org/10.1109/TIP.2003.819861
37. Wang, N., et al.: A comprehensive survey to face hallucination. Int. J. CV **106**(1), 9–30 (2014)

38. Yang, C., Lu, X., Lin, Z., Shechtman, E., Wang, O., Li, H.: High-resolution image inpainting using multi-scale neural patch synthesis. In: CVPR, vol. 1, p. 3 (2017)
39. Yeh, R.A., Chen, C., Lim, T.Y., Schwing, A.G., Hasegawa-Johnson, M., Do, M.N.: Semantic image inpainting with deep generative models. In: CVPR, vol. 2, p. 4 (2017)
40. Yu, J., Lin, Z., Yang, J., Shen, X., Lu, X., Huang, T.S.: Generative image inpainting with contextual attention

An Enhanced Louvain Based Image Segmentation Approach Using Color Properties and Histogram of Oriented Gradients

Thanh-Khoa Nguyen[1,2]([✉]), Jean-Loup Guillaume[1], and Mickael Coustaty[1]

[1] L3i Laboratory, Faculty of Science and Technology, University of La Rochelle,
Avenue Michel Crepeau, 17042 La Rochelle Cedex 1, France
{thanh_khoa.nguyen,jean-loup.guillaume,mickael.coustaty}@univ-lr.fr
[2] Ca Mau Community College, Ca Mau, Vietnam

Abstract. Segmentation techniques based on community detection algorithms generally have an over-segmentation problem. This paper then propose a new algorithm to agglomerate near homogeneous regions based on texture and color features. More specifically, our strategy relies on the use of a community detection on graphs algorithm (used as a clustering approach) where the over-segmentation problem is managed by merging similar regions in which the similarity is computed with Histogram of Oriented Gradients (named as *HOG*) and Mean and Standard deviation of color properties as features. In order to assess the performances of our proposed algorithm, we used three public datasets (Berkeley Segmentation Dataset *(BSDS300 and BSDS500)* and the Microsoft Research Cambridge Object Recognition Image Database *(MSRC)*). Our experiments show that the proposed method produces sizable segmentation and outperforms almost all the other methods from the literature, in terms of accuracy and comparative metrics scores.

Keywords: Image segmentation · Complex networks · Modularity · Superpixels · Louvain algorithm · Community detection

1 Introduction

Image segmentation is one of the main steps in image processing. It divides an image into multiple regions in order to analyze them. Image segmentation is also used to distinguish different objects in the image. For these reasons, the literature abounds with algorithms for achieving various image segmentation tasks. They are widely employed in several image processing applications including object detection [23], object tracking [41], automatic driver assistance [7], and traffic control systems [18], *etc..* These methods can be divided into some main groups according to the underlying approaches, such as feature-based clustering, spatial-based segmentation methods, hybrid techniques and graph-based approaches.

© Springer Nature Switzerland AG 2020
A. P. Cláudio et al. (Eds.): VISIGRAPP 2019, CCIS 1182, pp. 543–565, 2020.
https://doi.org/10.1007/978-3-030-41590-7_23

Recently, complex networks have been interested in many researchers both theories and applications as a trend of developments. To be inspired, image segmentation techniques based on community detection algorithms have been proposed and have become an interesting discipline in the literature [1,6,20–22,27,39,40]. A community is a group of nodes with sparse connections with members of other communities and dense internal connections. The general idea of those techniques is to highlight the similarity between the modularity criterion in network analysis and the image segmentation process. In fact, the larger the modularity of a network is, the more accurate the detected communities, *i.e.*, the objects in the image, are [1,6,27,39]. If the modeling of the image in a graph is well done then we can expect that a good partition in communities corresponds to a good segmentation of the image. The modularity of a partition is a scalar that measures the density of links inside communities as compared to links between communities, and its value falls into the interval $[-0.5, 1]$ [29].

Many algorithms have been proposed for detecting communities, but the Louvain method [5] has received a significant attention in the context of image segmentation [6,20,40]. However, it is still facing many problems such as how to generate an appropriate complex network that propose efficient image segmentation. Another concern of researchers is that community detection based image segmentation leads to over-segmented results. To overcome this issue, we propose a new segmentation approach based on the Louvain method and a merging step in order to gather homogeneous regions in a larger one and then to overcome the problem. Each sub-segment obtained during the Louvain method phase represents a region. We compute a histogram of oriented gradients (HOG) [12], and the values of mean and standard deviation are computed from the three color channels RGB individually. Then, the proposed algorithm operates by considering the similarity value between two adjacent regions based on combining HOG and color features in order to control the aggregation processes.

In this paper we propose a new feature, namely BMHF for "Balancing HOG and MeanSD Features". This new feature combines both previous features in a more meaningful way that gives more weight to color properties. This paper also offers a more comprehensive survey of existing works.

The rest of this paper is organized as follows. In Sect. 2, we briefly review graph-based image segmentation methods using superpixels and the complex networks analysis principle. In Sect. 3, we introduce complex networks, the concept of community detection and Louvain algorithm to point out how community detection algorithms can be applied in image segmentation efficiently. In Sect. 4, we give details of our method for implementation and performance. Experiments on three publicly available datasets are reported in Sect. 5. Finally, our conclusions are presented in Sect. 6.

2   Related Work

In this Section, we briefly review some graph-based image segmentation methods based on superpixels and complex networks analysis.

Ren and Malik [35] introduced superpixels in 2003 on the research of Learning a Classification Model for Segmentation. This research is very important in image segmentation discipline. Especially, with the development of complex networks both theories and applications, the cooperation between superpixels and complex networks display as an absolutely perfect partner-pair in graph-based image segmentation domain. Hence, image segmentation techniques based on community detection algorithms have been proposed and have become an interesting discipline in the literature. Complex networks analysis domain has been considered to segment images and has achieved outstanding results [1, 20–22, 38, 40].

Abin *et al.* [1] propose a new image segmentation method for color images which involves the ideas used for community detection in social networks. In this method, an initial segmentation is applied to partition input image into small homogeneous regions. Then constructing a weighted networks in which the small homogeneous regions obtained initial segmentation processes are nodes of graph. Every pairs of vertices are connected through an edge and the similarity between these regions are computed through Bhattacharyya coefficient [4]. The RGB color space is used to compute the color histogram. Each color channel is uniformly quantized into l levels and then the histogram of each region is calculated in the feature space of $l \times l \times l = l^3$ bins. The similarity distance between two region R_i and R_j is the Bhattacharyya coefficient which is defined as 1.

$$B(R_i, R_j) = \sum_{u=1}^{l^3} \sqrt{H_{R_i}^u . H_{R_j}^u} \tag{1}$$

where H_{R_i} and H_{R_j} are the normalized histograms of region R_i and region R_j, respectively. The superscript u represents the uth element of them. Bhattacharyya coefficient is actually the cosine of the angle between two vectors $(\sqrt{H_{R_i}^1}, \cdots, \sqrt{H_{R_i}^{l^3}})^\mathsf{T}$ and $(\sqrt{H_{R_j}^1}, \cdots, \sqrt{H_{R_j}^{l^3}})^\mathsf{T}$. If two regions have similar contents, their histograms will be also similar, and hence the Bhattacharyya coefficient will be high. The higher Bhattacharyya coefficient, the higher similarity is. In this method, the weight of the edge that corresponds to two regions R_1 and R_2 is assigned a weight corresponding to the normalized Bhattacharyya similarity measure, defined as 2:

$$W(R_1, R_2) = \frac{exp(-\frac{1}{B(R_1,R_2)})}{\sum_{i \neq j \in R} exp(-\frac{1}{B(R_i,R_j)})} \tag{2}$$

where R is the set of all regions.

After the creation of the network, one community detection method is applied to extract communities as segments. In this work, they have used the weighted Newman-Fast community detection algorithm [30] to extract the communities. Besides, the authors have suggested many existing community detection algorithm, for instance, the algorithm of Girvan and Newman [17] and the Louvain method [5]. The results of the proposed method is compared to three well-known methods, JSEG [13], EDISON [8] and MULTISCALE [37], and achieved much better results than others subjectively and in many cases. However, this method uses only the color histogram data as a measure of similarity, so it is an inadequate feature properties if we consider repetitive patterns of different colors in some homogeneous object. For example, in the case of zebra, cheetah and tiger.

Li [21] proposed a graph-partition segmentation method based on a key notion from complex network analysis namely *modularity segmentation*. This approach exploits the alternative concept of modularity [29] which has been successfully applied to community detection problems in complex networks [14,16,28]. The method automatically determines the number of segments by optimizing a natural and theoretically justified criterion, eliminating the need for human intervention. In graph-partition based segmentation, each image is represented as an undirected graph $G = (V, E)$, where each vertex in $V = (v_1, v_2, ..., v_n)$ corresponds to an individual pixel, and each edge in E connects pairs of vertices. The weight on each edge, w_{ij}, is a non-negative value that measures the affinity between two vertices v_i and v_j. A higher affinity indicates a stronger relation between the associated pixels. Let $d_i = \sum_j w_{ij}$ denote the sum of affinities associated with vertex v_i, and $m = \frac{1}{2} \sum_i d_i = \frac{1}{2} \sum_{ij} w_{ij}$ denote the total sum of edge weights in the graph. Given a candidate division of vertices into disjoint groups, the modularity is defined to be the fraction of the affinities that fall within the given groups, minus the expected such fraction when the affinities are distributed randomly. The randomization is conducted by preserving the total affinity d_i of each vertex. Under this assumption, the expected affinity between two vertices v_i and v_j is $d_i d_j / 2m$, hence the corresponding modularity is $w_{ij} - d_i d_j / 2m$. Summing over all vertex pairs within the same group, the modularity, denoted Q, is defined:

$$Q = \frac{1}{2m} \sum_{ij} \left[w_{ij} - \frac{d_i d_j}{2m} \right] \delta(c_i, c_j) \tag{3}$$

where c_i denotes the group to which v_i belongs and $\delta(c_i, c_j)$ is 1 if $c_i = c_j$ and 0 otherwise. An equivalent formulation can be given by defining s_{ik} to be 1 if vertex v_i belongs to group k and 0 otherwise. Then $\delta(c_i, c_j) = \sum_k s_{ik} s_{jk}$ and hence

$$Q = \frac{1}{2m} \sum_{ij} \sum_k [w_{ij} - \frac{d_i d_j}{2m}] s_{ik} s_{jk} = \frac{1}{2m} tr(S^\mathsf{T} B S) \tag{4}$$

where S is a matrix having elements s_{ik} and B is a modularity matrix with elements $b_{ij} = w_{ij} - d_i d_j / 2m$.

The modularity value Q always ranges in $[-0.5, 1]$. All rows and columns of the modularity matrix B sum to zero. It means that the modularity value of unsegmented graph is always zero. The value Q is positive if the intra-group affinities exceed the expected affinities achieved at random. Thus seeking for a partition that maximizes the modularity automatically determines the appropriate number of segments to choose as well as their respective structure, and does not require the number of segments to be pre-specified.

Li *et al.* [20] proposed to used some superpixels and features to solve the over-segmentation problem. This strategy starts with an (over-segmented) image segmentation in which each sub-segment is represented as a superpixel, and treats the over segmentation issue by proposed a new texture feature from low level cues to capture the regularities for the visually coherent object and encode it into the similarity matrix. In addition, the similarity among regions of pixels is constructed in an adaptive manner in order to avoid the over-segmentation. The main contribution of this research is the proposal of an efficient agglomerative segmentation algorithm incorporating the advantage of community detection and the inherent properties of images. A new feature is developed, Histogram of state (HoS), together with an adaptive similarity matrix to avoid over-segmented problem. The authors use a publicly available code [26] to generate super-pixel initialization. The L*a*b* color space is chosen to achieve good segmentation performance, and one neighborhood system has been constructed by considering the adjacent regions of this region to be its neighbors and store its neighboring regions using an adjacent list. The adjacent regions are regions that share at least one pixel with the current region. The method considers features for similarity based on color property and Histogram of State (HoS) of each region. The color feature obtained by computing Mean Distance, used the mean of the pixel value for the two regions to approximate the Earth Mover's distance, which neglects the difference of the co-variance matrices as shown in Eq. 5.

$$d_{DM}\Big(N(\mu_1, \Sigma_1), N(\mu_2, \Sigma_2)\Big)^2 = (\mu_1 - \mu_2)^\mathsf{T}(\mu_1 - \mu_2) \qquad (5)$$

To transform the above distribution distance into similarity measure, the authors used a radial basis function (Gaussian-like) presented in Eq. 6:

$$W_{ij}(color) = exp\Big(\frac{-dis(R_i, R_j)}{2\sigma^2}\Big) \qquad (6)$$

where $dis(R_i, R_j)$ measures the distance between the pixel value distributions for regions R_i and R_j. Empirically, they set $\sigma = 0.08$ and use the Mean Distance Eq. 5.

Cosine similarity measure 7 is used to measure the HoS texture feature that is represented by a 256-dimensional vector for each region, $i.e.$ for each region R_i and R_j, the HoS texture feature are h_i and h_j, $(h_i, h_j \in R^{256})$, respectively.

$$W_{ij}(texture) = cos(h_i, h_j) = \frac{h_i^\mathsf{T} h_j}{\|h_i\|.\|h_j\|} \qquad (7)$$

The adaptive similarity matrix is constructed during each iteration by recomputing the similarity between regions according to Eqs. 6 and 7. The reason for this is that during the merging process, the region keeps expanding and the similarity measure computed from the previous iteration might not be consistent for the current iteration. This adaptive similarity matrix is constantly updated and is the clue to overcome the problem of the non-uniformly distributed color or texture which should be grouped altogether. Empirically, the color feature and HoS texture feature have been combined using Eq. 8:

$$W_{ij} = \alpha \times \sqrt{W_{ij}(texture) \times W_{ij}(color)} + (1 - \alpha) \times W_{ij}(color) \qquad (8)$$

where $\alpha \in [0, 1]$ is a balancing parameter.

Verdoja [38] analyzed the benefits of using superpixels on a simple merging regions graph-based approach. A weighted undirected graph whose nodes are initialized with superpixels was built, and proper metrics to drive the regions merging were proposed in this research. They noticed that superpixels can efficiently boost merging based segmentation by reducing the computational cost without impacting on the segmentation performance. The proposed method provided accurate segmentation results both in terms of visual and objective metrics. Starting from an image I with over-segmented partition L^m composed of m regions, an undirected weighted graph $G^m = \{L^m, W^m\}$ is constructed over the superpixel set L^m, and W^m as defined in Eq. 9:

$$W^m = \{w_{ij}^m\}, \forall i \neq j \mid l_i^m, l_j^m \in L^m \wedge A(l_i^m, l_j^m) = 1 \qquad (9)$$

where A is the adjacency function. Note that G^m is an undirected graph, so $w_{ij}^m = w_{ji}^m$; the weights represent the distance (or dissimilarity measure) between pair of regions $w_{ij}^m = \delta(l_i^m, l_j^m)$. At each iteration, the algorithm picks the pair of labels $l_p^k, l_q^k \in L^k$ having $w_{pq}^k = min\{W^k\}$ and merge them (*i.e.* it generates a new partition $L^{k-1} = L^k - \{l_q^k\}$ having all the pixels $x \in l_p^k \cup l_q^k$ assigned to the label l_p^{k-1}). L^{k-1} contains now just $k - 1$ segments. Then, edges and corresponding weights need to be updated as well. W^{k-1} is generated according to the following formula 10:

$$w_{ij}^{k-1} = \begin{cases} \delta(l_p^{k-1}, l_j^{k-1}) & if \; i = p \vee i = q \\ w_{ij}^k & otherwise \end{cases} \qquad (10)$$

Note that w_{pq}^k is not included in W^{k-1} since it does not exist anymore. When $k = 2$ the algorithm stops and returns the full dendrogram $D = \{L^m, \dots, L^2\}$ that can be cut to obtain the desired number of regions.

Youssef *et al.* [40] also proposed the use of a community detection using complex networks for image segmentation purposes. They investigated a new graph-based image segmentation and compared it to other methods. In this research, the authors used the Infomap algorithm, Louvain algorithm, Fast multi-scale detection of communities based on Local Criteria (FMD), Multi-scale detection of communities using stability optimization (MD) and Stability optimization based on the Louvain method. These studies pointed out the potential perspective and prospect of community detection based image segmentation domain.

Finally, the image segmentation approach of Oscar A. C. Linares *et al.* [22] was based on the idea to construct weighted networks in which the smallest homogeneous regions *(super-pixels)* obtained by initial segmentation processes are nodes of graph, and the edge weighs represent the similarity distance between these regions computed by an Euclidean distance. In this strategy, a 3-dimensional descriptor formed by the components of the CIELAB color space was used and a link between two superpixels was created if the similarity distance was below or equal a certain threshold. This community detection method was applied to extract communities as segments based on the maximization of the modularity measure [9] and was improved by using a greedy optimization procedure proposed by Newman [30].

3   Description of the Approach

The global approach used in this paper consists in modeling an image as a graph where each node is a pixel and each edge between two nodes measures the similarity of the corresponding pixels. The network is then decomposed in communities. Since communities cannot be directly converted to regions, a noise removal and a merge of homogeneous regions is performed.

3.1   From Images to Complex Networks

One image can be represented as an undirected graph $G = (V, E)$, where V is a set of vertices ($V = \{v_1, v_2, ..., v_n\}$) and E is a set of edges ($E = \{e_1, e_2, ..., e_k\}$). Each vertex $v_i \in V$ corresponds to a pixel and an edge $e_{ij} \in E$ connects vertices v_i and v_j. For a given pixel, links towards other pixels are created if and only if the two pixels are close. In this paper we limit ourselves to a $N = 15$ distance for rows and columns directions (see Fig. 1). Empirically, the $N = 15$ value shows

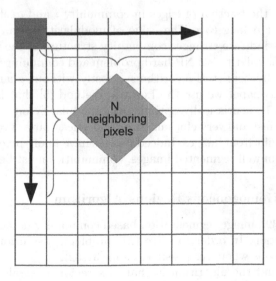

Pattern A

Fig. 1. Illustration of pattern for complex network.

a relatively good performance: the networks have few edges ($2.N$ per pixel) which helps Louvain algorithm to compute communities faster. It also allows to generate edges for distant pixels which again can help the Louvain method.

Furthermore, edges are weighted and the weight w_{ij}, is a non-negative value that measures the similarity/closeness between v_i and v_j. Many weighting schemes could be used but in this paper they are defined as:

$$w_{ij} = \begin{cases} 1 & if\ d_{ij}^c \leq t\ \text{for all color channels } c \\ nil & otherwise \end{cases} \tag{11}$$

where t is a threshold, d_{ij}^c is a measure of the similarity of pixels i and j intensity for color channel c (among R, G and B). It is defined by $d_{ij}^c = \left| I_i^c - I_j^c \right|$ where I_i^c and I_j^c represent the intensity of pixel i and j respectively for channel c.

3.2   From Complex Networks to Communities

The problem of community detection is usually defined as finding the best partition (or covering) of a network into communities of densely connected nodes, with the nodes belonging to different communities being only sparsely connected. Many algorithms have been proposed to find good partitions in a fast way [16]. The qualities of partitions resulting from these methods are often measured by the Newman-Girvan modularity [29]:

$$Q = \sum_i (e_{ii} - a_i^2) \tag{12}$$

where e_{ii} denotes the fraction of edges in community i and a_i if the fraction of ends of edges that belong to i. The value of modularity Q ranges in $[-0.5, 1]$ and higher values indicate stronger community structure of the network.

Maximizing modularity is a NP-hard problem and community detection algorithms are generally heuristics algorithms (mostly without guarantee) and not exact ones. In this paper we use the Louvain method [5] that is a hierarchical greedy algorithm to heuristically optimize the modularity on weighted graphs. This algorithm is fast and very efficient, two properties that are very necessary to work on graphs built from images. Indeed images have many pixels so graphs are large and to obtain well segmented images, communities must be well identified.

3.3   Merging Homogeneous Regions Algorithm

As shown in [31,32], image segmentation based community detection often leads to over-segmentation. In order to solve this problem, a solution is to combine homogeneous regions whenever possible, as in [31,32].

The idea behind the algorithm is that if a region is small or similar to a neighbouring region then both are merged. More formally we use the algorithm described as *Pseudo-code* below. It takes a set of homogeneous regions R, a threshold *regthres* that defines the notion of small region below which regions are merged and similarity threshold *simthres* that defines the limit above which similar regions are merged whatever their size. Furthermore, $C(R_i)$ returns the number of pixels in region R_i, and $\delta(R_i, R_j)$ returns 1 if R_i and R_j are adjacent and 0 otherwise.

Algorithm MHR
Input: A set of regions $R = \{R_1, R_2, ..., R_n\}$
01: **for** $(R_i \in R)$ **do**
02:     **for** $(R_j \in R;\ i \neq j)$ **do**
03:         **if** $(\delta(R_i, R_j) = 1)$
04:             **if** $((C(R_i) < regthres)\ OR\ (C(R_j) < regthres))$
05:                 Merge region R_i and region R_j
06:             **else**
07:                 Compute similarity distance $d(R_i, R_j)$
08:                 **if** $(d(R_i, R_j) >= simthres)$
09:                     Merge region R_i and region R_j
10:                 **end if**
11:             **end if**
12:         **end if**
13:     **end for**
14: **end for**
Output: The set of image segmentation result $R = \{R_1, R_2, ..., R_k\}; \forall k \leq n$

4   Implementation and Performance

In this Section, we detail our implementation strategy and study the effects of various choices on performances.

4.1   Features for Similarity

In the algorithm MHR, the similarity between region R_i and region R_j is computed as using Eq. 21 that we will detail below. The primary straightforward feature for image segmentation is color [20,27,31,32] which is essential when segmenting images using community detection. However, the color feature alone cannot achieve good segmentation if the image is composed of repetitive patterns of different colors in many homogeneous objects. In the proposed algorithm, we incorporate the histogram of oriented gradients (HOG) and the color features into a so-called similarity feature vector that represents each region.

The HOG Feature. The histogram of oriented gradients is computed based on a grayscale image: given a grayscale image I, we extract the gradient magnitude and orientation as using the 1D centered point discrete derivative mask (13), (14) in the horizontal and vertical directions to compute the gradient values.

$$D_X = \begin{bmatrix} -1 & 0 & 1 \end{bmatrix} \tag{13}$$

and

$$D_Y = \begin{bmatrix} -1 \\ 0 \\ 1 \end{bmatrix} \tag{14}$$

We obtain the x and y derivatives by used a convolution operation (15), (16).

$$I_X = I * D_X \tag{15}$$

and

$$I_Y = I * D_Y \tag{16}$$

In this paper, we decided to use the oriented gradient given by Eq. 17 to build the similarity feature vectors.

$$\theta = \arctan \frac{I_Y}{I_X} \tag{17}$$

In the implementation, we use the function *atan2* that returns a value in the interval $(-\pi, \pi]$. The orientation of gradient at a pixel is $\theta = atan2(I_Y, I_X)$ radians. The angle degrees are transformed by $\alpha = \theta * 180/\pi$, that give values in the range (-180, 180] degrees. To shift into unsigned gradient we apply formula 18 and obtain the range of the gradient [0, 360). The orientation of gradient is put into 9 bins that represent 9 elements in the similarity feature vectors. For each region, we compute the HOG feature by the statistic of the percentage of oriented gradient bins.

$$\alpha = \begin{cases} \alpha, & if \quad \alpha \geq 0 \\ \alpha + 360, & if \quad \alpha < 0 \end{cases} \tag{18}$$

The Color Feature. For the color feature, we consider color images (with RGB color space, but other color spaces could be used as the principle remains generic) on individually color channels. For each region, we compute Mean and Standard deviation for every channel of colors as formulas (19), (20) which contribute 6 elements in the similarity feature vectors.

$$Mean(R) = \frac{\sum_{i=1}^{n} C_i}{n} \tag{19}$$

$$SD(R) = \sqrt{\frac{\sum_{i=1}^{n} (C_i - Mean(R))^2}{n}} \tag{20}$$

where C_i is the color value channel of pixel i in image and n is the number of pixels in the set R.

The Similarity Feature Vectors. In the MHR algorithm, the similarity of two adjacent regions R_i and R_j is computed by cosine similarities of a pair of vectors u_i, u_j that represent the two considered regions, as indicated in equation (21).

$$d(R_i, R_j) = Cosine(u_i, u_j) = \frac{u_i^T u_j}{\|u_i\| . \|u_j\|} \qquad (21)$$

However, the feature vectors used in the similarity measure remain a parameter and in this study we propose some possible feature vectors. These proposed features rely on a distance using the mean and the standard deviation of color and texture information that is based on the Histogram of Oriented Gradients (HOG) feature.

4.2   Combination of Features Strategies

In this paper, we present two combinations of color feature and texture feature properties that are (i) Mean and Standard deviation of color and Histogram of oriented gradients (HOG), and (ii) Balancing Mean and Standard deviation of color and Histogram of oriented of gradients (HOG)features as detailed below:

Mean and Standard Deviation Plus HOG Feature. In this method, we build an 15-dimensional vector feature (HOGMeanSD) for each region, consists of 6 elements from Mean and Standard deviation (MeanSD feature) and 9 elements provided by HOG feature, detailed in [33].

Balancing HOG and MeanSD Features (BHMF Feature). It is noticed that the HOGMeanSD method above, the total of elements in each representative vector consist of 15 elements including three-fifth come from HOG feature (9 elements) and two-fifth obtained by MeanSD feature. According to many researches, the color property is a worthwhile feature for image segmentation [20,27,31,32]. Especially, it is more essential when segmenting images using community detection (Fig. 2). In this strategy, we make a combination three-times of Mean and Standard deviation and HOG into one feature, namely BHMF feature. To do that, we build a 27-dimensional feature vector for every region including 18 elements from Mean and Standard deviation ($2(Mean\ and\ SD)\ for\ each\ channel \times 3\ channels \times 3\ times$) and 9 elements that coming from HOG feature. It is illustrated in Fig. 3.

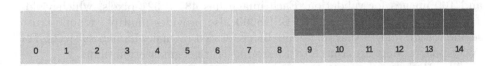

Fig. 2. Creating 15-dimensional vector for each region.

Mean and Standard deviation (18 elements)   Histogram of Oriented Gradients (9 elements)

Fig. 3. Creating 27-dimensional vector for each region.

4.3 Noise Removal

In the implementation, a primary technique that must be pointed out is the noise removal process. As mentioned above, the results obtained from Louvain processes consist of over-segmented results, which decrease the quality when evaluated. In this paper, we recommend applying a noise removal strategy that offers better results and obtains higher evaluation scores. The removing noise process is a crucial part of our algorithm because it merges the small regions that remain after Louvain process. Empirically, we tried different values to set the threshold $regthres = \{100, 200, ..., 600\}$ on our sample dataset (a part of BSDS500) and obtained some potential insights: it is stable in terms of PRI score when the threshold $regthres$ is in the range $\{100, 200, 300\}$. Therefore, we set the threshold for small regions to be $regthres = 200$ pixels for testing and evaluating on datasets.

5 Experimental Evaluation

This Section provides experiments that were performed to assess our algorithm. To evaluate the proposed model, we used three publicly available datasets for image segmentation: Berkeley Segmentation Data Set 300 *(BSDS300)* [24], Berkeley Segmentation Data Set 500 *(BSDS500)* [3] and MSRC object Recognition Data Set *(MSRC)* [36]. Three widely used evaluation segmentation metrics: Variation of Information *(VI)* [25], Segmentation Covering *(SC)* [2] and Probabilistic Rand Index *(PRI)* [34] have been applied to measure the accuracy of proposed algorithm. The qualitative and quantitative evaluation are presented below in Tables 1, 2 and 3.

5.1 Datasets

The Berkeley Segmentation Data Set 300 (BSDS300) has been built with the aim of providing an empirical basis for research on image segmentation and boundary detection. This dataset comprises 300 images, including 200 images for training and 100 images for validation. Each image has 481×321 pixels, which yields a graph of 154401 vertices. The BSDS300 also provides multiple ground-truth segmentation images that are manually generated by many human subjects. For every image, there are from 5 to 10 ground-truth segmentation maps.

The Berkeley Segmentation Data Set 500 (BSDS500) is an extension of BSDS300. This dataset comprises 500 images, including 200 images for training, 200 new testing images and 100 images for validation. Each image has

481×321 pixels and has in average 5 ground-truth segmentation maps. Supplying a benchmark for comparing different segmentation and boundary detection algorithms.

The Microsoft Research Cambridge Object Recognition Image Database (MSRC) contains a set of 591 natural images of size 320×213 with one ground-truth per image grouped into categories. Its intended use is research, in particular object recognition research.

5.2 Evaluation Metrics

In general, evaluation segmentation metrics have been used to evaluate different image segmentation algorithms in the literature. Some common one include Variation of Information *(VI)* [25], Segmentation Covering *(SC)* [2] and Probabilistic Rand Index *(PRI)* [34]. Especially, PRI brings exceedingly benefit of evaluation on BSDS300 and BSDS500 datasets which provide multiple ground-truth.

The Probabilistic Rand Index (PRI) [34] is a classical evaluation criterion for clustering. The PRI measures the probability that pair of pixels have consistent labels in the set of manual segmentation maps (ground-truth). Given a set of ground-truth segmentation images $\{S_k\}$, the Probabilistic Rand Index is defined as:

$$PRI(S_{test}, \{S_k\}) = \frac{1}{T} \sum_{i<j} [c_{ij} p_{ij} + (1 - c_{ij})(1 - p_{ij})] \qquad (22)$$

where c_{ij} is the event that the algorithm gives the same label to pixels i and j, and p_{ij} corresponds to the probability of the pixels i and j having the same label, and is estimated by using sample mean of the corresponding Bernoulli distribution on the ground-truth dataset. T is the total number of pixel pairs. The PRI values range in [0,1] in which a larger value likely indicates a greater similarity between these segmentation images.

The Variation of Information (VI) metric was introduced for the evaluation of clustering [25]. It measures the distance between two clusterings in terms of the information difference between them. VI is defined by:

$$VI(C, C') = H(C) + H(C') - 2I(C, C') \qquad (23)$$

where $H(C)$ and $H(C')$ are the entropy of segmentation image C and ground-truth C', respectively and $I(C, C')$ is the mutual information of two segmentation image C and ground-truth image C'. Let segmentation image C and ground-truth image C' have N levels of gray and distributions are uniform, *i.e.* $PN = 1/N$. The maximal values of entropies $H(C) = logN$ and $H(C') = logN$, and let mutual information $I(C, C')$ be equal to zero. Hence, the range of this metric is $[0, 2logN]$, and the smaller value is the better segmentation results.

The Segmentation Covering (SC) metric that measures averaged matching between proposed segment with a ground-truth labeling was introduced by Arbelaez *et al.* [2]. It is defined by:

$$SC(S, S_g) = \frac{1}{N} \sum_{R \in S} |R| \cdot \max_{R' \in S_g} O(R, R') \tag{24}$$

where N denotes the total number of pixels in the image and the *overlap* between two regions R and R', defined as:

$$O(R, R') = \frac{|R \cap R'|}{|R \cup R'|} \tag{25}$$

5.3   Results

For qualitative evaluations, we present image segmentation results obtained by using HOGMeanSD and BHMF features in Figs. 4, 5, 6 and 7 collected randomly from the dataset BSDS300, MSRC dataset and the dataset BSDS500. For qualitative assessment perspective, we can see that the proposed algorithm offers good results and produces sizable regions for all selected images. Our algorithm can aggregate homogeneous neighboring regions successfully even if pixels inside each region are dissimilar. Besides the success of our method, it remains a challenge for segmenting images whose colors contained are quite different in parts of an object as we point out in Fig. 8. It is seen that images in the second line which using HOGMeanSD feature [33] are worse than the images in the third line whose using BHMF feature in terms of qualitative results. In this paper, our proposed BHMF feature is not only better HOGMeanSD feature in both qualitative and quantitative segmentation results *(see in Tables: 1, 2 and 3)*, but also enhanced segmentation results significantly in cases of colors contained are quite different in parts of an object such as zebra, cheetah and tiger.

From a quantitative point of view, we evaluated the segmentation results using evaluation metrics presented in Sect. 5.2 (PRI, VI, SC) by comparing a test segmentation with multiple ground-truth images. We applied these evaluation metrics on the MSRC dataset, detailed results are given in Table 1. We run MHR algorithm on the validation set from the Berkeley segmentation data set 300 (BSDS300) and the test data set BSDS500, detailed results are given in Tables 2 and 3, respectively.

The evaluation results give the successful roof for our algorithm. Our method exceeds all previous graph-based algorithms in terms of PRI scores. Empirically, the threshold range for the agglomeration process is only taking range from 0.940 to 0.999 *(with 0.005 intervals)*. The best results are recorded when the value of cosine similarity distance equal to 0.995. Cosine similarity distance domain that offers best results in our algorithm fall into [0.990, 0.999]. Note that the regions belong to one segment have HOG and color features properties in common to each other.

Fig. 4. *Top:* Original images. *Second line:* Segmentation results obtained by the Louvain method. *Third line:* Segmentation results with the proposed algorithm. *Fourth line:* Ground-truth.

Table 1. Quantitative comparisons on MSRC Object Recognition Data set using the proposed algorithm and other methods.

Methods	PRI	VI	SC
IS4(MCG) [42]	**0.800**	**1.15**	0.690
gPb-owt-ucm [3]	-	-	0.750
Our algorithm (BHMF-feature method)	0.743	1.61	**0.754**
Our algorithm (HOGMeanSD-feature method) [33]	0.739	1.63	0.752
Canny-owt-ucm [3]	-	-	0.680

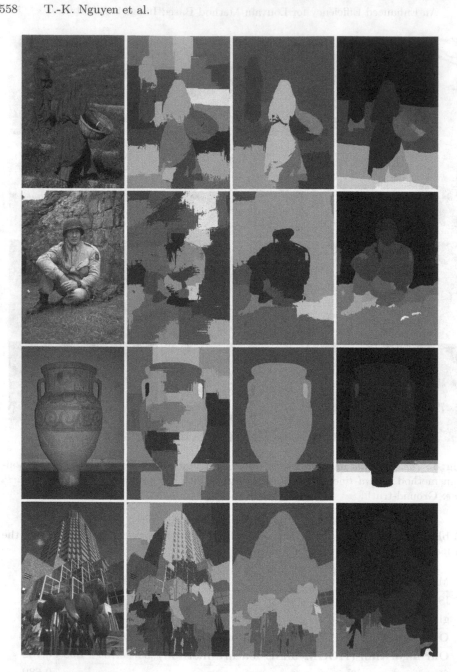

Fig. 5. *Left:* Original images. *The second:* Segmentation results obtained by the Louvain method. *The third:* Segmentation results with the proposed algorithm. *Right:* Ground-truth.

Fig. 6. *Top:* Original images. *Second line:* Segmentation results obtained by the Louvain method. *Third line:* Segmentation results with the proposed algorithm. *Fourth line:* Ground-truth.

Table 2. Quantitative comparisons on BSDS300 validation set of the proposed algorithm and other methods.

Methods	PRI	VI	SC
Human	0.870	1.16	-
Our algorithm (BHMF-feature method)	**0.828**	**1.33**	**0.77**
Our algorithm (HOGMeanSD-feature method) [33]	0.822	1.40	0.74
Lv-ara [31]	0.819	-	-
Youssef Mourchild's *(FMS(HOG))* [27]	0.811	-	-
gPb-owt-ucm [3]	0.810	1.47	0.75
IS4(MCG) [42]	0.810	1.54	0.61
Youssef Mourchild's *(HOG))*	0.803	-	-
Lv-ahr [32]	0.800	-	-
Mean Shift [10]	0.780	1.63	0.66
Shijie Li's method *(L*a*b (HoS))*	0.777	1.879	-
Felz-Hutt	0.770	1.79	0.68
Canny-owt-ucm [3]	0.770	1.81	0.66
NCuts [11]	0.750	1.84	0.66
Shijie Li's method *(RGB (HoS))* [20]	0.749	2.149	-

Fig. 7. *Top:* Original images. *Second line:* Segmentation results obtained by the Louvain method. *Third line:* Segmentation results with the proposed algorithm. *Fourth line:* Ground-truth.

Table 3. Quantitative comparisons on BSDS500 test set of the proposed algorithm and other methods.

Methods	PRI	VI	SC
Human	0.870	1.17	-
Learned STLD [19]	**0.860**	1.54	0.67
Our algorithm (BHMF-feature method)	0.838	**1.28**	**0.77**
Our algorithm (HOGMeanSD-feature method) [33]	0.835	1.30	0.74
gPb-owt-ucm [3]	0.830	1.48	0.74
IS4(MCG) [42]	0.830	1.35	0.63
Felz-Hutt [15]	0.800	1.87	0.69
Mean Shift [10]	0.790	1.64	0.66
Canny-owt-ucm [3]	0.790	1.89	0.66
NCuts [11]	0.780	1.89	0.67

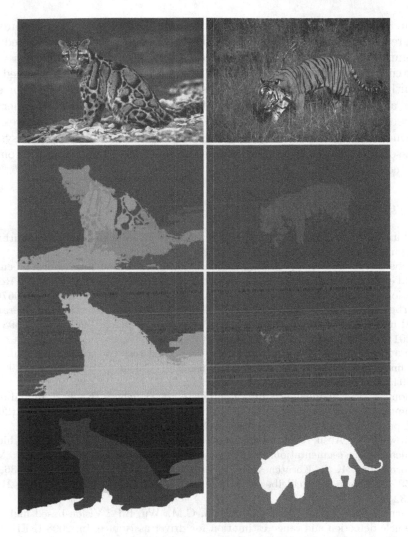

Fig. 8. *Top:* Original images. *Second line:* Segmentation results obtained by using the HOGMeanSD feature [33]. *Third line:* Segmentation results obtained by using the BHMF feature. *Fourth line:* Ground-truth.

6 Conclusion

In this paper we proposed a new and efficient algorithm to merge homogeneous regions issuing from the Louvain Method for segmenting the content of images. The Louvain method is a community detection techniques in graphs used here in order to segment images in homogeneous regions. By cooperating with our dedicated texture and color features, the classical over-segmentation problem is removed. Our method obtained some accurate and efficient results for image segmentation purposes. More specifically, this paper detailed how we merged similar and/or homogeneous regions using some texture and color features to measure the similarity. The second contribution of this paper is that we reduced its complexity as it does not need to recompute the feature properties when operated merging processes. Hence, the time complexity has been reduced significantly compared with the classical use of a 256-dimensional vector for each region and the recomputed processes of feature properties whenever merging processes implemented in some other techniques. Many experiments have been performed, and the results highlight the performances of our proposed system and assess that we produced more accurate objects while enhancing the computation complexity.

References

1. Abin, A.A., Mahdisoltani, F., Beigy, H.: A new image segmentation algorithm: a community detection approach. In: IICAI (2011)
2. Arbelaez, P., Maire, M., Fowlkes, C., Malik, J.: From contours to regions: an empirical evaluation. In: 2009 IEEE Conference on Computer Vision and Pattern Recognition, pp. 2294–2301, June 2009. https://doi.org/10.1109/CVPR.2009.5206707
3. Arbelaez, P., Maire, M., Fowlkes, C., Malik, J.: Contour detection and hierarchical image segmentation. IEEE Trans. Pattern Anal. Mach. Intell. 33(5), 898–916 (2011). https://doi.org/10.1109/TPAMI.2010.161
4. Bhattacharyya, A.: On a measure of divergence between two statistical populations defined by their probability distributions. Bull. Calcutta Math. Soc. 35, 99–109 (1943)
5. Blondel, V.D., Guillaume, J.L., Lambiotte, R., Lefebvre, E.: Fast unfolding of communities in large networks. J. Stat. Mech. Theory Exp. 10, 10008 (2008). https://doi.org/10.1088/1742-5468/2008/10/P10008
6. Browet, A., Absil, P.-A., Van Dooren, P.: Community detection for hierarchical image segmentation. In: Aggarwal, J.K., Barneva, R.P., Brimkov, V.E., Koroutchev, K.N., Korutcheva, E.R. (eds.) IWCIA 2011. LNCS, vol. 6636, pp. 358–371. Springer, Heidelberg (2011). https://doi.org/10.1007/978-3-642-21073-0_32
7. Chen, Y.L., Lin, C.T., Fan, C.J., Hsieh, C.M., Wu, B.F.: Vision-based nighttime vehicle detection and range estimation for driver assistance. In: 2008 IEEE International Conference on Systems, Man and Cybernetics, pp. 2988–2993, October 2008. https://doi.org/10.1109/ICSMC.2008.4811753
8. Christoudias, C.M., Georgescu, B., Meer, P.: Synergism in low level vision. In: Object Recognition Supported by User Interaction for Service Robots, vol. 4, pp. 150–155, August 2002. https://doi.org/10.1109/ICPR.2002.1047421
9. Clauset, A., Newman, M.E.J., Moore, C.: Finding community structure in very large networks. Phys. Rev. E Stat. Nonlinear Soft Matter Phys. 70, 066111 (2005). https://doi.org/10.1103/PhysRevE.70.066111
10. Comaniciu, D., Meer, P.: Mean shift: a robust approach toward feature space analysis. IEEE Trans. Pattern Anal. Mach. Intell. 24(5), 603–619 (2002). https://doi.org/10.1109/34.1000236
11. Cour, T., Benezit, F., Shi, J.: Spectral segmentation with multiscale graph decomposition. In: 2005 IEEE Computer Society Conference on Computer Vision and Pattern Recognition (CVPR 2005), vol. 2, pp. 1124–1131, June 2005. https://doi.org/10.1109/CVPR.2005.332

12. Dalal, N., Triggs, B.: Histograms of oriented gradients for human detection. In: 2005 IEEE Computer Society Conference on Computer Vision and Pattern Recognition, CVPR 2005, vol. 1, pp. 886–893, June 2005. https://doi.org/10.1109/CVPR.2005. 177

13. Deng, Y., Manjunath, B.S.: Unsupervised segmentation of color-texture regions in images and video. IEEE Trans. Pattern Anal. Mach. Intell. **23**(8), 800–810 (2001). https://doi.org/10.1109/34.946985

14. Easley, D., Kleinberg, J.: Networks, Crowds, and Markets: Reasoning about a Highly Connected World. Cambridge University Press, Cambridge (2010). https:// doi.org/10.1017/CBO9780511761942

15. Felzenszwalb, P.F., Huttenlocher, D.P.: Efficient graph-based image segmentation. Int. J. Comput. Vis. **59**(2), 167–181 (2004). https://doi.org/10.1023/B:VISI. 0000022288.19776.77

16. Fortunato, S.: Community detection in graphs. Phys. Rep. **486**(3–5), 75–174 (2010). https://doi.org/10.1016/j.physrep.2009.11.002. http://www.sciencedirect. com/science/article/pii/S0370157309002841

17. Girvan, M., Newman, M.E.J.: Community structure in social and biological networks. Proc. Natl Acad. Sci. **99**(12), 7821–7826 (2002). https://doi.org/10.1073/ pnas.122653799. https://www.pnas.org/content/99/12/7821

18. Junwei, L., Shaokai, L.: A novel image segmentation technology in intelligent traffic light control systems. In: 2013 3rd International Conference on Consumer Electronics, Communications and Networks. pp. 26–29, November 2013. https://doi.org/ 10.1109/CECNet.2013.6703263

19. Khan, N., Sundaramoorthi, G.: Learned shape-tailored descriptors for segmentation. In: 2018 IEEE/CVF Conference on Computer Vision and Pattern Recognition. pp. 666–674, June 2018. https://doi.org/10.1109/CVPR.2018.00076

20. Li, S., Wu, D.O.: Modularity-based image segmentation. IEEE Trans. Circuits Syst. Video Technol. **25**(4), 570–581 (2015). https://doi.org/10.1109/TCSVT. 2014.2360028

21. Li, W.: Modularity segmentation. In: Lee, M., Hirose, A., Hou, Z.-G., Kil, R.M. (eds.) ICONIP 2013, Part II. LNCS, vol. 8227, pp. 100–107. Springer, Heidelberg (2013). https://doi.org/10.1007/978-3-642-42042-9_13

22. Linares, O.A.C., Botelho, G.M., Rodrigues, F.A., Neto, J.B.: Segmentation of large images based on super-pixels and community detection in graphs. CoRR abs/1612.03705 (2016), http://arxiv.org/abs/1612.03705

23. Liu, D., Chen, T.: DISCOV: a framework for discovering objects in video. IEEE Trans. Multimedia **10**(2), 200–208 (2008). https://doi.org/10.1109/TMM.2007. 911781

24. Martin, D., Fowlkes, C., Tal, D., Malik, J.: A database of human segmented natural images and its application to evaluating segmentation algorithms and measuring ecological statistics. In: Proceedings of the 8th International Conference on Computer Vision, vol. 2, pp. 416–423, July 2001

25. Meilă, M.: Comparing clusterings: an axiomatic view. In: In ICML 2005: Proceedings of the 22nd International Conference on Machine Learning, pp. 577–584. ACM Press (2005)

26. Mori, G.: Guiding model search using segmentation. In: Tenth IEEE International Conference on Computer Vision (ICCV 2005) Volume 1, vol. 2, pp. 1417–1423, October 2005. https://doi.org/10.1109/ICCV.2005.112

27. Mourchid, Y., El Hassouni, M., Cherifi, H.: An image segmentation algorithm based on community detection. COMPLEX NETWORKS 2016 2016. SCI, vol. 693, pp. 821–830. Springer, Cham (2017). https://doi.org/10.1007/978-3-319-50901-3_65

28. Newman, M.: Networks: An Introduction. Oxford University Press Inc., New York (2010)

29. Newman, M.E., Girvan, M.: Finding and evaluating community structure in networks. Phys. Rev. E **69**(2), 026113 (2004)

30. Newman, M.: Fast algorithm for detecting community structure in networks. Physical Review E **69**, (2003). http://arxiv.org/abs/cond-mat/0309508

31. Nguyen, T., Coustaty, M., Guillaume, J.: A new image segmentation approach based on the Louvain algorithm. In: 2018 International Conference on Content-Based Multimedia Indexing (CBMI), pp. 1–6, September 2018. https://doi.org/10.1109/CBMI.2018.8516531

32. Nguyen, T.-K., Coustaty, M., Guillaume, J.-L.: An efficient agglomerative algorithm cooperating with louvain method for implementing image segmentation. In: Blanc-Talon, J., Helbert, D., Philips, W., Popescu, D., Scheunders, P. (eds.) ACIVS 2018. LNCS, vol. 11182, pp. 150–162. Springer, Cham (2018). https://doi.org/10.1007/978-3-030-01449-0_13

33. Nguyen, T., Coustaty, M., Guillaume, J.: A combination of histogram of oriented gradients and color features to cooperate with louvain method based image segmentation. In: Proceedings of the 14th International Joint Conference on Computer Vision, Imaging and Computer Graphics Theory and Applications - Volume 4: VISAPP, pp. 280–291. INSTICC, SciTePress (2019). https://doi.org/10.5220/0007389302800291

34. Pantofaru, C., Hebert, M.: A comparison of image segmentation algorithms. Technical report. CMU-RI-TR-05-40. Carnegie Mellon University, Pittsburgh, PA, September 2005

35. Ren, X., Malik, J.: Learning a classification model for segmentation. In: Proceedings Ninth IEEE International Conference on Computer Vision, vol. 1, pp. 10–17, October 2003. https://doi.org/10.1109/ICCV.2003.1238308

36. Shotton, J., Winn, J., Rother, C., Criminisi, A.: *TextonBoost*: joint appearance, shape and context modeling for multi-class object recognition and segmentation. In: Leonardis, A., Bischof, H., Pinz, A. (eds.) ECCV 2006, Part I. LNCS, vol. 3951, pp. 1–15. Springer, Heidelberg (2006). https://doi.org/10.1007/11744023_1

37. Sumengen, B., Manjunath, B.S.: Multi-scale edge detection and image segmentation. In: 2005 13th European Signal Processing Conference, pp. 1–4, September 2005

38. Verdoja, F., Grangetto, M.: Fast Superpixel-Based Hierarchical Approach to Image Segmentation. In: Murino, V., Puppo, E. (eds.) ICIAP 2015, Part I. LNCS, vol. 9279, pp. 364–374. Springer, Cham (2015). https://doi.org/10.1007/978-3-319-23231-7_33

39. Mourchid, Y., El Hassouni, M., Cherifi, H.: A new image segmentation approach using community detection algorithms. In: 15th International Conference on Intelligent Systems Design and Applications, Marrakesh, Marocco, December 2015

40. Mourchild, Y., El Hassouni, M., Cherifi, H.: Image segmentation based on community detection approach. Int. J. Comput. Inf. Syst. Ind. Manage. Appl. **8**, 195–204 (2016)

41. Zhou, J.Y., Ong, E.P., Ko, C.C.: Video object segmentation and tracking for content-based video coding. In: 2000 IEEE International Conference on Multimedia and Expo. ICME2000. Proceedings. Latest Advances in the Fast Changing World of Multimedia (Cat. No. 00TH8532), vol. 3, pp. 1555–1558 (2000). https://doi.org/10.1109/ICME.2000.871065
42. Zohrizadeh, F., Kheirandishfard, M., Kamangar, F.: Image segmentation using sparse subset selection. CoRR abs/1804.02721 (2018). http://arxiv.org/abs/1804.02721

Vehicle Activity Recognition Using DCNN

Alaa AlZoubi[1(✉)] and David Nam[2]

[1] School of Computing, The University of Buckingham, Buckingham, UK
alaa.alzoubi@buckingham.ac.uk
[2] Propelmee Ltd., Milton Keynes, UK
david@propelmee.com

Abstract. This paper presents a novel Deep Convolutional Neural Network (DCNN) method for vehicle activity classification. We extend our previous approach to be able to classify a larger number of vehicle trajectories in a single network. We also highlight the flexibility of our approach in integrating further scenarios to our classifier. Firstly, a spatiotemporal calculus method is used to encode the relative movement between vehicles as a trajectory of QTC states. We then map the encoded trajectory to a 2D matrix using the one-hot vector mapping, this preserves the important positional data and order for each QTC state. To do this we associate the QTC sequences with pixels to form a 2D image texture. Afterwards, we adapted trained CNN architecture into our vehicles activity recognition task. Two separate types of driving data sets are used to evaluate our method. We demonstrate that the proposed method out-performs existing techniques. Along with the proposed approach we created a new dataset of vehicles interactions. Although the focus of this paper is on the automated analysis of vehicle interactions, the proposed technique is general and can be applied for pairwise analysis for moving objects.

Keywords: Vehicle activity classification · Spatiotemporal calculus · Trajectory texture · Transfer learning · Deep convolutional neural networks

1 Introduction

Vehicular collisions are typically the result of a deficient situational understanding of the surrounding scene, and therefore being unable to identify and act upon dangerous situations, before there are consequences. Being able to understand the relative interaction between a vehicle and its surroundings is imperative in defining the situation a vehicle is in, or about to enter, and hence to plan accordingly and do appropriate decision making [16]. The objective of vehicle activity recognition is to classify a set of actions (from one of more vehicles) for a sequence of observations. During dynamic traffic scenarios complex interactions can occur, either between vehicles themselves, or even a possible broken

© Springer Nature Switzerland AG 2020
A. P. Cláudio et al. (Eds.): VISIGRAPP 2019, CCIS 1182, pp. 566–588, 2020.
https://doi.org/10.1007/978-3-030-41590-7_24

or stalled vehicles; in this situation we refer to the stationary vehicle as an obstacle. Larger group interactions can then be constructed from the pair-wise interactions—vehicle-vehicle or vehicle-obstacle.

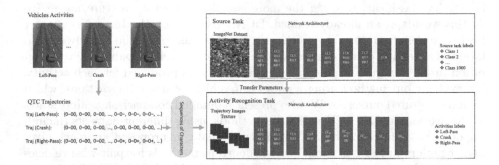

Fig. 1. Overview of presented approach. Representation of trajectories using QTC (left). Customised CNN for activity classification (right) [4].

There are two main approaches for trajectory analysis, quantitative and qualitative approaches. Traditionally there has been significant research within the quantitative approach; which uses sequences of real-valued features (trajectories) [10,12,13,15,25,27], however, recent interest has grown around qualitative approaches. Qualitative methods use a symbolic representation of scenarios of moving objects, and have shown superior performance for vehicle activity analysis when compared with quantitative methods [2]. It has found applications in vehicle interaction and human behavior analysis [2], and human-robot interaction [9]. Along with improved performance there are also other benefits for a qualitative approach, such as having a more compact representation (and therefore is more computationally efficient). Qualitative representations are also a more natural way of describing interactions [7]; where it is more flexible to variations in trajectories, but still captures the overall behaviour.

Within the context of activity recognition having a compact and informative representation for encoding trajectories, of moving objects, can be beneficial. [12] encodes trajectories within a two dimensional matrix, while [20] encodes them in a trajectory texture image. Both approaches were shown to be successful in different application domains, human activity recognition [19,20] and pair-wise vehicles activity recognition [12]. Based on this the 2D representations are then used to train a classifier to perform the activity recognition task. Also, with the recent success of deep learning methods for image classification, representing trajectories within a 2D image enable the capabilities of previously built and trained networks. In particular neural networks are well suited for learning features based on the shape and texture of an object.

In this series, we present our novel approach for vehicle pair-activity recognition and classification, based on QTC and DCNN. Our method consists of two stages, firstly we employ QTC as a means to, compactly, represent the relative motion between pairs of objects (vehicle-vehicle or vehicle-obstacle). We then encode their interactions as a trajectory of QTC states. To convert this representation as a 2D image we use one-hot vectors to convert QTC sequences to a two dimensional matrix (or image texture). The second stage of our approach is activity classification, using the more encoded trajectory texture image. To do this we adapt, an already trained, DCNN (trained on the ImageNet dataset for image classification). The motivation for this was to use the already trained layers as a starting point to transfer knowledge to the vehicle pair-activity recognition task. A unique image signature (or texture) is produced for each activity. We evaluate our method using a dataset of vehicle-obstacle interactions, which we have captured ourselves. We also present a detailed comparison against state-of-the-art quantitative and qualitative methods, using different datasets (including our own). Moreover, results demonstrate that our proposed approach gives higher performance when compared with current methods for pair-wise vehicles trajectory classification. An overview of the method and main contributions of our work is shown in Fig. 1.

This article is an extension to our work previously presented [4]. We have evaluated our approach on a new dataset which combines vehicles activity trajectories from different data sources. We have extended the categories of our activity classifier, while still achieving high performance. Further detailed evaluations on this dataset are given in Sect. 4.4; where our approach is now able to accurately distinguish between eight different pairwise activities, without compromising on accuracy. This allows for a more general method, which can be used for applications with a wide range of scenarios. To highlight the main contributions made within this paper, we demonstrate, for the first time, a novel approach for classifying pair-wise vehicle activities using QTC with DCNNs.

Our work is primarily motivated by our interest in the automated recognition vehicle's scenario or activity, however, being able to predict a vehicle's future trajectory can help in avoiding any potential collisions. To be able to gain traction and boost usability as a main-stream analysis method, we present a novel driver model for predicting a vehicle's future trajectory, from its partially observed prior trajectory. Main novel aspects of this work are as follows: proposing a new CNN tuned for pair-wise vehicle activity recognition, (using a modified version of AlexNet [11]). The second contribution is a novel method for vehicle activity classification, based on a vehicle's potential future scenario; to achieve this we utilise a driver model to produce likely trajectories. Lastly, through experimentation we demonstrate that our proposed CNN out-performs current approaches (which includes [2,12,13,15,27]). Additionally, we include a new, open-source, dataset of pairwise vehicle-obstacle interactions, (with associated ground-truth). It consists of 554 vehicle scenarios (complete and incomplete scenarios) for three different types of interactions.

2   Background

In this section we give a brief overview of the state-of-the-art in trajectory analysis, qualitative trajectory calculus, and deep learning.

2.1 Trajectory Analysis Techniques

In order to do trajectory analysis a spatial-temporal representation of motion information has to be defined. Currently there are numerous techniques for trajectory representation; in order to encode a sequence of continuous states in an efficient and fast way. A long-term motion descriptor given the name, sequential deep trajectory descriptor (sDTD), was presented in [19]. Their technique initially determines the simplified dense trajectories of a single object and then converts these trajectories into 2D images. Chavoshi et al. [5] presented a visualization technique, sequence signature (SESI), to convert a basic variation of QTC (QTC_B) movement patterns of moving point objects into a 2D indexed rasterized matrix. The approach in [12] represents trajectories of pair-vehicles as a series of heat sources, where a thermal diffusion process creates an activity map as a 2D matrix. The above techniques either encode trajectories of a single object, or utilise traditional similarity methods (e.g. Euclidean distance) which unable to cope with varying lengths trajectories and compound behaviors. The vehicle activity analysis tackled in this paper requires a more general technique, invariant to trajectory length and compound behaviors. It has been demonstrated that quantitative [12] and qualitative [2] methods are the most efficient for encoding vehicle pair-wise activity. We thus utilise both as the standard against which we evaluate our own work.

Approaches for trajectory analysis are grouped into three categories: single-role activities [26], pair-wise-activities [2], and group-activities [12]. Typically these techniques were focused on specific types of behaviours: human behavior recognition [2,12,27], human-robot interaction [9], animal behavior clustering [2], or vehicles interaction recognition [2,10,12,25]. Quantitative features were mainly used in [12] for modelling traffic behaviours, and the same was done for autonomous driving applications in [24]. [12] proposed an algorithm for vehicle pair activity recognition using a heat map. There vehicle trajectories were represented as an activity map and then a Surface-Fitting method was used to group the resulting vehicle activities.

An approach using qualitative features for traffic activity recognition was proposed in [2]. There a Normalized Weighted Sequence Alignment technique was introduced, to calculate the similarity between QTC sequences. They evaluated their approach using three different datasets: human generated trajectories, trajectories from animals, and trajectories from vehicle interactions. It was demonstrated that their techniques provided improved performance over other state-of-the-art quantitative methods [12,13,15,27]. The latter are more closely associated with the work proposed in this paper. Hence, we use these approaches to compare our method. We also do a full comparison using the vehicle-interaction dataset [12], provided with ground truth, to give a complete evaluation of our recognition and classification approach (Sect. 4.1). Within this work we will focus on pair-wise vehicle-to-vehicle and vehicle-to-obstacle interactions. For a deeper understanding of trajectory analysis we refer to reader to [1]; where further review is provided.

2.2 Trajectory Representation

We have now established that for robust and efficient trajectory analysis being able to represent trajectories, in a compact and meaningful way, is essential. Qualitative spatial-temporal reasoning is a class of techniques which processes trajectory information in a similar manner to the human perception system, in terms of relative interactions. It does so by using symbolic representations, of specific classes of interactions, rather than numeric measurements [23]. For activity analysis, approaches using QTC representations have demonstrated a higher level of robustness and performance, when contrasted against quantitative methods; in different application areas, including: human activity analysis and recognition of vehicle interaction [2].

Four features are used as a base unit for describing more complex pair-wise trajectory information. Given the coordinates of two objects, k and l, their interactions can be described using QTC as follows:

1. **Distance Features**
    - S_1: distance of k with respect to l: "$-$" indicates decrease, "$+$" indicates increase, "0" indicates no change.
    - S_2: distance of l with respect to k.
2. **Speed Features**
    - S_3: Relative speed of k with respect to l at time t (which dually represents the relative speed of l with respect to k).
3. **Side Features**
    - S_4: Displacement of k with respect to the reference line L connecting the objects: "$-$" if it moves to the left, "$+$" if it moves to the right, "0" if it moves along L or not moving at all.
    - S_5: Displacement of l with respect to L.
4. **Angular Features**
    - S_6: The respective angles between the velocity vectors of the objects and vector L: "$-$" if $\theta_1 < \theta_2$, "$+$" if $\theta_1 > \theta_2$ and "0" if $\theta_1 = \theta_2$

here S_i represents the qualitative relations in QTC. Figure 2 demonstrates the concept of qualitative relations within QTC for two disjoint objects, (k and l). Three main calculi were defined [23]: QTC_B, QTC_C and QTC_{Full}. Where QTC_C (S_1, S_2, S_4 and S_5) and the combination of the four codes results in $3^4 = 81$ different states.

2.3 Deep Neural Networks

The increasing popularity of convolutional neural networks have demonstrated their applicability for object recognition; in particular when it comes to very large-scale visual recognition problems [18]. Much of its success is attributed to the ability for CNNs to learn base image features and then combine them to form, discriminative and robust, object features (such as shapes and textures) [17]. Some successful CNNs for image classification are presented in AlexNet [11] and GoogLeNet [22], which were both designed within the context of the "Large Scale

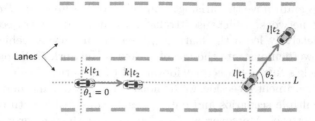

Fig. 2. Diagrammatic representation showing QTC relations between two moving vehicles: $QTC_C = (-, +, 0, +)$ [4].

Visual Recognition Challenge" (ILSVRC) [18] for the ImageNet dataset [6]. For a further in-depth analysis we refer the reader to [14], which provides a review of deep neural network architectures and their applications for object recognition.

Within our proposed approach for activity recognition we utilise AlexNet [11]. This model provided a good foundation for activity recognition, as it was trained on approximately 1.2 million labeled images, consisting of 1,000 different categories from the ILSVRC dataset [18]; as a point to note, each image in this dataset has one centrally located object, which occupies a significant portion of the image and there is also limited background clutter. This has allowed the network to learn a wide a robust set of base features. AlextNet uses the entirety of the image as an input and produces probabilities for each class. In terms of the network structure there are 650,000 neurons and 60 million parameters in AlexNet. Its architecture is made up of two normalisation layers, three fully-connected layers, three max-pooling layers, five convolutional layers, and a linear layer with softmax activation, to produce the probabilistic classification outputs. Dropout regularization [21] is employed as a means to reduce overfitting in the fully connected layers. Rectified Linear Units (ReLU) are also used as the activation of the layers, and to provide non-linearity to the system.

A novel aspect of this work involved customizing AlexNet for the activity recognition task, and evaluating the performance of our improved DCNN architecture on multiple datasets.

3   Theoretical Formulation

In Fig. 1 we overview the key steps and contributions for our vehicle activity recognition method; where the right side demonstrated our novelty in our deep learning framework. Within our first step QTC trajectories are computed from consecutive observations and then projected onto 2D matrices, (here observations consist of the 'x' and 'y' coordinates). At this point the resulting 2D matrices (or trajectory texture images) now encode the relative motions' of pairs of vehicles at consecutive time frames, $[t_1, t_N]$.

One of the main novelties arising from this work is given in *TrajNet*; a new deep learning network, which uses transfer learning, using trajectory texture images, to effectively learn the features presented in varying vehicle activities. Additionally, we demonstrate the use of a Neural Network (NN) model for predicting a vehicle's future trajectory, for our set of scenarios, from an incomplete one. In our experimental section we demonstrate that our method is generalizable across multiple scenarios and data types, is compatible with complete and predicted trajectories. Moreover, we also demonstrate that the proposed method, consistently, out-performs other state-of-the-art approaches.

We summarize the key novel contributions of this work as follows: Representing specific scenarios using sequences of QTC states, for a pair of vehicles' relative movements. We introduce a technique to represent QTC sequences as an image texture, based on the one-hot vector encoding. A novel CNN model is proposed for vehicle activity recognition; it utilizes AlexNet and the ImageNet dataset, but improves upon the base network, for vehicles pair-activity recognition task. This results in our new network *TrajNet*. For predicting a complete trajectory from a partially observed one, a NN based human driver model is presented.

3.1   Encoding Pair-Wise Vehicle Interactions

Representing Vehicle Behaviour with QTC. Given x, y, in the top-down coordinate space, as the two-dimensional position of the centroid of a vehicle, we employ QTC for representing the behaviour of a pair of interacting vehicles. Here we consider interacting vehicles to be in close enough proximity that its presence could potentially influence the actions and decision making of the other vehicle. Changes in their relative positions, (hence their interactions) are expressed as a sequence of QTC states. We use the common QTC variant (QTC_C) within our approach. The common QTC variant is represents the vehicle pair's relative two-dimensional movement qualitatively.

Definition: Given two interacting vehicles, or vehicle/obstacle pair, interacting—in this case the presence of an obstacle in the path of a vehicle will cause an action, likely to avoid it, to be taken—with one another and their centroid x, y coordinates, we define:

$$V1_i = \{(x_1, y_1), ..., (x_t, y_t), ..., (x_N, y_N)\}, \tag{1}$$

$$V2_i = \{(x_1', y_1'), ..., (x_t', y_t'), ..., (x_N', y_N')\}, \tag{2}$$

here (x_t, y_t) and (x_t', y_t') are the centroids of the first interacting vehicle and second centroids of the interacting vehicles, at time t, respectively. The pairwise trajectory is therefore represented as a sequence of ordered QTC_C states: $Tv_i = \{S_1, ..., S_t, ..., S_N\}$, where S_t is the QTC_C state representation of the relative movement of the two vehicles (x_t, y_t) and (x_t', y_t') at time t in trajectory Tv_i. N is the number of observations in Tv_i.

Transforming Sequential QTC States to a 2D Image Texture. The time varying sequence of QTC_C states, described in Sect. 3.1, is a one-dimensional succession of QTC_C states. It can be seen as analogous to both vehicles trajectories. When drawing a comparison with text information, there are limitations; such as where are no spaces between QTC_C states and there are no concepts of words. In order to be able to decode a higher level of information from these succession of states, we translate QTC_C trajectories into sequences of characters; with the aim of applying the same representation technique for text data without losing location information of each QTC_C state in the sequence. We then represent this sequence numerically, so that it is able to be used as an input for our CNN. Further discussion on our contribution to this is covered in Sect. 3.2.

To be able to achieve this we first represent the QTC_C states using the a symbolic representation Cr: cr_1, cr_2, ..., cr_{81}. We then map our representation, the one-dimensional sequence of characters (or QTC_C trajectory), onto a two-dimensional matrix (image texture). This is done using one-hot-vector representation, to efficiently evaluate the similarity of relative movement. This results in images texture which we use as an input to train our network ($TrajNet$), with the goal of being able to distinguish between different vehicle behaviours.

Definition: Given a set of trajectories $\zeta = \{Tv_1, ..., Tv_i, ...Tv_n\}$ where n is the total number of trajectories in ζ, we convert each QTC trajectory; calculated from ζ, to form a sequences of characters $\zeta_C = \{Cv_1, ..., Cv_i, ...Cv_n\}$. Following this, we represent each sequence Cv_i in ζ_C as an image texture I_i using one-hot vector representation. Here the columns represent Cr (or QTC_C states) and the rows indicate the presence of a unique character (or QTC_C state) at a specific time-stamp. Algorithm 1 describes mapping QTC_C trajectories into image texture. For example, the relative motion between two vehicles, as seen in Fig. 7(a), one vehicle is passing by the other vehicle (or obstacle) towards the

---

Algorithm 1. Image representation of QTC trajectory.

1: Input: set of trajectories $\zeta = \{Tv_1, ..., Tv_i, ...Tv_n\}$ where n is the number of trajectories in ζ
2: Input: QTC_C states Cr: $cr_1(- - - -), ..., cr_{81}(+ + + +)$
3: Output: n 2D matrices (images I) of movement pattern
4: Extract: sequences of characters $\zeta_C = \{Cv_1, ..., Cv_i, ...Cv_n\}$ from QTC_C trajectories ζ
5: Define: a 2D matrix (I_i) with size ($N \times 81$) for each sequence in ζ_C, where 81 is the number of characters in Cr and N is the length of Cv_i
6: Initialise: set all the elements of I_i into zero
7: Update: each matrix in I:
8: **for** $i = 1$ to n **do**
9:     **for** $j = 1$ to N **do**
10:         $I_i(j, Cv_i(j)) = 1$
11:     **end for**
12: **end for**
13: return I

left during the time interval t_1 to t_e. This interaction is described using QTC_C:
$(0 - 0\,0, 0 - 0\,0, ..., 0 - 0 -, 0 - 0 -, ...)_{t_1-t_e}$ or $(cr_{32}\ cr_{32}\ ...\ cr_{31}\ cr_{31}\ ...)_{t_1-t_e}$.
This trajectory can be represented as an image texture I_i using our Algorithm 1.

3.2   CNN Based Activity Classification

Upon creating our two-dimensional image representation of QTC sequences, from the pair-wise vehicle trajectories, we are able to train our proposed CNN, using these images as inputs. Classification of the generated image textures is still a challenging task; hence our motivation for transfer learning. Given a scenario between two vehicles, despite the overall activity between two vehicles being the same, there can be countless variations in the trajectories and the resulting activity image. These variations can be due to multiple factors: variations in environmental conditions, influences from other actors, and the different driving styles of people. Considering all these factors, the types of variations are not tractable, and to be able to efficiently model this we adopted a CNN based approach for our activity recognition algorithm.

We decided to use AlexNet; as it is a proven and well established classification network. Its structure is utilised, which has shown to be able to classify many images with complex structures and features, as a base layer for our approach. Transfer learning is done on this network, as completely learning the parameters of this network from start would be unrealistic, do to the relatively small images texture of QTC trajectories available in our dataset (where AlexNet was trained on the order of a million images). As a result of AlexNet being trained on such a large number of images for a general classification task, the earlier layers are well tuned for extracting a wide array of basic features (edges, lines, etc.). We take advantage of this by only replacing and fine-tuning the final convolutional layer ($CL5$), the last three fully connected layers ($FL6$, $FL7$ and $FL8$), softmax (SL) and the output layer (OL), this results in our new network *TrajNet* (Fig. 1).

For the final convolutional layer we uses a smaller layer CL_n, which consists of 81 convolutional kernels. This was preceded by ReLU and max pooling layers (using the same parameters as in [11]). We then incorporated one fully connected FC_{n1}, with 81 nodes. This is used in place of the last two fully connected layers ($FL6$ and $FL7$), of 4096 nodes each. The number of nodes used in the final layers is correlated with the reduction in higher level features in our trajectory texture images (as opposed to [6]). This gives more tightly coupled responses. After our new FC_{n1} we include a ReLU and dropout layer (50%). Based on the number of vehicle activities (a) defined in the dataset, we add a final new fully connected layer (FC_{n2}) to match the a classes. A softmax layer (SL_n), and a classification output layer (OL_n) were also added to reflect the number of classes. The output of the final fully-connected layer is passed to an a-way softmax (or normalized exponential function) which produces a probability distribution over the a class labels. The following are training and testing procedures using our image texture I. Each image texture (I_i) is used as input for the data layer *data* for the network. The network parameters were initialised as follows: iteration number $= 10^4$, initial learn rate $= 10^{-4}$ and mini batch size $= 4$. These parameters were chosen empirically and were based on fine tuning the later layers for the activity recognition task. The other network parameters were set according to [11].

3.3 Extending Vehicle Activity Classification with Predictive Trajectory Modelling

In this section we introduce a novel approach to predicting a vehicle's future trajectory, given its and its pair's states. Ideally, the earlier we are able to identify which scenario a vehicle is in the more time there is to take appropriate action. This step would act as a prerequisite for pair-wise vehicle activity classification, in order to better identify scenario sooner.

To be able to predict the future pathway a vehicle may take, within the context of a specific scenario, we propose a Feed Forward Neural Network (FFNN). Details of our driver model, to predict full vehicle trajectories using partially observed (or incomplete trajectories), are shown in Fig. 3. The proposed model architecture is made up of 9 hidden layers; where each hidden layer has z, such that $\{z \in \mathbf{Z} : \mathbf{Z} = [10, 10, 20, 20, 50, 20, 20, 20, 15]\}$. Given a hidden layer H^i with z nodes, that layer is defined such that $z \hat{=} i^{th}$ element in $\mathbf{Z}$. This configuration was determined empirically, and was well suited to model the complex and intricate decisions a human driver a human can make. Our trajectory prediction model is unidirectional, where inputs are fed sequentially first through the input layer, then processed in hidden layers and finally passed to the output layers. Each layer is made up of nodes, where these nodes have weights and biases associated with each input to the node. To determine the output for a given node the sum of the weighted inputs, along with addition of the bias value, are calculated and then passed through an activation function; here we use a hyperbolic tangent function. The FFNN can be conceptualised as a way to approximate a function, where the values of each node's weight and bias are learned through training. For training data we use prior (incomplete) trajectories with their correct or desired output trajectory. Training is done through Levenberg-Marquardt backpropagation with a mean squared normalized error loss function.

Definition: Given x, y and x', y' as centroid positions of the ego-vehicle and obstacle, respectively, we calculate the relative changes in the ego-vehicle's heading angle and translation of the ego-vehicle between times $(t - 1)$ and t follows:

$$\theta(t) = \tan^{-1} \frac{(y_t - y_{t-1})}{(x_t - x_{t-1})}, \tag{3}$$

$$\nabla(t) = \sqrt{(x_t - x_{t-1})^2 + (y_t - y_{t-1})^2}, \tag{4}$$

Here $\theta(t)$ is the ego-vehicle heading angle and $\nabla(t)$ is the magnitude of the change in the ego-vehicle's motion. Both features ($\theta(t)$ and $\nabla(t)$) were used as prior information in the training of our FFNN. To account for the effects of noise, (mainly associated with tiny fluctuations in the ego-vehicle's heading angles, caused by the driver), we calculate a moving average of the heading angle

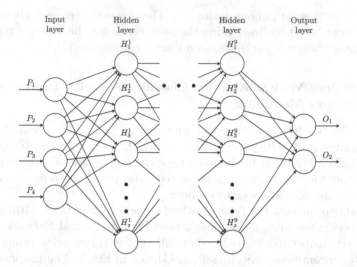

Fig. 3. Network architecture for trajectory prediction model. Layers with their associated nodes are shown. The inputs, outputs and hidden states, are labelled P, O and H, respectively. Note that we use 9 hidden layers, each with z hidden states [4].

and translation over the previous 0.5 s. With this prior information we can define the inputs to our FFNN as:

$$\Theta(t) = \frac{1}{5} \sum_{i=(t-5)}^{t} \theta(i), \tag{5}$$

$$\mathcal{R}(t) = \frac{1}{5} \sum_{i=(t-5)}^{t} \nabla(i), \tag{6}$$

$$\beta(t) = \sum_{j=1}^{t} \left(\nabla(j) \sin \left(\sum_{i=1}^{j} \theta(i) \right) \right), \tag{7}$$

$$\lambda(t) = \sqrt{(x_t - x_t')^2 + (y_t - y_t')^2}. \tag{8}$$

Where $\lambda(t)$ is the distance between the ego-vehicle and the object. To help bound the driver model from going off the road when avoiding the obstacle we introduced $\beta(t)$, which is the lateral shift from the centre of the road. Θ and $\mathcal{R}$ are directly related the ego-vehicle movement, β relates the vehicle location to the road, and λ corresponds to the vehicle and obstacle interaction. The inputs are further visualised in Fig. 3, where $[P_1, P_2, P_3, P_4] \equiv [\Theta(t), \mathcal{R}(t), \beta(t), \lambda(t)]$ and outputs are $[O_1, O_2] \equiv [\theta(t+1), \nabla(t+1)]$. To determine the future trajectory the algorithm in run iteratively, producing a future motion at each time-step. This is described in Algorithm 2.

Turn   Follow   Pass   Overtake   Bothturn

Fig. 4. Example pair-wise manoeuvres for traffic dataset used in [4], based on work presented in [12].

Algorithm 2. Vehicle trajectory prediction.

▷ A distance of ε meters between the vehicle and the obstacle was chosen to start prediction.

1: **if** $(\lambda(t) < \varepsilon)$ **then**
2:     **while** $((x_t < x'_t) \wedge (y_t < y'_t))$ **do**
3:         Inputs: $[\Theta(t), \mathcal{R}(t), \beta(t), \lambda(t)]$, using (5)-(8), respectively;
4:         Outputs: $[\theta(t + 1), \nabla(t + 1)]$;
5:
$$\begin{cases} x_{t+1} = \nabla(t + 1)\cos(\sum_{i=1}^{t+1}\theta(i)) + x_t, \\ y_{t+1} = \nabla(t + 1)\sin(\sum_{i=1}^{t+1}\theta(i)) + y_t. \end{cases}$$
6:         Increment time-step: $t = t + 1$;
7:     **end while**
8: **end if**

---

4   Evaluation

Within our experimental section we explore the performance of our vehicle activity recognition approach by doing multiple comparisons, on different datasets and against other methods. We contrast the performance of our classification approach (Sect. 3.2) with state-of-the-art quantitative and qualitative approaches to activity classification [2,12,13,15,27]. We also evaluate our driver model by comparing with our manually obtained ground truth. We demonstrate the generality of our approach and explain how to can be applied to other scenarios. We also show that gives the best performance, compared with the other approaches; making it a suitable choice for time and safety critical applications.

4.1   Datasets and Experimental Set-up

In this section we explain our evaluation methodology, firstly we present our experimental approach and data used, after which we present a detailed evaluation of our approaches. We performed our tests on two publicly available datasets. The two different datasets represent different application domains, motion and interaction from vehicle traffic (obtained from surveillance cameras [12]) and vehicle-obstacle interactions, for modelling more scarce scenario of a potential collision [3]. Two state-of-the-art pair-wise activity recognition approaches [2,12] have been demonstrated to give superior performance (on the [12] dataset) when compared with a number of other methods [13,15,27].

Table 1. Definitions of vehicle manoeuvres as described in [4].

Activity	Definition
Turn	One vehicle moves straight and another vehicle in another lane turns right
Follow	One vehicle followed by another vehicle on the same lane
Pass	One vehicle passes the crossroad and another vehicle in the other direction waits for green light
Bothturn	Two vehicles move in opposite directions and turn right at same time
Overtake	A vehicle is overtaken by another vehicle

For consistency this has motivated up to compare our approach using these algorithms, the traffic dataset, and ground truth, as a base for evaluating our own activity classification technique. [12] is also, to best of our knowledge, the only pair-wise traffic surveillance dataset publicly available. All experiments were conducted on an Intel Core i7 desktop, CPU@3.40 GHz with 16.0 GB RAM.

Traffic Motion Dataset. The traffic dataset was obtained by extracting trajectories from 20 surveillance videos; see [12] for more details. Five unique vehicle activities, *Turn*, *Follow*, *Pass*, *BothTurn*, and *Overtake*, are defined and their corresponding annotations are provided. 175 clips are presented in total, each activity having 35 clips. Here clips are composed of segments with 20 frames each The dataset provides x, y coordinates associated with the centroid of each vehicle, for each frame and time-stamp t. Figure 4 gives representative frames from the dataset. Table 1 details the definitions of each activity in the dataset.

Vehicle Obstacle Dataset. For our second dataset [3] we focused on close proximity manoeuvring for vehicle/obstacle interactions. These types of scenarios can be potentially very dangerous for the vehicles involved, hence, they happen rarely and there is not much available data—real testing would also not be a viable option for our crash scenario. To address this we developed our own data through a simulation environment, developed using Virtual Battlespace 3

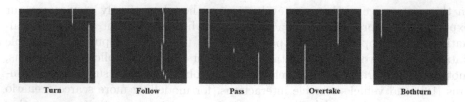

Turn   Follow   Pass   Overtake   Bothturn

Fig. 5. Our image texture representation of QTC encode trajectories [4]. Scenarios from traffic dataset [12].

Table 2. Description of pair-wise vehicle activities in vehicle/obstacle dataset [3].

Scenario	Description
Left-Pass	The ego-vehicle successfully passes the object one the left
Right-Pass	The ego-vehicle successfully passes the object one the right
Crash	The ego-vehicle and the obstacle collide

(VBS3), with the Logitech G29 Driving Force Racing Wheel and pedals. For realism, our simulated tests were carried out on a realistic model of a Dubai highway. A six lane highway was used, with the obstacle placed in the center lane. A total of 40 participants were used in our trials, all of varying ages, genders and driving experiences. Participants were encouraged to drive naturally and to use their driving experience to avoid the obstacle. Within the simulation environment a Škoda Octavia was used, and a maximum speed of 50 km/h was adhered to. We label the manoeuvres manually into three groups: pass left, pass right, and crash. The Cartesian centroid positions for the obstacle and ego-vehicle's was recorded, along with their velocity, yaw angle, and Euclidean distance from each other. Data was recorded at 10 Hz. This data was used for pair-wise vehicle-obstacle activity classification but also for our trajectory prediction model.

Within our simulation trails, drivers completed a full manoeuvre, until they had passed the obstacle. However, we part of the motivation of this work, our approach is focused around recognising events early within the manoeuvre. Hence, our new dataset for vehicle-obstacle interaction recognition task [3], was be partitioned into three subsets, to demonstrate different aspects of our method.

- Our first subset (SS_1) consists of 122 vehicle-obstacle trajectories of about 600 m each (43,660 samples). We used this group to train our driver trajectory prediction model.
- Our second group (SS_2) contains complete trajectories. The activity breakdown within this group is as follows: 67 crash, 106 left-pass, and 104 right-pass trajectories. We consider this group to be our ground truth, and used it to evaluate the accuracy of our recognition method and our trajectory prediction model. The initial distance between the driven vehicle and the obstacle (i.e. the total distance travelled in each trial) was 50 m. Here two experts labelled each trial as one of the three activities. We summarise the definitions of each scenario in Table 2. Examples of each scenario are shown in Fig. 7; where the three scenarios are given from left to right, Left-Pass, Right-Pass and Crash, respectively.
- The third subset (SS_3) consisted of 277 incomplete trajectories. This group is taken as partial trajectories from this trajectories derived in SS_2. We use these partially observed trajectories to evaluate our recognition method, for predicting the events in advance; with the use of our trajectory prediction model. This subset is composed of: 67 crash, 106 left-pass, and 104 right-pass incomplete trajectories, of 25 m in length each.

When using SS_3 we selected an observed distance of $\varepsilon = 25$ m between the ego-vehicle and the obstacle. This distance represented 50% of the full distance of the manoeuvre. Figure 7(d) gives an example of an incomplete trajectory for Right-Pass scenario. Here the solid red line is the ground truth trajectory (taken by the human participant, and is a distance of 25 m) and the dashed line is the predicted trajectory.

4.2 Validation of Driver Model

To be able to classify a scenario earlier within the manoeuvre, and with a higher degree of accuracy, we utilised our driver model to complete the remainder of a manoeuvre. It is therefore imperative to demonstrate the accuracy of the driver model. Firstly, we trained our driver model (Sect. 3.3) using the SS_1 dataset, (which consisted of 43,660 data points). We separated SS_1 into training, validation and test sets, in a 70%, 15% and 15% split, respectively. The training subset is used by our FFNN to learn the network parameters (weights and biases), the purpose of the validation subset was to obtain, unbiased, network parameters for the training process. Lastly the test subset was used to evaluate the network performance, to ensure correct functioning. Figure 8 demonstrates that our network parameters were learned correctly.

We used SS_3, to perform a more in-depth analysis of our driver model. This dataset was also not seen during network training. It consisted of 277 incomplete trajectories, where SS_2 (the completed version) is considered as the ground truth. SS_3 consists of 67 crash, 106 left-pass, and 104 right-pass scenarios. Figure 9 shows samples of the input trajectory, predicted trajectory using our approach, and the ground truth trajectory, for our three scenarios. Our trajectories are positioned so that they begin closer to the origin and move from left to right, with increasing 'x' and 'y' values. The trajectories, created by the trail participants, are in green and red (where the red section is the ground truth), and the trajectories from our driver model are in blue. We measure the error from our FFNN based driver model using the Modified Hausdorff Distance (MHD) [8]. The selection of our error metric MHD, was motivated by its property of increasing monotonically as the amount of difference between the two sets of edge points increases. It is robust to outlier points. Given the driver model generated trajectory as $\mathbf{T_v}$ and the ground-truth trajectory as $\mathbf{T_v}^{gt}$, we determine our error measure as follows:

$$MHD = min(d(\mathbf{T_v}, \mathbf{T_v}^{gt}), d(\mathbf{T_v}^{gt}, \mathbf{T_v})). \tag{9}$$

Here $d(*)$ is the average minimum Euclidean distances between points of predicted and ground-truth trajectories. The error across SS_3 is shown in Fig. 6. Here the red line shows the average error and the bottom and top edges of the box give the 25th and 75th percentiles, respectively. The whiskers of each box extend to cover 99.3% of the data. Red pluses represent outliers. Across all three manoeuvres a mean error of 0.4 m was seen, this amount of error is negligble for activity recognition, because the overall characteristic of the trajectory is

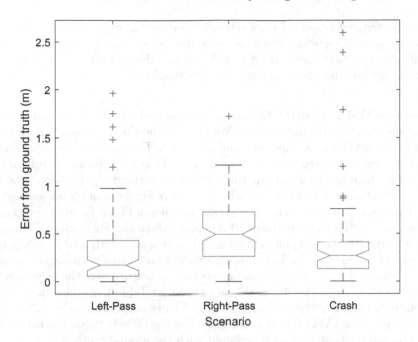

Fig. 6. Errors ($\mathcal{MHD}$) from scenarios, across $SS3$ [4].

Table 3. Comparison of proposed approach with state-of-the-art algorithms on the traffic dataset [12].

Type	TrajNet	NWSA [2]	Heat-Map [12]	WF-SVM [27]	LC-SVM [15]	GRAD [13]
Turn	2.9%	2.9%	2.9%	2.0%	16.9%	10.7%
Follow	0.0%	5.7%	11.4 %	22.9%	38.1%	15.4%
Pass	0.0%	0.0%	0.0%	11.7%	17.6%	15.5%
Bothturn	0.0%	2.9%	2.9%	1.2%	2.9%	4.2%
Overtake	2.9%	5.7%	5.7%	47.1%	61.7%	36.6%
Average Error	**1.16%**	**3.44%**	**4.58%**	**16.98%**	**27.24%**	**16.48**

still captured. Moreover, a standard highway lane can be over 3 m, therefore variations will still be within lane.

4.3 Results on Vehicle Activity Recognition

The main objective of this work is to be able to perform supervised classification of pair-wise vehicle activity, between the ego-vehicle and its surroundings. Having prior information about the type of event a vehicle is in, or about to enter, can have benefits for path planning and decision making. As part of this work we have collected and labelled a comprehensive dataset of high speed vehicle object interactions, in simulation. With this new dataset of complex scenarios unique classification problems arise. Here we will examine the performance and accuracy of our algorithm and provide insight into its results.

Results on Traffic Mostion Dataset. We first evaluate the performance of our algorithm on the traffic dataset [12]. We utilised the Cartesian coordinate pairs, given for each vehicle, as inputs to our algorithm. For each clip the coordinates were then processed into their corresponding QTC_C trajectories. The final pre-processing stage involved constructing our image texture (I_i) for corresponding QTC_C trajectories (done with Algorithm 1). Figure 5 gives example images texture for five different interactions along with the samples in Fig. 4. In order thoroughly evaluate our approach we perform 5-fold cross validation. For each fold we separate the images texture (I) into training and testing sets, with a 80% to 20% split for each class, respectively. The training sets were used to determine the weights of our network ($TrajNet$). The unseen test images texture were then classified by our trained $TrajNet$. Results of this are shown in Table 3. We also incorporate comparative results obtained from [2,12], and three other approaches [13,15,27]. The average error (AVG Error) is calculated as the ratio between the total number of incorrect classifications (compared with the ground truth labels) and the total number of activity sequences within the test set. We show that our approach gives better performance when compared with the five state-of-the-art approaches (Table 3). It is able to classify the dataset with errors rate 1.16% and with a standard deviation 0.015.

Results on Vehicle Obstacle Dataset. Our second set of experiments for activity recognition algorithm was done on our simulation based dataset [3]. We test our classification approach using SS_2 for complete trajectories and SS_3 for partial trajectories. The pair-wise centroid positions of each vehicle x, y in SS_2 were given as inputs. This was preprocessed and encoded to their QTC_C trajectory representation for each scenario. The next step in preprocessing was to transform the trajectory representations to image texture (I_i). One image was created for each QTC_C trajectory, based on Algorithm 1. To evaluate our classification approach we used 5-fold cross validation on the SS_2 dataset. For each fold we separated the images texture, generated from SS_2, into two sets, representing 80% to 20% of the dataset, for training and testing respectively. The network weights were trained using the training subset $TrajNet$, and then tested on the test subset. This approach showed the robustness of our technique, and helps to reduce any bias the network may have received from a training set. Results of the complete manoeuvre classification are given in Table 4. The network trained on full trajectory manoeuvres is represented as "*Complete Traj*". Here the error ratio (AVG Error) is defined as the total number of false classifications, as obtained from the ground truth labelling, divided by the total number of sequences in the test subset.

(a) Left-Pass   (b) Right-Pass

(c) Crash   (d) Driver Model

Fig. 7. Representation of manoeuvres (shown through VBS3) displayed in red (a–c). Here the green car represents an ego-vehicle and the white car a possible obstacle. Results of applying our driver model are shown in (d), where the solid red line is the previously driven path and the dashed line is the predicted trajectory [4].

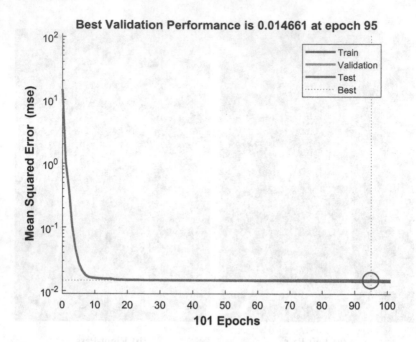

Fig. 8. Results of training for FFNN based driver model. Note that the error rates for the training and testing set are similar [4].

While we have demonstrated the efficacy of our classification method on full manoeuvres, we will now evaluate results on partial trajectories; in order to make pre-emptive decisions. The first step for evaluating the incomplete trajectories was to train our FFNN based driver prediction model (Fig. 3), using SS_1. Using our fully trained driver model, we now perform a prediction step (on the SS_3 dataset) to generate future trajectories and a complete manoeuvre. We refer to this augmented dataset as SS_4.

Similarly to evaluating the full trajectories, we utilise the x, y centroid pairs in the Cartesian space, for both the vehicle and the obstacle from SS_4, as inputs to our classifier. We then generate QTC_C trajectories and associated images texture (I) for all runs. A sample trajectory is shown in Fig. 7(d). We calculated the classification accuracy of our algorithm on SS_4, using a 5-fold cross validation. For each fold, the images texture in SS_4 were separated into training and testing subsets, 80% to 20% respectively. The network weights were optimised for $TrajNet$ using the training subset and test subset was classified by the optimised network $TrajNet$. Results of our testing on incomplete trajectories are presented in Table 4. We differentiate classifications from our network trained on incomplete trajectories by labelling it "*Predicted Traj*". It can be seen that our approach still delivers a high level of performance, despite not being given the complete trajectory. This can be seen a strong motivator for coupling this approach with potential decision making and path planning algorithms.

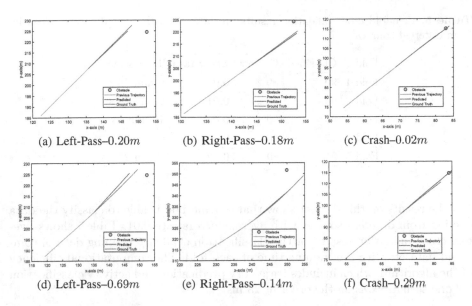

(a) Left-Pass–0.20m (b) Right-Pass–0.18m (c) Crash–0.02m

(d) Left-Pass–0.69m (e) Right-Pass–0.14m (f) Crash–0.29m

Fig. 9. Example trajectories for simulation dataset [3]. Error rates are based on, $\mathcal{MHD}$. Two examples from each scenario are given (top and bottom rows), covering all three scenarios. (a)–(c) and (d)–(f) represent left-pass, right-pass, and, crash scenarios, respectively [4].

Table 4. Classification error for full and partially observed trajectories, SS_2 and SS_3, respectively [4].

Type	Complete Traj (SS_2)	Predicted Traj (SS_4)
Crash	0.0%	0.0%
Left-Pass	0.0%	0.0%
Right-Pass	0.0%	1.0%
AVG Error	**0.0%**	**0.3%**

4.4 Results on Datasets from Different Sources

To demonstrate the robustness and generality of our method, we conducted similar classification experiment using a challenging dataset which combines the two datasets ([3] and [12]). This results in a new dataset which contains 452 scenarios for eight different activities, *Turn, Follow, Pass, BothTurn, Overtake, Crash, Left-Pass* and *Right-Pass*. This is to show that our single network is able to accurately distinguish between an even larger number of driving tasks. Again, we calculated the classification accuracy of our algorithm using a 5-fold cross validation with training and testing subsets, 80% to 20% respectively. The network weights were optimised for *TrajNet* using the training subset and test subset was classified by our generic *TrajNet*.

Table 5. Five-fold cross validation results on datasets with eight different activities from different sources.

Fold	Classified	Correct	Error rate	Train time
Fold 1	90	89	0.01	9 min 50 s
Fold 2	90	89	0.01	9 min 58 s
Fold 3	90	90	0.0	10 min 13 s
Fold 4	91	90	0.01	9 min 36 s
Fold 5	91	89	0.02	9 min 23 s

The results on this dataset show that our method is able to classify the data coming from different sources with average errors rate 0.01. Table 5 shows five-fold cross-validation results for the classification challenge training data of combined datasets with the training time of each fold. The computational efficiency of the algorithm, which includes trajectory predication and activity classification for one scenario was on the order of 28 ms.

5 Conclusion

A new approach based on a deep neural networks for vehicles activity recognition was presented. We have built upon our previous approach by extending the number of trajectories classes our network is able to classify. This further demonstrates the capabilities of our approach to interpret multiple types of vehicle interactions. We utilise a QTC representation to capture invariant interaction features. We then detailed steps on constructing a corresponding image textures, for a more interpretable representation of the QTC trajectories; based on a one-hot vector encoding. Our technique is able to efficiently capture multiple scenarios of vehicles interactions. We describe the reasoning behind our architecture and how we efficiently used a limited amount of data to train TrajNet, while not compromising on a high level of classification accuracy.

The method has been tested on two different challenging datasets: traffic motion and vehicles interaction datasets. The experimental results showed that the method is robust and performs better than existing methods with 1.16% error rate, as opposed to 3.44%, 4.58%, 16.98%, 27.24%, and 16.48% of [2,12,15,27] and [13], respectively.

A separate contribution is our vehicle-obstacle interaction dataset (VOIDataset), which includes complete and incomplete trajectories. VOIDataset is publicly available for download. Our method obtained high accuracy with error rates 0.0% and 0.3%, for complete and incomplete scenarios in VOIDataset, respectively. Through combining the two datasets we showed the robustness of our activity classification method.

To be able to pre-empt a future scenario from a partial observed example we introduced a FFNN method for trajectory predication. The trajectory prediction model was also evaluated on VOIDataset and the average error was 0.4 m.

This approach can have potential impacts in future driving technologies, whether in safety related applications or improving path planning. As a step towards future work we would aim to tackle more complex manoeuvres, involving multiple vehicles and infrastructure.

References

1. Ahmed, S.A., Dogra, D.P., Kar, S., Roy, P.P.: Trajectory-based surveillance analysis: a survey. IEEE Trans. Circuits Syst. Video Technol. **29**, 1985–1997 (2019)
2. AlZoubi, A., Al-Diri, B., Pike, T., Kleinhappel, T., Dickinson, P.: Pair-activity analysis from video using qualitative trajectory calculus. IEEE Trans. Circuits Syst. Video Technol. **1850–1863**, 28 (2018)
3. AlZoubi, A., Nam, D.: Vehicle Obstacle Interaction Dataset (VOIDataset). https://figshare.com/articles/Vehicle_Obstacle_Interaction_Dataset_VOIDataset_/6270233 (2018)
4. AlZoubi, A., Nam, D.: Vehicle activity recognition using mapped QTC trajectories. In: Proceedings of the 14th International Joint Conference on Computer Vision, Imaging and Computer Graphics Theory and Applications, VISAPP, INSTICC, vol. 5, pp. 27–38. SciTePress (2019). https://doi.org/10.5220/0007307600270038
5. Chavoshi, S.H., De Baets, B., Neutens, T., Delafontaine, M., De Tré, G., de Weghe, N.V.: Movement pattern analysis based on sequence signatures. ISPRS Int. J. Geo-Inf. **4**(3), 1605–1626 (2015)
6. Deng, J., Dong, W., Socher, R., Li, L.J., Li, K., Fei-Fei, L.: ImageNet: a large-scale hierarchical image database. In: 2009 IEEE Conference on Computer Vision and Pattern Recognition, CVPR 2009, pp. 248–255. IEEE (2009)
7. Dodge, S., Laube, P., Weibel, R.: Movement similarity assessment using symbolic representation of trajectories. Int. J. Geogr. Inf. Sci. **26**(9), 1563–1588 (2012)
8. Dubuisson, M.P., Jain, A.K.: A modified Hausdorff distance for object matching. In: Proceedings of 12th International Conference on Pattern Recognition, pp. 566–568. IEEE (1994)
9. Hanheide, M., Peters, A., Bellotto, N.: Analysis of human-robot spatial behaviour applying a qualitative trajectory calculus. In: 2012 IEEE RO-MAN, pp. 689–694. IEEE (2012)
10. Khosroshahi, A., Ohn-Bar, E., Trivedi, M.M.: Surround vehicles trajectory analysis with recurrent neural networks. In: 2016 IEEE 19th International Conference on Intelligent Transportation Systems (ITSC), pp. 2267–2272. IEEE (2016)
11. Krizhevsky, A., Sutskever, I., Hinton, G.E.: ImageNet classification with deep convolutional neural networks. In: Advances in Neural Information Processing Systems, pp. 1097–1105 (2012)
12. Lin, W., Chu, H., Wu, J., Sheng, B., Chen, Z.: A heat-map-based algorithm for recognizing group activities in videos. IEEE Trans. Circuits Syst. Video Technol. **23**(11), 1980–1992 (2013)
13. Lin, W., Sun, M.T., Poovendran, R., Zhang, Z.: Group event detection with a varying number of group members for video surveillance. IEEE Trans. Circuits Syst. Video Technol. **20**(8), 1057–1067 (2010)
14. Liu, W., Wang, Z., Liu, X., Zeng, N., Liu, Y., Alsaadi, F.E.: A survey of deep neural network architectures and their applications. Neurocomputing **234**, 11–26 (2017)

15. Ni, B., Yan, S., Kassim, A.: Recognizing human group activities with localized causalities. In: 2009 IEEE Conference on Computer Vision and Pattern Recognition, CVPR 2009, pp. 1470–1477. IEEE (2009)
16. Ohn-Bar, E., Trivedi, M.M.: Looking at humans in the age of self-driving and highly automated vehicles. IEEE Trans. Intell. Veh. 1(1), 90–104 (2016)
17. Oquab, M., Bottou, L., Laptev, I., Sivic, J.: Learning and transferring mid-level image representations using convolutional neural networks. In: 2014 IEEE Conference on Computer Vision and Pattern Recognition (CVPR), pp. 1717–1724. IEEE (2014)
18. Russakovsky, O., et al.: Imagenet large scale visual recognition challenge. Int. J. Comput. Vision 115(3), 211–252 (2015)
19. Shi, Y., Tian, Y., Wang, Y., Huang, T.: Sequential deep trajectory descriptor for action recognition with three-stream CNN. IEEE Trans. Multimedia 19(7), 1510–1520 (2017)
20. Shi, Y., Zeng, W., Huang, T., Wang, Y.: Learning deep trajectory descriptor for action recognition in videos using deep neural networks. In: 2015 IEEE International Conference on Multimedia and Expo (ICME), pp. 1–6. IEEE (2015)
21. Srivastava, N., Hinton, G., Krizhevsky, A., Sutskever, I., Salakhutdinov, R.: Dropout: a simple way to prevent neural networks from overfitting. J. Mach. Learn. Res. 15(1), 1929–1958 (2014)
22. Szegedy, C., et al.: Going deeper with convolutions. In: Proceedings of the IEEE Conference on Computer Vision and Pattern Recognition, pp. 1–9 (2015)
23. Van de Weghe, N.: Representing and reasoning about moving objects: a qualitative approach. Ph.D. thesis, Ghent University (2004)
24. Xiong, X., Chen, L., Liang, J.: A new framework of vehicle collision prediction by combining SVM and HMM. IEEE Trans. Intell. Transp. Syst. 19(3), 699–710 (2018). https://doi.org/10.1109/TITS.2017.2699191
25. Xu, D., et al.: Ego-centric traffic behavior understanding through multi-level vehicle trajectory analysis. In: 2017 IEEE International Conference on Robotics and Automation (ICRA), pp. 211–218. IEEE (2017)
26. Xu, H., Zhou, Y., Lin, W., Zha, H.: Unsupervised trajectory clustering via adaptive multi-kernel-based shrinkage. In: Proceedings of the IEEE International Conference on Computer Vision, pp. 4328–4336 (2015)
27. Zhou, Y., Yan, S., Huang, T.S.: Pair-activity classification by bi-trajectories analysis. In: 2008 IEEE Conference on Computer Vision and Pattern Recognition, CVPR 2008, pp. 1–8. IEEE (2008)

Quantifying Deformation in Aegean Sealing Practices

Bartosz Bogacz[1(✉)], Sarah Finlayson[2], Diamantis Panagiotopoulos[2], and Hubert Mara[1]

[1] Forensic Computational Geometry Laboratory, Heidelberg University, Heidelberg, Germany
{bartosz.bogacz,hubert.mara}@iwr.uni-heidelberg.de
[2] Heidelberg Corpus der Minoischen und Mykenischen Siegel, Heidelberg University, Heidelberg, Germany
{sarah.finlayson,diamantis.panagiotopoulos}@zaw.uni-heidelberg.de

Abstract. In Bronze Aegean society, seals played an important role by authenticating, securing and marking. The study of the seals and their engraved motifs provides valuable insight into the social and political organization and administration of Aegean societies. A key research question is the determination of authorship and origin. Given several sets of similar impressions with a wide geographical distribution on Crete, and even beyond the island, the question arises as to whether all of them originated from the same seal and thus the same seal user. Current archaeological practice focuses on manually and qualitatively distinguishing visual features. In this work, we quantitatively evaluate and highlight visual differences between sets of seal impressions, enabling archaeological research to focus on measurable differences. Our data are plasticine and latex casts of original seal impressions acquired with a structured-light 3D scanner. Surface curvature of 3D meshes is computed with Multi-Scale Integral Invariants (MSII) and rendered into 2D images. Then, visual feature descriptors are extracted and used in a two-stage registration process. A rough rigid fit is followed by non-rigid fine-tuning on basis of thin-plate splines (TPS). We compute and visualize all pairwise differences in a set of seal impressions, making outliers easily visible showing significantly different impressions. To validate our approach, we construct a-priori synthetic deformations between impressions that our method reverses. Our method and its parameters is evaluated on the resulting difference. For testing real-world applicability, we manufactured two sets of physical seal impressions, with a-priori known manufactured differences, against which our method is tested.

Keywords: Feature extraction · Image registration · Visualization · Computational humanities

1 Introduction

Seals are artifacts mostly made of stone and other materials, such as bone/ivory, metal, and various artificial pastes. They display engraved motifs, ranging from

A. P. Cláudio et al. (Eds.): VISIGRAPP 2019, CCIS 1182, pp. 589–609, 2020.
https://doi.org/10.1007/978-3-030-41590-7_25

simple geometrical patterns to complex figurative scenes. Seals are generally perforated, enabling them to be worn as necklaces or at the wrist supporting - beyond their significance as insignia, prestige objects, and amulets - their primary administrative purposes: securing, authorizing and marking. Their study provides important insight into the Aegean and its socio-political organization and administration.[1]

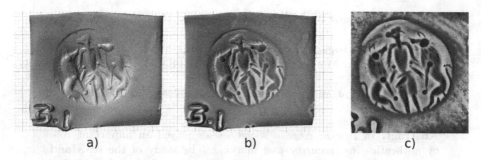

a)   b)   c)

Fig. 1. Source data and pre-processing steps are shown. (a) Original 3D seal impressions are scanned with a structured-light scanner resulting in a 3D object with color. (b) MSII curvature computation is applied to colorize areas with their local surface curvature. (c) The 3D object is rendered into a raster image and local histogram scaling is applied to maximize enhance local visual features.

One key question arising in archaeological research on Aegean seals is the authorship, origin of impressions, and sealing practice. In particular, researchers are interested in determining whether a set of seemingly identical impressions has been stamped by one or more seals, and also whether the same person made each impression or not. Typical approaches focus on qualitative comparisons of salient visual features manually ascertained among a set of impressions. Their presence or absence determines the possibility of originating from a specific seal.

In our work, we enable quantitative archaeological reasoning by automatically aligning deformed impressions, highlighting differences and computing their impact. Our data are modern casts of ancient seal impressions that are acquired with a high-resolution structured light 3D scanner. The data and their preprocessing is shown in Fig. 1.

We address the challenges of analyzing commonalities and differences among a set of impressions too large for detailed manual inspection. This work is largely based on material published in [3]. Contributions new to this work are as follows: (1) space of parameters explored has been increased, (2) visualizations of (a) overlaid impressions, (b) deformation grids, and (c) distance matrices have been revised for greater legibility, (3) an Autoencoder trained on our data replaces the convolutional network pre-trained on general images for greater visual feature

---

[1] Further introductory information can be found at https://www.uni-heidelberg.de/ fakultaeten/philosophie/zaw/cms/seals/sealsAbout.html.

accuracy, (4) and, most importantly, two sets of seal impressions for testing have been manufactured to better validate real-work applicability of our method.

This paper is structured as follows: In Sect. 2 we present related work on registration of images of different content. Section 3 introduces our historic data and our manufactured data for testing. We describe the steps necessary from acquisition of the physical objects to rendering into images. Then, in Sect. 4 we describe three feature extraction methods used to generate correspondences. In Sect. 5, based on the proposed correspondences, first a rigid mapping is estimated. The residuals are fine-tuned bases on thin-plate splines. In Sect. 6, we evaluate our method (i) on a synthetic dataset and estimate optimal parameters, (ii) on the manufactured dataset, visually connecting physical differences to computed highlights, and in (iii) we apply our method on our historic data to gain insights into the archaeological context. A summary and conclusion is given in Sect. 7.

2 Related Work

Image registration is a well-known and common challenge in many settings, including medical [13], geographical [4, 25, 29], and stereo vision [30]. Typically, registration is performed rigidly, using either a Euclidean transform or a projective transform. Non-rigid approaches use flow-fields [2], diffeomorphisms [1] (continuous and topology preserving transforms), and radial basis functions which thin-plate splines are a part of. In all cases, correspondence is established by a set of landmarks (requiring visual feature extraction), borders (requiring a-priori segmentation) or by intensity of pixel values (requiring a-priori meaning e.g. density in medical imaging). In our setting, both seal impressions differ in the depicted content and singular pixel values have no semantic meaning beyond denoting surface curvature. We rely on extracting salient visual landmarks using feature descriptors.

In [12] Ham et al. introduce *Proposal Flow*, an approach generating a flow field to match two images of significantly differing content. The authors proceed in two steps to match two images. First, a set of multi-scale region proposals are generated and matched. The matching considers the spatial support of the region pair and its visual similarity. Visual similarity is computed with Spatial-Pyramid Matching (SPM) [16] or a convolutional neural network (CNN) [15] enabling matching semantically similar regions in addition to pure pixel-wise similarity. The spatial support is computed and weighted considering neighboring matches, i.e. support is low if neighbors disagree. This scheme enforces local smoothness of matches. In the second step, a dense flow field is estimated from the matches, and used to transform the source image onto the target image.

In [28] Rocco et al. revisit the classical computer vision registration pipeline, of (i) extracting descriptors and (ii) estimating a transform based on the computed landmarks. The authors construct a fully differentiable and end-to-end trainable network imitating these stages. A pre-trained CNN, employing all but its classification layer, is used for extracting visual features from the source (to be transformed) and target image. Then, the extracted visual features are cross-correlated and a regression CNN estimates rigid transformation parameters. This two stage approach is repeated again to regress non-rigid transformation parameters for a thin-plate spline based deformation. The main advantage is the capability of full end-to-end training of the pipeline and its ability to predict rigid and non-rigid transformation parameters using only one inference step, i.e. no optimization of parameters is necessary for prediction as the neural network predicts the parameters in one forward pass.

Classic transformation estimation is performed with RANSAC [10]. Due to the adaptable complexity and high count of parameters of thin-plate spline based deformation, a minimal subset of points cannot be defined, i.e. a minimal deformation can be found for any pair of point-sets. However, Tran et al. show in [31] that in the space of correspondences a hyperplane can be fit to inlying matches and separate them from outlying matches. That is, the set of inlying matches is a lower dimensional manifold in the higher dimensional correspondence space.

In our setting, no training data is available, making training of transformation networks not feasible. We adopt a two-stage approach of first performing a rigid registration followed by a non-rigid fine-tuning based on a thin-plate spline deformation model.

3   Dataset

Our dataset originates from the CMS project [22] (Corpus der Minoischen und Mykenischen Siegel) in Heidelberg, Germany. It is an archive of approx. 12.000 impressions of ancient seals and sealings, with the goal of systematically documenting and publishing the entire Aegean material. The archive contains impressions of the seals and sealings, copies manufactured in plasticine, silicon and gypsum, as well as photographs, tracings, and the meta-data associated with current research into Aegean glyptic. The impressions of the seals and sealings are digitized, by structured-light 3D acquisition, in the course of an on-going interdisciplinary project.

The impressions for this work have been acquired with 550DPI resolution which corresponds to 47.000 vertices per square centimeter. While photographs and manual tracings of the seal impressions are available, we only rely on the 3D data. Therefore our method is independent of any artist bias in the tracings, and of any lighting effects or occlusion by shadows in the photographs.

In addition to our previously published work [3], we manufactured two supplementary sets of completely new seal impressions. Thus, our dataset consists of (i) 4 modern casts of ancient seal impressions denoted as the *HR* group, (ii) 12 modern handmade impressions of a dice denoted as the *Dice* group, and (iii) 20 modern handmade impression of a seal denoted as the *Motif* group. We make the 3D scans and pre-processed raster images used for this work available online as Open-Access at DOI 10.11588/data/UMJXI0. Two samples from all datasets are shown in Fig. 2.

At the time of experiments only a subset of impressions were fully acquired pre-processed. The first two sets of impressions has been made with a dice by the same person, with *Dice 1.1, 1.2, 1.3, 1.4, 1.6* impressed with a vertical hand movement pressing the dice upwards from bottom to top, and *Dice 2.1, 2.2, 2.5, 2.6*, with a horizontal movement pressing the dice from right to left against the clay.

The second two sets of impressions, using a replica seal with a complex motif, were made by two different people. Impressions *Motif 3.1, 3.2, 3.3, 3.4, 3.5, 3.7, 3.10* were made by a left-handed person holding the seal in the left-hand, and impressions *Motif 4.2, 4.3, 4.5, 4.8, 4.9* a right-handed person holding the seal in the right-hand. The gestures used in stamping the seal were kept as similar as possible. The modern casts of ancient seal impressions consists are denoted as *HR591, HR632, HR634, HR635*.

3.1   Pre-processing

The texture data accompanying the acquired surface data of the seal and sealing impressions does not contain any relevant information. It is discolored and weathered through the centuries, not representative of the original artifacts color. Therefore, we rely only on measured surface and its local curvature. The acquired 3D is cleaned and its surface curvature is computed in the GigaMesh[2] software framework. For very small and fine surface gradients, such is the case for the

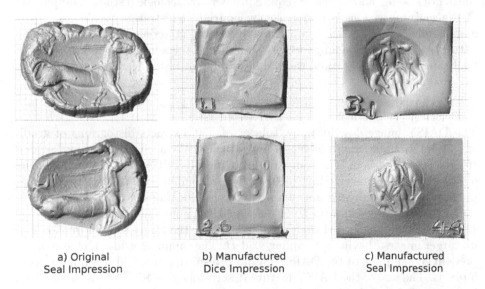

| a) Original | b) Manufactured | c) Manufactured |
| Seal Impression | Dice Impression | Seal Impression |

Fig. 2. Two samples each from the three datasets used in our work. In addition to the (a) modern casts of ancient Aegean seal impressions, we manufactured two test sets of seal impressions: (b) using a common dice to create regular impressions and (c) a custom seal face to create impressions similar to ancient ones. Shown here are 3D objects with lighting only.

---

[2] https://gigamesh.eu.

Dice impressions, we employ multi-scale integral invariants (MSII) [19,20]. MSII intersects, at multiple scales, a sphere (in our work, of 1 mm radius) with the scanned seal impression surface and computes the difference of volumes below and above the surface. Large differences of volumes indicate a high curvature. For large surface structures, e.g. for round seals with motifs cf. Fig. 2(a), we compute the Ambient Occlusion (AO) [24] of the 3D objects. AO computes the amount of ambient light that can possibly reach points on the surface.

Then, the surface enhances with the computed features, either MSII or AO, is orthogonally projected onto a raster image of size 1400 × 1000 pixels. To enhance local features further and to use the full domain of color values, we apply local histogram scaling with a disk sized 100 pixels. The process, from an actual manufactured seal impression, its 3D acquisition to the input into our method pipeline is shown in Fig. 1.

4   Correspondence

Our main motivation for this work is the assumption that a set of seal impressions originating from the same seal ought to share highly similar visual features. Any deviations are either accidental, for example resulting from poor preservation of some seal impressions, or are potentially caused by differences in practice, i.e. the act of sealing, and therefore require further archaeological study. Comparison of the seal impressions is performed by nonrigid registration to abstract away global influences from hand movement and soft materials. Further, repeating patterns of deformation may indicate styles of gestures used to impress seals. To facilitate precise registration, we evaluate different visual feature descriptors.

4.1   DAISY

The DAISY image descriptor by Tola et al. [30] extracts histograms of gradient orientations at multiple scales and arranged in a circular fashion around the reference position. It is comparable to the SIFT [17] and GLOH [23] in its descriptiveness, while enabling highly efficient computation by sharing overlapping and computed histograms among reference locations.

In this work, we define the to be deformed seal impression as the source image I having a sampling grid A that has to align with the target seal impression and its target image J having a sampling grid B. The sampling grids A, B match the predefined sampling of the DAISY method, resulting in a grid of samples every 5 pixels. The size of the DAISY feature descriptors $f_i \in \mathbb{R}^{8*(1+6r)}$ is determined by count of rings r having 6 histogram probes in one ring, with one additional central histogram, and eight orientations per histogram.

$$d_{ij} = \frac{\langle f_i, g_j \rangle}{\|f_i\| \|g_j\|} \tag{1}$$

The distance between the descriptors is computed with the cosine distance.

4.2 BOVW

The bag-of-visual words (BOVW) image descriptor, first introduced by Fei-Fei et al. [8], aggregates local visual words, low level image descriptors such as SIFT or DAISY, and computes a histogram of their occurrences. Since this descriptor models its neighborhood by the counts of visual words only, it is robust against any movement, rigid and nonrigid, of its low-level words, within its local aggregation radius. This makes the BOVW descriptor particularly suited for our task of describing image regions irrespective of their local deformation.

We re-use the previously computed DAISY descriptors as low-level visual words for the high-level description with BOVW. Here, we denote the sets of extracted DAISY descriptors from the images I, J at locations $\tilde{a} \in \tilde{A}, \tilde{b} \in \tilde{B}$ with $g_{\tilde{a}}, g_{\tilde{b}} \in \mathbb{R}^{8*(1+6r)}$. The sampling grids of the DAISY descriptors $\tilde{A}, \tilde{B}$ differ (enforced by the DAISY method) from the sampling grids for the BOVW descriptors $a \in A, b \in B$. These sampling points are arranged as regular grids overlaid on the images and define the level of accuracy of the correspondence computation. We denote BOVW feature descriptors sampled on the respective grid points as $f_a, f_b \in \mathbb{R}^k$ with k words used to describe an location. Our experiments determined 1024 to be the optimal count of words.

The computation proceeds by clustering the continuous low-level DAISY descriptors $g_{\tilde{a}}, g_{\tilde{b}}$ into discrete words. Given a clustering, in this work with k-means [18], with cluster centers $\{v_1 ... v_k\}$ the discrete words $\tilde{g}_{\tilde{a}}, \tilde{g}_{\tilde{b}}$ are defined as follows.

$$\tilde{g}_{\tilde{a}} = arg \min_k \|g_{\tilde{a}} - v_k\| \tag{2}$$

Then, for each sample on the BOVW grids A, B we count the occurrences of visual words within a radius θ. In our experiments a radius of $\theta = 20$ has proven to produce the most accurate results. We record the count of visual words, the cardinality of the set $|\{...\}|$ of visual words, in the components of a BOVW feature vector $f_{a,i}$ with $i \in \{1...k\}$.

$$f_{a,i} = |\{\tilde{g}_{\tilde{a}} = i \text{ and } \|\tilde{a} - a\| < \theta\}| \tag{3}$$

On basis of these feature vectors we compute similarity by computing their distance.

$$d_{ab} = \frac{\langle f_a, f_b \rangle}{\|f_a\| \|f_b\|} \tag{4}$$

We use the cosine distance, as the absence of a visual word should not count towards the distance. Otherwise, we will miss salient visual features in the present of significant noise. Our experiments also confirmed the use of the cosine distance over the Euclidean distance.

4.3 Convolutional Autoencoder

To capture high-level structure and visual features of our objects at multiple levels, i.e. from small features such as eyes or hands to large feature as bodies, we employ a convolutional neural network for feature extraction. In our previous work [3], we extracted features following the approach of Razavian et al. [27] by removing the classification layer from a pre-trained convolutional network, in our case the DenseNet architecture [14] pre-trained on ImageNet [6]. However, our previous experiments have shown that the image domain of ImageNet, photographs of natural scenes, is significantly different from our image domain, projected 3D objects colored with surface curvature. The performance of the pre-trained network was worse than of the classical methods.

Compared to ImageNet, out dataset is very limited and unlabeled, precluding the training of a supervised architecture. Therefore, in this work, we use a convolutional Autoencoder [21] to learn an embedding of image patches of our data into a low-dimensional space, our feature space. The architecture of the Autoencoder is stack of convolutional and pooling layers reducing the initial patch size 16-fold into a bottleneck with 16 feature maps. Then, a stack of up-sampling and convolutional layers increases the filter map sizes back to its original extents. The topology of the network is also illustrated in Fig. 3.

The network learns to embed an image patch $g_a, g_b \in \mathbb{R}^{64 \times 64}$ at sampling positions a, b of the grids A, B into an $4 \times 4 \times 16$ dimensional feature space. Then, the same image patch is reconstructed and compared to the original image with mean squared error loss. After training, the feature space models the repeating high-level visual patterns in the images. Similar patterns are close in feature space. The subsequent registration make use of this property. Therefore, the extract feature descriptors f_a, f_b at sampling grid positions $a, b \in A, B$ are computed from the trained encoder part of the network.

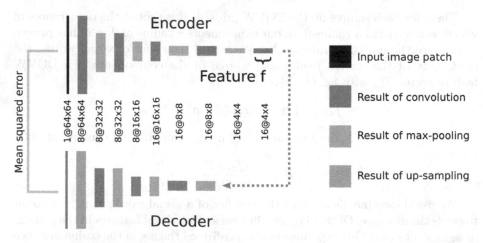

Fig. 3. Convolutional neural network topology of the autoencoder used for feature extraction. Rectangle width denotes count of feature maps. Rectangle height denotes the length of each side of the feature maps. Operations are shown indirectly by denoting which layers are results of which operations. The feature describing the visual contents of the image patch, subsequently used for alignment, is taken after the last convolutional operation in the encoder part.

$$f_a \in \mathbb{R}^{256} = \text{encode}(g_a) \tag{5}$$

Parameters are optimized with AdaGrad [7] with mean squared error loss.

5   Registration

Registration of images proceeds typically [11] by identifying and aligning land-marks. The landmarks, a set of labeled points, from the source image is trans-formed to match the point set of the target image. Unlike aligning point-clouds, landmarks have distinct one-to-one correspondences since each landmark repre-sents a specific visual feature.

From the grids $a, b \in A, B$ of all visual features f_a, f_b of the images we com-pute a set of correspondences with a greedy approach. Each feature on the source image is matched to the closest on the target image, based on the computed dis-tances $d_a b$.

$$m_{ab} = \begin{cases} 1 & \text{if } \underset{a}{\text{argmin}} \, d_{ab} = \underset{b}{\text{argmin}} \, d_{ab} \\ 0 & \text{otherwise} \end{cases} \tag{6}$$

Cross-checking makes sure that a correspondence is optimal from source to target and vice-versa. In Fig. 4 we visualize the matching process and in Fig. 5 the computed deformation grid.

5.1   Rigid Pre-alignment

The first stage of alignment is performed with a rigid transformation model, i.e. translation and rotation only. To enable non-rigid fine-tuning in the second stage, small differences, within an error of ϵ to the rigid model, are kept. Estimation of the rigid model is performed with the RANSAC [10] method. The method randomly chooses a subset of correspondences, fits a rigid model to the chosen subset, and counts the amount of remaining correspondences agreeing with the chosen model. RANSAC minimizes the squared error between the two point-sets of source and target.

$$\min_{(R,t)} \sum_{i}^{M} \sum_{j}^{N} m_{ij} \|(Ra_i + v) - b_j\| \tag{7}$$

The rotation matrix is denoted by $R \in \mathbb{R}^{2 \times 2}$ and translation is denoted by $v \in \mathbb{R}^2$. The set of inlying correspondences is given by $\bar{m}_a b$.

$$\bar{m}_{ij} = \begin{cases} m_{ij} & \text{if } \|(Ra_i + v) - b_j\| < \epsilon \\ 0 & \text{otherwise} \end{cases} \tag{8}$$

From the set of all correspondences m_{ab} the remaining inlying correspon-dences $\bar{m}_{ab}$ are used for non-rigid fine-tuning. The subsets of grid points of the inlying correspondences are denoted as $\bar{A} \subseteq A$ and $\bar{B} \subseteq B$.

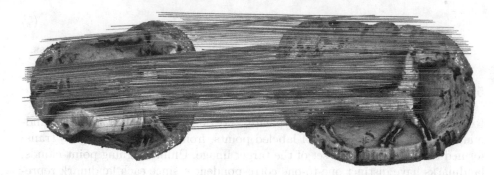

Fig. 4. Left: to be deformed source seal impression *HR591*. Right: target seal impression *HR635*. Correspondence between key-points is shown with colored lines. Line color has no meaning beyond increasing legibility. (Color figure online)

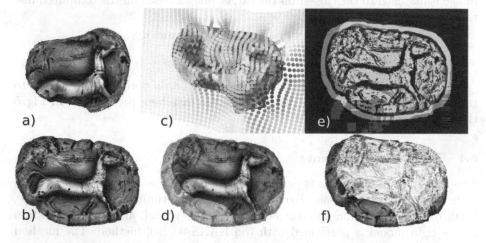

a)

b)

c)

d)

e)

f)

Fig. 5. To be deformed source seal impression *HR591* (a) and target seal impression *HR635* (b). Deformation grid (c) necessary to deform (a) to match (b). Compressed regions are shown with hot (red) colors and stretched regions are shown with cool colors (blue). Displayed grid only cover image extents of the original source image. Regions with no grid points are interpolated from border reflection. (d) Additive overlay of deformed source image from (c) and target image from (b). (e) Overlay of Niblack [26] binarized deformed source and target images for improved legibility of common and different regions. (f) Inverted difference overlay of deformed source and target image. Similarities are shown as bright colors while differences as dark colors. (Color figure online)

5.2   Non-rigid Finetuning

We estimate the necessary non-rigid transformation of source point-set $\bar{A}$ to match the target point-set $\bar{B}$ on basis of thin-plate splines (TPS). Our approach closely follows that of Chui et al. in [5]. While Chui et al. re-computes correspondences in each step, all pairs of feature-less points are eligible, in our approach correspondences $\bar{m}_{ab}$ are fixed and determined by visual features and the rigid model. Therefore we skip the expectation maximization procedure outlined by Chui et al. and directly estimate the necessary transformation.

In the TPS model we minimize the bending energy, outlined in the following equation, to find a transfer function $t_{d,w} : \mathbb{R}^3 \to \mathbb{R}^3$, mapping source points to target points.

$$\min_{t_{d,w}} \sum_i \sum_j \|\bar{a}_i - s_{d,w}(\bar{b}_j)\| + \lambda\|Lt_{d,w}\| \qquad (9)$$

The regularization operator L controls the amount allowed non-rigidity and is determined by the parameter λ. The choice of the parameter λ controls the amount of allowed bending energy, between nearly rigid transformations with $\lambda > 10000$ and unconstrained transformation with $\lambda = 0$. We set the parameter to $\lambda = 1000$ for experiments, as the initial rigid matching already constrains possible deformations. The definition of the regularization operator L and the minimization of the transfer function parameters d, w follows definition outlined in Chui et al.

The parameter d denoted the rigid transformation matrix $d \in \mathbb{R}^{3\times3}$ and the parameter w the spline transformation parameters for each correspondence $w \in \mathbb{R}^{\mathrm{card}(\bar{A})\times 3}$. The spatial relationships, radial basis functions, of the grid points are given in the TPS kernel Φ.

$$\Phi = \|a_i - \bar{a}_j\|^2 \log \|\bar{a}_i - \bar{a}_j\| \qquad (10)$$

Then, a point x in homogeneous coordinates $x \in \mathbb{R}^4$ can be mapped from source to target, i.e. applying the TPS transformation, as follows.

$$t_{d,w}(x) = x * d + \Phi * w \qquad (11)$$

We warp the source image to the target image by re-using and transforming the regular grid A on top of the source image. We compute pixel values by linear interpolation in the rectangles spanned by neighboring points. Difference of grid resolution has a visually imperceptible impact on interpolation fidelity.

6   Evaluation

The choice of the dataset and the goal of the research project do not admit to a typical evaluation against a manually labeled ground-truth dataset. However, to enable researchers to draw conclusions from our work we evaluate our approach w.r.t three different criteria.

1. A qualitative evaluation of the feature descriptors. A point-query response heat-map visualizes the accuracy and bias of the correspondences generated by a feature descriptor.
2. A quantitative evaluation of the complete method pipeline with synthetic generated data. We deform seal impression images with random a-priori known deformations and use our method to recover this deformation. The difference of the known and recovered deformation gives us a measure of its performance.

3. A quantitative evaluation with hand-made seal impressions. We manually created seal impressions to form a dataset of known hand movements and authors. We compute and visualize difference to measure accuracy of our method in separating hand-movement and authors.

Finally, in the results section, we compute distance matrices of all seal impressions of our three datasets, to compare the output of our method to the modes of manufacture of the seal impressions.

6.1 Visualization of Feature Descriptors

We qualitatively evaluate the suitability of our feature descriptors by visualizing a heat-map of regions on the target image similar to a point on the source image.

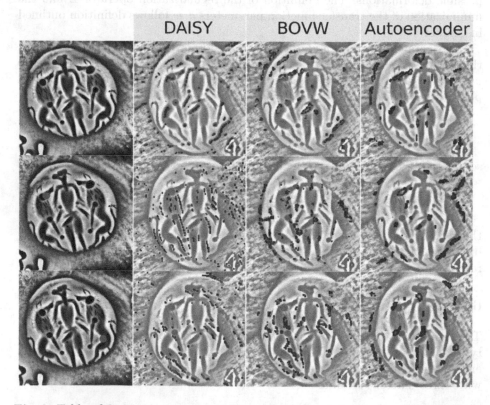

Fig. 6. Table of 3 point queries for each of the presented descriptors. The left-most column shows the source impression with a query point. All other columns show visually similar regions highlighted on a different seal impression. Shown are the 500 most similar sampling points on the sampling grid of the target seal impression. Hot colored sample points denote high descriptor similarity, cold colored sample points denote low descriptor similarity. Point queries on seal impression *Motif 03.01* are compared to regions in seal impression *Motif 04.02*. (Color figure online)

As a point on the heat-map on target image and the query-point on the source image may produce a correspondence mapping, by inspecting the heat-map we can draw conclusions of the fidelity of the feature descriptor.

The heat-map should (i) accurately highlight semantically similar regions, i.e., in the *HR* group of seals, highlight horse-heads on the target if a horse-head was sampled on the source, c.f. Figs. 4 and 6, and (ii) be unimodal, i.e. only the region and all of the region in the immediate vicinity of horse-head on the target should have an high similarity score while other regions should have a low score.

Figure 6 shows similarity heat-maps of the target images for a query-point on a source image for our introduced feature descriptors. Visually closest points as determined by the DAISY descriptor are scattered with no clean hot-spot. Such a pattern makes it challenging to find a good set of correspondences. While the BOVW approach shows more focused regions of visually similar points, the most favorable behavior is shown by the Autoencoder, with one or two very focused hot-spots of highly similar values. Choice of correspondences is thereby limited and focused on the most similar regions.

6.2   Evaluation of Synthetic Deformations

To enable a quantitative evaluation of our methods without ground truth we computationally create an artificial test dataset, separate from our physically manufactured seal impressions. The original images of the rendered seal impressions are deformed with randomly created deformation grids. Then, we employ our method to recover the deformation from the synthetically deformed images. By comparing the a-priori created deformation to the recovered deformation, and the original image to the recovered image, we are able to measure the accuracy of our methods.

The synthesis of deformations is based on the same TPS model as used to fine-tune the estimated rigid transformations. A 4×4 point grid of control points on the original image is randomly perturbed with a uniform distribution. The warped image is created by linear interpolation between the perturbed grid points. We choose a low count of control points and high perturbation vectors to simulate the deformation expected in real data. We assume that hand movement and clay weathering introduced large and global deformations instead of small and local errors.

We use two evaluation metrics to assess the accuracy of the recovered deformation: (i) the point-wise error between a grid transformed with a-priori known parameters d, w and the estimated parameters $\hat{d}, \hat{w}$, and (ii) the pixel values error between two images, warped with known and with estimated parameters.

While homogeneous coordinates $x \in \mathbb{R}^4$ are necessary for the estimation of an optimal deformation, here, we consider the transformation function $t'(d, w) : \mathbb{R}^3 \rightarrow \mathbb{R}^3$ which maps $x \in \mathbb{R}^3$ into homogeneous coordinates, performs the transformation, and maps its back into $\mathbb{R}^3$. We re-use the sampling grid A used for extracting patches as the evaluation grid to measure errors.

$$E_{\text{grid}} = \frac{1}{|A|} \sum_a \|t'_{\hat{d},\hat{w}}(a) - t'_{d,w}(a)\| \tag{12}$$

The pixel-wise distance is computed from the difference of pixel values of the deformed images. With $I(p), J(p)$ we denote the pixel value at position $p \in I$ in the respective images, and $|I|$ denotes the count of pixels in image I. Pixel values of continuous positions in the discrete grid of the raster images are retrieved by linear interpolation. Values outside the raster images are use reflected images as continuation.

$$E_{\text{pixel}} = \frac{1}{|I|} \sum_{p \in I} \|I(t'_{\hat{d},\hat{w}}(p)) - I(t'_{d,w}(p))\| \tag{13}$$

The target image J is not stated in the equations above, as it is I deformed with a-priori known parameters $J(p) = I(t_{d,w}(p))$ with $p \in I$. The errors computed by the two measures are predictive of the accuracy of our methods. Since pixel values of the images denote surface curvature, not texture or light intensity, no undue biases are introduced. Figure 7 shows an original seal impression image and its deformation. The a-priori known deformation grid is re-estimated and the original seal impression is deformed to match the a-priori known deformation. This evaluation is repeated for a set of different deformations strengths. Figure 8 shows the growing differences of grids and pixel values with increasing deformation strength.

6.3   Evaluation of Hand-Made Impressions

While the comparison against a-priori known deformations measures our method's ability to recover deformations, we are also interested in separating seal impressions by their deformation characteristics. Unique deformation characteristics imply a different technique of impressing seals, e.g. using a different hand, a different author, or a different material.

We compute the similarity measure E_{pixel} on each pair of impressions in the sets of handmade seal impressions. Additionally, we compare the deformations grids themselves, by the amount of deformations that was necessary to match a pair of impressions with the following equation.

We transform a regular grid of points and label grid points by the difference of distances e_i to their neighbors, weighted by the Gaussian distribution, before and after transformation. The value σ controls the fuzziness of the coloring of the deformations and is set to $\sigma = 0.01$.

$$E_{\text{warp}}(a) = \sum_i \sum_j (\|\hat{a}_j - \hat{a}_i\| - \|a_j - a_i\|) e^{\frac{-\|a_j - a_i\|}{\sigma}} \tag{14}$$

From these two measures we compute two distance matrices, comparing each pair of seal impressions. Figure 9 shows two distance matrices of pairwise comparisons of pixel distances and grid deformation energy of all three sets of seal impressions.

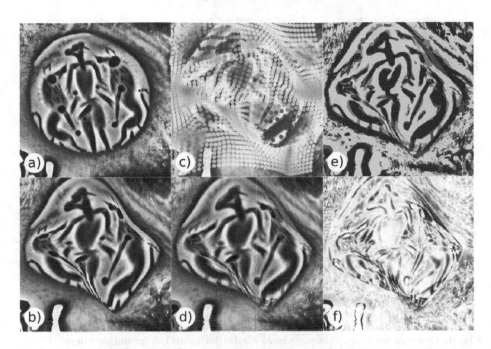

Fig. 7. Registration result of an original image to an a-priori deformed image. Our method successfully recovered the initial deformation, the differences in the overlays are minimal. This figure follows the conventions of Fig. 5. (a) Source image to be deformed to match (b) a-priori deformed target image. (c) Recovered deformation grid and (d) overlay of source deformed with inferred deformation and a-priori deformed target. Overlay of (e) binarizations and (f) pixel-wise differences.

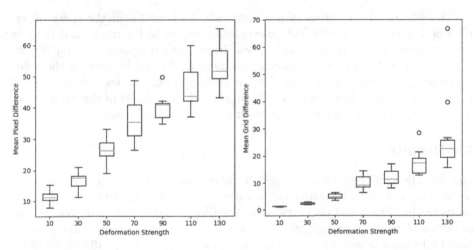

Fig. 8. Distributions of pixel differences E_{pixel} and grid differences E_{grid} between a-priori deformed image and a-posteriori recovered deformation and deformed image. Synthetic deformations have been applied to the set of manufactured *Motif* seal impressions.

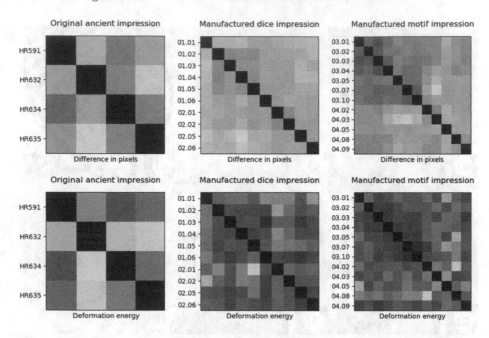

Fig. 9. Distance matrices showing visual dissimilarity and deformation energy for all pairs of seal impressions in our three sets. Motif impressions *Motif 03.01 - Motif 03.10* have been created by a different person than *Motif 04.02 - Motif 04.09*. Dice impressions *Dice 01.01 - Dice 01.06* have been created by the same person but using a different gesture than *Dice 02.01 - Dice 02.06*. The commonalities of manufacture of the ancient seal impressions *HR* are currently unknown.

To optimize the accuracy of our approach, we search over the space of our parameters used to tune the feature extraction methods. Figures 10 and 11 show the distribution of pixel distances of seal impression comparisons from the set of manufactured *Motif* seals. Evaluations were performed by varying the value under study and holding all other fixed at their optimum values. Optimal values for parameters not shown are 3 Daisy rings, and 1024 words in the visual word dictionary. All other parameters were set to their optimal values.

7   Results

Our work provides two directly applicable visualizations for the work of archaeologists: (i) a comparison table of pair-wise transformed and matched impressions as shown in Fig. 12 and close-ups with differences highlighted and deformation direction plotted as shown in Fig. 5, allowing a direct visual comparison of common and unique features across the set of impressions, and (ii) a distance matrix of differences across the set of impressions, as shown in Fig. 9, allowing the identification of clusters of similar seals.

The distance matrices in Fig. 9 are computed from alignment similarity E_{pixel} and necessary deformation energy E_{warp}. While the respective technique of

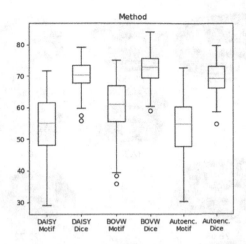

Fig. 10. Evaluation of visual feature extraction methods split for the two manufactured datasets, Motif and Dice. All Y-axis denotes mean difference in pixels, lower values Orange lines denotes medians, box extents denote quartiles and whiskers denote 95-percentiles. (Color figure online)

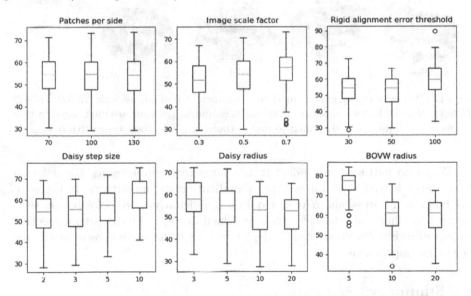

Fig. 11. Grid search evaluation of a subset parameters influencing the results. All Y-axes denote mean difference in pixels, lower values are better.

impressing the seals cannot be ascertained, it can be seen in Fig. 9(a) that impressions *Motif 03.01, 03.02, 03.03, 03.04*, to some degree also *Motif 03.06, 03.07, 03.08*, have lower pixel-wise distance among each other than to the remaining impressions. These sets coincide with the two different persons that created sets *Motif 03.01 - Motif 03.10* and *Motif 04.02 - Motif 04.09*.

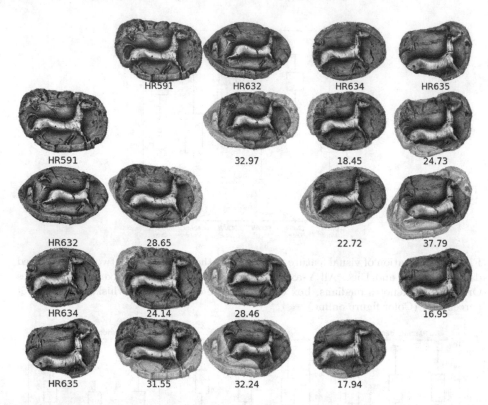

Fig. 12. Pair-wise comparison of modern casts ancient seal impressions *HR591, HR632, HR634, HR635*. Rows are deformed to match columns, i.e. seals are unchanged in their respective columns, and deformed to match their target in their respective row.

When no patterns are evident in the pixel-wise comparisons, the deformation energy matrix is complementary visualization to find outliers. In the set of original seal impressions, *HR632* requires significantly more deformation to be aligned to other seals and vice versa, as shown in Fig. 9. From visual inspection, as shown in Fig. 12, we see that it depicts a larger and more diverse motif than the other impressions.

8   Summary

In this work, we presented an approach to quantify deformation and similarity in a set of seal impression images. In the analysis of authorship and origin of Aegean seals and sealings, archaeologists are interested in the visual commonalities and specific differences between seals and sealings. However, due to different manufacturing technique and weathering of the material, the artifacts are deformed. We approach this challenge by 3D scanning the seal and sealing impressions with a structured-light 3D scanner. Then, we compute surface curvature with MSII (Multi-Scale Integral Invariants) and render the impression meshes into raster images. On these images we evaluate different visual feature descriptors, including an Autoencoder, to extract visually similar regions. Given the set of corresponding regions we compute, first, a rigid transformation with RANSAC, and, secondly, estimate a non-rigid fine-tuning with TPS (thin-plate splines). We evaluate our approach on the basis of an a-priori known and synthetically deformed set of images and comparing the recovered deformations to the known deformations. Further, we manufactured a set of seal impressions with known gestures, vertical and horizontal, and recover these differences by computing pairwise distances between the aligned impression images. Finally, we visualize the differences and deformations on a set of modern casts of ancient seal impressions.

Our overlay visualization is based on a smooth and natural, i.e. differentiable and regularized, deformation to match two seal impression images. The remaining visual differences are from actual semantically relevant content of the seal impression, instead of large-scale deformations caused by weathering or material difference. Further, by computing the pixel-wise error on the surface curvature values, we provided a quantifiable measure of similarity independent of texture and lighting. Our visualizations enable archaeologists to identify in different sets of seal impressions the common or unique gestures used in handling seals during the act of sealing.

The precise attribution of visual similarity of the Autoencoder in our qualitative comparison, yet its middling performance in the quantitative evaluations, implies that our approach is currently limited by the first rigid pre-registration step. In future work we will evaluate diffeomorphisms [1] based deformations. Further, we intend to enable the computations of our method to be performed directly on the 3D data itself, without the need of rendering the objects to raster images. This requires the use of expressive feature descriptors of 3D surfaces [9], which are less wide-spread than 2D image descriptors.

Acknowledgments. We genuinely thank and greatly appreciate the efforts of Maria Anastasiadou co-supervisor of the "Corpus der Minoischen und Mykenischen Siegel" (CMS). We sincerely thank Markus Kühn for his contributions to our tooling. Katharina Anders for her feedback on related work We gratefully thank BMBF eHeritage II for funding this work and ZuK 5.4 for additional funding for this work.

References

1. Ashburner, J.: A fast diffeomorphic image registration algorithm. NeuroImage **38**, 95–113 (2007)
2. Bartoli, A.: Groupwise geometric and photometric direct image registration. Pattern Anal. Mach. Intell. **30**, 2098–2108 (2008)
3. Bogacz, B., Papadimitriou, N., Panagiotopoulos, D., Mara, H.: Recovering and visualizing deformation in 3D aegean sealings. In: International Conference on Computer Vision Theory and Applications (2019)
4. Chen, C., Li, Y.: A robust method of thin plate spline and its application to DEM construction. Comput. Geosci. **48**, 9–16 (2012)

5. Chui, H., Rangarajan, A.: A new point matching algorithm for non-rigid registration. Comput. Vis. Image Underst. **89**, 114–141 (2003)
6. Deng, J., Dong, W., Socher, R., Li, L.J., Li, K., Fei-Fei, L.: ImageNet: a large-scale hierarchical image database. In: Computer Vision and Pattern Recognition (2009)
7. Duchi, J., Hazan, E., Singer, Y.: Adaptive Subgradient methods for online learning and stochastic optimization. J. Mach. Learn. Res. **12**, 2121–2159 (2011)
8. Fei-Fei, L., Perona, P.: A Bayesian hierarchical model for learning natural scene categories. In: Computer Vision and Pattern Recognition (2005)
9. Fey, M., Lenssen, J.E., Weichert, F., Müller, H.: SplineCNN: fast geometric deep learning with continous B-spline kernels. In: Computer Vision and Pattern Recognition (2018)
10. Fischler, M.A., Bolles, R.C.: Random sample consensus: a paradigm for model fitting with applications to image analysis and automated cartography. SRI International (1980)
11. Goshtasby, A.A.: 2-D and 3-D Image Registration: for Medical, Remote Sensing, and Industrial Applications. Wiley, Hoboken (2005)
12. Ham, B., Cho, M., Schmid, C., Ponce, J.: Proposal flow. In: Computer Vision and Pattern Recognition (2016)
13. Holden, M.: A review of geometric transformations for nonrigid body registration. Trans. Med. Imaging **27**, 111–128 (2008)
14. Huang, G., Liu, Z., van der Maaten, L., Weinberger, K.Q.: Densely connected convolutional networks. In: Computer Vision and Pattern Recognition (2017)
15. Krizhevsky, A., Sutskever, I., Hinton, G.E.: ImageNet classification with deep convolutional neural networks. In: Proceedings of the 25th International Conference on Neural Information Processing Systems, pp. 1097–1105 (2012)
16. Lazebnik, S., Schmid, C., Ponce, J.: Beyond bags of features: spatial pyramid matching for recognizing natural scene categories. In: Computer Society Conference on Computer Vision and Pattern Recognition (2006)
17. Lowe, D.: Distinctive image features from scale invariant keypoints. Comput. Vis. **60**, 91–110 (2004)
18. MacQueen, J.B.: Some methods for classification and analysis of multivariate observations. In: Berkeley Symposium on Mathematical Statistics and Probability (1967)
19. Mara, H.: Made in the humanities: dual integral invariants for efficient edge detection. J. IT Inf. Technol. **58**, 89–96 (2016)
20. Mara, H., Krömker, S.: Visual computing for archaeological artifacts with integral invariant filters in 3D. In: Eurographics Workshop on Graphics and Cultural Heritage (2017)
21. Masci, J., Meier, U., Cireşan, D., Schmidhuber, J.: Stacked convolutional auto-encoders for hierarchical feature extraction. In: Honkela, T., Duch, W., Girolami, M., Kaski, S. (eds.) ICANN 2011. LNCS, vol. 6791, pp. 52–59. Springer, Heidelberg (2011). https://doi.org/10.1007/978-3-642-21735-7_7
22. Matz, F., Biesantz, H.: Die minoischen und mykenischen Siegel des Nationalmuseums in Athen. Propylaeum (2016). https://doi.org/10.11588/propylaeum.93.112
23. Mikolajczyk, K., Schmid, C.: A performance evaluation of local descriptors. Pattern Anal. Mach. Intell. **27**, 1615–1630 (2005)
24. Miller, G.: Efficient algorithms for local and global accessibility shading. In: Computer Graphics and Interactive Techniques (1994)
25. Mongus, D., Žalik, B.: Parameter-free ground filtering of LiDAR data for automatic DTM generation. J. Photogramm. Remote. Sens. **67**, 1–12 (2012)

26. Niblack, W.: An Introduction to Digital Image Processing. Prentice-Hall, Upper Saddle River (1986)
27. Razavian, A.S., Azizpour, H., Sullivan, J., Carlsson, S.: CNN features off-the-shelf: an astounding baseline for recognition. In: Computer Vision and Pattern Recognition (2014)
28. Rocco, I., Arandjelović, R., Sivic, J.: Convolutional neural network architecture for geometric matching. In: Computer Vision and Pattern Recognition (2017)
29. Tennakoon, R.B., Bab-Hadiashar, A., Suter, D., Cao, Z.: Robust data modelling using thin plate splines. In: Digital Image Computing: Techniques and Applications (2013)
30. Tola, E., Lepetit, V., Fua, P.: DAISY: an efficient dense descriptor applied to wide-baseline stereo. Pattern Anal. Mach. Intell. (2010)
31. Tran, Q.-H., Chin, T.-J., Carneiro, G., Brown, M.S., Suter, D.: In defence of RANSAC for outlier rejection in deformable registration. In: Fitzgibbon, A., Lazebnik, S., Perona, P., Sato, Y., Schmid, C. (eds.) ECCV 2012. LNCS, vol. 7575, pp. 274–287. Springer, Heidelberg (2012). https://doi.org/10.1007/978-3-642-33765-9_20

Author Index

Printed in the United States
By Bookmasters